MOLECULAR ELECTRONICS: SCIENCE AND TECHNOLOGY

ANNALS OF THE NEW YORK ACADEMY OF SCIENCES
Volume 852

MOLECULAR ELECTRONICS: SCIENCE AND TECHNOLOGY

Edited by Ari Aviram and Mark Ratner

The New York Academy of Sciences
New York, New York
1998

♾*The paper used in this publication meets the minimum requirements of the American National Standard for Information Sciences—Permanence of Paper for Printed Library Materials, ANSI Z39.48-1984.*

Library of Congress Cataloging-in-Publication Data

Molecular electronics: science and technology / edited by Ari Aviram and Mark Ratner.
p. cm. — (Annals of the New York Academy of Sciences, ISSN 0077-8923 ; v. 852)
Includes bibliographical references and index.
ISBN 1-57331-155-3 (cloth : alk. paper). — ISBN 1-57331-156-1 (pbk. : alk. paper)
1. Molecular electronics. I. Aviram, Ari. II. Ratner, Mark A., 1942– . III. Series.
TK7874.8.M665 1998
621.381—dc21 98-9927
CIP

GYAT / PCP

Printed in the United States of America
ISBN 1-57331-155-3 (cloth)
ISBN 1-57331-156-1 (paper)
ISSN 0077-8923

ANNALS OF THE NEW YORK ACADEMY OF SCIENCES

Volume 852
June 30, 1998

MOLECULAR ELECTRONICS: SCIENCE AND TECHNOLOGY[a]

Editors
ARI AVIRAM AND MARK RATNER

Conference Organizers
DAVID ALLARA, PHILLIP CHANG, JAMES ELLENBOGEN, CHRISTIAN JOACHIM, MARK REED, JAMES TOUR, AND YASUO WADA

CONTENTS

[a]This volume is the result of a conference entitled **Molecular Electronics: Science and Technology** held by the Engineering Foundation in Puerto Rico on December 14–18, 1997.

Financial assistance was received from:

- COLEGIO DE INGENIEROS Y AGRIMENSORES DE PUERTO RICO
- DEFENSE ADVANCED RESEARCH PROJECT AGENCY
- ENGINEERING FOUNDATION
- NATIONAL SCIENCE FOUNDATION
- OFFICE OF NAVAL RESEARCH

Preface

The trend of miniaturization of electronic circuit elements has progressed unabated since the invention on the integrated circuit (IC) by Noyce and Kilby. The number of transistors on a state-of-the-art microchip has steadily grown, and by the end of the century, it will near one billion per chip. At the same time the size of the features on a microchip has declined to 0.175 microns and the Moore's Law trend suggests that it will shrink a lot further. However, it is also clear that once the size of the printed lines shrinks below 0.03 microns, the performance of the silicon device will suffer because of cross-talk and quantum limitations. It is entirely possible that at those small dimensions and below, a new technology will emerge based on properties of some molecules that act as electronic components. This technology is Molecular Electronics.

The first theoretical proposals for molecular-scale devices were published as early as 1974. Since then, hundreds of publications (experimental, theoretical, and speculative) have appeared. The attractive features of molecular electronics are that it offers an opportunity to study a new group of materials on the molecular level, while offering engineers a whole new microelectronics technology. The unique properties of molecular devices include structural control, high density, reduced heat dissipation, and high switching speeds. The high density is inherent in the size of molecules and the speed is implied by the short electronic-transition times that occur in molecules (picoseconds or below); and reduced thermal dissipation arises in coherent transport.

Another advantage inherent in molecular electronics is in the mode of fabrication of the molecular IC. In conventional fabrication, every transistor has to be built on the surface of the microchip. This is accomplished by a large number of *lithographic steps* (more than 30 for the current generation). In contrast, the transistors and the other devices in molecular electronics are built by synthetic means in the bulk and are connected to the circuit by self-assembly, thus reducing the number of lithographic steps to only a few. In itself this is a substantial advantage.

Recent developments in science and technology have a direct bearing on molecular electronics and have contributed immensely to the advancement of the field. Among these developments are the discovery of the scanning tunneling microscope and the atomic force microscope that made access to single molecules possible and new advances in nanolithography that make it possible to produce reproducibly nanoscale structures. Also new synthetic and surface chemistry allows the fabrication of many new molecular devices and their attachment to specific electrodes. In addition, there have been advances that are aimed directly as solutions to problems posed uniquely by molecular electronics. An example is the development of break junctions that permits measurement of the conductivity of a single small molecule such as benzene dithiol. Larger molecular wires such as nanotubes have been also studied and the actual current that passes through one single molecular wire has been measured.

The basic premise of molecular electronics is that molecular species, organic or organometallic, can be used for electronic applications. This thesis has been substantiated by various theoretical models. These models span the spectrum of electronic components from theoretical studies of molecular wires to molecule-metal electrode interface and includes studies of components like rectifiers and transistors. The actual experimental demonstration of many of these features is recent and exciting, but still incomplete. During the last year we witnessed the demonstration of measurements of single molecular wires connected to two metallic electrodes, the demonstration of the rectification of a single molecule, and the synthetic preparation of single-molecule field-effect transistors (though the transport measurements have not been done yet). By all standards, these are remarkable advances.

The conference *Molecular Electronics — Science and Technology* was held in Humacao, Puerto Rico, on December 14–18, 1997 under the sponsorship of the Engineering Foundation, a subsidiary of United Engineering Trustees. The conference covered recent developments in the field and discussed future expectations. This book, which represents the proceedings of the conference, constitutes an excellent representation of the state of the art of molecular electronics today.

— ARI AVIRAM
— MARK RATNER

Optimization and Chemical Control of Porphyrin-Based Molecular Wires and Switches

NOEL S. HUSH,[a,b] JEFFREY R. REIMERS,[a] LACHLAN E. HALL,[a] LESLEY A. JOHNSTON, AND MAXWELL J. CROSSLEY[a]

[a]*School of Chemistry and* [b]*Department of Biochemistry, University of Sydney, N.S.W. 2006 Australia*

ABSTRACT: The design rationale is described as are some experimental and theoretical developments concerning rigidly fused porphyrin molecular wires. These materials consist of porphyrin units fused to acene-type bridges and have been synthesized in a range of topologies including linear porphyrin octamers of length ca. 120 Å. Next we demonstrate, for some linear oligoporphyrins, how the electronic coupling between the end porphyrin units can be modulated by simple (possibly *in situ*) chemical modulation of the bridging units. Specifically, the chemical systems considered involve either pH-controlled protonation of bridge azines or conversion of bridge quinone or quinone dioxime rings to or from benzenoid or hydroquinone rings. In the most general terms, the electronic coupling through oligoporphyrin molecular wires is discussed in terms of a simple model in which complete end-to-end π electronic delocalization is required in order to provide strong long-range interactions. Computationally, we monitor interorbital coupling using an appropriate mixture of density functional (B3LYP) and *ab initio* SCF computational schemes. Finally, we examine bridge modulation of the intermetallic coupling in three homovalent bis-metallic oligoporphyrin systems, two in which $Ru(CO)_2$ is complexed within terminal porphyrin rings, the other in which $[Cu(I)Cl_2]^-$ is tethered using phenanthroline end groups. Results are obtained both using an effective two-level model, appropriate for spectroscopic properties, and using a more general scheme, appropriate for molecular conduction.

1. INTRODUCTION

While the field of through-molecule long-distance electron transfer is very well established, a new area involving steady-state molecular conduction through single molecules is emerging (for reviews see, e.g., Refs. 1 and 2; key works include Refs. 3–5). Fused oligoporphyrin molecular wires offer many potential and/or demonstrated advantages as basic materials for the construction of nanodevices exploiting these phenomena. In general, desired properties include:

(1) Synthesizability. There must be practical methods available for the controlled synthesis of molecular electronic materials. Synthetic strategies for porphyrins are well established, and new and innovative solutions to forthcoming problems are likely to be found. To date, the longest linear oligoporphyrin molecular wire produced is an octamer of length ca. 120 Å, a length suitable for nanoelectrode spanning. But, because the synthetic strategy employed doubles the length with each step, much longer molecules could, in principle, be produced.

(2) Stability. Once made, the materials must be chemically stable *in situ* and not decompose when used. Porphyrinic materials form the basis of many essential biological functions such as blood oxygen transport and photosynthesis. While the lifetime of these molecules in harsh biological conditions is at least of the order of days, lifetimes under controlled conditions could well be expected to meet requirements for nanodevices stable for several years.

(3) Synthetic flexibility. It should be possible, by making small changes to the molecules, to tailor molecular properties to match precisely those required in a particular application. Oligoporphyrins may be substituted at the β, *meso* and central porphyrin sites or converted to chlorins, bacteriochlorins, etc., and the composition of the bridging material may be varied, substituted, or chemically controlled. In all, about 100 different oligoporphyrins have been synthesized in our laboratories, and a small selection of these are shown in FIGURE 1; some plausible, simpler model compounds used herein for theoretical investigation are shown in FIGURE 2.

(4) Solubility. The materials and their precursors must remain soluble so that they may be manipulated chemically and oriented properly in a device. By exploiting the synthetic flexibility of the oligoporphyrins, solubility (or, if required, the lack of solubility) can be readily modified to meet requirements. The large *meso* substituents shown on molecules in FIGURE 1 serve this purpose, as well as the purpose of forming an insulating sheath somewhat analogous to the β-sheet surrounding nerves to isolate the wire from undesired or uncontrolled chemical and electrical effects.

(5) Rigidity. The oligoporphyrins consist of long chains of fused rings producing a rigid structural backbone. Hence, relative orientations and distances are nicely controlled, and one may design and analyze systems with confidence. Many popular alternate systems, including polyacetylene wires and acetylene-containing wires do not have fused rings and, while these materials may have structures which are nominally planar, large torsional motions about *each* link in the chain are possible making the overall structure somewhat problematic.[6]

(6) Self-assembly. While delicate nanofabrication of porphyrins is possible,[7] self-assembly is a very desirable property. Preliminary studies have shown that oligoporphyrin systems can self-assemble, for example, on gold surfaces[8,9] and in Langmuir-Blodgett films.[10]

(7) Attachment. It must be possible to attach a molecular wire to terminal or other devices. The synthetic flexibility of the oligoporphyrins allows for the possibility of attachment by a multitude of means including the phenanthroline[11] end groups shown in FIGURES 1 and 2 designed to tether oligoporphyrin molecular wires to electrodes or redox centers. Thiophene end groups have also been synthesized.[9]

(8) Junctions. An essential property in a nanoelectronic device is the ability to fabricate junctions through which signals may be combined or split. Oligoporphyrins intrinsically offer the possibility of this since they are two-dimensional molecules, a porphyrin having four pyrrole rings and can hence form from one to four inter-unit connections per porphyrin. The 90° angle between the pyrroles allows for geometrically simple grid-type design layouts for large devices, and offers the possibility of layer-by-layer assembly of a three-dimensional device using, for example, the out-of-plane coordination sites of a metal porphyrin or spiro-type switches[12] to provide inter-layer communication.

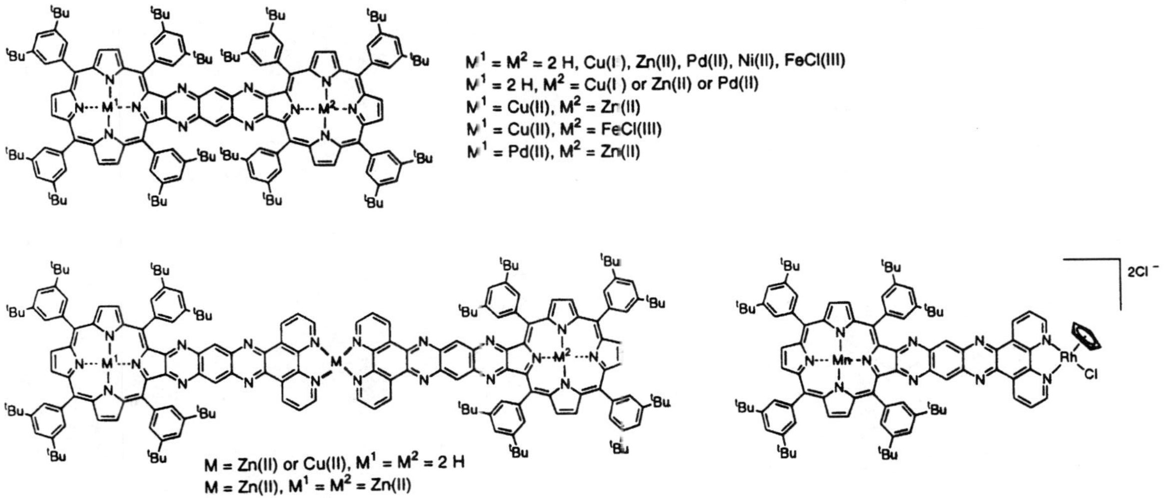

FIGURE 1. Some of the ca. 100 oligoporphyrins which have been synthesized.

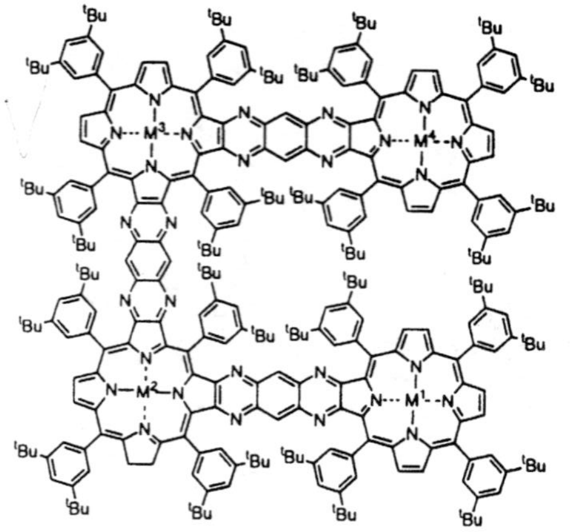

$M^1 = 2$ H, $M^2 = Zn(II)$, $M^3 = Zn(II)$, $M^4 = 2$ H

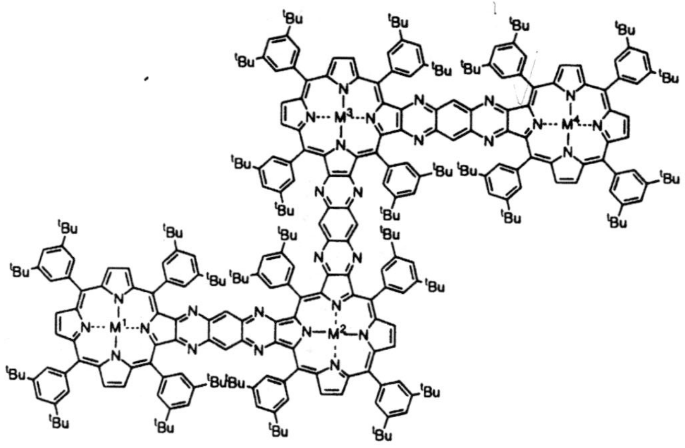

$M^1 = 2$ H, $M^2 = Zn(II)$, $M^3 = Zn(II)$, $M^4 = 2$ H

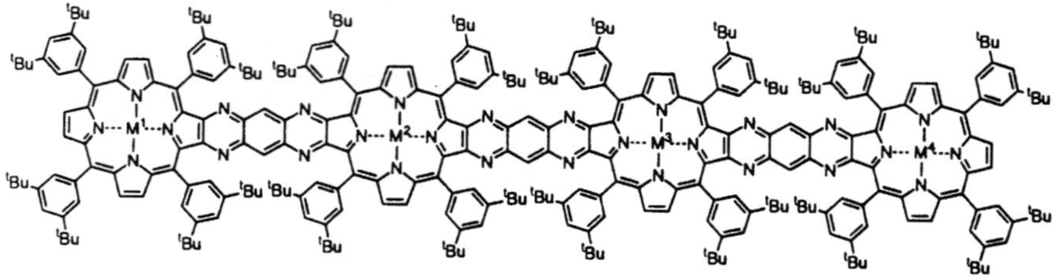

$M^1 = M^2 = M^3 = M^4 = 2$ H, Ni(II)
$M^1 = 2$ H, $M^2 = Ni(II)$, $M^3 = Ni(II)$, $M^4 = 2$ H

FIGURE 1. Continued.

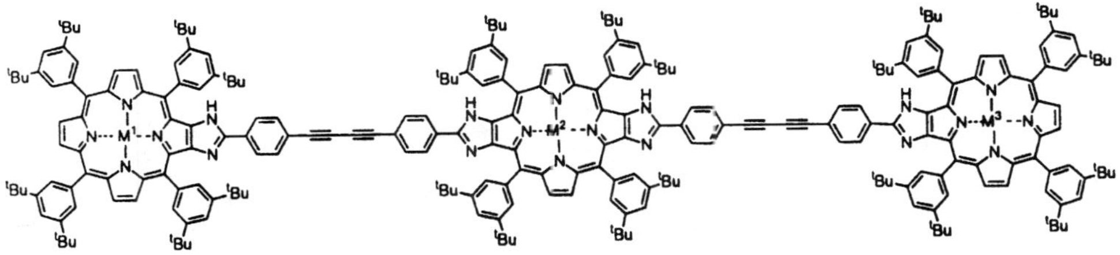

$M^1 = M^2 = M^3 = Zn(II)$

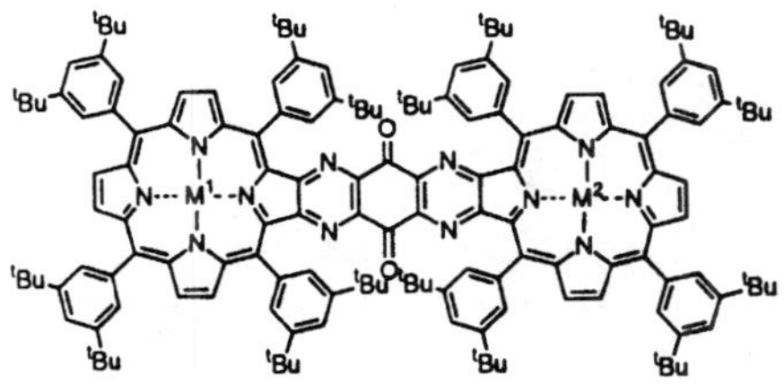

$M^1 = M^2 = 2\,H, Cu(II), Zn(II)$

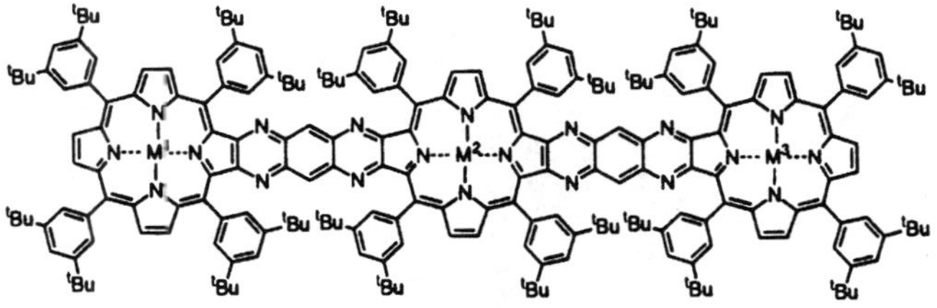

$M^1 = M^2 = M^3 = 2\,H, Ni(II)$
$M^1 = Pd(II), M^2 = 2\,H, M^3 = Pd(II)$
$M^1 = Pd(II), M^2 = 2\,H, M^3 = Zn(II)$
$M^1 = Pd(II), M^2 = 2\,H, M^3 = Ni(II)$
$M^1 = Pd(II), M^2 = Cu(II), M^3 = Pd(II)$
$M^1 = Pd(II), M^2 = Cu(II), M^3 = Zn(II)$

FIGURE 1. Continued.

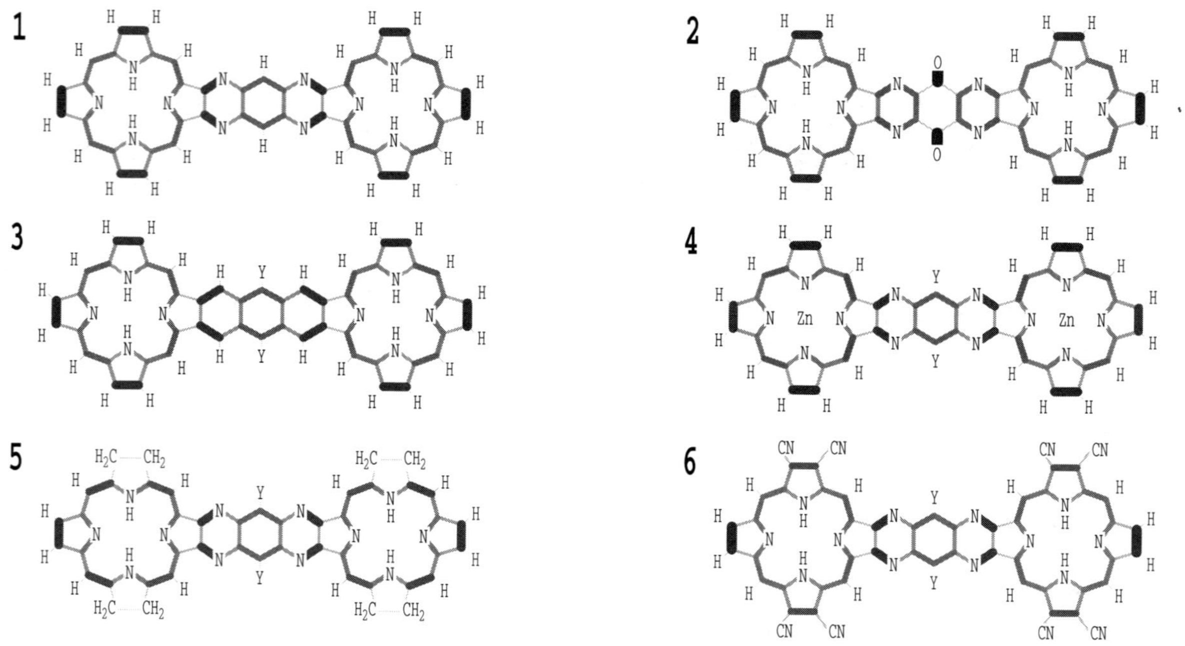

FIGURE 2. SCF/3-21G atomic bond orders[28] at B3LYP/3-21G structures for the benzenoid/quinonoid pair **1** and **2**, and those for the benzenoid form of the remaining oligoporphyrins studied theoretically.[13,23] X = $Ru(CO)_2$, Y = H for benzenoid form, Y = O for quinonoid form, Z = OCH_3. All calculations are performed for closed-shell singlet wavefunctions using D_{2h} point-group symmetry; all molecules are neutral except for **8** (charge = 1 *e*) and **9-10** (charge = −2 *e*). Full π bonds are indicated by thick black lines while weak π bonds are depicted by thin, faint lines.

7

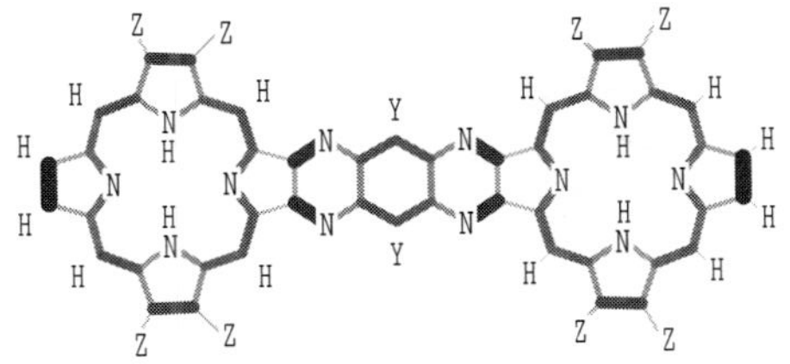

8

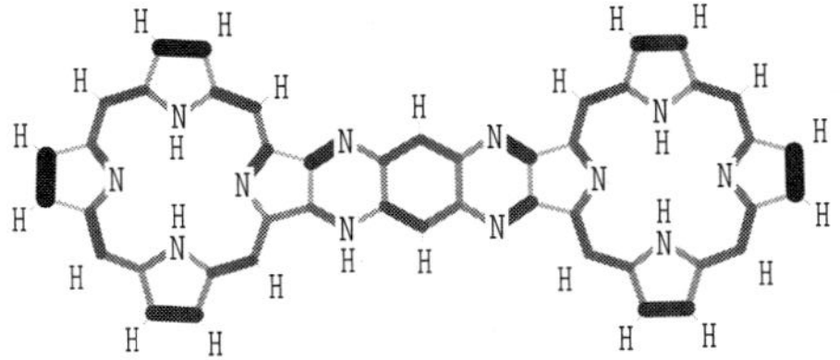

9

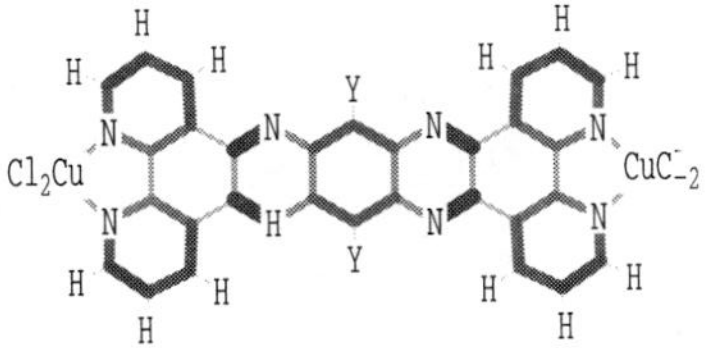

10

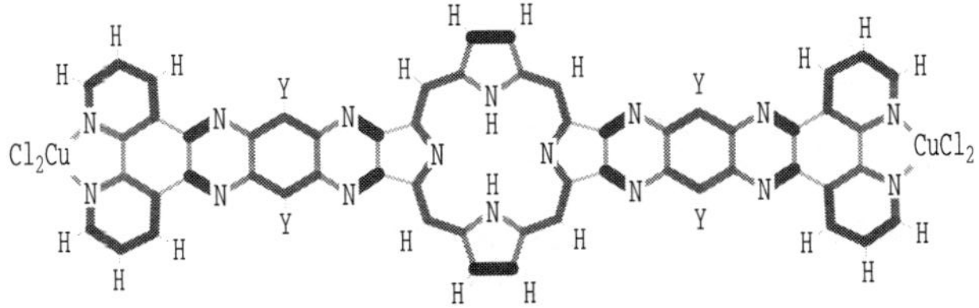

11

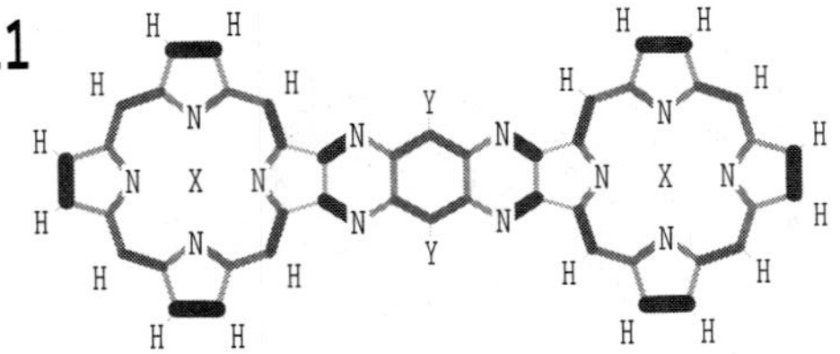

12

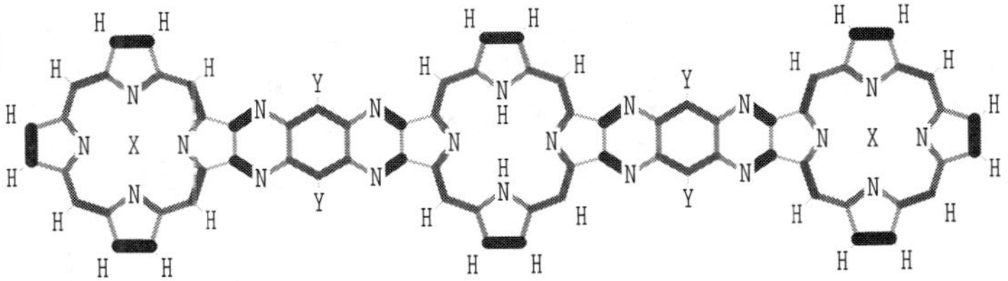

FIGURE 2. Continued.

13 14 15 16

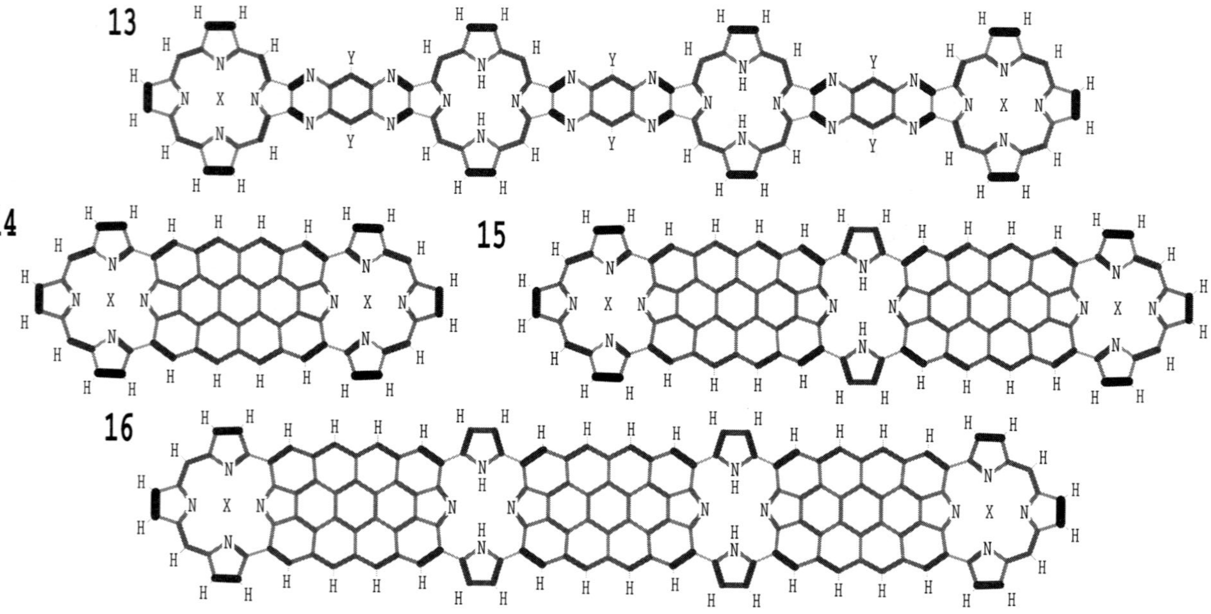

FIGURE 2. Continued.

(9) Functionality. An important property of a molecular wire is that it must be able to transmit information over long distances. This may be done by a variety of means including polarization or exciton transfer, electron or hole transfer via direct electronic coupling (superexchange), or electron or hole transfer via the propagation of geometric disturbances such as solitons, or, more probably, polarons. While each particular mechanism has its own intricacies, all ultimately rely for efficiency on the long-range coupling between the π electrons in the oligoporphyrins, and this is explored most conveniently through the study of superexchange phenomena. For the oligoporphyrins, significant superexchange effects have been demonstrated experimentally[13] through analysis of the rather large changes induced in the porphyrin Soret electronic absorption band structure.

(10) Control of conductivity: rectification. An essential property of material used in a nanodevice is the ability to control its function. A simple and powerful model for how such control could be exercised is the field-effect transistor (FET) in which a large electronic current is controlled by a small applied external field, as well as simpler devices such as rectifying diodes. In principle, oligoporphyrins could be designed with the ability to rectify through the use of asymmetric substitution to control the occupied-virtual band gap and locate localized orbitals at one particular end of the porphyrin within the "band gap." This functionality is yet to be demonstrated, but appropriate compounds have been synthesized.[14]

(11) Control of conductivity: switching. A FET is, in effect, usually employed in electronic circuitry simply as a switch: the current from the source to the drain is either allowed or forbidden to flow. Logic elements can be built from such switches without the need for rectification (see, e.g., Refs. 15–17). Ideally, a dynamical switch is desired, one which can be switched many times at a fast rate by, say, the application of an external electric field. Oligoporphyrin molecular wires have been shown to be switchable between low and high conductance states by chemical and electrochemical means.[13] This provides an important first step towards the demonstration of fast dynamical switching, the length of time required for especially the chemical process at first thought limiting the application to static switching. The demonstration of static switching is, however, a significant achievement with possible application, for example, to the development of write-rarely read-only memory devices such as video tapes.

Our work in this field has two aspects: synthetic development and testing, and theoretical modelling and prediction. Recent synthetic activity has concentrated on the production of new oligoporphyrins, the exploration of the available synthetic flexibility,[18,19] self-assembly,[8] attachment,[11] and switching.[13] Earlier theoretical work concentrated on the description of the basic functionality and on how the synthetic flexibility may be exploited to optimize this.[20,21] This involved the use of semiempirical methods for determining the geometric and electronic structure of the oligoporphyrins, the only methods feasible at the time. Since then we performed calibrations of the theoretical techniques used to model porphyrins[22], and very recently have developed a combined *ab initio* — density functional approach.[13,23,24]

The molecules discussed in Section 2 are all realistic model compounds for known or readily producible oligoporphyrins. Calculations are performed for experimentally observable quantities and clearly demonstrate chemical control of inter-porphyrin coupling. The properties considered are relevant in that they are highly indicative of individual components of oligoporphyrins in a functioning molecular electronic device, but they do not probe properties of more complete systems. One reason for this is that more complete systems are yet to be synthesized and characterized experimentally so it is not possible to verify that the results from more elaborate calculations are reliable (although there is no reason to suspect that they should not be at least qualitatively indicative). Not surprisingly, the experimental synthesis and characterization of more complete functioning units is, however, a priority in our laboratories. Another reason is that the properties of more complete devices, especially those containing metals, will be strongly influenced by solvent effects. Hence, we also have active research programs investigating solvent effects (see, e.g., Refs. 25–27), especially those involving strong solvent–solute interaction, as is likely to occur in and be modulated by charge-transfer processes in molecular electronic devices. *A priori* calculations of this type are, at the moment, at the limit of feasibility. While early results[24] are encouraging, accuracy is not yet quantitative, and since experimental data is not currently available to test the results, we do not pursue such approaches herein. Instead, in Section 3, we introduce an adjustable parameter, an energy-level shift e_0, into otherwise *ab initio* calculations and determine the molecular electronic properties as a function of it. The energy-level shift e_0 is used to include implicitly a range of possible substituent, solvation, and applied electric field effects. What emerges is a more global picture of the superexchange electronic coupling in the oligoporphyrin systems considered.

2. CHEMICAL CONTROL OF INTER-PORPHYRIN COUPLING IN SIMPLE MODEL OLIGOPORPHYRINS

We consider two methods by which the coupling through an oligoporphyrin molecular wire could be modulated *in situ* by chemical means. These are (1) pH control, discussed briefly later, and (2) benzenoid/quinonoid conversion of a bridge ring, discussed in detail. Viewed at the simplest and most general level, benzenoid/quinonoid conversion of the bridge unit in an oligoporphyrin dimer modulates the coupling because the π bonds in the benzenoid formed are delocalized across the bridge, facilitating long-range electronic coupling, while cross-conjugation through the quinonoid link interrupts the delocalization, effectively electrically isolating the two ends of the dimer. It is illustrated in FIGURE 2, which shows the atomic π bond orders[28] for the corresponding benzenoid and quinonoid oligoporphyrin dimers **1** and **2**, respectively. The central ring is clearly strongly delocalized for the benzenoid (even grey shading), while the quinonoid contains isolated double and single bonds (black and light shading, respectively). The phenomenon is actually not restricted to oligoporphyrins, but is a general feature of electron transfer[23,29] through benzenoid/quinonoid bridges.

Interporphyrin coupling can manifest itself in a large number of ways. We choose to model theoretically a property which is important in both spectroscopic applications and to hole transport through oligoporphyrin bridges — the coupling between localized porphyrin HOMO levels, *J*. This property is in fact more precisely defined as half of the energy difference between the first and second ionization potentials of the dimer. Ionization energies can be explicitly evaluated from SCF or density-functional total energies, and inter-HOMO couplings obtained using these two approaches for oligoporphyrins agree quite well.[13] Ionization-energy calculations are expensive and difficult, but reliable SCF results may also be obtained with ease through the application of Koopmans' theorem. This theorem allows J to be determined simply from the differences in the energies of paired molecular orbitals, and we take this approach herein. It is important to perform all calculations at accurate molecular geometries, and to obtain these we use the density-functional B3LYP method since it is both efficient and accurate;[13] a 3-21G basis set is used throughout. Note that, as is essential in quantitative studies (see e.g., Refs. 30,31), our calculations explicitly include all non-nearest-neighbor interactions.

Calculated couplings J between localized porphyrin HOMO levels for the benzenoid oligoporphyrins of FIGURE 2 are shown in TABLE 1, as is the attenuation to this caused by quinonoid conversion. This chemical modulation is shown to reduce the coupling by typically an order of magnitude. If, for example, a hole-transfer process through the oligoporphyrin was being monitored in the Golden-Rule limit, this would imply an attenuation of the conductance of the wire by the square of this ratio, some two orders of magnitude. Clearly, this is much more than would be adequate to provide properly defined switch-on and switch-off states, as would be desired in a molecular electronics application. However, the single-step interconversion of benzenoid and quinonoid forms thus envisaged is somewhat impractical, but the corresponding hydroquinones have benzenoid structures and may be readily converted using wet-chemical or electrochemical techniques to or from either the quinone or the corresponding dioxime, both of which have the quinonoid structure.[13] Indeed, interconversion of benzenoid and quinonoid forms may be feasible using very-low-energy hydrogen tautomerization reactions between pyridinones and pyridinols; our other article in this volume[29] is concerned with the details of such reactions and their possible application to fast color switching devices.

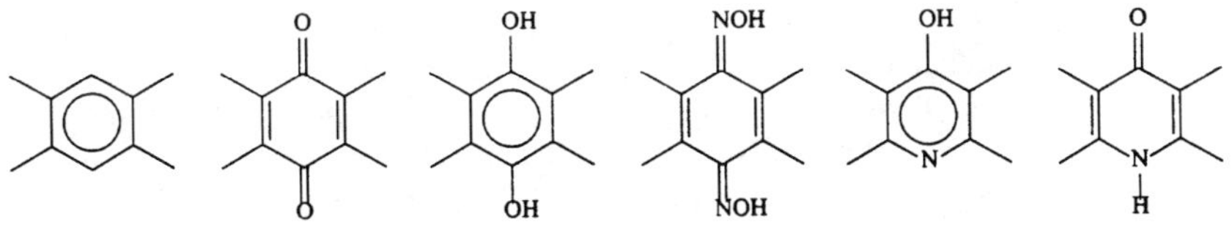

Chemical control of oligoporphyrin molecular wires could also be achieved by introducing asymmetry into the system. This could, for example, place the energy levels at one end of the molecule much higher than those at the other, allowing for the possibility of rectification. A gross but easily chemically controllable effect of this type is monoprotonation of the (basic) azine nitrogens in tetraazaanthracene bridging units, for example, forming molecule **8** from FIGURE 2. The π energy levels on the porphyrin closest to the charge would then be lowered, with SCF calculations at B3LYP geometries for **8** indicating that the magnitude of the induced porphyrin-HOMO level splitting is 1.2 eV. This splitting is much larger that the superexchange coupling (cf. TABLE 1), effectively isolating the ends of the molecule.

TABLE 1. Porphyrin-localized HOMO orbital interactions for some of the oligoporphyrin dimers shown in FIGURE 2 and the attenuation in this coupling when the bridge unit is converted to quinonoid form[a]

Molecule (modification)	Coupling/meV	Attenuation
1	119	17
3 (anthracene)	196	10
4 (Zn)	107	13
5 (bacteriochlorin)	–57	3.4
6 (cyano)	–137	53
7 (methoxy)	132	16
11 ($Ru_2(CO)_2$)	114	16
14 (*f*-coronene)	–101	–

[a]For example, the absolute value of the ratio of the coupling calculated for the benzenoid **1** to that calculated for its related quinonoid **2**.

3. INTER-METALLIC ELECTRONIC COUPLING THROUGH OLIGOPORPHYRINS

A significant number of oligoporphyrins containing metal atoms have now been synthesized,[13] but none have yet been examined to determine the intermetallic coupling strength. A series of spectroelectrochemical and related studies are planned for the forthcoming year, but at this stage we wish to examine theoretically the type of results that are expected- and the related issues of just what constitutes the best possible system for the initial studies. There are two basic ways in which metals may be connected: chelated within a porphyrin ring, or attached to an end group such as phenanthroline. Here, we consider model compounds for each case, inner-ring chelation of $Ru(II)(CO)_2$, and attachment of phenanthroline chelated to $[Cu(I)Cl_2]^-$. Specifically, the molecules considered are structures **9–16** shown in FIGURE 2. For the inner-ring chelated complexes, two types of bridging materials are used: 1,4,5,8-tetraazaanthracene (TAA), a linker commonly used in oligoporphyrins,[13] and a coronene bridge with additional CH groups to fuse rings between the bridge and the porphyrin meso sites (*f*-coronene). In a recent study of through-bridge coupling in oligoporphyrins,[23] the *f*-coronene bridge was found to provide strong coupling between porphyrin orbitals concentrated on the long-axis pyrrole nitrogen atoms, atoms that interact strongly with chelated metals; more common bridging materials such as TAA, by contrast, produce very weak coupling between pyrrole orbitals. We thus expect to find weak intermetallic coupling for TAA bridges but strong coupling for *f*-coronene.

In a long oligoporphyrin chain it is possible to chelate metals in every porphyrin, or, for example, only in the terminal porphyrins. Resonance effects imply that the end-to-end coupling would vary considerably as a function of the chelation pattern; here we study only the terminally chelated porphyrins as, currently, these are easier to produce synthetically.

For the molecules studied we evaluate the inter-porphyrin coupling between the metal d_π localized molecular orbitals at the SCF/3-21G level of theory. As previously described, such calculations *a priori* are rather difficult. However, the most significant is that the metal ions are strongly solvated, and this effect must be taken into account in the calculations. Treatment of solvation is possible[27] either by calculating explicitly the solvent structure via evaluation of the intermolecular interactions, or implicitly by assuming that the solvent forms a dielectric continuum in which the molecule is embedded. Such calculations[24] are, however, just at the level of current technical feasibility, and we choose instead to proceed by a simpler method in which the effect of the solvent is treated empirically through the introduction of a single fittable parameter. We proceed by:

(1) determining the gas-phase geometric and electronic structure for the metallic oligoporphyrin (for small molecules this is done by direct B3LYP optimization, while for larger molecules the geometry is estimated very accurately by splicing together the B3LYP structures of some basic building blocks[23]);

(2) evaluating the Fock matrix in the Löwdin-orthogonalized atomic orbital basis from the SCF molecular orbital coefficients, molecular orbital energies, and atomic orbital overlap matrix;

(3) adding an (arbitrary) energy shift e_0 to the appropriate (see below) diagonal elements of the Fock matrix in order to represent the effect of interactions of the metal, and also possibly its non-porphyrin ligands, with its surroundings (e.g., solvent, applied electric field);

(4) diagonalizing the modified Fock matrix to obtain revised molecular orbital energies and coefficients, and

(5) extracting the appropriate intermetallic couplings.

This procedure is somewhat simplistic, but it does manage to capture many of the key physical features, although it does not account for, amongst other effects, solvent-induced polarizability changes. Some time ago[6,32] we modelled the properties of inorganic complexes using a much simpler approach than this, one that augmented metal orbitals to free-ligand Fock matrices. This approach, which involved two adjustable parameters, was quite successful in modelling a large amount of experimental data. Our present method has been tested by the evaluation of intermetallic couplings between ruthenium d_π orbitals in bis-pentammineruthenium complexes of α,ω-dipyridyl *trans*-polyenes $Ru^{2+}(NH_3)_5$–C_5H_5N–$(C_2H_2)_n$–C_5H_5N-$Ru^{2+}(NH_3)_5$, comparing the results to those observed[6,33] for the corresponding II-III mixed-valence species. This comparison is only possible since the appropriate value of e_0 can be determined chemically, and the results obtained[24] show agreement to within 20% between the calculated and observed values for bridges of length $n = 0$ to 4. Further, we found[24] that these result do not alter significantly when more sophisticated models of solvation are used. We thus expect the results presented here to be at least qualitatively indicative of those that may be observed for oligoporphyrins.

Because e_0 does not explicitly correspond to an observable quantity, its definition is somewhat arbitrary. We have considered three alternatives, all of which achieve the desired qualitative features but they treat minor aspects differently. These are: (1) assign e_0 as an energy shift for the metal d_π orbitals only, (2) assign e_0 to all metal orbitals, and (3) assign e_0 to all metal orbitals plus all orbitals on out-of-plane ligands (CO or Cl^-). We adopt method (1) here because it best allows us to focus on the major features. As we employ a double-ζ basis, we define the key metal d_π orbitals based on their gas-phase representation and do not allow the metal's intra-atomic coefficients to change with e_0.

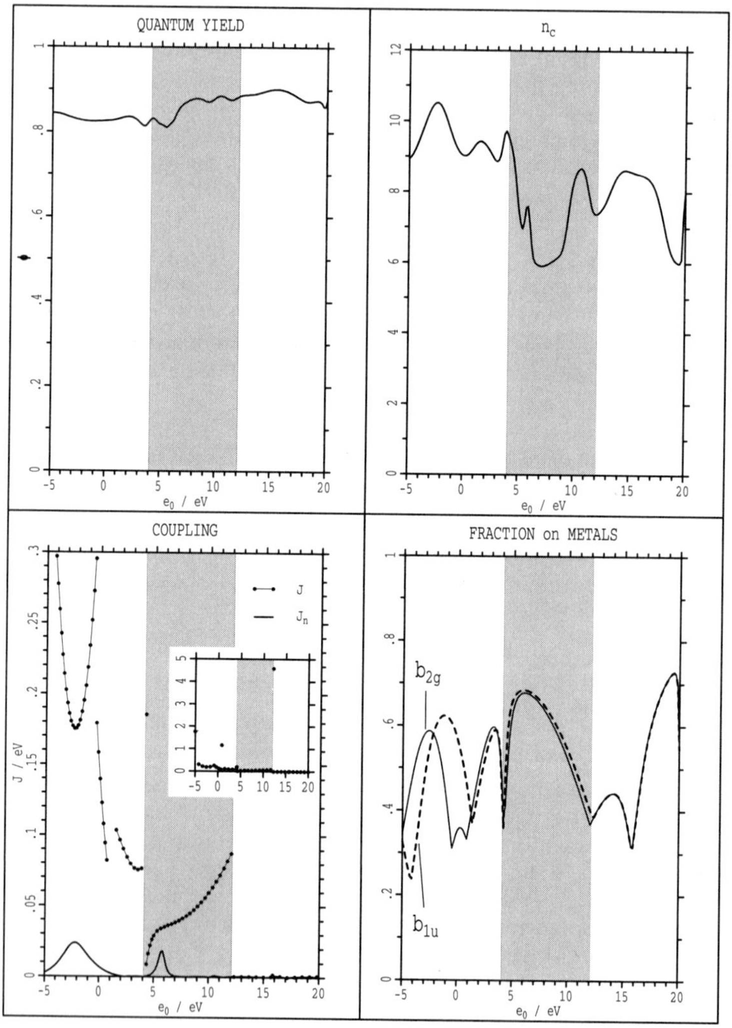

FIGURE 3. The effective end-to-end coupling between the long-axis metal d_π orbitals for the ruthenium oligoporphyrin dimer **11** as a function of e_0, evaluated with (J) and without (J_n), the use of an effective two-level model; the inset shows the data on a reduced scale. J is evaluated at the center of the band gap while J_n is evaluated at its maximum. Also shown are the fractional metal d_π character of the molecular orbitals (of b_{1u} and b_{2g} symmetry), which best singly represent the d_π orbitals, the quantum yield ϕ, and the minimum number of states in an effective n-level model required to adequately describe the kinetics of hole-transfer, n_c. Shown shaded is the oligoporphyrin HOMO-LUMO band gap.

Technically, SCF and B3LYP calculations on dimeric inorganic complexes such as **9-16** are quite difficult because of the large number of low-lying electronic states, states separated in energy by order 1 part in 10^5 of the total energy, or less. The copper complexes are all have a d^{10} configuration, while the ruthenium complexes are all low-spin d^6. A useful technique for obtaining the starting guess for the wavefunction or density is to break complex molecules up into overlapping parts (say central bridge benzene rings), and to average a molecular Fock or density matrix from the individual fragment matrices. From these, initial orbitals can be found which impart the desired electronic structure to the full molecule, and subsequent SCF or B3LYP calculations then rapidly converge.

Step 5, extracting the intermetallic end-to-end couplings, is also not completely straightforward. The simplest method is to assume an effective two-level model and write a two-state Hamiltonian which reproduces the energies associated with the "donor" and "acceptor" states, and this indeed was the approach taken in Section 2. Results obtained for the ruthenium dimer **11** are shown in FIGURE 3 which displays, as a function of e_0, J obtained as half of the difference between the two molecular-orbital energies associated with the localized Ru long-axis d_π levels, as well as the fractional contribution of those metal orbitals to the molecular orbitals. Away from resonances, this fraction would approach unity and indicate that effective two-level models are appropriate. On resonance, this fraction falls to at most 50%, indicating that at least a three-level model is required in order to describe the basic physics; however, FIGURE 3 clearly indicates that even more than three levels are quite often required in order to describe this system. Even when the d_π orbitals are placed in the middle of the oligoporphyrin HOMO-LUMO gap (which extends from $e_0 = 4 - 12$ eV), at least three levels are clearly required. As a resonance is crossed, the identity of the molecular orbitals which most closely resemble the original metal atomic orbitals changes discontinuously. Hence J, when evaluated using an effective two-level model, is a discontinuous function[34] of e_0.

Some experiments, such as, spectroscopic ones, involve molecular eigenstates and directly probe J as evaluated from the effective two-level model, and, despite the obvious limitations, it is rare for theoretical treatments of electronic coupling to proceed beyond this level. Experiments relevant to molecular electronics often involve non-equilibrium electron transport, however, and it will become increasingly necessary to go beyond it. As an alternative to the introduction of an effective two-level model, one can[35] solve directly for the rate constant k for the process in which a signal of some kind is injected at the donor and drained from the acceptor to a bath. The Hamiltonian describing this process is non-conservative, the total energy decaying to zero as the signal is removed to the bath. While this rate depends on the drain rate of the acceptor (and possibly also the injection rate, but here we assume that injection is instantaneous), the maximum possible rate, the *Rabi* rate, can be readily related to J when two-level models are appropriate: $\hbar k = 2^{-1/2} |J|$ (in most experimental situations, rates as large as this are not observed and the kinetics is best described by the Golden Rule limit, in which the rate becomes proportional to J^2). By inverting this equation, an effective coupling J_n can be obtained which is appropriate in the resonance region as well as the non-resonance region. When only two levels are involved, the kinetics of the envisaged process is simply exponential decay, and this is trivially modeled by a single rate constant k. In general,[35] the kinetics is described as the real part of the sum of many complex exponentials; we have reduced this kinetics to a single parameter by defining k based on the total area under the decay curve. Small amounts of the input signal readily get trapped in poorly connected states of the oligoporphyrin bridge, residing there for extremely long times before finally being transferred to the acceptor and escaping to the bath. This dramatically increases the area under the decay curve and hence greatly reduces k and J_n. To circumvent this problem, we have introduced[35] a damping parameter σ (the value herein is set to 0.03 throughout) to remove the contributions of weak components from the decay curve. The net result is that we deem the transfer process to be complete at a cut-off time when a postdetermined fraction, the *quantum yield* ϕ, of the input signal is recovered from the acceptor, and evaluate k from the data obtained up to this point in time. When effective two-level models are appropriate (e.g., for bis-ruthenium complexes of α,ω-dipyridyl trans-polyenes[24]), the quantum yield will easily exceed 90% and J_n obtained in this fashion will agree well with that determined directly using a two-level model. This approach allows an explicit definition[35] of the number of coupled modes n_c as the minimum number of levels needed in an effective Hamiltonian to correctly describe the kinetics of electron transfer up until the cut-off time. For the ruthenium oligoporphyrin dimer **11**, results for the J_n, ϕ, and n_c are also shown in FIGURE 3 for the hole-transfer process from one ruthenium d_π orbital to the other. They indicate that effective two-level models are *nowhere* appropriate for the description of this process; even if the metal levels fell in the middle of the oligoporphyrin HOMO-LUMO gap then a minimum of six levels would be necessary to describe the kinetics when the quantum yield is just 80%. As a consequence, the effective coupling J_n is much less than the two-level result J. It peaks near $e_0 = -2$ eV, when the d_π orbitals are placed in the middle of the oligoporphyrin occupied band. This result is, incidently, quite different from that obtained for ruthenium pyridyl complexes, for which the coupling is strongest to the close lying unoccupied ligand orbitals.[6,24]

Without prior knowledge of the actual value of e_0, it is not possible to predict results appropriate for a given experiment, although we know from chemical evidence that the d_π orbitals reside in the oligoporphyrin band gap and so 4 eV < e_0 < 12 eV. In order to compare results for compounds **9-16** we identify key aspects of FIGURE 3. These properties are shown in TABLE 2. For J, we choose the coupling evaluated in the center of the band gap. This is indicative of results that could be obtained spectroscopically, for example, from the intensity of the intervalence transitions in the corresponding Ru (II)/Ru (III) or Cu (I)/Cu (II) mixed-valence compounds. A shift of e_0 away from the center of the band gap will, in general,[6,35,17] result in increased coupling and decreased attenuation of the coupling with increasing bridge length. This is not the case for **12**, however, as the coupling changes sign within the band gap, thwarting simplistic analysis. For the kinetic coupling J_n we choose a different property and tabulate the maximum value of J_n obtained at any e_0 for each molecule (TABLE 2). This is indicative of what, under ideal energy-matching conditions, would be the maximum coupling available to drive hole transport.

Results for the TAA-bridged complexes **9-13** shown in TABLE 2 indicate that incorporation of a quinonoid linkage considerably reduces the coupling between the metal centers: switching based on quinonoid conversion thus remains a robust effect. However, the overall couplings are quite weak in magnitude. The inner-ring ruthenium complexes display very strong d_π–pyrrole interactions, but these interactions are not conveyed through the bridge to the other end of the molecule. Qualitatively, this arises due to the weak conjugation through the fused pyrrole rings: long-distance interactions are strong when the π electrons are delocalized through *all* intermediate bonds. Indeed, the same behavior is seen in the phenanthroline complexes, the bond-order diagrams of FIGURE 2 indicating that delocalization is lost through the central phenanthroline ring. With the relatively weak coupling comes also a relatively large attenuation of the coupling with increasing bridge length R. Measured between the metal centers, R is given in TABLE 2 for molecules **9-13**, and the coupling is fitted to an exponential decay

$$J = A e^{-\alpha R}$$

with the results given in TABLE 3. Shown for comparison in this table are also results calculated by similar means for the ruthenium α,ω-dipyridyl trans-polyenes.[24] Clearly, the two-level model coupling J decays much faster with increasing bridge length for these oligoporphyrin bridges, although the hole-transfer response is competitive.

Experimentally, single-molecule conduction has only been observed for relatively short molecules, through distances of order 10 Å. The rapid decay of the coupling with increasing length depicted for the porphyrins **9-13** and the ruthenium polyenes in TABLES 2 and 3 is the primary factor responsible for this. Investigations of the coupling through free-base oligoporphyrins[23] indicate that some localized porphyrin orbitals are quite strongly coupled (J = 200 meV in dimers) whilst others remain very weakly coupled. Further, the orbitals with large coefficients on the pyrrole nitrogens were regularly found to be very weakly coupled, leading to the relatively

TABLE 2. Metal-localized long-axis-oriented d_π orbital interactions (at intermetallic separation R) for the inner-ring chelated ruthenium (II) complexes and phenanthroline-linked copper (I) complexes shown in FIGURE 2 and the attenuation in this coupling when the bridge unit is converted to quinonoid form

		J from two-level model		J_n for hole transport	
Molecule	R/Å	Coupling/ meV	Attenuation	Coupling/ meV	Attenuation
9 (Cu, TAA)	17.24	18	3.8	24	0.99
10 (Cu, TAA)	32.69	0.22	110	2.8	4.2
11 (Ru, TAA)	15.51	43	3.1	24	2.6
12 (Ru, TAA)	31.00	0.39[a]	3.5	1.1	11
13 (Ru, TAA)	46.50	0.0058	2.0	0.15	6
14 (Ru, *f*-coronene)	15.24	14	—	13	—
15 (Ru, *f*-coronene)	30.13	2.9	—	2.0	—
16 (Ru, *f*-coronene)	45.22	4.4	—	6.0	—

[a]Difficult to obtain as the coupling passes through a node within the oligoporphyrin HOMO-LUMO gap.

TABLE 3. Bridge-length dependence of the calculated intermetallic coupling for the copper–phenanthroline–oligoporphyrin system **9-10**, the ruthenium-oligoporphyrin systems **11-13**, and the ruthenium α,ω-dipyridyl trans-polyenes[24]

	J from two-level model		J_n for hole transport	
System	*A*/eV	α/Å^{-1}	*A*/eV	α/Å^{-1}
Cu	2.44	0.29	0.26	0.14
Ru-porphyrin	1.17	0.21	0.25	0.16
Ru-polyene	0.28	0.13		

weak couplings and rapid fall-offs seen here. In an attempt to change this picture, *f*-coronene bridges were conceptualized and found to provide strong coupling between pyrrole-nitrogen-based orbitals. The results shown in TABLE 2 for the coupling between the complexed ruthenium-d_π orbitals indicate strong coupling, of order 5 meV, for the tetramer **16**, a molecule whose intermetallic separation is 45 Å. Indeed, the calculated couplings are larger for the tetramer than for the trimer **15**, indicative of bridge resonance. Both for the TAA and *f*-coronene bridges, the effective number of resonant modes n_c is high; only for *f*-coronene do we see the expected consequence of non-exponential bridge-length dependence of the coupling, however.

CONCLUSIONS

We have considered desirable properties for materials to be used in molecular electronic applications and find that oligoporphyrins offer many natural advantages. In particular, we concentrate on a proposed mechanism for chemical control of the switching of the conductivity of oligoporphyrin wires through the conversion of benzenoid rings to quinonoid form. Because this conversion introduces a region of localized single and double bonds into the wire it reduces the π-electron delocalization and hence electrically isolates both sides. The effect is investigated theoretically by consideration of localized porphyrin HOMO orbital interactions and oligoporphyrin-mitigated intermetallic interactions. For intermetallic interactions we find that effective two-level models are inadequate for describing the kinetics of hole transfer. In magnitude, couplings through distances of order 16 Å are predicted to be quite large (0.02 – 0.2 eV) and readily usable. Although for the commonly used TAA bridges the attenuation of superexchange coupling with increasing bridge length is calculated to be quite rapid, possibly measurable (order 10 – 100 μeV) chemically controllable couplings through oligoporphyrin tetramers are anticipated. Nevertheless, for trimeric and larger molecules with linear-acene-derived bridges, such as TAA, the superexchange coupling will most certainly be sufficiently low that hole- or electron-transfer processes will proceed via activated non-resonant (inelastic scattering) processes in which ions are localized on individual porphyrin rings, most probably as polarons. Clearly, considerable experimental and theoretical work is needed to characterize such processes.

Much larger intermetallic couplings, of order 5 meV over distances of 45 Å, are predicted when *f*-coronene bridges are used instead of TAA bridges. This arises because the strongly interacting porphyrin orbitals have large coefficients on the pyrrole nitrogens, and the interactions are so large as to place the coupling in the resonance region. Conduction through single molecules of this type and length should be readily detectable.

ACKNOWLEDGMENT

We thank the Australian Research Council for financial assistance to M.J.C., L.A.J., J.R.R., and L.E.H.

REFERENCES

1. ROTH, S. & C. JOACHIM. 1997. Atomic and molecular wires. Kluwer. Dordrecht.
2. RATNER, M.A., B. DAVIS, M. KEMP, V. MUJICA, A. ROITBERG & S. YALIRAKI. 1998. Ann. New York Acad. Sci. This volume.
3. MUJICA, V., M. KEMP, A. ROITBERG & M.A. RATNER. 1996. Current-voltage characteristics of molecular wires: Eigenvalue staircase, Coulomb blockade, and rectification. J. Chem. Phys. **104:** 7296–7306.
4. BUMM, L. A., J.J. ARNOLD, M.T. CYGAN, T.D. DUNBAR, T.P. BURGIN, L. JONES II, D.L. ALLARA, J.M. TOUR & P.S. WEISS. 1996. Are single molecular wires conducting?. Science **271:** 1705–1707.
5. REED, M.A., C. ZHOU, C.J. MULLER, T.P. BURGIN & J.M. TOUR. 1997. Conductance of a molecular junction. Science **278:** 252–254.
6. REIMERS, J.R. & N.S. HUSH. 1990. Electron and energy transfer through bridged systems. V. The electronic structure of the bis-pentaammineruthenium complexes of α,ω-dipyridyl trans-polyenes in nitrobenzene. Inorg. Chem. **29:** 4510–4513.
7. JUNG, T.A., R.R. SCHLITTLER, J.K. GIMZEWSKI, H. TANG & C. JOACHIN. 1997. Controlled room-temperature positioning of molecules: Molecular flexure and motion. Science **271:** 181.
8. CROSSLEY, M.J., S. MARTIN, L.A. JOHNSTON & C.W. TANSEY. 1998. Structural and electrical characterization of self-assembled monolayers of porphyrin-based molecular wires. Adv. Mater. In preparation.
9. CROSSLEY, M.J. & J.K. PRASHAR. 1997. Thiophene-appended porphyrin systems. Tetrahedron Lett. **38:** 6751–6754.
10. AZUMI, R., M. MUTSUYOSHI, S. KURODA, L.G. KING & M.J. CROSSLEY. 1995. Orientation change of dimer-type porphyrin in Langmuir-Blodgett films caused by a trigger molecule. Langmuir **11:** 4056–4060.
11. CROSSLEY, M.J., P.L. BURN, S.J. LANGFORD & J.K. PRASHAR. 1995. Porphyrins with appended phenanthroline units: A means by which porphyrin π systems can be connected to an external redox centre. J. Chem. Soc. Chem. Commun. 1921–1923.
12. HUSH, N.S., A.T. WONG, G.B. BACSKAY & J.R. REIMERS. 1990. Electron and energy transfer through bridged systems. VI. Molecular switches: The critical field in electric-field activated bistable molecules. J. Am. Chem. Soc. **112:** 4192–4197.
13. CROSSLEY, M.J., L.A. JOHNSTON, J.R. REIMERS & N.S. HUSH. 1998. A laterally extended porphyrin switch incorporating a switchable unit. J. Am. Chem. Soc. In preparation.
14. CROSSLEY, M.J. & C.S. SHEEHAN. 1998. A new approach to molecular-sized rectifiers based on the Aviram-Ratner proposal. Synthesis of thienopyrazinobisporphyrins with an inbuilt electronic bias. J. Chem. Soc., Perkin Trans. I. Submitted for publication.

15. HOPFIELD, J.J., J.N. ONUCHIC & D.N. BERATAN. 1989. Electronic shift register memory based on molecular electron-transfer reactions. J. Phys. Chem. **93**: 6350–6357.

16. GILAT, S.L., S.H. KAWAI & J.-M. LEHN. 1993. Light-triggered electrical and optical switching devices. J. Chem. Soc. Chem. Commun.: 1439–1442.

17. LEHN, J.-M. 1995. Supramolecular Chemistry. VCH. Weinheim.

18. CROSSLEY, M.J. & P.L. BURN. 1991. An approach to porphyrin-based molecular wires: Synthesis of a bis(porphyrin)tetraone and its conversion to a linearly conjugated tetrakisporphyrin system. J. Chem. Soc. Chem. Commun.: 1569–1571.

19. CROSSLEY, M. J., L.J. GOVENLOCK & J. PRASHAR. 1995. Synthesis of porphyrin-2,3,12,13- and 2,3,7,8-tetraones: Building blocks for the synthesis of extended porphyrin arrays. J. Chem. Soc. Chem. Commun.: 2379–2380.

20. LÜ, T.X., J.R. REIMERS, M.J. CROSSLEY & N.S. HUSH. 1994. Rigid fused oligoporphyrins as potential molecular wires: I. Geometry and connectivity of 1,4,5,8-tetraazaanthracene-bridged oligoporphyrins. J. Phys. Chem. **98:** 11878–11884.

21. REIMERS, J.R., T.X. LÜ, M.J. CROSSLEY & N.S. HUSH. 1996. The geometric and electronic structures of some simple porphyrinic molecular wires. Nanotechnology **7:** 424–429.

22. REIMERS, J. R., T.X. LÜ, M.J. CROSSLEY & N.S. HUSH. 1995. The mechanism for inner-hydrogen migration in free base porphin: *Ab initio* MP2 calculations. J. Am. Chem. Soc. **117:** 2855–2861.

23. REIMERS, J. R., L.E. HALL, M.J. CROSSLEY & N.S. HUSH. 1998. Rigid fused oligoporphyrins as potential versatile molecular wires: II. B3LYP and SCF calculated geometric and electronic properties of 98 oligoporphyrin and related molecules. J. Phys. Chem. B. Submitted for publication.

24. REIMERS, J.R. & N.S. HUSH. 1998. Density functional modelling of the electronic structure of the bis-pentaammineruthenium complexes of α,ω-dipyridyl trans-polyenes. Chem. Phys. Lett. In preparation.

25. ZENG, J., N.S. HUSH & J.R. REIMERS. 1996. Solvent effects on molecular and ionic spectra. IX. The change in dipole moment accompanying metal to ligand charge transfer in pentaamminopyridylruthenium(II). J. Phys. Chem. **100**: 19292-19294.

26. HUSH, N.S. & J.R. REIMERS. 1995. Vibrational Stark spectroscopy 1. Basic theory and application to the CO stretch. J. Phys. Chem. **99:** 15798–15805.

27. BACSKAY, G.B. & J.R. REIMERS. 1998. Solvent effects. *In* Encyclopaedia of Computational Chemistry. P. von R. Schleyer, Ed. Wiley. London. In press.

28. VILLAR, H.O. & M. DUPUIS. 1987. Bond order and valencies from ab initio wavefunctions: Application to prototypical molecules and to the characterization of solitons in polyenes. Chem. Phys. Lett. **142:** 59–66.

29. REIMERS, J.R., L.E. HALL, N.S. HUSH & K. SILVERBROOK. 1998. Chemical control of tautomerization-based molecular electronic and color switches. Ann. N. Y. Acad. Sci. This volume.

30. PADDON-ROW, M.N. & K.D. JORDAN. 1993. Analysis of the distance dependence and magnitude of the $\pi+,\pi-$ and $\pi+^*,\pi-^*$ splittings in a series of dietynyl[n]staffanes: An ab initio molecular orbital study. J. Am. Chem. Soc. **115:** 2952–2960.

31. CURTISS, L.A., C.A. NALEWAY & J.R. MILLER. 1995. Superexchange pathway calculation of electronic coupling through cyclohexane spacers. J. Phys. Chem. **99:** 1182–1193.

32. REIMERS, J.R. & N.S. HUSH. 1990. Electron and energy transfer through bridged systems. IV. The electronic structure of the bis-pentaammineruthenium complexes of α,ω-dipyridyl trans-polyenes in D_2O. Inorg. Chem. **29:** 3686–3697.

33. WOITELLIER, S., J.P. LAUNAY & C.W. SPANGLER. 1989. Invervalence transfer in pentaammineruthenium complexes of α,ω-dipyridyl polyenes. Inorg. Chem. **28:** 758–762.
34. REIMERS, J.R. & N.S. HUSH. 1990. Electron and energy transfer through bridged systems. II. Tight binding linkages with zero asymmetric band gap. Chem. Phys. 146: 89–103.
35. REIMERS, J.R. & N.S. HUSH. 1989. Electron and energy transfer through bridged systems. I. Formalism. Chem. Phys. **134:** 323–354.
36. REIMERS, J.R. & N.S. HUSH. 1994. Electron and energy transfer through bridged systems. III. Tight binding linkages with finite asymmetric band gap. J. Photochem. Photobiol. A **82:** 31–46.

Molecular Wires: Charge Transport, Mechanisms, and Control

MARK A. RATNER,[a,d] BILL DAVIS,[a] MATHIEU KEMP,[a] VLADIMIRO MUJICA,[b] ADRIAN ROITBERG,[c] AND SOPHIA YALIRAKI[a]

[a]*Department of Chemistry, Northwestern University, Evanston, Illinois 60208, USA*
[b]*Universidad Central de Venezuela, Caracas 1020A, Venezuela*
[c]*National Institute of Standards & Technology, Gaithersburg, Maryland 20899, USA*

ABSTRACT: By molecular wires, one generally means molecular structures that transmit a signal between two termini. We discuss some theoretical models and analysis for electronically conductive molecular wires in which a single molecule conducts charge between two electrodes. This situation resembles both intramolecular non-adiabatic electron transfer, in which electronic tunneling between donor and acceptor is seen, and mesoscopic quantum transport.

We discuss formal methods for predicting conductance in molecular wire circuits. The critical component that differs from the usual conductivity is the interface between electrode continuum and the discrete levels of the molecule. This can be described in several ways. We present an analysis based on the Bardeen tunneling formula. Specific problems (electron polarization, disorder, nuclear scattering, charge distribution) are discussed. Finally, the differing mechanisms expected for the conductance, ranging from ballistic tunneling to gated transfer, are outlined.

1. INTRODUCTION

Molecular electronics[1–9] is of interest because of some enticing properties of molecules: perhaps the most important of these is their size. Molecules make up structures with characteristic lengths from sub-nanometer to hundreds of nanometers. They therefore bridge the dimensional regime between lithographic structures and atomic sizes. Other attractive properties of molecules include their remarkable synthetic adaptability, light weight, relative stability, nontoxicity, and low cost. Perhaps most intriguing of all is their novelty: while response properties of molecules have been of interest for a hundred years or more, assembling those molecules into structures with nanoscale dimension, and investigating the properties of such structures, is a very new and exciting enterprise — it is this enterprise that, in general, constitutes molecular electronics.[1–9]

Perhaps the simplest component of an integrated electronic system is an interconnect, or wire. In the simplest instance, this merely serves to convey either charge or excitation energy from one active unit to another. The two "flavors" of molecular wires that have been investigated are photonic molecular wires, which transfer excitation,[10] and electronic molecular wires, which move either electron or hole

[d]Address for telecommunication: Phone, 847-491-5652; fax, 847-491-7713; e-mail: ratner @ chem.nwu.edu

TABLE 1. Similarities and differences between intramolecular electron transfer and wire conductance

Aspect	Molecular Wire	Nonadiabatic Intramolecular Electron Transfer
Potential Energy Diagram	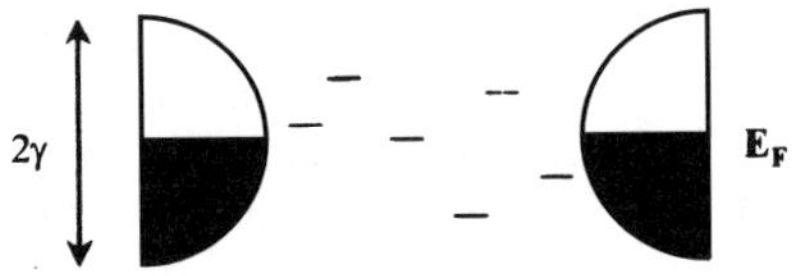	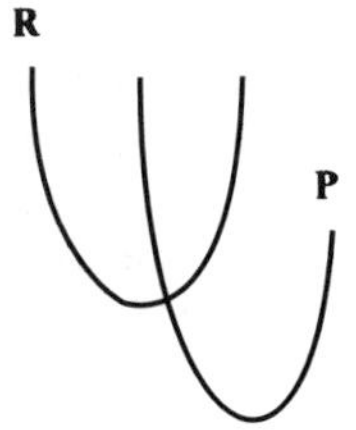
Observable	Current	Rate Constant
Continuum	Electronic States	Nuclear Motions
Initial State	Electrode	Vibronic Levels
Process	Electron Tunneling	Electron Tunneling
Theory	$\frac{2\pi}{\hbar} e \sum f(E_1)[1-f(E_N+eV)]T_{1N}^2\delta(E_1-E_N)$	$k = \frac{2\pi}{\hbar} T_{DA}^2(DWFC)$

charge.[11,12] Our discussion here will be limited to charge transport (although many of the same ideas are applicable to photonic wires).

Charge transport in molecular wires is clearly closely related to intramolecular electron transfer; these are reactions in which either photo-induced or initially deposited non-equilibrium electron density transfers, in a rate process, between sites linked by a molecular bridge.[13–22] Electron-transfer reactions of this kind are very important in areas ranging from corrosion to photosynthesis, and their understanding has been a major theme of physical chemistry for the past half century.[13,14] The link to molecular wire interconnects is a direct one, since in both cases tunneling occurs from an initial state to a final state through a bridging medium. Some differences and similarities between the intramolecular electron transfer and the molecular wire situation are sketched in TABLE 1.

Biological systems require electron transfer for energy transduction, signaling, charge neutralization, and enzyme function.[11–22] Thus the electron-transfer process constitutes a bridge from synthetic molecular electronics to molecular rate reactions to biological structures.

The molecular wire differs from macroscopic conductive wires in its size, and therefore in the nature of its energy levels and transport processes. Synthetic techniques based on molecular self-assembly,[6,12,20] and measurement techniques based on scanning probe structures[23] and nanoassembled circuitry[24–27] have led to the study of the conductance in molecular wires under varying conditions. The current paper is devoted to discussion of some computational models for molecular wire conductance, some of the expected phenomena in molecular wire conductance, and some of the challenges that remain in this area. Section II describes theoretical and computational approaches to the conductance problem, some recent results, and some of the outstanding problems that remain to be effectively treated. Section III overviews some categories of mechanisms expected for charge transport in molecular wire interconnects, and some remarks are made in Section IV.

II. FORMAL ANALYSES OF MOLECULAR CONDUCTANCE

1. Formalism

Landauer Formula

The Landauer approach[28,29] is a simple and powerful way of addressing the problem of computing the conductance across a molecular wire. In essence, Landauer argues that in the steady state, the conductance is proportional to the transmission probability t^2 of the electronic states around the Fermi level:

$$g \propto \frac{2e^2}{\pi\hbar}t^2 \tag{2.1}$$

This result presupposes that an electron in the donor electrode proceeds to the acceptor electrode through an elastic scattering medium. The assumption of elasticity is crucial to Landauer's analysis, and a point of major departure from previous theories of electronic conduction. Whereas other theories assume that the same mechanisms that limit the current also cause energy dissipation, Landauer observes that even in the absence of inelastic mechanisms, the system has a well-defined conductance. He points out that the electron arrives in the acceptor electrode in a hot state, and once there thermalizes to the Fermi level of the second electrode through e-e scattering.

The importance of Landauer's formula lies in its expressing an explicit connection between the transmission probability and the conductance. It shows that the best conductance, or minimal resistance, is finite. It also shows, by acknowledging that the number of conduction channels is discrete, that the maximum conductance is an integer multiple of the quantum of conductance. The Landauer approach has been used in several very successful computational[6,29,30] studies of wire conductance.

Time-Independent Scattering

Landauer's analysis is fundamental, because it expresses the basic physics of conduction through mesoscopic media. From a computational standpoint, however, other approaches (such as ones based on cluster analysis)[30] may be more useful or direct.

We have addressed the problem of developing a formalism applicable to molecules, based on Bardeen's analysis of tunneling.[31] In the case of small voltage, we obtained an expression for the conductance similar in form to Landauer's:

$$g \propto \frac{2e^2}{\pi\hbar}\Delta_D\Delta_A|G_{1N}|^2 \tag{2.2}$$

where the Δs are chemisorption couplings to the electrodes, and G_{1N} is the 1,*N* element of the total molecular Green's function.

As with Landauer's expression, Equation 2.2 shows that the conductance depends on the transmission probability, but it makes explicit the factors that influence the transmission probability. First, the chemisorption couplings Δ_D and Δ_A enter separately, and multiplicatively. Second, the conductance maxima correspond to the poles of the molecular Green's function, and therefore to the MO energies of the isolated molecule (up to electrostatic screening and lifetime corrections).

Another extension allowed by this formalism is to the important case of finite voltage. When the voltage is finite, the current is given by:

$$I = \int_{-eV}^{0} dE f_D(E)(1 - f_A(E + eV))g(E)/e \tag{2.3}$$

where $g(E)$ is the conductance corresponding to transmission at an energy E, and where the Fermi level of the donor reservoir defines the zero of energy. The result is straightforward to interpret: the total current is the sum of all the conductances coming from all energy states between the two Fermi levels. A direct consequence of this result is that the current is linear for small voltages. A second consequence arises from the resonance-like structure of the conductance spectrum $g(E)$: indeed, when any of the molecular energies crosses one of the Fermi levels, the current increases, and remains constant until the next energy crossing. The resulting current-voltage curve is staircase-like[34] (an example is shown in FIGURE 1), and very similar to what is observed experimentally.

Even using a simple Hückel-like model for the molecule, surprisingly accurate results can be obtained: for conductance through an alkanethiol structure and through a buckyball, the calculation using Equation 2.2, with the band structure parameters for gold, agrees with measurements to within a factor of 2.[32] The resistance is of order 10 MΩ, so that these molecules can indeed act as wires (1 volt of potential gives a current of roughly 0. 1 μA).

Keldysh Formalism

The previous two approaches are rigorously justified for one-electron, time-independent Hamiltonians. This is appropriate in some situations, such as for small voltages, but it breaks down in other important cases. The optical response of a wire, the charging of the molecule, and the charge transfer occurring at the interface between the wire and the electrodes, are three cases where either because of dynamical or of many-electron aspects, it is necessary to go beyond the simple formalism.

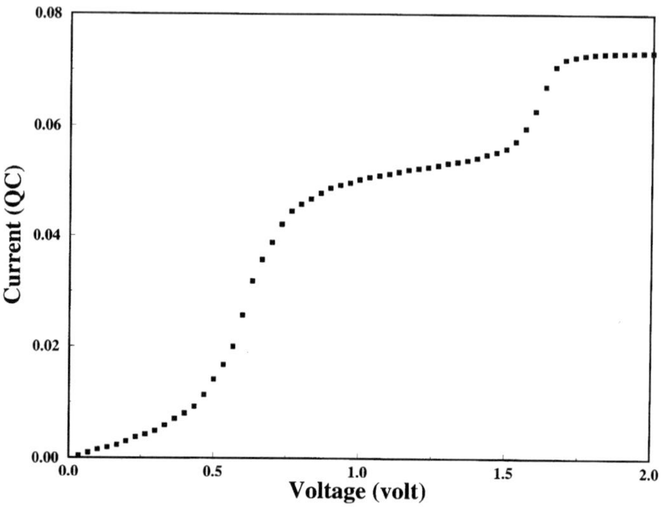

FIGURE 1. Calculated current-voltage characteristic of a molecular wire. The molecule is described by a simple four-site tight-binding Hamiltonian with equal site energies and site couplings. The current is given in units of the quantum of current. The current increases whenever one of the molecular MO energies crosses the donor/acceptor reservoir Fermi levels. Adapted from Reference 36.

The Keldysh formalism[29] provides a general method for dealing with these more complex situations. In essence, it tells how to compute the density matrix. Clearly, this is crucial for understanding the photoresponse, where many changes in the conduction properties can be traced to changes in the density matrix, or the charging of the molecule associated with coulomb blockade, which signals a change in state occupancy.

Recent important applications of the Keldysh formalism include the study of the transient behavior following application of the voltage, and that of the transient photoresponse.[33] In both these cases, the results are somewhat unexpected, and include photon sidebands, nonlinear transient response, and higher harmonic generation. It is probably fair to say that new conceptual developments are likely to come out of these studies, developments that could influence our understanding of what controls transport in wires.

2. Recent Developments

The study of current in molecular wire circuits involves the composite system — reservoirs and molecule. In the analysis that leads to Equation 2.2, the interaction between electrode and wire is modeled as a one-dimensional continuum coupled to the molecule at one point. In a real system, however, this is too simple a view. To make the theory more realistic requires generalization to account for the many-orbital nature of the contact.

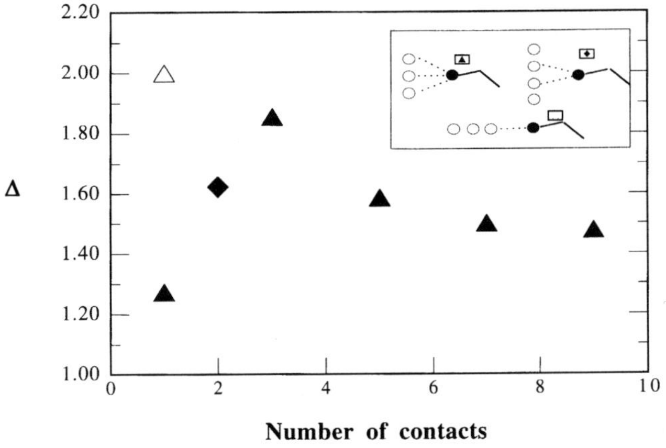

FIGURE 2. Spectral density Δ in units of the one-dimensional electrode half-bandwidth 2γ, as a function of number of contacts between molecule and electrode, and for various geometries. The *open triangle* shows the spectral density of a one-dimensional electrode connected to the molecule by a single orbital (lower part of the *inset*). The *solid triangles* correspond to a two-dimensional metal multiply connected to a single orbital (upper left part of the *inset*). The *solid diamond* shows the spectral density of a two-dimensional metal, with two atoms connected to a single orbital (upper right part of the *inset*).

This generalization has been realized recently. Although we only present results for two-dimensional metallic electrodes, the level of generalization permits the study of different contact geometries, the effect of electrode dimensionality, and more complex molecular structures, such as monolayers of molecules.

The generalization amounts to writing the conductance as:

$$g = \frac{2e^2}{\pi\hbar}\sum_{l,r}\sum_{l',r'} G_{lr}G^*_{l'r'}\Delta_{ll'}\Delta_{rr'} \qquad (2.4)$$

where the sums run over all the contacts orbitals l (left) and r (right), and where the spectral density matrices Δ are proper generalization of Newns's chemisorption spectral density.[31] A typical result is shown in FIGURE 2. For a single wire, the single contact chemisorption parameter decreases from the one-dimensional to the two-dimensional case, for the same unit of energy. The effect of the number of contacts and of geometry on Δ is shown in FIGURE 3.

The formalism also permits the study of multiple wires with different connectivities between electrodes. Through interaction with the electrodes, the molecules acquire an indirect intermolecular coupling, reflected in the spectral densities. We observe that in the case of single interchain coupling — molecules with one particular active side group — the conductance as a function of position of the active site is maximal when the interchain coupling takes place in the middle of the chain. Fi-

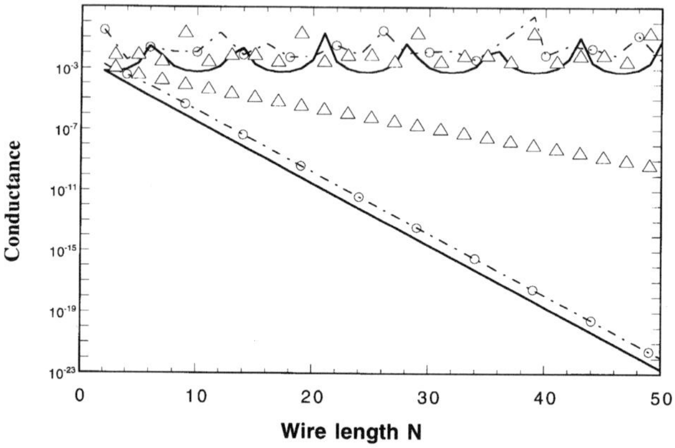

FIGURE 3. Length dependence of the conductance of a system of two identical wires attached between two-dimensional metallic electrodes. The wires are attached to two metallic atoms, as shown in the *inset* of FIGURE 2 (*solid diamond*). The conductance is given in units of the quantum of conductance. The *thick unbroken line* corresponds to two non-interacting wires, infinitely separated; the *dashed line with circles* represents two wires connected by a single orbital; the *open triangles* correspond to wires connected at every point. The length dependence is computed at two injection energies: Injection below the molecular energy band produces exponential decay. The magnitude and the decay rate of the conductance are very sensitive to the coupling between wires; injection in the molecular energy band produces similar conductance oscillations in all cases.

nally, we observe that the maximal conductance increase occurs for a fully connected network.

3. Outstanding Problems

Effect of Molecular Vibrations

A remarkable aspect of molecular wires is that their response is dominated by quantum coherence effects, sometimes even at high temperature. The possibility of having a high temperature quantum device is certainly very attractive. At a very basic level, high temperature coherent operation relies on the conjunction of two factors: first, the coupling between electrons and molecular vibrations must be small, and second, the thermal motion of the wire atoms must also be small. As studies of static disorder have shown,[34] the second condition is required because the conductance is exponentially sensitive to variations in the molecular Hamiltonian (practically speaking, a variation of 20% in the overlap between two atoms is enough to change the conductance by one order of magnitude).

The first factor is not completely understood, and will have to be addressed in the near future. The second factor has been studied recently by Roitberg *et al.*[34] As FIGURE 4 shows, thermal disorder is not as important as might have been anticipated.

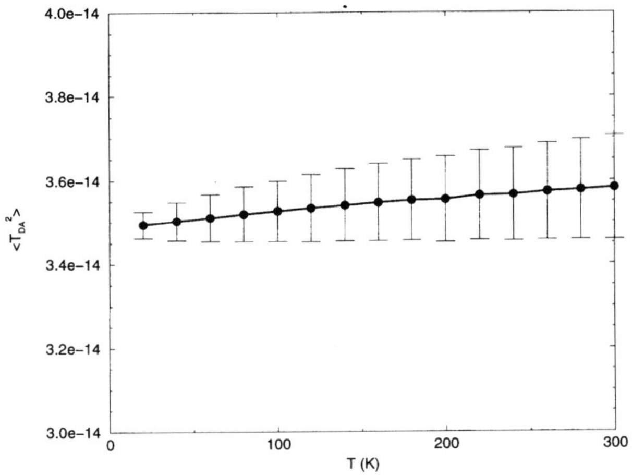

FIGURE 4. Effect of thermally induced disorder, for a model polypeptide bridge. The conductance fluctuations are small, of order 5% at room temperature. (From Roitberg *et al.*[35] Reproduced by permission.)

Thermal disorder does broaden the distribution of conductance values, but even at room temperature, the fluctuations are only of order 5%.

The system studied by Roitberg *et al.* is entirely made of strong covalent bonds. It is not clear at present whether their prediction of small thermal influence can be extended to systems where gating is expected to be important.

Electrostatics

The last aspect we discuss is the inclusion of electrostatic effects in the modeling of molecular wires. At a basic level, a molecular wire is made up of two parts, the molecule and the two electrodes. The problem we are confronted with is that, in order to make it from one electrode to the other, the charge carrier must go through the small molecular bottleneck. At the simplest level, one can assume that the molecule is not affected by the presence of the electrodes. This is what is usually done. At a second level of complexity,[36] one assumes that the electrostatic potential varies linearly between the electrodes. As a result, the electronic structure of the molecule is Stark-modified, and the current-voltage response of the device changes. A third level of complexity comes from the recognition that a change in the electronic structure due to the external field produces redistribution of the charge on the molecule. As the molecule loses charge neutrality, it produces an electrostatic potential. This extra potential affects the molecular Hamiltonian, which in turn affects the charge distribution, etc. As a result, the electrostatic potential acting on the molecule is not fixed externally, but must be solved self-consistently along with Schrödinger's equation.[37]

The problem is, in fact, slightly more complex. Classical electrostatics teaches that a charge in front of a plate maintained at a constant potential produces a charge rearrangement on the metal (image charge). Quantum mechanically, the same effect takes place. From the molecular wire perspective, this causes all sorts of difficulties.

The most important of these is that the partitioning of the electrode into an effective molecular Hamiltonian, which is at the heart of all present calculations, is now a serious task. At present, this problem is still open.

A second facet of the electrostatic problem is the following: when computing current in a wire using low-level theories, there is an ambiguity that comes from the positioning of the electrode Fermi level. Charge transfer can produce a dipole layer at the interface, whose effect is to push the Fermi level until the system becomes stable. By analogy with metal-semiconductor interfaces, the system is expected to be stable when the Fermi level falls in the HOMO-LUMO gap. The difficulty here is twofold: first, the gap is typically 1–2 eV wide, which leaves a lot of freedom for placing the Fermi level. Second, studies of the dependence of the conductance on the Fermi level position show that the conductance changes by orders of magnitude with small changes in the Fermi level position. From a predictive standpoint, this combination is disastrous. In order to make accurate conductance predictions, this problem has to be solved.[29,30]

Some of these difficulties can be solved using the methodology suggested by Lang.[37] He takes into account all electrostatic and exchange-correlation effects, using DFT. Further, he is able to find the precise location of the Fermi level. It is not clear how DFT must be used when the voltage is finite, which is required to understand coulomb blockade, but certainly DFT looks very promising as a tool for understanding the full complexity of the problem.

III. CHARGE TRANSPORT MECHANISMS FOR MOLECULAR WIRES

Because actual measurements on individual molecular wires are still rare, not all of the transport regimes have yet been investigated. Striking similarities with heterostructures, quantum point defect and mesoscopic systems,[38] suggest particular transport mechanisms, as do comparisons with conductive polymers[39] and long-range electron transfer. On the basis of computation, modeling, measurement, and analogy, one might suggest that there are, in fact, five appropriate mechanisms for charge transport in molecular wires between metallic contacts:

1. Coherent Electron Motion: Nonresonant

This is the most common situation for long-range electron transfer reactions, and should hold for molecular wire conduction in the situation where the Fermi level of the contact occurs in the middle of the broad HOMO/LUMO gap for the molecular wire. Under these conditions, a superexchange-type mechanism should hold:[40–43] In the absence of strong dissipation or trapping effects, the site orbitals associated with the molecular wire provide a superexchange pathway for the electron to transfer down the wire. Because the local mixing matrix elements are small compared to the gap, simple perturbation theory suggests that the behavior should be exponential.[38–43] Under these conditions, then, the conductance in the linear regime of Equation 2.2 will be proportional to the square of the molecular Green's function evaluated[31] at the Fermi energy; this in turn is given (in the simplest case) by

$$|tG_{1N}|^2 = |t/\Delta E_{\mathrm{BW}}|^N \quad \mathbf{(3.1)}$$

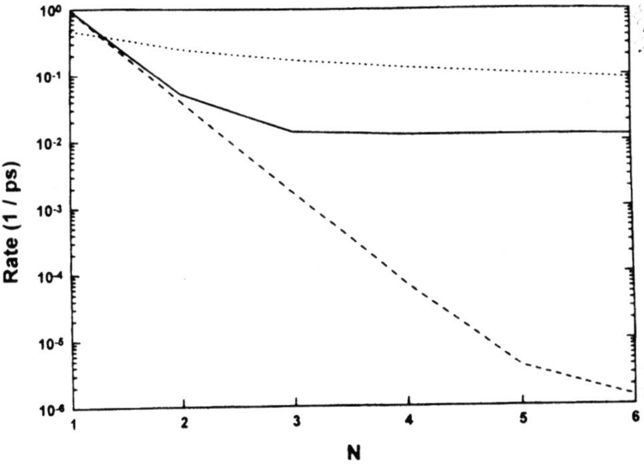

Figure 5. Evolution of the electron transfer rate as N, the number of bridge sites, is increased. $\Delta E_{BW} = 1500\ cm^{-1}$ and $t = 300\ cm^{-1}$. The plots correspond to systems where the magnitude of the bridge dephasings are $\gamma = 1\ cm^{-1}$ (*unbroken line*), $10^{-3}\ cm^{-1}$ (*dashed line*), and $10^4\ cm^{-1}$ (*dotted line*). Even for moderate values of the dephasing, the rates are predicted to quickly become nearly distance independent. (From Davis *et al.*[49] Reproduced by permission.)

Here t, ΔE_{BW}, and N are, respectively, the matrix element between neighbor sites on the molecular wire, the gap energy between the Fermi energy on the electrode and the site orbital energy on the wire, and the number of sites on the molecular wire. This formula (which holds for the situation $|t/\Delta E_{BW}| \ll 1$, for a linear, N-site equivalent chain) yields exponential behavior[34,44] of the conductance $g = A\ \exp(-\beta N)$, where N indicates the number of sites in the wire (proportional to wire length) and β is the characteristic decay parameter, $\beta = 1/2 \ln |t/\Delta E_{BW}|$. This exponential decay should hold for short wires with large HOMO/LUMO gaps, such as oligoalkanes or very short conjugated wires. The exponential length dependence is very widely found in long range electron-transfer reactions, and is sketched in FIGURE 5.

2. Coherent Electron Tunneling: On Resonance

When the Fermi levels become resonant or near resonant with site levels or molecular orbital energy levels on the bridge, the superexchange development no longer holds. Indeed, under such conditions, a chemical rate analogy fails, because it is not clear how to differentiate reactant and product. One can do simple calculations, and the result is that any resistance must arise from contact behavior; this is one of the important messages of the Landauer formula.[28,29] In the absence of dissipative effects, the conductance should scale linearly with the transmission coefficient and therefore with the number of transverse eigenmodes in the wire. This suggests that longer wires might actually suffer no loss in conductance (in the absence of any weak quantization due to the surface or any structural trapping or scattering). The linear increase in conductance with a number of modes, given by the Landauer formula, has been spectacularly observed in atomic wires,[30,45] where the scattering or the probability of transmission, T, is essentially unity for each available channel; the Landauer formula then suggests that, indeed, the conductance will increase with the number of modes.

Such ideal behavior is improbable for molecular wires, given the nature of the molecular eigenstates. Nevertheless, injection at resonance will result in only very weak length-dependence (in the absence of inelastic scattering, trapping, dissipation or surface quantization). For symmetric molecules, Datta[46] has developed a very elegant result for the conductance, based on the Landauer formula and the Friedel sum rule. The low-temperature conductance in metallic carbon nanotubes apparently is in this regime of behavior.[26]

3. Incoherent Transfer — Ohmic Behavior

If the electronic levels in the molecular wire couple either to nuclear motions (vibration, libration) in the wire, or to modes of the environment, then the electronic states in the wire will develop an effective intramolecular lifetime, and inelastic scattering can become important. Under these conditions, the wire should behave very much like the usual sort of electrical wire, with the conductivity now a well-defined property. The conductivity itself will depend on inverse length, as suggested by Ohm's law.

The transition between coherent transport for short wires at low temperatures with large gaps, and incoherent conductivity behavior for longer wires at higher temperatures with smaller gaps can be derived phenomenologically, using a density matrix analysis.[47–49] Results for a four-site wire[49] are shown in FIGURE 5. Note that for very short chains, the rate constant (these calculations were done for rate constants rather than for conductances, but the results should be transferrable) decays exponentially with chain length, as suggested in Section I above. For longer chains, however, the incoherent, frictional channel becomes dominant. It can be shown,[49] using a very simple single-site model, that the rate constant for small friction, γ, behaves like

$$k = k_0 \left| t/\Delta E_{\mathrm{BW}} \right|^N + \gamma t^2/\Delta E_{\mathrm{BW}}^2 \qquad \textbf{(3.2)}$$

The second term is the incoherent term, the first term the coherent one. In the case of large friction, a slightly different limit occurs, such that

$$k = k_0 \left| \Delta E_{\mathrm{BW}}^2 \right| / \gamma \qquad \textbf{(3.3)}$$

The characteristic turnover, from rate constant scaling like friction at low friction to rate constant scaling like inverse friction at high friction, is reminiscent of the Kramers escape problem in one-dimensional potentials,[50] and to the expected result for electron-transfer reactions.[51] The important insight, however, is that for long wires, an incoherent channel may well be more efficient than a coherent one, if the gap between the Fermi level and the HOMO or LUMO is sufficiently large.

4. Quasiparticle Formation and Diffusion

In conductive polymer systems[39] the primary charge carriers at ambient temperature consist of quasiparticles. For degenerate electronic ground states (polyacetylene), the carriers are charged solitons, corresponding to a structural defect, with (formally) an extra electron sitting on a free radical site. These carriers transport electrical charge through the conductive polymer; the carrier consists of the electron plus a lattice distortion which corresponds to a local kink in the bond length displacement pattern. Solitons extend over a dozen or so lattice sites, and acquire effective masses that correspond to self-trapped electrons.

In non-degenerate conductors, such as poly *p*-phenylene or poly *p*-phenylenevinylene, two carriers will trap nearby one another, giving rise to the so-called bipolaron effect. Once again, in these systems charge is carried not by individual electrons but rather by charged structural defects.[39]

Both solitons and bipolarons have been extensively studied in isolated systems and in the bulk. They exhibit ohmic transport behavior, but the resistivity is not proportional to temperature (as it is in a good metal). Indeed, electronically conductive polymers exhibit highly unusual charge-transport processes; nevertheless, one expects that defect motions of this kind might well provide effective conduction in longer molecular wires.

5. Gated Electron Transfer Mechanisms

One substantive difference between molecules and other nanoscale structures such as atomic wires, quantum dots, or mesoscopic junctions is that molecular stereochemistry is dynamic: that is, molecules can change their shape from one local minimum geometry to another. These different geometries have different electronic structures, and therefore different conductances should be expected. Several examples involve, for example, the difference between a chain of cyclopropane structures and a chain of open shell propenes, or all *cis* vs. *trans* polyacetylene. More generally, the conductance can be gated by change of the nuclear geometry of the molecular wire.

For example, computational study and electron-transfer rate measurements show that difference in local stereochemistry can substantially change intramolecular electron transfer rates. A very simple example is shown in TABLE 2, from work of Closs *et al.*,[52] who observed that, as a substituent on the cyclohexane ring changed position from equatorial to axial, the rate of electron transfer changes by roughly a factor of ten.

While the electron-transfer process itself, and therefore the conductance, will not substantially modify the geometry, geometric modification can substantially change the tunneling current. In this sense, electron-transfer processes through molecules capable of undergoing dynamic stereochemical change can be considered gated — that is, by changing the local molecular geometry, one can change the conductance. Geometric change can be caused by chemical reaction, thermally, or (perhaps most interestingly) by photoisomerization. Gating reactions of this sort are of substantive importance in biological systems, and might well be important in the biological analogues of molecular wires (that is, long-distance electron-transfer schemes in proteins).

TABLE 2. Effect of difference in stereochemistry on the rate of electron transfer. The compounds are all based on cyclohexane. D and A are, respectively, 4-biphenylyl and 2-naphthyl. (From Closs *et al.*[52] Reproduced by permission.)

COMPOUND	k	k_{corr}
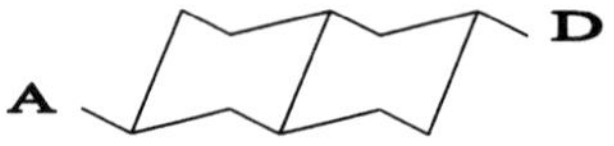	3.1×10^6	5.0×10^7
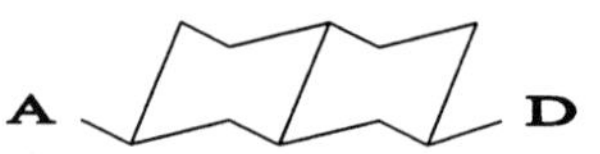	9.1×10^7	1.7×10^8
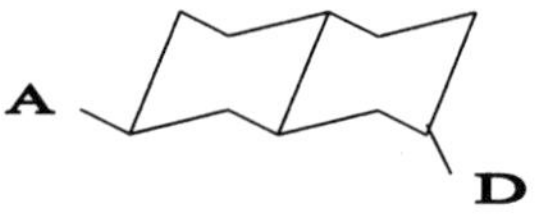	1.1×10^7	7.6×10^7

These five mechanisms appear to be the important, limiting behaviors that can characterize conduction in molecular wire interconnects. Clearly, by changing such parameters as the molecular and electronic structure of the bridge, the electrode attachments, the temperature, and environmental parameters such as solvent or coating, one can modify the nature of the conductance of these molecular bridges.

IV. REMARKS

Possible use of molecules as interconnect structures in nanostructured circuitry raises an interesting set of issues that differ fundamentally from those involved with atomic wires or mesoscopic solid-state materials. Perhaps the most promising of these are the possibilities for self-assembly schemes to create the structures and circuits themselves, and the possibility of different mechanisms modulating the charge transport in the molecular wires, as opposed to hard or more traditional materials. Conceptually, the focus must be on the nature of the contact/wire coupling: What is the effective capacitance of such a junction? How do different molecular linkage and chemical functionalization structures determine that capacitance, as well as the resistance of the interface? The second major issue is the actual transport through the wire itself, and the mechanisms that determine it. On the other hand, some of the possible operative mechanisms for molecular-charge transfer, including defect diffusion, dissipative diffusion, and particularly gated transfer may represent differing ways in which molecular interconnects are actually superior to ballistic transport structures. The gating possibility, in particular, is intriguing: It suggests that wires could be used not only as passive interconnects, but also as switchable interconnects, whose conductance could be varied either by applied electromagnetic radiation or by molecular recognition. Such molecular recognition could be used for sensor and activator purposes or to provide a link to a biological environment.

The gating by molecular stereochemistry can be complemented by gating of the sort generally found in solid-state circuitry — that is, gating controlled by external charge densities on a third terminal. In the simplest situation, that third terminal could be a molecular structure such as a quinone, which can accept charge, become anionic, and thereby act as a gate electrode.

Much of the theoretical framing of the molecular wire conductance problem is now complete. Some difficulties do remain in allowing for dissipative mechanisms (the Green's function for the molecular wire needs to be replaced by an effective Liouville operator which includes dissipation or relaxation terms), and in the proper solution of the self-consistency issue (as discussed briefly in Section II, the simple tight-binding representation of the metal is incorrect, and the issue of how the applied voltage is translated into effective fields across the molecule, with appropriate molecular charge distributions, remains to be fully solved). Nevertheless, both experimental and theoretical advances clearly demonstrate the fascination and promise of molecular wire interconnects.

ACKNOWLEDGMENTS

We are grateful to the Chemistry Division of the National Science Foundation, to CONICIT and the International Division of the National Science Foundation, to the Chemistry Division of the Naval Research Laboratory, and to the MURI Program of the DOD for generous support of this research over the past five years. We also thank Ari Aviram, Abe Nitzan, Michael Wasielewski, Richard vanDuyne, Mark Reed, and Supriyo Datta for very helpful and informative conversations and collaborations.

REFERENCES

1. CARTER, F.L. 1982. Molecular Electronic Devices. Marcel Dekker. New York.
2. CARTER, F.L., R.E. SIATKOWSKI & H. WOHLTGEN. 1982. Molecular Electronic Devices. Elsevier. New York.
3. AVIRAM, A. 1988. Molecular Electronics — Science and Technology. Engineering Foundation. New York.
4. AVIRAM, A. 1992. Molecular Electronics — Science and Technology. American Institute of Physics. Washington, DC.
5. METZGER, R.M., P. DAY & G.C. PAPAVASSILIOU. 1991. Lower Dimensional Systems and Molecular Electronics. Plenum. New York; BORISSOV, M. 1987. Molecular Electronics. World Scientific. Singapore.
6. JORTNER, J. & M.A. RATNER. 1997. Molecular Electronics. Blackwell, London.
7. BIRGE, R.R. 1994. Molecular and Biomolecular Electronics. American Chemical Society. Washington, DC.

8. HONG, F.T. 1988. Molecular Electronics. Plenum. New York; PETTY, M.C., M.R. BRYCE & D. BLOOR. 1995. Introduction to Molecular Electronics. Oxford University Press. London.
9. MILLER, J.S. 1990. Adv. Mater. **2:** 378 and 601.
10. WAGNER, R.W. & J.M. LINDSEY. 1994. J. Am. Chem. Soc. **116:** 9759; LINDSEY, J.M. 1996. J. Am. Chem. Soc. **118:** 3996.
11. RATNER, M.A. & J. JORTNER. In Ref. 6, p.5.
12. MIRKIN, C.A. & M.A. RATNER. 1992. Annu. Rev. Phys. Chem. **43:** 719.
13. ULSTRUP, J. 1997. Charge Transfer Processes in Condensed Media, Lecture Notes in Chemistry, Vol. 10. Springer Verlag. New York.
14. TAUBE, H. 1970. Electron Transfer Reactions of Complex Ions in Solution. Academic Press, New York.
15. BILLING, G.D. & K.V. MIKKELSEN. 1997. Advanced Molecular Dynamics and Chemical Kinetics. John Wiley. New York.
16. FOX, M.A. & M. CHANON. 1988. Photoinduced Electron Transfer, Vols. A-D. Elsevier. Amsterdam and New York.
17. BARBARA, P.F. *et al.* 1996. J. Phys. Chem. **100:** 13148.
18. JORTNER, J. & B. PULLMAN. 1990. Perspectives in Photosynthesis. Kluwer. Dordrecht.
19. See: 1994. J. Photochem. Photobiol. A **82**.
20. See: 1998. Adv. Chem. Phys. In press.
21. NEWTON, M.D. & N. SUTIN. 1984. Annu. Rev. Phys. Chem. **35:** 437.
22. MICHEL-BEYERLE, M.E. 1995. The Reaction Center of Photosynthetic Bacteria: Structure and Dynamics. Springer Verlag. Berlin.
23. CHEN, C.J. 1993. Introduction to Scanning Tunneling Microscopy. Oxford University Press. New York.
24. REED, M.A. *et al.* 1997. Science *278:* 252.
25. ROTH, S. & C. JOACHIM. 1997. Atomic and Molecular Wires. Kluwer, Dordrecht.
26. TANS, S.J. *et al.* 1997. Nature **386:** 474.
27. ZHOU, C. *et al.* 1997. Appl. Phys. Lett. **71:** 611.
28. LANDAUER, R. 1957. IBM J. Res. Dev. 1: 223; LANDAUER, R. 1981. Phys. Lett. **8a:** 91.
29. DATTA, S. 1995. Electronic Transport in Mesoscopic Systems. Cambridge University Press. Cambridge, England; Samanta, M.P. *et al.* 1996. Phys. Rev. B **53:** 7626; JOACHIM, C. & J.F. VINUESA. 1996. Europhys. Lett. **33:** 635; ANDRES, R.P. *et al.* *In* Handbook of Nanostructured Materials and Nanotechnology. H.S. Nalwa, Ed. Academic Press (1998).
30. DATTA, S. *et al.* 1997. Phys. Rev. Lett. **79:** 2530.
31. MUJICA, V. *et al.* 1994. J. Chem. Phys. 101: 6849; MUJICA, V. *et al.* 1994. J. Chem. Phys. **101:** 6856.
32. KEMP, M. *et al.* In Ref. 25, p.203; KEMP, M. *et al.* In Ref. 20.
33. WINGREEN, N.S. *et al.* 1993. Phys. Rev. B *48:* 8487; KEMP, M. Submitted to Phys. Rev. B.
34. KEMP, M. *et al.* 1994. J. Phys. Chem. **101:** 5172; KEMP, M. *et al.* 1996. J. Phys. Chem. 100: 8349; Filali-Mouhim, A. *et al.* 1996. J. Phys. Chem. **100:** 12311.
35. ROITBERG, A. *et al.* 1998. J. Phys. Chem. In press.
36. MUJICA, V. *et al.* 1996. J. Chem. Phys. **104:** 7296.
37. LANG, N.D. 1995. Phys. Rev. B **52:** 5335; JOVANOVIC, D. & J.P. LEBURTON. 1994. Phys. Rev. B **49:** 7474; Macucci, M. *et al.* 1993. Phys. Rev. B **48:** 17354.
38. GRABERT, H. & M.H. DEVORET. 1992. Single Charge Tunneling. Coulomb Blockade Phenomena in Nanostructures. Plenum Press. New York; BEENAKKER, C.W.J. & H. VAN HOUTEN, 1995. Solid State Phys. **44:** 1.

39. SALANECK, W.R., 1. LUNDSTROM & B. RANBY. 1993. Conjugated Polymers and Related Compounds. Oxford University Press, London; BREDAS, J.L. & R. SILBEY. 1991. Conjugated Polymers. Kluwer. Dordrecht.
40. MCCONNELL, H.M. 1961. J. Chem. Phys. **35:** 508.
41. RATNER, M.A. 1990. J. Phys. Chem. **94:** 4877.
42. DEGAMA, A.A.S. 1985. Theor. Chim. Acta *68:* 159; EVENSON, J.W. & M. KARPLUS. 1993. Science **262:** 1247; MAGARSHAK, Y. *et al.* 1991. J. Chem. Phys. **95:** 418; Gehlen, J.N. *et al.* 1996. Inorg. Chim. Acta **243:** 271; GOLDMAN, C. 1991. Phys. Rev. A **43:** 4500; REIMERS, J. & N.S. HUSH, 1994. J. Photochem. Photobiol. A **82:** 31.
43. MAGOGA, M. & C. JOACHIM. 1997. Phys. Rev. B **56:** 4722.
44. SAMANTA, M.P. *et al.* 1996. Phys. Rev. B **53:** R7626.
45. ZHOU, C. *et al.* In Ref. 6, p. 191.
46. DATTA, S. & W. TIAN. 1997. Phys. Rev. B **55:** R1914.
47. SKOURTIS, S. & S. MUKAMEL. 1995. Chem. Phys. **197:** 367.
48. POLLARD, W.T. *et al.* 1996. Adv. Chem. Phys. **XCIII:** 77.
49. DAVIS, W.B. *et al.* 1997. J. Phys. Chem. A **101:** 6158.
50. KRAMERS, H.A. 1940. Physica **4:** 284.
51. Hush, N.S. *et al.* This volume; REIMERS, J.R. & N.S. HUSH. 1989. Chem. Phys. **134:** 323; KOSLOFF, R. *et al.* In press.
52. CLOSS, G.L. *et al.* 1988. J. Am. Chem. Soc. **110:** 2652.

Chemical Control of Tautomerization-Based Molecular Electronic and Color Switches

JEFFREY R. REIMERS,[a] LACHLAN E. HALL,[a] NOEL S. HUSH,[b] AND KIA SILVERBROOK[c]

[a]*School of Chemistry, University of Sydney, NSW Australia*
[b]*Department of Biochemistry and School of Chemistry, University of Sydney, N.S.W. Australia*
[c]*Molecular Electronics Research, Balmain, N.S.W. Australia*

ABSTRACT: Many schemes which have been suggested for the chemical control of molecular electronic devices are impractical because of the large energy changes associated with bond breaking and reforming: these would result in unacceptably large amounts of energy being required for device operation. We have chosen to look in detail at a class of chemical reactions prominent in biological systems that, on energetic grounds, could produce a feasible device: tautomerization reactions of heterocyclic bases. In particular, we concentrate on a model reaction, the 4-pyridinol/4(1H)-pyridinone tautomerization, through which an aromatic pyridinol unit is converted into a π-localized pyridinone unit. In these molecules the functionalities are arranged *para*, allowing the possibility that the molecules align in a solid phase in linear hydrogen-bonded chains. The change in the π electronic structure associated with reactions of this type is quite large, as illustrated, for example, in the conversion of quinacridone, a brightly colored pigment, to quinacridol, an uncolored species with strong UV absorption. Our primary purpose is to calculate energy changes and reaction barrier heights to see whether such tautomerization reactions could provide the basis for an operational chemically controlled molecular electronic device. We also consider, briefly, other required properties in a practicable device.

INTRODUCTION

The aim of this exploratory work is to investigate molecular systems whose properties make them candidates for functional molecular electronic devices. An extended discussion of desirable properties of such molecules is provided in a companion article in this volume[1] which considers chemical and electrochemical switching of conductance properties of oligoporphyrin molecular wires. The system considered therein has possible applications to the construction of storage devices, but is limited in that the control is exercised using wet-chemical processes, processes which are inherently slow. Here we explore a system in which molecular properties relevant to molecular electronics could possibly be controlled by chemical or electrochemical means without the use of wet-chemistry techniques, systems in which the cycle time could be short (μsec or better) and the device cycled many many times. A possible application of such a device, under electronic or optical control, is in RAM memory or computer logic functions; applications to display devices are also possible.

A variety of different systems have been proposed over the years as chemically controllable molecular electronic devices. Often, however, these are centred around standard laboratory wet chemistry, involving reactions with large activation energy barriers and typically also large energy changes per cycle. Just one example of this is the soliton switch system operated via heterocyclic ring closure/opening reactions proposed by Groves.[2] Such devices could be useful in applications such as long-term information storage, but they are *not* suited to RAM or fast computer logic functions.

To find a system which involves chemical bonding changes with low barriers and low energy changes, we considered ways in which this process has been optimized biologically and chose to examine tautomerization reactions amongst the conjugate acids of heterocyclic nitrogen bases[3] including cytosine[4] and many others. Such systems have already been shown by Aviram *et al.*[5] to have interesting molecular electronic properties. Specifically, we choose to study tautomerization between 4-pyridinol and 4(1H)-pyridinone, molecules for which the geometry and electronic structure, as well as those of their conjugate base and doubly protonated acid, are shown in FIGURE 1; these are known as the oxo (OXO) and hydroxy (HYDROXY) tautomers, respectively. These tautomers are also known as the lactam and lactim forms, respectively, but this notation is somewhat misleading as the OXO form remains aromatic in nature and its chemical properties are distinct from those of standard unsaturated lactams.[6] Quite frequently (see, for example, Refs. 4,7–14) pyridinol/pyridinone tautomerization is used as a model for biological systems, a key feature of which is that the relative energies of the OXO and HYDROXY tautomers are very similar so that large energy penalties are not encountered.

Of specific importance here is also the large change in chemical properties which accompanies OXO/HYDROXY tautomerization. As shown in FIGURE 1, the HYDROXY form is aromatic with highly delocalized π electrons, the electronic structure of 4-pyridinol closely resembling that of its parent, pyridine. On the other hand, the OXO form has a bonding pattern indicating alternating localized single and double π bonds. One consequence of this is that, fused in a molecular wire, the HYDROXY form would be conductive while the OXO form would be insulating. Our other article in this issue[1] examines this aspect in detail. Another consequence is that the OXO form absorbs light at much longer wavelengths than does the HYDROXY form. Indeed, the OXO form is the active unit in the archetypal organic pigment quinacridone,[15] while the corresponding HYDROXY tautomer does not absorb visible light at all. By tailoring substitution and fusing of pyridinol/pyridinone units, it would be possible to design systems we believe would have very useful switchable optical and electronic properties. Synthetic investigation of the properties of such molecules has been suprisingly lacking, but a major program is currently under way in the laboratories of Müllen[16] with quinacridones having, amongst other properties, demonstrated photovoltaic[17] and photoconductive[18] activity.

The condensed-phase structure of materials of this type is controlled by intermolecular hydrogen bonding. We envisage a scenario, in a molecular electronic device, in which pyridinol/pyridinone tautomerization (possibly in wider fused molecules such as quinolines[19,20] or quinacridones[15]) could occur in the solid state as a result of the transfer of hydrogens from one side of each hydrogen bond to the other, for example, converting a chain of OXO tautomers with NH—O intermolecular hydrogen bonds to a chain of HYDROXY tautomers with N—HO hydrogen bonds. In order to produce such a device, a variety of significant obstacles must be overcome:

(1) We must be able to synthesize materials with the desired geometrical properties. The molecules must form hydrogen-bonded chains, but these chains could be

arranged in a variety of ways into, for example, three-dimensional crystallites or effectively 2-dimensional layers on top of a surface (e.g., silicon) .

(2) The material must not mechanically destruct after the solid-state reaction is complete. The hydrogen-bond topology of the HYDROXY and OXO tautomers as well as the reaction intermediates will differ (hopefully only slightly), and hence switching will induce internal stresses. The material must be strong enough to tolerate these, and they must not shear the sample from its supports.

(3) The material must have the correct molecular electronic properties, e.g., the desired linear or nonlinear optical properties for each tautomer.

(4) The tautomerization reaction must be inducible by some electrochemical process, the reaction being reversed when the applied voltage is switched. A simple hydrogen-bonding topology which would allow for the possibility of this would be one in which the molecules stacked in long electrode-spanning chains in a head-to-tail configuration. Some chemical paste between the electrodes and the sample may be necessary to initiate the reaction at one or both ends of each chain.

(5) Once initiated, the reaction must propagate with minimal energy change until all molecules have been isomerized.

A block diagram of the envisaged operation of a device through one cycle for one chain is shown in FIGURE 2. The reaction may, in principle, proceed via either proton transfer or hydrogen-atom transfer chemistry. Both intuition and our calculations (see later) indicate that a proton-transfer mechanism would be more likely to work than a free-radical hydrogen-atom transfer mechanism; FIGURE 2 indicates only the proton transfer mechanism. In this figure at time 0 the system is in its OXO tautomeric form; at time step 1 the electrode potentials are set so as to induce protonation and deprotonation of the end OXO groups forming ACID and BASE ions of the type shown in FIGURE 1. This step will cost significant energy ΔE_{AB} which would be provided from the electrode current. A possible set of operating conditions could be an applied electrode voltage of 10 V with an electrode spacing of 100–1000 nm (say 160–1600 molecules in each chain, using our calculated spacing of 6.3 Å per link). Our calculated energy cost for the initiation step, assuming a dielectric constant of 4 as described later, is 1.7 eV, leaving 8.3 eV or 190 kcal mol^{-1} available to drive the propagation reaction. Hence, the maximum energy change allowed for an individual OXO-HYDROXY interchange is just 1.2 kcal mol^{-1} for a chain of 160 molecules. Propagation of the ions, perhaps under the influence of the applied electric field, is illustrated at time step 2 in FIGURE 2. At time step 3, the counter-travelling ACID and BASE ions meet and annihilate, liberating their excess energy as heat. Application of a reversed voltage, with initiation depicted at time step 4, propagation at time step 5, and annihilation at time step 6, returns the system to its initial state.

This project is, at present, in an exploratory stage. Issues (1)–(4) above are all, in principle, solvable, given a sufficiently intense experimental research program. A possible starting molecule would be *cis*-quinacridone, a variant of the popular *trans*-quinacridone[15] (pentacene with its second and fourth rings as in FIGURE 1 with antiparallel orientation). *Trans*-quinacridone does form a crystal with the topology of linear chains of pyridinone units, but there are really two sets of such chains oriented approximately perpendicular to each other. In a *cis*-quinacridone, the pyridinone units lie parallel and are more likely to align in the desired fashion. Nevertheless, at this stage, the key question is not related to issues (1)–(4), but is rather the feasibility of the overall scenario depicted in FIGURE 2: This is determined by the number of molecules required in a functioning device, the need to choose (suppliable) voltages and length scales in order to avoid dielectric breakdown, and the energy supply required and the related issue of the maximum amount of heat that the device could successfully dissipate. We do not address these aspects explicitly here, but the most important issues turns out to be in fact the feasibility of the propagation step, issue (5). This paper addresses basic questions as to the energetics of the interconversion of OXO and HYDROXY chains. Specifically, we investigate properties of the model compounds 4-pyridinol and 4-pyridinone.

MONOMER PROPERTIES

We have considered the properties of 4-pyridinol and 4-pyridinone (in the gas phase) in detail elsewhere.[21] There we employed a range of *ab initio* and density-functional computation schemes using small, medium, and large basis sets. It was found that, for quantitative accuracy, methods using large basis sets with sophisticated post Hartree-Fock treatments of electron correlation are essential. However, of the more economical methods, Hartree-Fock self consistent field (SCF) theory using a small (3-21G) basis set is sufficient to provide qualitatively correct descriptions of the monomer energetics as well as hydrogen-bond topologies and conformer energetics. Optimized geometries and their associated π atomic bond orders are shown in FIGURE 1 while the relative energy of the OXO compared to HYDROXY tautomer and the individual dipole moments are given in TABLE 1.

Also shown in this table are values obtained using the sophisticated CCSD(T) method with a cc-pVTZ basis set; the differences in the calculated results are small compared to changes that could possibly be induced by chemical substitution effects or by solvation effects, and hence we conclude that SCF/3-21G is appropriate in an exploratory study such as this. Results obtained using the advanced B3LYP density-function method are also included for comparison; note, however, that the reliability of density-functional methods for treating proton-transfer problems is variable.[22,23]

The computed results predict that 4-pyridinol is more stable in the gas phase than is its OXO tautomer. This is borne out by experimental results which to date have failed to detect gas-phase 4-pyridinone at equilibrium, suggesting[14] that the energy difference ΔE is at least 3.8 kcal mol^{-1}; extrapolation of liquid-phase data suggests[12] that $\Delta E = 7 \pm 2$ kcal mol^{-1}, but one suspects that the CCSD(T)/cc-pVTZ value of 3.9 kcal mol^{-1} is the most reliable estimate currently available. The hydroxy group in the HYDROXY and ACID forms can undergo hindered rotation, the lowest-energy configuration being a planar structure. The computed rotation barrier is given in TABLE 1 as 4.7 kcal mol^{-1}. Because this barrier is of the order of hydrogen bond strengths, significant conformational flexibility in hydrogen-bonded complexes is anticipated. The barrier for this rotation in the ACID form increases significantly to 8.7 kcal mol^{-1}, however, indicating that reactive intermediates may have tighter geometrical requirements.

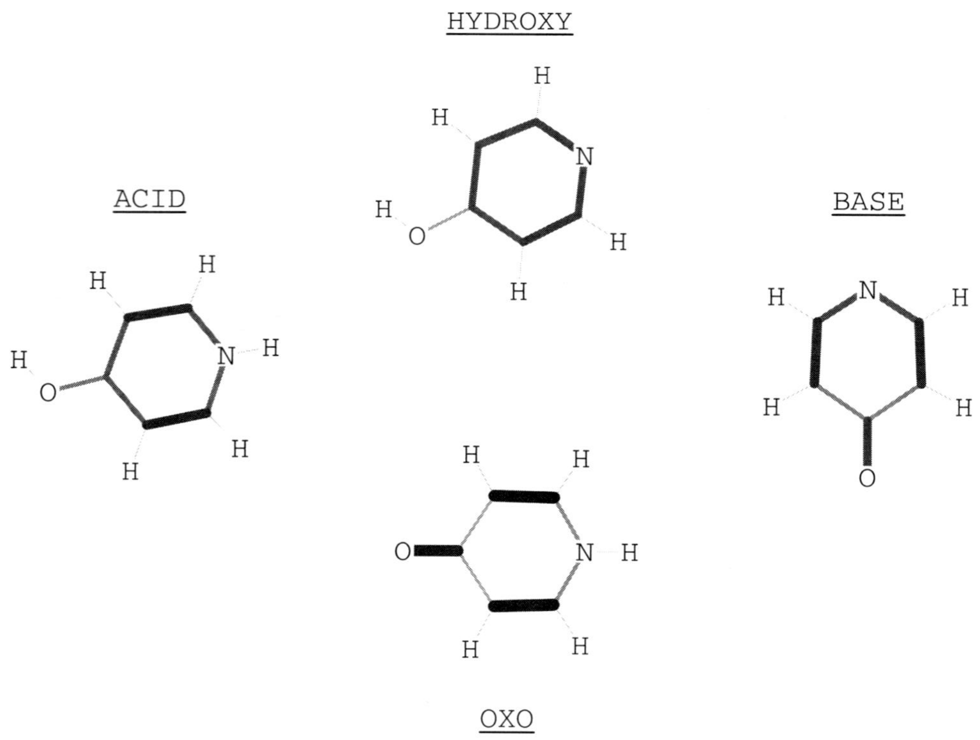

FIGURE 1. SCF/3-21G geometries and π atomic bond orders of the planar aromatic molecules 4-pyridinol (HYDROXY) and 4(1H)-pyridinone (OXO) as well as their conjugate base (BASE) and doubly protonated acid (ACID). Full π bonds (double bonds) are indicated by thick, dark lines, while single bonds are indicated by thin, faint lines; molecules with a delocalized π system have an even gray shading throughout.

TABLE 1. Calculated properties on 4-pyridinone (OXO), 4-pyridinol (HYDROXY), and their acid and base conjugates (see FIG. 1)

Property	SCF/3-21G	B3LYP/cc-pVDZ	CCSD(T)/cc-pVTZ[a]
ΔE	0.7	2.1	3.9
ΔE_{rot}	3.5	5.5	4.7
ΔE_{AB}	119.1	115.4	117.2
μ_{OXO}	7.01	6.54	6.33
$\mu_{HYDROXY}$	2.69	2.71	2.62

NOTE: ΔE is the energy of 4-pyridinone with respect to 4-pyridinol, in kcal mol^{-1}.
ΔE_{rot} is the OH rotational energy barrier for 4-pyridinol in kcal mol^{-1}.
ΔE_{AB} is the energy of infinitely separated ACID and BASE ions with respect to that of two infinitely separated 4-pyridinol molecules, in kcal mol^{-1}.
μ_{OXO} and $\mu_{HYDROXY}$ are the magnitudes of the dipole moments, in D, for 4-pyridinone and 4-pyridinol, respectively. The SCF and B3LYP calculations were performed using GAUSSIAN-94[33] while the CCSD(T) calculations were performed using MOLPRO-97.[34]
[a]At B3LYP/cc-pVDZ geometry.

Also shown in TABLE 1 is the energy cost for the production ΔE_{AB} of the ACID and BASE forms infinitely separated in the gas phase from two 4-pyridinol molecules by the process

$$2NC_5H_4OH \rightarrow [HNC_5H_4OH]^+ + [NC_5H_4O]^- \quad (1)$$

For this reaction, the best available energy estimate is ΔE_{AB} = 117 kcal mol^{-1} or 5.1 eV. This quantity is highly relevant since energy must be provided in the initialization step, issue (4), to induce the acid/base tautomerization reaction. In condensed phases, the ions will be solvated and hence the calculated gas-phase energy cost significantly reduced. Crudely, this effect can be estimated using the Born equation[24] to evaluate the solvation energy

$$\Delta E_{SOLV} = -\frac{\varepsilon - 1}{2\varepsilon}\frac{q^2}{a} \quad (2)$$

for each charge of magnitude q located in a cavity of radius a in a dielectric continuum of dielectric constant ε. Using a radius of a = 2.8 Å for both the ACID and BASE species obtained by fitting the van der Waals molecular volume, for ε = 2, 4, and 30 the net energy cost $\Delta E_{AB} + \Delta E_{SOLV}$ reduces to 61, 32, and 6 kcal mol^{-1}, respectively, or 2.7, 1.4, and 0.3 eV.

The calculated dipole moments for the neutral molecules are also shown in TABLE 1. While 4-pyridinol has a dipole moment, normally thought of as being quite large, typical of that of pyridine and other azines, the dipole moment of 4-pyridinone is nearly three times larger. This feature is very important in determining intermolecular interactions and solvation properties, and arises from the admixture of ionic forms into the ground-state electronic structure of the OXO form:

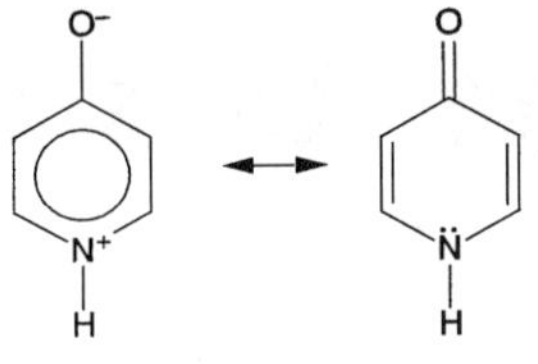

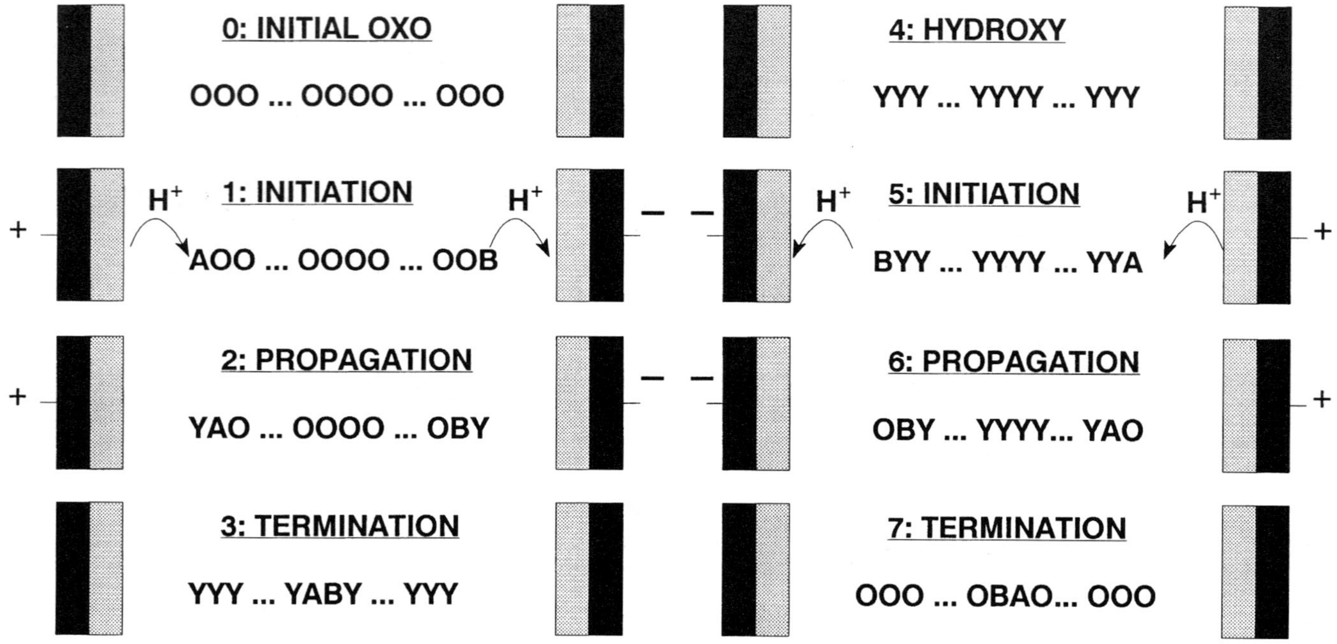

FIGURE 2. Operation cycle of a proposed molecular electronic device based on pyridinol/pyridinone tautomerization. The *solid areas* are electrodes while *shaded areas* represent the proton source/sink paste between the electrodes and the sample; O = OXO molecule, Y = HYDROXY, A = ACID, and B = BASE. At stage 0 the device is in its fully OXO state while, at the half-way point in stage 4, it is in its fully HYDROXY state. Initiation transfers protons to and from the ends at stages 1 and 5; the ions propagate down the chain during stages 2 and 6, meet in the center and terminate at stages 3 and 7.

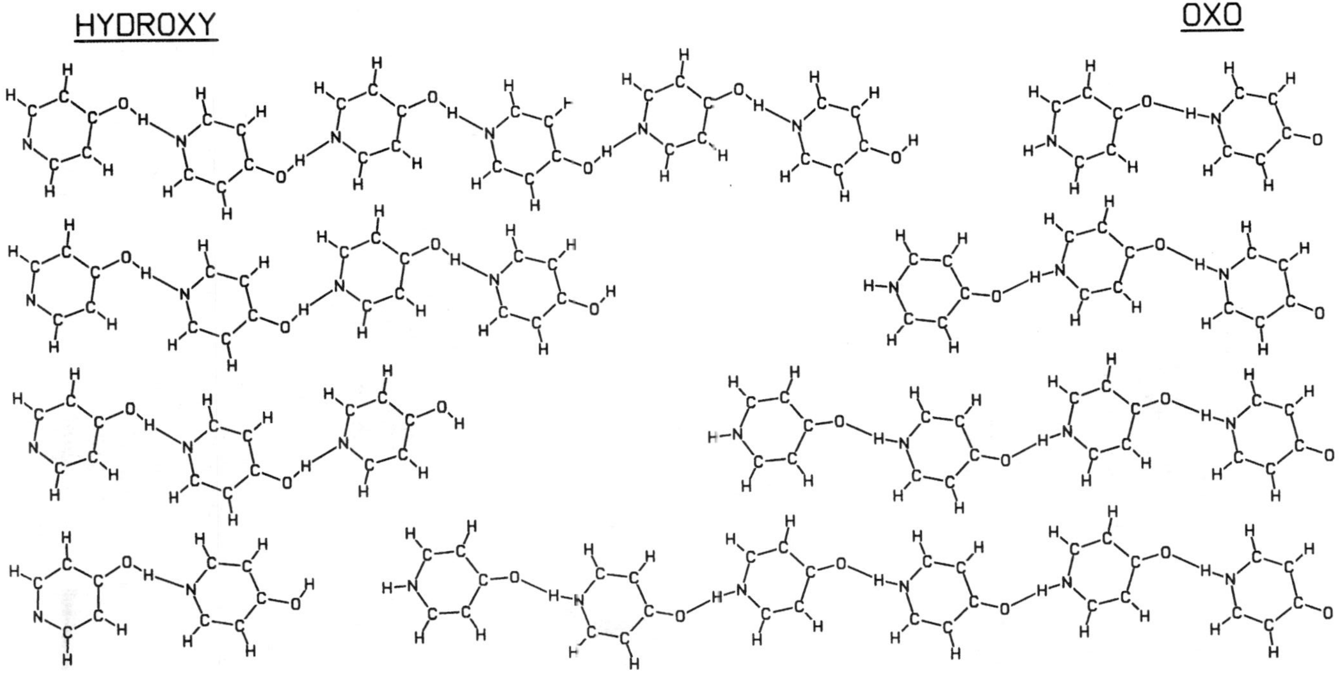

FIGURE 3. SCF/3-21G optimized planar structures for linear chains of HYDROXY and OXO molecules.

CHAIN PROPERTIES

Dimers of 4-pyridinol and 4-pyridinone may one day be observed in molecular beam or matrix isolation studies. Elsewhere[21] we have exhaustively searched using SCF/3-21G, MP2/cc-pVDZ, and B3LYP/cc-pVDZ optimizations for different isomers and have identified 12 key structures. The geometries of these structures are reasonably independent of the method used, as are the energies of the hydrogen-bonded complexes (2 of the 12 structures are relatively high-energy π-stacked structures for which only MP2 correctly represents the interaction energy). For the pyridinol dimer and the pyridinone dimer, the lowest-energy structures are twisted, but the intermolecular torsional potentials are very flat and only small energy penalties are encountered on making the dimers planar. Planar structures can be more easily moulded into the types of topologies of interest to us here, and henceforth we consider only SCF/3-21G results obtained for planar geometries.

SCF/3-21G planar optimized structures for individual HYDROXY and OXO linear dimers, trimers, tetramers, and hexamers have been determined[21] and are shown in FIGURE 3. The charge distribution within these clusters can be succinctly described by fitting atomic charges to represent the molecular electrostatic potential outside the van der Waals surface, constrained to reproduce the ground-state dipole moment. Net charges and net dipole moments (corrected for the net charges) for each molecule in the chain can then be evaluated; the results are given in TABLE 2. This reveals that strong polarization effects occur, reducing the polarity of 4-pyridinol, but significantly increasing that of 4-pyridinone. On the basis of this electrostatic picture, one would expect that the OXO-OXO hydrogen-bond strength would be much greater than that for HYDROXY-HYDROXY, but other contributions are also important[21] and negate this effect considerably.

In TABLE 3 we analyze the calculated hydrogen-bond strengths after basis-set-superposition-error (BSSE) correction in HYDROXY and OXO chains, considering the terminal molecules to interact differently to internal molecules. Hence we obtain, for each chain type, terminal-terminal, terminal-internal, and internal-internal hydrogen-bond strengths. The terminal-terminal bond strengths (i.e., those for the dimers) are large and roughly equal for HYDROXY and OXO chains, the value increasing by 22% for internal-internal interactions of 4-pyridinol, but by a massive 47% for 4-pyridinone. Indeed, it is well known and understood[14,25–28] that, while 4-pyridinol is more stable in the gas phase as solution in low dielectric materials, in condensed phases, including the neat solid and solution in polar liquids, 4-pyridinone becomes much more stable than 4-pyridinol. These calculations indicate that linear chain production favors 4-pyridinone by 2.8 kcal mol^{-1} per link whereas our best estimate of the relative monomer energies favors 4-pyridinol by 3.9 kcal mol^{-1} (TABLE 1) so that the net result still favors 4-pyridinol by 1.1 kcal mol^{-1}. In a three-dimensional sample, interactions with molecules located above and below the molecular planes, as well as longer distance interactions, establish the OXO form as the most stable.[13] Simply by control of substitution and the dielectric constant of the material it should be possible to design a system in which the HYDROXY and OXO forms have equal free energy, allowing for the conversion of one tautomer to the other at minimal energy cost.

TABLE 2. Net charges, in *e*, and dipole moment magnitudes, in D, partitioned to each molecule in linear chains of length *n* of 4-pyridinol (HYDROXY) and 4-pyridinone (OXO) molecules

		HYDROXY		OXO	
n	Ring	*q*	μ	*q*	μ
2	1	−.17	1.5	.05	7.9
	2	.17	2.0	−.05	7.9
3	1	−.17	1.8	.09	7.8
	2	.00	1.5	−.02	8.4
	3	.17	2.1	−.06	8.2
4	1	−.18	1.7	.05	8.5
	2	.00	1.6	.02	9.4
	3	−.01	1.4	−.01	9.2
	4	.19	2.1	−.06	8.3
6	1	−.18	1.8	.08	8.1
	2	.00	1.6	.00	9.1
	3	.00	1.6	.03	9.3
	4	.00	1.6	−.03	9.3
	5	.01	1.5	.02	9.1
	6	.18	2.1	−.09	8.0

NOTE: The rings are numbered from left to right in FIGURE 3.

TABLE 3. Hydrogen-bond strengths involving terminal or internal molecules in linear chains of 4-pyridinol (HYDROXY) or 4-pyridinone (OXO) molecules

Position	HYDROXY	OXO
Terminal-terminal	−10.5	−10.7
Terminal-internal	−11.5	−13.2
Internal-internal	−12.8	−15.6

NOTE: These energies, in kcal mol^{-1}, are evaluated from results for the molecules shown in FIGURE 3 using SCF/3-21G with corrections for BSSE.

One qualitative feature of FIGURE 3 is important in connection with issue (2) discussed in the introduction: the material must not mechanically self-destruct as a result of a solid-state tautomerization. The OXO and HYDROXY chains have similar overall topologies, occupying boxes of very similar dimensions, but the chains differ by small rotations about the centers of each molecule. It is thus conceivable that the tautomerization reaction could proceed without the need to change significantly the overall shape and structure of the solid. Other solid-state hydrogen-transfer tautomerization reactions are known to display this property.[29]

THE TAUTOMERIZATION PROPAGATION REACTION

As stated earlier, the *a priori* most worrisome aspect of the operation of the tautomerization switch is the energy per cycle that would be required to convert a chain of one tautomer to the other and back again. This may occur by one of two possible mechanisms — hydrogen-atom transfer and hydrogen-ion transfer. Hydrogen atom transfer involves free radicals and is naively expected to require more energy and have greater risk of side reactions. For both cases we proceed by assuming that the initiation step near the electrodes can be performed for a reasonable energy cost (e.g., 2 eV per chain) so that the crucial aspect becomes the energy cost required to propagate the tautomerization reaction along a chain.

To investigate the hydrogen-atom transfer (acid/base) mechanism, we take the optimized chain structures shown in FIGURE 3 and add or remove a proton to or from each, as appropriate. The response of the clusters to this perturbation is the determined, not using molecular dynamics simulations at constant energy, as would seem most appropriate, but more simply by relaxing the molecular coordinates along the path of steepest descent in order to find stationary points which characterize the overall reaction profile. For initial OXO and HYDROXY dimers, this process is represented, respectively, by

$$HNC_5H_4O\cdots HNC_5H_4O \overset{\text{initiation}}{\Rightarrow} [NC_5H_4O]^-\cdots[HNC_5H_4OH]^+ \overset{\text{termination}}{\rightarrow} NC_5H_4OH\cdots NC_5H_4OH$$

and (3)

$$NC_5H_4OH\cdots NC_5H_4OH \overset{\text{initiation}}{\Rightarrow} [HNC_5H_4OH]^+\cdots[NC_5H_4O]^- \overset{\text{termination}}{\rightarrow} HNC_5H_4O\cdots HNC_5H_4O$$

The electronic structure calculations indicate that, in both cases, the termination step proceeds spontaneously without activation, releasing the appropriate initiation energy ΔE_{AB} less the appropriate tautomerization energy change ΔE. Similar calculations were also performed for the atom-transfer mechanism, with similar results.

This type of calculation could be repeated for our trimers, tetramers, and hexamers, but it is not felt to be one which provides the most informative results: even the hexamers are quite short and ΔE_{AB} provides a large driving force for the reaction. Rather, we choose to model the situation in which just one of the ACID or BASE disturbances (or the corresponding radical intermediates produced by the atom-transfer mechanism) is propagating through a chain, the other group assumed to be infinitely far away. In this fashion we use our calculated clusters to model isolated propagation reactions, and, as it should be possible to produce isoenergetic propagation by control of substituents and the dielectric constant of the material, we concentrate on estimating bounds for the activation energy of propagation.

For hydrogen-atom transfer, we found that ACID-HYDROXY dimers spontaneously rearrange to give OXO-ACID dimers without activation, while HYDROXY-

BASE dimers similarly rearrange to give BASE-OXO dimers. This is encouraging in that no large activation barrier must be surmounted. For hydrogen-atom transfer we found that SCF/3-21G calculations predict a barrier for the corresponding reactions of ca. 25 kcal mol^{-1}, evaluated using a seam search algorithm[30] as the two electronic states involved interact extremely weakly. This decreases by a small amount when MP2 corrections are made, but decreases by a large amount to 8 kcal mol^{-1} using the B3LYP density-functional technique. The large values for this energy, combined with the lack of solvation energy stabilization from Eq. (2) strongly suggest that hydrogen-atom transfer mechanisms would not be useful, and we consider them no further.

Next we consider trimers made up of either one ACID group or one BASE group. At the SCF/3-21G level, the linear ACID-HYDROXY-HYDROXY spontaneously rearranges to form the OXO-ACID-HYDROXY trimer, the charged species clearly preferring to be in the center rather than at the end of the chain. A second reaction to produce the OXO-OXO-ACID trimer

$$\mathrm{HNC_5H_4O \cdots [HNC_5H_4OH]^+ \cdots NC_5H_4OH} \rightarrow \mathrm{HNC_5H_4O \cdots HNC_5H_4O \cdots [HNC_5H_4OH]^+} \quad \textbf{(4)}$$

costs $\Delta E = 0.6$ kcal mol^{-1}, however, and has an activation energy of $\Delta E^{\ddagger} = 0.7$ kcal mol^{-1} after BSSE. This provides an indication that the activation energy for transfer of a proton along a chain is going to be rather small. Indeed, an analogous situation is found for transfer of a proton-hole along the chain, with HYDROXY-HYDROXY-BASE spontaneously forming HYDROXY-BASE-OXO, but the reaction

$$\mathrm{NC_5H_4OH \cdots [NC_5H_4O]^- \cdots HNC_5H_4O} \rightarrow \mathrm{[NC_5H_4O]^- \cdots HNC_5H_4O \cdots HNC_5H_4O} \quad \textbf{(5)}$$

to produce the BASE-OXO-OXO species costs $\Delta E = 0.9$ kcal mol^{-1} and has an activation energy of $\Delta E^{\ddagger} = 1.4$ kcal mol^{-1} after BSSE.

An analogous situation is found for the tetramer, with the ACID group in ACID-HYDROXY-HYDROXY-HYDROXY and the BASE group in HYDROXY-HYDROXY-HYDROXY-BASE spontaneously moving past all but the last group in the chain. The most relevant step is when the reactive group passes through the centre of the chain, that is,

$$\mathrm{HNC_5H_4O \cdots [HNC_5H_4OH]^+ \cdots NC_5H_4OH \cdots NC_5H_4OH} \rightarrow \mathrm{HNC_5H_4O \cdots HNC_5H_4O \cdots [HNC_5H_4OH]^+ \cdots NC_5H_4OH} \quad \textbf{(6)}$$

and

$$\mathrm{NC_5H_4OH \cdots NC_5H_4OH \cdots [NC_5H_4O]^- \cdots HNC_5H_4O} \rightarrow \mathrm{NC_5H_4OH \cdots [NC_5H_4O]^- \cdots HNC_5H_4O \cdots HNC_5H_4O} \quad \textbf{(7)}$$

Both reactions are spontaneous, both before and after BSSE correction. We estimate the energy change by constraining the appropriate reactant OH bond lengths at 1.05 Å and 1.04 Å, respectively, while optimizing all other (in plane) nuclear coordinates. Though the potential energy surfaces decrease monotonically as these bond lengths increase, they are very flat in the critical region and the results obtained are independent of the details of this clamping procedure. After BSSE, they indicate ΔEs of -3.1 and -2.7 kcal mol^{-1} for reactions 6 and 7, respectively. On the basis of the simple idea that the groups in a chain can be classified as being either central or terminal, these reactions correspond to the conversion of a central HYDROXY to a central OXO, the expected energy for which is -2.1 kcal mol-^{1}.

Hexamers were investigated since clearly end and polarization effects are still important for the tetramers. Without correction for BSSE, the ACID and BASE groups again spontaneously move past HYDROXY groups till the penultimate unit, but, with BSSE, the last step becomes activated. The most important reactions are

$$HNC_5H_4O\cdots HNC_5H_4O\cdots[HNC_5H_4OH]^+\cdots NC_5H_4OH\cdots NC_5H_4OH\cdots NC_5H_4OH$$
$$\rightarrow$$
$$HNC_5H_4O\cdots HNC_5H_4O\cdots HNC_5H_4O\cdots[HNC_5H_4OH]^+\cdots NC_5H_4OH\cdots NC_5H_4OH \qquad (8)$$
$$\rightarrow$$
$$HNC_5H_4O\cdots HNC_5H_4O\cdots HNC_5H_4O\cdots HNC_5H_4O\cdots[HNC_5H_4OH]^+\cdots NC_5H_4OH$$

for which (after BSSE) $\Delta E = -3.6$ and -1.2 kcal mol^{-1}, respectively, with $\Delta E^\ddagger$ = 1.7 kcal mol^{-1} for the last reaction, and

$$NC_5H_4OH\cdots NC_5H_4OH\cdots NC_5H_4OH\cdots[NC_5H_4O]^-\cdots HNC_5H_4O\cdots HNC_5H_4O \rightarrow$$
$$NC_5H_4OH\cdots NC_5H_4OH\cdots[NC_5H_4O]^-\cdots HNC_5H_4O\cdots HNC_5H_4O\cdots HNC_5H_4O \rightarrow \qquad (9)$$
$$NC_5H_4OH\cdots[NC_5H_4O]^-\cdots HNC_5H_4O\cdots HNC_5H_4O\cdots HNC_5H_4O\cdots HNC_5H_4O$$

for which (after BSSE) $\Delta E = -4.0$ and -0.9 kcal mol^{-1}, respectively, with $\Delta E^\ddagger$ = 2.0 kcal mol^{-1} for the last reaction. The calculated ΔEs for the reaction at the chain centre are even further away from the value of -2.1 kcal mol^{-1} expected for the conversion of a central HYDROXY to a central OXO group, indicating again the extent of the polarization and end effects that are operative, especially when ions are involved. Of most interest, however, are the activation energies, which again remain very low. Values of this order are less that errors associated with the treatment of electron correlation, basis set incompleteness, and basis set superposition, and much smaller than solvation effects. Hence, within the accuracy of the calculations, the propagation reaction requires almost no activation energy.

CONCLUSIONS

Tautomerization reactions involving the acids of nitrogen bases are very important biologically as large changes in molecular electronic and reactivity properties can be induced through reactions which are easily controlled, do not involve large energy changes, and have low activation energies. We have demonstrated, using the model compounds 4-pyridinol and 4-pyridinone at useful geometries for operation in a molecular electronic device, that these properties could, in principle, also be exploited in molecular electronics applications. Our calculations reproduce the experimentally known facts about tautomerization reactions of this type concerning the delicate balance that exists between the energies of different tautomers, and how this energy difference can be readily modulated through the control of the polarity of the medium, etc. They indicate that, from an energetic standpoint, it is feasible to interconvert tautomers using nano-electrodes to initial a solid-state chemical reaction. In particular, we calculate that the activation energy barrier to the propagation of an interconversion reaction initiated on one side of a solid sample is very small, of the order of a few kcal mol^{-1}. This number is in fact below the error bounds of the computational schemes used and it is 1 to 2 orders of magnitude smaller than that for many reactions that have been suggested as being useful in molecular electronic devices. This indicates that proton-transfer tautomerization reactions may indeed be very useful. Further, reactions of this type can proceed rapidly through long-range effects such as concerted tunnelling,[31] an effect recently described as being "not hindered by the relatively large number of moving protons."[31] Hence, even if a reasonably large activation energy is needed, key propagation reactions could still be quite fast, e.g., of the order of a few psec.[32]

ACKNOWLEDGMENTS

We thank the Australian Research Council for financial assistance to J.R.R. and L.E.H. We also thank Dr. Gabin Treboux for useful discussions.

REFERENCES

1. Hush, N. S., J.R. Reimers, L.E. Hall, L.A. Johnston & M.J. Crossley. 1998. Optimization and chemical control of porphyrin-based molecular wires and switches. Ann. N.Y. Acad. Sci. This volume.
2. Groves, M.P. 1997. Soliton circuit design using molecular gate arrays. Aust. Comp. Sci. Commun. **19:** 245–252.
3. Katritzky, A. R., M. Karelson & P.A. Harris. 1991. Prototropic tautomerism of heteroaromatic compounds. Heterocyclics **32:** 329–369.
4. Smets, J., L. Adamowicz & G. Maes. 1994. Combined matrix-isolation FTIR and ab initio study of H-bonded complexes between water and molecules modelling cytosine and isocytosine. J. Molec. Struct. **322:** 113–130.
5. Aviram, A., C. Joachim & M. Pomerantz. 1988. Evidence of switching and rectification by a single molecule effected with a scanning tunneling microscope. Chem. Phys. Lett. **146:** 490–495.
6. Hall, G. G., A. Hardisson & L.M. Lackman. 1963. Quantum chemical interpretation of the chemical shift in aromatic molecules. Tetrahedron **19S2:** 101–109.
7. Buyl, F., J. Smets, G. Maes & L. Adamowicz. 1995. Matrix isolation FTIR and ab initio study of H-bonded complexes between water with 4-hydroxypyridine and 3-hydroxypyridine in Ar matrices. J. Phys. Chem. **99:** 14967–14979.
8. Les, A., L. Adamowicz, M.J. Nowak & L. Lapinski. 1994. Concerted biprotonic tautomerism of 2-hydroxypyridine. Theochem. **118:** 157–166.

9. PERSON, W.B., J.E. DEL BENNE, W. SZAJDA, K. SZCZEPANIAK & M. SZCZESNIAK. 1991. Matrix isolation studies and ab initio calculations of the vibrational spectra of complexes of water with 3-hydroxypyridine. J. Phys. Chem. **95:** 2770–2780.
10. KWIATKOWSKI, J. S., R.J. BARTLETT & W.B. PERSON. Contributions from electron correlation to the relative stabilities of the tautomers of nucleic acid bases. J. Am. Chem. Soc. **110:** 2353–2358.
11. SORDO, J. A., M. KLOBUKOWSKI & S. FRAGA. 1985. Theoretical studies of the tautomeric equilibrium of 4(1H)-pyridinone in solution. J. Am. Chem. Soc. **107:** 7569–7572.
12. SCHLEGEL, H. B., P. GUND & E.M. FLUDER. 1982. Tautomerization of formamide, 2-pyridone, and 4-pyridone. J. Am. Chem. Soc. **104:** 5347–5351.
13. CORONGIU, G., E. CLEMENTI, M. DAGNINO & L. PAOLONI. 1979. Interaction of water with 4-hyrdoxy-pyrisine and 4-pyridone. Chem. Phys. **40:** 439–455.
14. BEAK, P., F.S. FRY JR., J. LEE & F. STEELE. 1976. Equilibrium studies. Protomeric equilibria of 2- and 4-hydroxypyridines, 2- and 4-hydroxypyrimidines, 2- and 4-mercaptopyridines, and structurally related compounds in the gas phase. J. Am. Chem. Soc. **98:** 171–179.
15. POTTS, G. D., W. JONES, J.F. BULLOCK & S.J. ANDREWS. 1994. The crystal structure of quinacridone: An archetypal pigment. J. Chem. Soc. Chem. Commun. 2565–2566.
16. KELLER, U., K. MÜLLEN, S. DE FEYTER & F.C. DE SCHRYVER. Hydrogen-bonding and phase-forming behaviour of a soluble quinacridone. Adv. Mater. **8:** 490–493.
17. TOMIDA, M., S. KUSABAYASHI & M. YOKOYAMA. 1987. Organic solar cell fabrication using quinacridone pigment. Chem. Lett. 609–612.
18. MANABE, K., S. KUSABAYASHI & M. YOKOYAMA. 1984. Long-life organic solar cell fabrication using quinacridone pigments. Chem. Lett. 1305–1308.
19. DE LA CRUZ, A., J. ELGUERO, P. GOYA & A. MARTINÍNEZ. 1992. Tautomerism and acidity in 4-quinolone-3-carboxylic acid derivatives. Tetrahedron **48:** 6135–6150.
20. BARTOLOMEI, M., M. CIGNITTI, M.C. RAMUSINO & G. LA MANNA. 1995. Ab initio study of the tautomeric forms of some quinolediones . Theochem **330**: 431-5.
21. REIMERS, J. R., L.E. HALL & N.S. HUSH. 1998. J. Phys. Chem. B. Submitted for publication.
22. ZHANG, Q., R. BELL & T.N. TRUONG. 1995. Ab initio and density functional theory studies of proton transfer reactions in multiple hydrogen bond systems. J. Phys. Chem. **99:** 592–99.
23. LI, G. -S., M.F. RUIZ-LÓPEZ, M.-S. ZHANG & B. MAIGRET. 1997. Ab initio calculations of tautomer equilibrium and protonation enthalpy of 4- and 5-methylimidazole in the gas phase: basis set and correlation effects. J. Molec. Struct. (Theochem). In press.
24. BORN, M. 1920. Z. Phys. **1:** 45–48.
25. BELLAMY, L.J. & P.E. ROGASCH. 1960. The use of solvent effects in the identification of carbonyl absorptions in pyridones. Spectrochim. Acta **16:** 30–34.
26. FRANK, J. & A.R. KATRITZKY. 1976. Tautomeric pyridines. XV. Pyridone-hydroxypyridine equilibria in solvents of different polarity. J. Chem. Soc. Perkin Trans. **II:** 1428–1431.
27. NOWAK, M. J., L. LAPINSKI & J. FULARA. 1989. Matrix isolation studies of cytosine: The separation of the infrared spectra of cytosine tautomers. Spectrochim. Acta A **45:** 229.
28. WANG, J. & R.J. BOYD. 1996. A theoretical study of proton transfers in aqueous para-, ortho-hydroxypyridine and para-, ortho-hydroxyquinoline. Chem. Phys. Lett. **259:** 647–653.

29. DE PAZ, J. L. G., J. ELGUERO, C. FOCES-FOCES, A.L. LLAMAS-SAIZ, F. AGUILAR-PARRILLA, O. KLEIN & H.-H. LIMBACH. 1997. Theoretical study of the structure and tautomerism of N1-unsubstituted pyrazoles in the solid state. J. Chem. Soc., Perkin Trans. **2:** 101–109.
30. FARAZDEL, A. & M. DUPUIS. 1991. On the determination of the minimum on the crossing seam of two potential energy surfaces. J. Computat. Chem. **12:** 276–282.
31. LIEDL, K. R., S. SEKUŠAK, R. T. KROEMER & B. M. RODE. 1997. New insights into the dynamics of concerted proton tunnelling in cyclic water and HF clusters. J. Phys. Chem. A **101:** 4707–4716.
32. BROUGHAM, D. F., A.J. HORSEWILL & R.I. JENKINSON. 1997. Proton transfer dynamics in the hydrogen bond: a direct measurement of the incoherent tunnelling rate by NMR and the quantum-to-classical transition. Chem. Phys. Lett. **272:** 69–94.
33. FRISCH, M. J., G.W. TRUCKS, H.B. SCHLEGEL, P.M.W. GILL, B.G. JOHNSON, M.A. ROBB, J.R. CHEESEMAN, T.A. KIETH, G.A. PETERSSON, J.A. MONTGOMERY, K. RAGHAVACHARI, M.A. AL-LAHAM, V.G. ZAKRZEWSKI, J.V. ORTIZ, J.B. FORESMAN, J. CIOSLOWSKI, B.B. STEFANOV, A. NANAYAKKARA, M. CHALLACOMBE, C.Y. PENG, P.A. AYALA, W. CHEN, M.W. WONG, J.L. ANDRES, E.S. REPLOGLE, R. GOMPERTS, R.L. MARTIN, D.J. FOX, J.S. BINKLEY, D.J. DEFREES, J. BAKER, J.J.P. STEWART, M. HEAD-GORDON, C. GONZALEZ & J.A. POPLE. 1995. GAUSSIAN 94 Gaussian Inc. Pittsburgh, PA
34. WERNER, H.-J. & P.J. KNOWLES. MOLPRO-97 program package.

Theory of Electrical Conduction Through a Molecule

ELDON G. EMBERLY[a] AND GEORGE KIRCZENOW

Department of Physics, Simon Fraser University, Burnaby, British Columbia, Canada V5A 1S6

ABSTRACT: We present a theoretical study of electronic transport through a 1,4 benzene-dithiolate (BDT) ion chemically bonded to two gold leads. The model consists of a single BDT attached to two Au clusters, which are in turn attached to ideal one-dimensional leads. Extended Hückel theory is used to model the electronic structure and the Lippmann-Schwinger equation is solved for the lead-to-lead transmission. The current-voltage characteristics of the system are calculated using the finite temperature, finite voltage Landauer formula. Different atomic configurations of the Au clusters are examined including the effects of thermally induced positional disorder of the Au atoms. We find that while in some cases there is conduction via states derived from the BDT HOMO and LUMO levels, conduction due to levels derived from hybrid Au cluster-BDT states often predominate. The current-voltage characteristics are found to be sensitive to the specific configuration of the Au clusters, BDT and ideal leads.

I. INTRODUCTION

There has been renewed interest in studying the electronic transport properties of molecular systems in recent years. The novel ideas first envisioned by Ratner and Aviram[1] are now being realized and characterized experimentally,[2–7] as new molecular devices are being fabricated. Theoretical ideas and calculations are also emerging motivated in part by the experimental findings.[8–15]

One of the devices is a molecular wire which may be studied by passing current through a molecule attached to two metallic leads. A molecule of current interest for use as a molecular wire is 1,4 benzene-dithiol, which when attached to two gold leads becomes 1,4 benzene-dithiolate (BDT).[2,10] It is a good candidate since the benzene ring offers delocalized states in the form of π bonds which would seem to be beneficial to electron transport. The sulphur end groups also allow for the desired functionality of bonding readily to the gold leads. Many other candidates exist, but the transport properties of this molecule have been measured experimentally[2] and thus data are available for comparison with calculations.

Many questions arise when considering the microscopic nature of the lead-molecule-lead system: What is the atomic configuration of the leads, and how does the molecule bond to the leads' surfaces? Should the coupling between the molecule and leads be regarded as strong or weak? What role does the structure and chemistry of the system play in transport?

[a]Address for telecommunication: phone, 604/291-3480; fax, 604/291-3592; e-mail: eemberly @sfu.ca

To address the issue of the microscopics, specifically the coupling between the molecule and metallic leads, we have modeled the system in a somewhat different fashion from that of previous studies. To complement previous work that has modeled molecules connected to semi-infinite macroscopic leads, we have chosen to represent the leads as finite gold clusters to which ideal leads are attached. The clusters bind to the BDT, and we treat the complex consisting of the BDT, and clusters as an integral chemical entity in order to address the chemical nature of the junctions between the molecule and leads as well as is possible at the level of our calculations. It seems reasonable that the binding of the metallic system to the molecule strongly affects the electronic nature of the molecule. We find that conduction through the molecule is not necessarily controlled by the highest occupied molecular orbital (HOMO) and lowest unoccupied molecular orbital (LUMO) when the molecule is bonded to leads. Thus it is important to focus on the states of the coupled molecular-metallic system rather than on those of the molecular subsystem alone. Because of the sensitivity to the nature of the coupling, the microscopic details of the structure have a strong influence on the resulting transport.

In Section II, we describe the scattering and transport theory used in this work. It is based on a direct numerical solution of the Lippmann-Schwinger equation. The evolution of the energy states of the system when Au clusters are attached to the BDT is discussed in Section III. The calculated transmission and conductance for various BDT-cluster systems is presented in Section IV. Finally, the effect of elastic scattering of electrons by thermal atomic vibrations in the Au clusters is considered in Section V.

II. THEORY

The transport of electrons through a molecule connected to two metallic leads can be modeled in many different ways. At its simplest level, this is a tunneling problem where an electron incident from the source with energy E has a probability $T(E)$ of being transmitted through the structure to the drain. By calculating this transmission probability for a range of energies around the Fermi energy, ϵ_F of the lead, the finite temperature, finite voltage, Landauer formula can be used to calculate the transmitted current

$$I(V) = \frac{2e}{h}\int_{-\infty}^{\infty} dET(E)\left(\frac{1}{\exp[(E-\mu_s)/kT]+1} - \frac{1}{\exp[(E-\mu_d)/kT]+1}\right) \tag{1}$$

The two chemical potentials μ_s and μ_d, refer to the source and drain, respectively. They are defined to be, $\mu_s = \epsilon_F + eV/2$ and $\mu_d = \epsilon_F - eV/2$, where V is the source-drain bias voltage.

We calculate the transmission probability $T(E)$ starting from the Lippmann-Schwinger (LS) equation,

$$|\Psi\rangle = |\Phi_0\rangle + G_0 W|\Psi\rangle \tag{2}$$

Here, both $|\Phi_0\rangle$, the state of the incident electron, and G_0, the free propagator, refer to the unperturbed system. We take the unperturbed system to consist of three non-

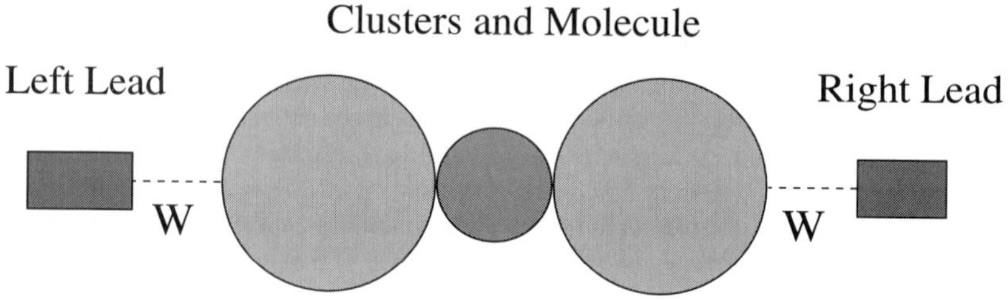

FIGURE 1. Schematic diagram of the three noninteracting systems: left lead, CMC, and right lead. The systems are coupled by the potential W.

interacting subsystems. These are the left lead, the right lead, and the clusters+molecule (CMC), as illustrated in FIGURE 1. These noninteracting subsystems are then coupled by an interaction Hamiltonian, W, which connects the free propagator G_0 between the subsystems and transforms the incident one electron state, $|\Phi_0\rangle$, into the scattering state $|\Psi\rangle$.

Each of the state vectors, $|\Phi_0\rangle$ and $|\Psi\rangle$, are represented using LCAO, where the basis is composed of the atomic orbitals on the CMC and the leads. The ideal leads attached to the CMC are modeled as semi-infinite 1-D chains of atoms with one orbital per site, and only nearest neighbor hopping. The incoming boundary conditions for $|\Phi_0\rangle$ are such that it is an incoming Bloch wave in the left lead, and has a value of 0 on the CMC and right lead systems. This allows for an exact analytical evaluation for the free propagator on the leads. The free propagator for the CMC is calculated using its energy eigenstates. These can be found via different quantum chemistry methods, but for our calculation, they were determined by using simple extended Hückel. Because initially the three subsystems are not coupled to each other, the free propagator connecting sites between subsystems is zero.

The matrix elements for the coupling potential W are defined to be non-zero only between the lead sites adjacent to the CMC and sites on the CMC. We approximate them by using extended Hückel to calculate the Hückel matrix elements between the single lead orbital and all orbitals on the CMC. Thus the coupling W depends on where the lead site is placed with respect to the CMC, and also on the nature of its atomic orbital.

We solve the LS equation, Eq. (2), for $|\Psi\rangle$ on the lead sites adjacent to the CMC and also for the sites on the CMC. The transmission, $T(E)$ is then just the square modulus of $|\Psi\rangle$ on the site of the right lead adjacent to the CMC.

III. ENERGY LEVEL SCHEMES

A possible controlling factor in understanding molecular wires is the nature of the coupling of the continuum of energy states in the lead to the discrete energy levels of the molecular system. We believe it to be important to study how the energy levels of the molecule change when placed in contact with the metallic leads. The nature of these changes will be discussed below for different configurations of the contacts and the BDT molecule.

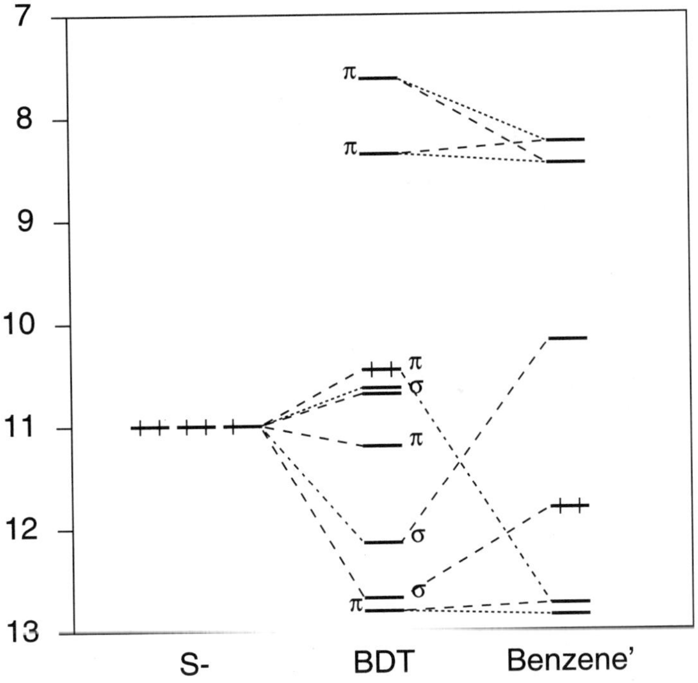

FIGURE 2. Energy-level diagram for BDT ion. Left diagram shows sulphur ion (S) p orbitals. Not shown is the sulphur S level, which is lower in energy. *Right diagram* is for benzene with the 1,4 Hs removed. The *dashed lines* connecting the diagrams indicate where the levels of a given segment get mixed into the BDT ion.

When 1,4 benzene-dithiol reacts with gold, it has been experimentally shown[17] that it loses a pair of H ions to become BDT. The H ions react with the gold by capturing a valence electron. In FIGURE 2, the energy level diagram of BDT is shown. To its right in the figure are the energy levels for benzene with the Hs at positions 1 and 4 removed (labeled Benzene'). On its left are shown the energy levels for the sulphur end groups of the BDT ion. These were calculated using extended Hückel. From a calculation for benzene (not benzene') the HOMO is at −12.814 eV with a LUMO of −8.248 eV. Both of these levels describe π bonds on the C ring, and it is shown on the diagram where the corresponding benzene' levels mix into the BDT. The HOMO-LUMO gap shrinks for BDT with the HOMO now at −10.470 eV, but with almost the same LUMO of −8.247 eV. The character of the HOMO is that of Cπ around the ring with some Sπ content. The other levels in the BDT spectrum between the benzene HOMO-LUMO also have S character, and as marked on the diagram, also contain Cπ and Cσ bonding states. The σ states are due to the two states in the benzene' spectrum that occur within the traditional HOMO-LUMO gap of benzene. It will be seen that these levels are significantly influenced when the S is bonded to the Au clusters.

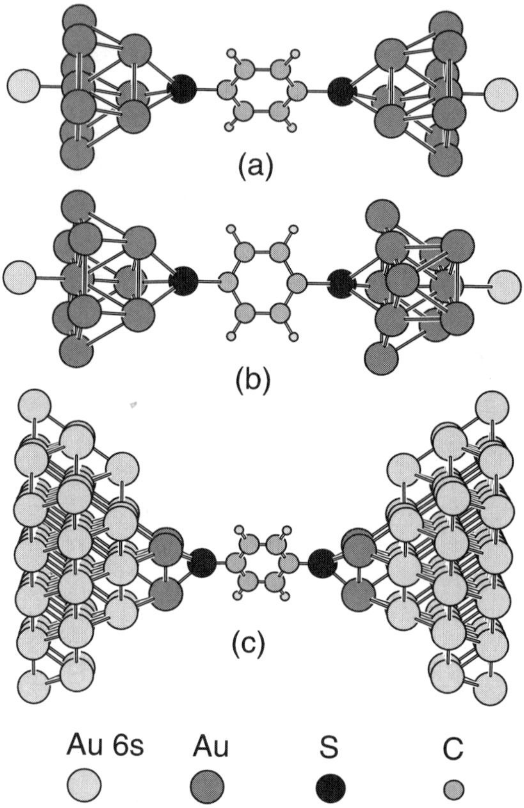

FIGURE 3. Atomic diagrams of CMC systems. **(a)** 10 Au (111) clusters attached to BDT. **(b)** 52 Au (111) clusters attached to BDT. **(c)** 10/11 Au (111) clusters attached to BDT. In **(b)**, the two lead atoms that bond to the left and right clusters are not shown.

No studies as yet have been done to characterize the nature of the interface between the sulfur and the gold nanowire. Recent studies of gold nanowires and break junctions of macroscopic gold wires[18,19] have shown them to be different from bulk gold; the region between two gold contacts is composed of filament structures. The reconstructions of the Au atoms for a nanobridge tend to form hcp structures. Though neither of these reports may characterize the exact nature of the interface, it seems reasonable to model the leads with different geometrical configurations that are consistent with the above findings for non-bulk gold. The cluster contacts to be considered are shown in FIGURE 3. In all the clusters, the gold atoms have been arranged in an ideal fcc (111) configuration with the lattice parameter of bulk gold. It has been assumed that the Ss of the BDT each bond to a triangle of Au atoms. An SCF calculation by Sellers[16] has shown this to be energetically favorable for an alkanethiolate. The perpendicular distance of the S from the Au triangle has been chosen to be 1.9 angstroms. The other important bond is the S—C bond, which has been chosen to be 1.795 angstroms.

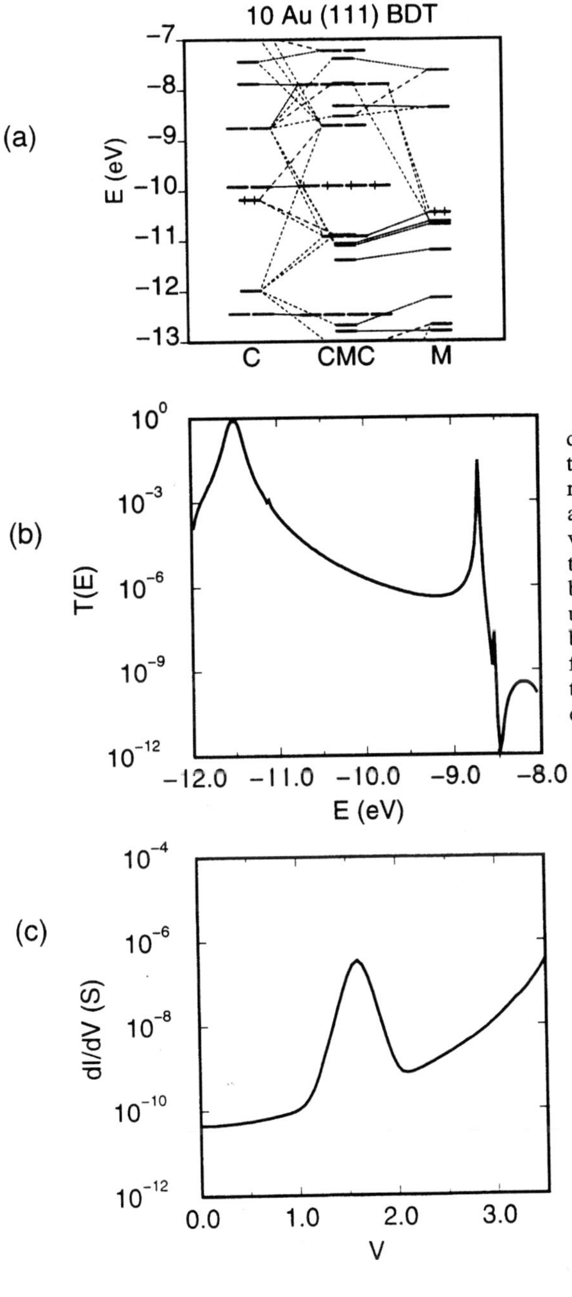

FIGURE 4. (**a**) Energy-level diagram for 10 Au (111) cluster and BDT system. (**b**) Transmission vs. electron energy E, and (**c**) conductance vs. bias voltage V, for this system. In the energy-level diagram, C labels the energy levels of the uncoupled left cluster, M labels the energy levels for the free molecule, and CMC labels the energy levels for the bonded clusters + molecule system.

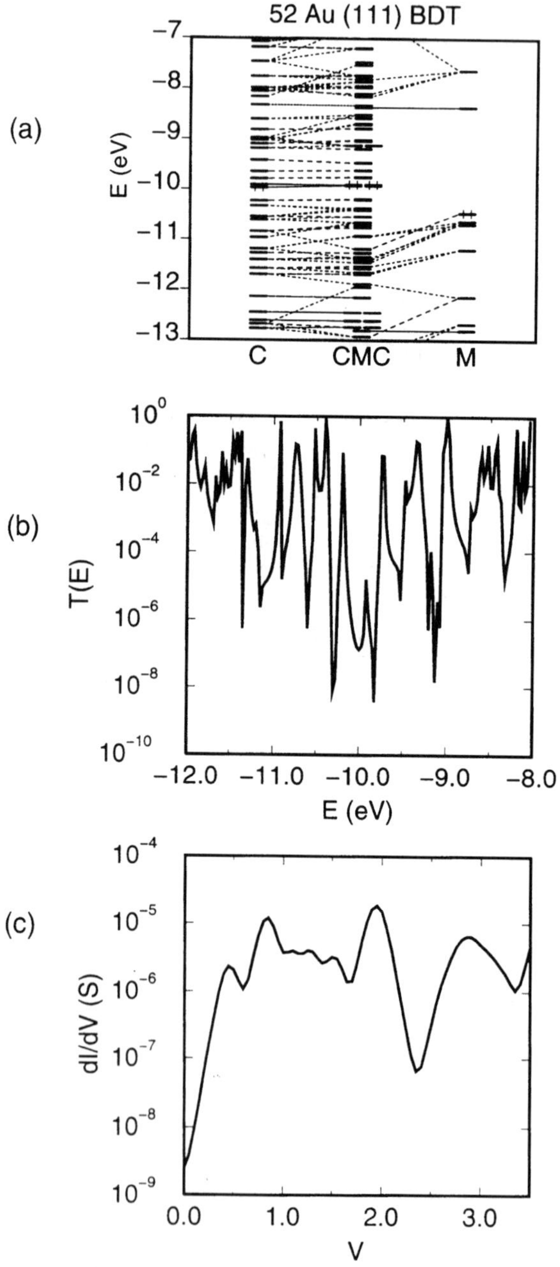

FIGURE 5. **(a)** Energy-level diagram for 52 Au (111) cluster and BDT system. **(b)** Transmission vs. electron energy E, and **(c)** conductance vs. bias voltage V, for this system.

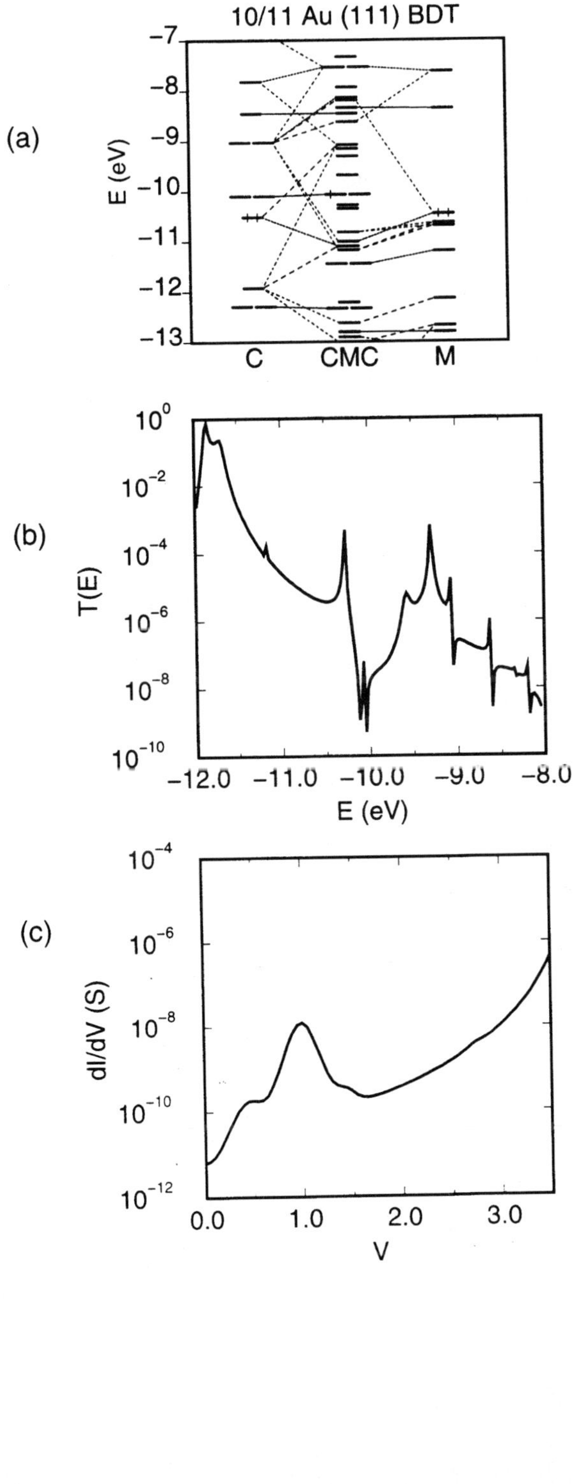

FIGURE 6. **(a)** Energy-level diagram for 10/11 Au (111) clusters and BDT system. **(b)** Transmission vs. electron energy E, and **(c)** conductance vs. bias voltage V, for this system.

For the clusters of 10 Au atoms (FIGURE 3a) it is seen in the energy spectrum (FIGURE 4a) that the CMC system is a mixture of molecular and cluster states. The BDT levels around its HOMO, which have large S content, mix into different levels of the combined system. The Fermi level for the cluster (approximated by the HOMO for the cluster) lies in the HOMO-LUMO gap of the molecule, and this has the effect of narrowing the gap in the combined system. Other geometries with similar numbers of atoms have also been studied, but are not presented here. The same general effects are noticed; the addition of the Au cluster alters the molecular states. It seems that the relevance of the BDT HOMO-LUMO to the combined system is questionable due to the mixing of states that occurs. Transport in such a system will be different from that through a free molecule.

The disadvantage of modeling with small clusters is that they do not fully capture the continuum character of the leads. Considerably bigger clusters were constructed in order to generate a denser spectrum. These larger clusters were simplified by using only 6s orbitals on most of the Au atoms. Only the Au atoms neighboring the molecule were modeled using all of the valence orbitals for Au. This is different from the smaller clusters, where all cluster atoms were modeled using the full valence orbital set. The incident electrons are from the 6s band, and although it is known that the 5d and 6p orbitals play a role in shifting the 6s energies, the inclusion of only 6s makes the calculation for large clusters feasible.

In the energy level diagram given in FIGURE 5a for the cluster consisting of 52 Au atoms (FIGURE 3c), the levels around the HOMO are predominantly shifted further below the Fermi energy, and the levels around the LUMO are mixed into levels that still lie above the Fermi energy. It is possible to argue that there is still a BDT HOMO-LUMO character to the resulting combined system, in that there is a gap between where most of the HOMO levels mix and where the LUMO levels mix. In the latter sections of this paper the usefulness of this notion will be seen to be limited since it will be shown that there can be transport through levels not connected with the molecule's HOMO or LUMO.

The other configuration to be presented is that of asymmetric binding — FIGURE 3b. It is reasonable to assume that the left part of the break junction will not be a mirror image of the right part. This is modeled by considering a CMC system with two different clusters. In the energy-level diagram for this system (FIGURE 6a) degeneracies are lifted, and as a result the levels are more uniformly distributed than in the case of two 10 Au clusters. The energy levels, to which neither the left cluster nor molecule connect are due to the right cluster.

It should be pointed out that all the energy diagrams presented have been calculated with the first atom of both ideal leads attached to the CMC. The lead atom is modeled using an Au 6s orbital. This coupling, W, can shift and split the energy levels. It has been found, though not shown here, that it is possible to attach the ideal leads in such a way that the energy spectrum changes quite significantly. Here the leads are attached on the central axis so as not to break the symmetry of the CMC, thereby preserving the general energy structure of the CMC.

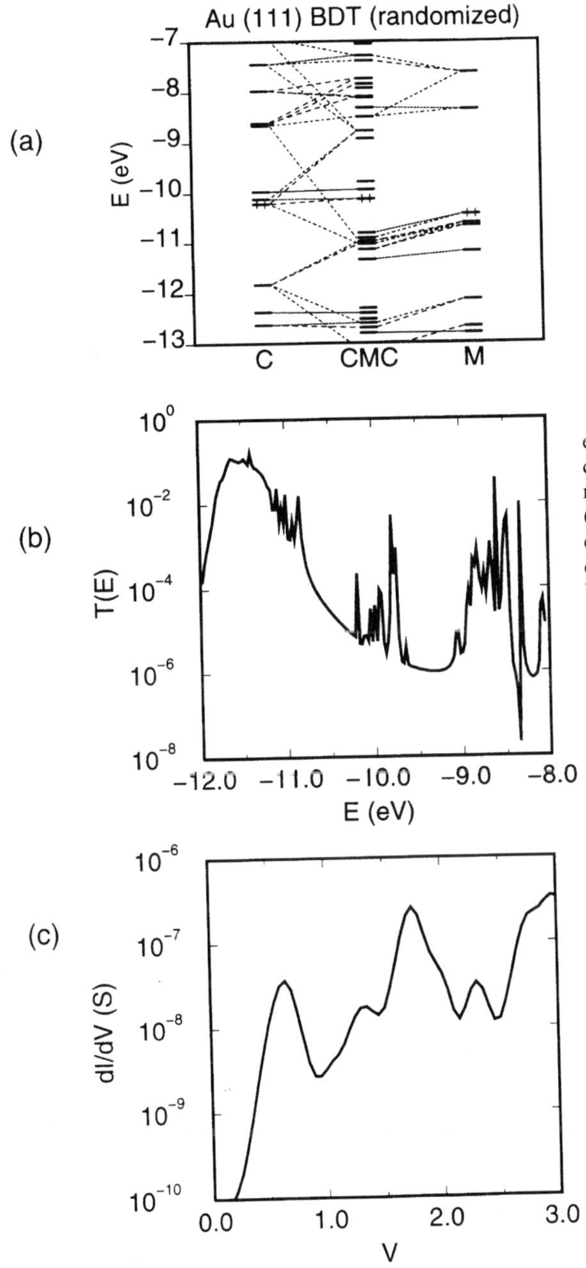

FIGURE 7. **(a)** Energy-level diagram for 10 Au (111) cluster and BDT system for randomized cluster geometry. **(b)** Averaged transmission vs. electron energy E, and (c) conductance vs. bias voltage V, for this system.

IV. TRANSMISSION AND CONDUCTANCE OF CMC SYSTEMS

Using the LS formalism, the transmission through the above systems was calculated for energies around the Fermi level of the Au leads. The Landauer formula was then used to evaluate the current vs. bias voltage curve for each system. The differential conductance was calculated from this data.

The transmission curve for the 10 Au cluster is shown in FIGURE 4b. It has a region of strong transmission below the Fermi energy, and peaks of weaker strength above. The transmission at –11.5 eV can be attributed to a Sπ– and Cπ-like state with some Au presence. The little peak at –11.1 eV is due to a CMC σ-like state. The transmission at –8.7 eV is due to a σ like state also. The smaller peak at –8.52 eV is due to the BDT LUMO which has been mixed with some cluster states. Nonetheless, the LUMO for the CMC is different from the LUMO for the BDT. It was mentioned previously that the coupling of the lead has the effect of shifting the energy levels of the CMC. The coupling of the entire lead to the CMC can further shift the levels, which is the case for the peak at –11.5 eV. This is due to the state at around –11.4 eV in the spectrum, which has strong binding to the lead.

The corresponding conductance curve is shown in FIGURE 4c. It has been calculated with a lead Fermi energy at –9.5 eV, which seems reasonable based on the energy level diagram presented. The lack of any transmission peak around the Fermi energy, leads to the valley around 0 applied bias. The two peaks at –11.5 and –8.7, and their spacing give rise to the structure in the conductance.

For the 52 Au cluster, there are many more CMC levels to tunnel through. One would expect though that only those levels that have a wavefunction which extends across the whole of the system would be conducting. This is found. There is transmission at energies which fall within the HOMO-LUMO gap of the molecule, and for these states there is enough presence on the molecule to transmit (see FIGURE 5b). These levels, though they have π character, are not connected strongly to either the HOMO or LUMO of the BDT, and yet still conduct. Though these levels have molecular content, it is not possible, as is shown in the energy diagram, to attribute this content to any particular molecular orbital. These states seem to be admixtures of many states from the molecule.

Because of the transmission over virtually all of the energy range, the corresponding conductance curve, FIGURE 5c, has much more structure. The Fermi energy was chosen to be –9.98 eV since for a larger cluster the HOMO of the cluster should approximate the Fermi level. Because there are many more peaks in the transmission spectra, the magnitude of the differential conductance has risen. The corresponding rise in the conductance also occurs at a voltage that is closer to 0 than what was found experimentally. Other theoretical work by Kemp *et al.* and Datta *et al.*[12] has also found that the calculated conductance for BDT to be greater than what was experimentally measured.

Finally, the last case to be considered is for the molecule binding to two dissimilar clusters. Notice that in the conductance curve (FIGURE 6c) the magnitude is down compared to the other two configurations. In the transmission diagram (FIGURE 6b) the common characteristic of the larger peaks is that the wavefunction of the CMC has significant presence on the cluster. The smaller peaks are due to states that are more localized on the molecule. The major transmission at –11.6 eV is due to predominantly σ bonds between the cluster and molecule. This state is also highly coupled to the lead. The same can be said about the peak at –9.3 eV, where the state is again σ like, with good coupling to the lead, though not as strong as the one at –11.6 eV.

The conductance curve is shown in FIGURE 6c. The Fermi energy was chosen to be –9.8 eV. With the two transmission peaks at –10.4 eV and –9.3 eV, the conductance rises at around 1 V. The large transmission peak at –11.6 eV contributes to the rise in the tail of the conductance curve at 3 V.

V. FINITE TEMPERATURE EFFECTS

The above calculations for the conductance have included some finite temperature effects by using the 300 K Fermi distribution for the source and sink electrons via the Landauer formula. However the CMC system has been assumed to be in a regular geometrical configuration. At room temperature the effects of thermal vibration on the configuration of the cluster will be significant. A simplified way to model these effects, ignoring inelastic scattering, is to examine the nature of positional disorder in the Au clusters at room temperature. This model does not seem unreasonable, since the gold clusters do not form a perfect lattice, and hence deviations from the regular cluster geometry could be important.

EXAFS measurements on small clusters have given values for the Debye-Waller factor,[20] which can be used to obtain the rms displacement of atoms in a solid at a temperature T. Different geometrical configurations of a given cluster were generated by randomly displacing each Au atom in the cluster according to a Gaussian distribution consistent with this value of the rms displacement. For each random configuration of the various clusters, the transmission was evaluated. FIGURE 7 shows the effect of randomizing the positions on the energy spectrum, the transmission, and the conductance. The transmission is an average of over 50 random configurations of the 10 Au clusters. All the energy-level degeneracies that exist in the regular geometry are broken due to the positional change. It is evident that this causes a general broadening in the transmission spectra, as peaks are shifted or new ones created due to the change in geometry. The transmission is also weaker than in the regular geometry. This smearing effect causes the conductance peaks to broaden, and in some cases flatten into plateaus as shown in FIGURE 7c. A sampling of only a few random configurations can not be expected to completely describe the physical situation, but it can be seen that the effect of geometric disorder does play a significant role.

VI. CONCLUSION

We have presented calculations for the room-temperature differential conductance of BDT. The role of the contacts has been studied by including them with the molecule in the quantum chemistry calculation at the extended Hückel level. This has shown that there is mixing between cluster and molecule states, which alters the energy levels of the free molecule. It has been found that the resulting transport and conductance is sensitive to the geometry of the Au cluster contacts. This is because the electronic structure of the system is sensitive to the CMC configuration.

It was also found that the size of the clusters has noticeable effects on the conductance. With the larger size, a denser spectrum of cluster states is achieved, which is a better approximation to the continuum of states that would exist in reality. Thus many CMC states occur between the BDT HOMO and LUMO, and many of these are transmitting. This leads to a rise in conductance at smaller voltages than were observed experimentally. The magnitude is also up by a few orders compared to experiment. Thus experimentally, the molecule is less transparent than what our theory would seem to suggest. Why this is the case is still an open question.

To include more finite temperature effects in the model, positional disorder within the gold clusters was considered. By changing the regular geometry of the cluster, the energy levels were shifted around, thereby moving or creating new transmission peaks. The effect of averaging over configurations broadened and lowered the magnitude of the transmission spectrum. The conductance was also broadened and flattened in some regions. This appears to be consistent with the experimental findings, where variations in the conductance are relatively smooth and less pronounced. Thus positional disorder does seem to be a factor that has a role in determining the experimentally observed conductance.

ACKNOWLEDGMENTS

We would like to thank Mark Reed, Ross Hill, and John Black for helpful discussions. For the extended Hückel calculations, we would like to acknowledge the EHC program, Yaehmop, by Greg Landrum. This work was supported by NSERC.

REFERENCES

1. Aviram, A. & M.A. Ratner. 1974. Molecular Rectifiers. Chem. Phys. Lett. **29:** 257.
2. Reed, M. A., C. Zhou, C. J. Muller, T. P. Burgin & J. M Tour. 1997. Conductance of a molecular junction. Science **278:** 252–253.
3. Zhou, C., M. R. Deshpande, & M. A. Reed. 1997. Nanoscale metal/self-assembled monolayer/metal heterostructures. Appl. Phys. Lett. **71:** 611.
4. Stipe, B.C., M.A. Rezaei, W. Ho, S. Gao, M. Persson & B.I. Lundqvist. 1997. Single-molecule dissociation by tunneling electrons. Phys. Rev. Lett. **78:** 4410-4413.
5. Andres, R.P., J.D. Bielefeld, J.I. Henderson, D. B. Janes, V.R. Kolagunta, C.P. Kubiak, W.J. Mahoney & R.G. Osifchin. 1996. Self-assembly of a two-dimensional superlattice of molecularly linked metal clusters. Science **273:** 1690–1693.
6. Mirkin, C. A., R. L. Letsinger, R.C. Mucic & J.J. Storhoff. 1996. A DNA-based method for rationally assembling nanoparticles into macroscopic materials. Nature **382:** 607–609.
7. Fischer, C.M., M. Burghard, S. Roth & K. von Klitzing. 1995. Novel tunneling experiments on organic heterostructures. Surf. Sci. **361/362:** 905–908.
8. Datta, S., W.Tian, S. Hong, R. Reifenberger, J.I. Henderson & C.P. Kubiak. 1997. Current-voltage characteristics of self-assembled monolayers by scanning tunneling microscopy. Phys. Rev. Lett. **79:** 2530–2531.

9. MAGOGA, M. & C. JOACHIM. 1997. Conductance and transparence of long molecular wires. Phys. Rev. B **56:** 4722–4729.
10. DATTA, S. & W.TIAN. 1997. Application of the Friedel sum rule to symmetric molecular conductors. Phys. Rev. B **55:** R1914–R1917.
11. JOACHIM, C., & J.F. VINUESA. 1996. Length dependence of the electronic transparence (conductance) of a molecular wire. Europhys. Lett. **33:** 635–640.
12. KEMP, M *et al.* & S. DATTA e*t al.* 1997. Molecular Electronics — Science and Technology Conference. Unpublished material.
13. MUJICA, V., M. KEMP, A. ROITBERG & M. RATNER. 1996. Current-Voltage characteristics of molecular wires: Eigenvalue staircase, Coulomb blockade, and rectification. J. Chem. Phys. **104:** 7296–7305.
14. SAMANTA, M.P., W. TIAN, S. DATTA, J.I. HENDERSON, & C.P. KUBIAK. 1996. Electronic conduction through organic molecules. Phys. Rev. B **53:** R7626–R7629.
15. KEMP, M., A. ROITBERG, V. MUJICA, T. WANTA & M. A. Ratner. 1996. Molecular Wires: Extended Coupling and Disorder Effects. J. Phys. Chem **100:** 8349–8355.
16. SELLERS, H., A. ULMAN, Y. SCHNIDMAN & J.E. EILERS. 1993. Structure and binding of alkanethiolates on gold and silver surfaces: implications for self-assembled monolayers. J. Am. Chem. Soc. **115:** 9389–9401.
17. LAIBINIS, P.E., G.M. WHITESIDES, D.L. ALLARA, Y.T. TAO, A.N. PARIKH, R.G. NUZZO. 1991. J. Am. Chem. Soc. **113:** 7152.
18. CORREIA, A. & N. GARCIA. 1996. Nanocontact and nanowire formation between macroscopic metallic contacts observed by scanning and transmission electron microscopy. Phys. Rev. B **55:** 6689–6692.
19. KONDO, Y. & K. TAKAYANAGI. 1997. Gold nanobridge stabilized by surface structure. Phys. Rev. Lett. **79:** 3455–3458.
20. PINTO, A., A. R. PENNISI, G. FARACI, G. D'AGOSTINO, S. MOBILIO & F. BOSCHERINI. 1995. Evidence for truncated octahedral structures in supported gold clusters. Phys. Rev. B **51:** 5315–5321.

Ab Initio Methods for the Study of Molecular Systems for Nanometer Technology: Toward the First-Principles Design of Molecular Computers

JORGE M. SEMINARIO[a] AND JAMES M. TOUR[b]

Department of Chemistry and Biochemistry, University of South Carolina, Columbia, South Carolina 29208, USA

ABSTRACT: A top-priority objective in the area of nanoelectronics is the design of logical circuits using single molecules so that the feature size of present computers can be dramatically reduced by several orders of magnitude, thus allowing an enormous increase in speed and leading us toward the design of a molecular CPU. Achieving this goal clearly requires a good understanding of the factors that determine single interactions between unitary molecules, as well as the means to predict these interactions for proposed target molecules prior to undertaking what are often lengthy and costly synthetic efforts. The use of *ab initio* techniques will help us to learn how structural, electronic and external factors influence the communication between molecules, and to develop computational approaches that will allow us to quantitatively predict the characteristics of molecular scale electronics device architectures.

INTRODUCTION

Extremely higher computational performances are already a need in science, industry, defense, and commerce. Presently, single CPU architectures hardly reach one gigaflops performance. Gigaflops performances are obtained with very high efforts using massively parallel architectures. We are approaching teraflops performances using massively parallel computers with hundreds of CPUs, however these performances are just mere estimates obtained by multiplying the ideal single-processor megaflops times the number of processors. Because of the present need to use vector and parallel schemes, thousands of researcher-hours are spent just in the algorithm preparation rather than in the science itself. Scientific problems are inherently and easily solved through serial processing not parallel, except when the components of the system under study are totally independent of each other, which is equivalent to solving them individually.

Petaflops would be an ideal accomplishment in the solving of several problems. Examples of present problems in need of petaflops include intelligent systems, precise weather predictions, design of nanotechnological systems, design of molecules

[a,b]Addresses for telecommunication: J.M.S.: Phone, 803/777-9567; fax, 803/777-9521, e-mail, jorge@psc.sc.edu J.M.T: Phone, 803/777-9517; fax, 803/777-9521; e-mail, tour @psc.sc.edu

TABLE 1. Main characteristics of bulk and molecular CPU circuits

Feature	Bulk Electronics	Molecular Electronics
Time cycle	1 ns	1 fs
Electrons needed for bit of information	16,000 electrons	Much less than one electron
Power sources to operate	Need external power supply	Molecules are always ready to operate
Integration scale	10^8 gates/cm^2	10^{13} gates/cm^2
Gate length of a transistor	300 Å	2 Å

that target cancer cells, food improvement, creating better and new living environments on earth and on other planets, and the design of new materials.

Strong efforts have been made to move toward parallel techniques; however, the transformation of programs is being made slowly. Only commercial and heavily used programs are taken as candidates for conversion to fully parallel algorithms. Researchers in science and technology need to make efficient new programs quickly on a day-to-day basis to test theories or ideas; however, they cannot afford a tedious paralellization process able to optimize a program that can be disposed of after one run. Research would be much easier and progress much quicker if there can be petaflops performance on single CPUs, which is a top-priority objective in the area of nanometer-electronics. Thus the design of logical circuits using single molecules, so that the feature size of present computers can be dramatically reduced by several orders of magnitude, would allow an enormous increase in speed.

The advantages that can be obtained with the design of a molecular CPU are illustrated in TABLE 1, which shows the main differences between the present bulk electronic devices and our proposed molecular electronic devices. On the basis of these estimates, an increase in performance of 10^{11} may be expected with molecular electronics. Therefore, petaflops performances would be a conservative prediction, and exaflops performances would be an optimistic one; meanwhile, massive use of these systems can take us to zetaflops performances. A molecular CPU would be several orders of magnitude more powerful than any contemporary device and will be able to accelerate development in scientific areas related to its own design, yielding an iterative progress in their design and study. This will have a tremendous positive impact on our society.

Despite our well-developed understanding of classical semiconductor-based digital systems, including the complexities of semiconductors, precise equations regarding their behavior are unattainable because the Schrödinger equation, an exact equation of nature and able to describe any real material, can be only solved with chemical accuracy for very small systems, for example, small molecules, leaving bulk systems to be guided by model approximations and experimentation. Alternatively, the use of single molecules to perform electronic or digital functions can be guided by nearly exact calculations since solutions for single molecules can be treated with very high degrees of precision when using first-principles (*ab initio*) techniques. TABLE 2 shows that electronic digital devices are governed by considerations of size and speed. Optimizing the size of the basic units (usually the transistors) and optimizing their speed (limited by their natural temporal responses) are normally conflicting design goals; therefore, several trade-offs have to be made.

TABLE 2. The hardware-software duality as a CPU-memory compromise

Characteristic	CPU	Memory
Basic units	Faster and bigger	Smaller and slower
Programming	Wired logic	Programmed logic
Size of problems	Very small: few bit adders and few logical operations	Very large: largest possible memory, programs and data
Rate of usage	Practically all circuits at a time	Less than one in 106 at a time
Power dissipation	Very high (design limit)	Very low
Real problem	Cannot get below 1 ns	Limited by speed, transfer
Present solution	Going parallel (several problems)	Except for cost, no problem

The most important compromise in computational technology involves the hardware-software duality, which materializes in the requirements of programmed logic (memory- or software-dominant) versus wired logic (CPU-, or hardware-dominant). Components of a programmed logic are smaller and able to handle larger problems than a wired-logic system; however, a wired logic is faster than a programmed logic. At one extreme, there can be a bit adder (a minimum logic unit able to sum) with a small number of logical gates that will require a large memory to obtain the results, while at the other extreme, there could be a large CPU with all specific functions wired into the system that will be able to process the entire problem, having only a small memory for the input and output data. Present technology is heavily inclined toward programmed logic, for example, a computer with a large memory and a fast but simple CPU. This requires very large programs to solve present problems. This software-tilted technology leaves a large margin for improvement on the hardware side.

A classical two-terminal semiconductor device is characterized by the ability to conduct charges (electrons or holes) in one direction with low impedance, and high impedance in the reverse direction. The simplest of these two-terminal devices is the diode where a *p*-type semiconductor is joined to an *n*-type semiconductor (*p-n* junction). Therefore, a potential in one direction will deplete the junction and will not allow a current to pass while a potential in the opposite direction will allow the crossing of charged carriers. A more advanced device is the transistor where an additional collector, *p*- or *n*-type semiconductor, is joined to the *n*- or *p*-side of the diode (or base), respectively. The main feature of this device is that a small current in the base is able to control a large current in the collector or emitter. This notable amplification can also be viewed as a device having a low-input impedance and a high-output impedance. A related device which can be made even smaller than the bipolar transistor is the field-effect transistor (FET). It is similar to the transistor, but the output current is controlled by an input voltage rather than an input current. It has the three terminal arrangements: gate, source and drain corresponding to the base, emitter, and collector of the bipolar transistor. These three terminal electronic devices can operate as amplifiers and switches. Normally good amplifiers are also good switches, but the opposite is not necessarily true. When the transistor is used as an amplifier, the goal is to increase the power (energy per unit of time) of an input signal. If the device has a small input resistance, it will have a small input power, and if it has a large output resistance, it will have a large output power. The power of the signal is amplified at the expenses of a power supply, thus an amplifier shapes the energy of a power supply according to the shape of a small input signal. Since the output must sustain a large amount of energy under amplification conditions, the use of single molecules as amplifiers is discouraged and is therefore not being considered within the paradigm described here. However, the other application of transistors is their use as digital systems for processing of information, such as in digital computers. This is the area where single molecules can be used with possibilities of making computers orders of magnitude more efficient.[1–7]

In bulk electronics, it is a simple task to convert a good amplifier to a switch in which a small signal controls the passage of current; the logic is therefore based on whether or not the current passed. A large output current provides the circuit with a capable fan out, namely the ability to drive subsequent circuits in a cascade with high reliability. Logical circuits, in addition to performing the logical operations, also need to have a good driving capability. The output of a gate needs to be powerful enough to excite the input of one or more gates. At the same time, this high output should not couple to the input of the same device. This implies that there must be a very high impedance between the input and output. These are problems that would limit the use of molecular devices if we use bulk device architectural philosophies. We probably will not be able to shape the energy of a power supply using a single molecule. Alternatively, simulating gates by neighboring electrostatic potentials is proposed (*vide infra*).

What is the difficulty in simply downsizing present silicon-based devices so that the CPU's wired-logic components could become smaller? The eventual limitation to downscaling of conventional devices will be brought about by a combination of difficulties in fabrication, reliability, yield, interconnects, economics of production, and ultimately fundamental limits. In this size regime, several fundamental properties of quantum systems have to be considered like superposition, interference, entanglement, nonclonability, and uncertainty. Therefore, extension of microelectronics beyond current size regimes requires exploration of nonconventional electronic structures which scale far beyond these limits.[4–7]

Since conventional devices would be difficult to downsize, novel approaches to logic circuit design have to be devised wherein transistors and integrated circuits are substituted by using simple molecular structures. Of course, this implies the consideration of nonstandard architectures and design philosophies. Entirely new logic architectures would be required from molecular-based electronic components which will be complementary yet nonidentical to their bulk counterparts. Hence, molecules that work like the basic electronic devices (transistor, diodes, etc.) may not be desired, except for purely testing purposes, and certain basic principles of present devices such as amplification may not be of paramount interest. However, as we consider novel designs, we should not discard some basic principles that underpin CPU construction. For example, proposals in molecular-scale electronics must regard the need for input/output signal homogeneity within devices. An electronic input and a photonic output (or vice versa) within a device would be difficult to consider since the second device in the series would then require photonic operation. We must maintain a homogeneity of input and output signal types (e.g., voltage in and voltage out) and magnitudes within a device so that the second device can be driven with the same signal type and signal size that operated the first device. Therefore, new architectural strategies must be proposed while considering fundamental needs of overall CPU operation.

A SOUND APPROACH FOR THE FUTURE

A molecular CPU would have to operate by electrostatic interactions rather than electron currents, as in present systems. Electron currents at the molecular level would produce energies per unit of area that are impossible to be dissipated, as we will show later in this section. On the other hand, the use of the electrostatic potentials around the atoms of the molecules is a perfect choice because it can be controllable (as the current in a transistor) just by the rearrangement of a small amount of the charge density equivalent to a charge several orders of magnitude smaller than the charge of one electron, and also because these small changes in electrostatic potential are easily detected by the neighboring molecules.

The molecular CPU would be formed by small molecular units making the CPU itself a supramolecular unit. These units, which can be studied and designed exactly, will require parallel developments in synthetic efforts. New areas of technology would have to be created and developed like molecular-control theory, molecular-circuit theory, and molecular-digital theory based on present *ab initio* theories of quantum chemistry. Given the tremendous computational power that can be obtained with molecular CPUs, a super high-level programming can be developed.

Molecular-based systems can offer distinct advantages in uniformity and potential fabrication costs. If devices were to be based upon single molecules, using routine chemical syntheses, one could prepare 6×10^{23} (Avogadro's number) devices in a single-reaction flask, hence more devices than are presently in use by all the computational systems combined, world-wide.[4,8] However, the real problem facing synthetic chemists is how to arrange such a vast amount of molecular units. This process must be guided by precise laws of nature using quantum mechanics. Additionally, the properties of quantum systems can be used in a favorable manner at the molecular level. Thus molecular-based systems can offer distinct advantages in the density of structures that can be prepared and in their stability. Hence, molecular-scale electronics would likely shift the software/hardware equilibrium in the direction of hardware, namely massively wired-logic or CPU-dominant. Molecular circuits offer the possibility of constructing large and fast CPUs with complicated functions. By the use of large molecular arrays, problems could be calculated within the CPU with minimal need for main or auxiliary memories. Wired-logic would supplant programmed-logic computing, thereby affording several additional orders of magnitude increase in performance.[9–21]

A major consideration in molecular devices is the energy consumption/dissipation needs. Transfer of electrons or electron currents is actually incompatible with the idea of using single molecules for improvements in present digital technology. Such a strategy may only be useful for testing purposes. Considering 10^8 gates/cm^2 (as in present-sized silicon-based systems) at the rate of 10^{-9} sec (present speeds) yields 10^{17} electrons/sec (~0.02 amperes/cm^2) if only one electron per gate is used to transport, indicate, fetch, or represent a binary digit. At this point heat considerations are already extreme if the average resistance of the circuit is 30 Ω, this represents 20 watts/cm^2. If an increase of several orders of magnitude (10^{11}) in performance is expected with molecular circuits, a proportional increase in power dissipation is implied. Such levels of power dissipation rule out most conventional current- or electron-transfer methods for practical molecular devices.

Using the new design paradigm, molecules can be viewed as inherently active devices rather than simply passive devices like a resistor. Molecules are quantum-mechanical entities whose behavior cannot be explained by classical methods which embody our macroscopic world. A molecular system with N electrons can be successfully approximated (in zero order) as a group of one-electron orbitals φ_i where every electron, with a determined energy ε_i, moves under the field of the nuclei and of the other N-1 electrons. This is the Hartree-Fock (HF) approximation. The errors introduced by this approximation can be systematically corrected to any degree of precision limited only by the available computational resources. Methods like Møller-Plesset (MP), configuration interaction (CI), coupled cluster (CC), and combinations of them are typical ways to upgrade results obtained by the HF approximation. Even more encouraging, the modern Hohenberg-Kohn-Sham-Levy density functional theory (DFT) allows an exact description of the molecular system using auxiliary orbitals of noninteracting electrons.[22] Therefore, even within the quantum-mechanical picture, one-electron orbitals can be successfully used to study any molecular system. This does not imply that electrons are assigned to a specific orbital, but that there is one indistinguishable electron per orbital. Interestingly, these movements of electrons (wave-particle entities) in the molecular orbitals (or circuits) are performed without any dissipation of energy (stationary states). Unlike standard classical electronic circuits, these devices can be regarded as having inextinguishable power supplies that maintain the movement of the electrons. In this way, we can regard a molecule as an electronic circuit with active devices ε_i (energy sources) and passive devices φ_i (the orbitals). There are additional orbitals, like the φ_{LUMO} (LUMO = lowest unoccupied molecular orbital), with energy $\varepsilon_{\mathrm{LUMO}}$, which can be connected to the circuit if an additional electron is available or can also exchange occupancy with any of the other occupied orbitals (in most cases with the highest occupied molecular orbital-HOMO) and thus take the system to an excited state.

As part of these design scenarios, molecular devices could function by electrostatic interactions produced by small reshapes of the electron density due to the input signals. In turn, electrostatic potential interactions between molecules would transport the information throughout the CPU. Using this approach, there is no need for electron currents or electron transfers as in present devices; a small change in the electrostatic potential of one molecule could be enough to send the information to another molecule. External fields or excitations are able to change the boundary conditions of the molecule producing a change in the electrostatic potential generated by these sources. This is an efficient way to transmit a signal. There is no need for electron transfer, just a charge reshape. If the excitation ceases, the charge shape comes back to its original form. Is this method of information transfer really detectable since the change in the electrostatic potential is minimal (~10 kcal/mole/electron, where the formal units of electrostatic potential are energy per unit of charge)? The changes observed in the electrostatic potentials are in the range of values of nonbonded interactions, such as van der Waals interactions, which are easily detected by neighboring molecules. In fact, these are precisely the range of signal energies needed if we will ultimately utilize large scale integration in very small areas of materials. If the signals were large, problems of energy consumption and dissipation, as discussed above, rule out the use of molecular-based electronics. Thus the use of electrostatics directly addresses energy concerns in downsizing.

Distinguishing between two types of gates in molecular circuits is also in order. There are those gates at the interface of the logical circuits or CPU. The gates at the input interface would have to be able to take signals from standard electronic circuits, lasers, scanning probe microscopy (SPM) tips, or in general, from well-localized electrical fields. Molecular gates at the output would have to drive or inject a current onto a minuscule tip or metal cluster, or be able to excite the input of standard bulk devices. The vast majority of molecular gates, however, would be within the molecular CPU and will thus perform the information processing by controllably affecting the electrostatic potential of neighboring molecules. The ideas described here belong to the large group of gates that would allow the processing of information and the interconnection between them. The bulk/molecular interfacing structures are not being considered here.

Present technologies are also beset with problems from working at very high frequencies.[1] The use of electrostatics also addresses this difficulty. Small spurious capacitances (C) usually appear between portions of devices and circuits that otherwise should be totally electrically disconnected, or at infinite impedance (Z). If the operational frequency of the device increased, the impedance would decrease proportionally with the frequency (f) of operation ($Z = 1/2\pi fC$), rendering the device useless after a characteristic cut-off frequency. This cut-off frequency constitutes an upper limit to the operational frequency which, in present technology, approaches 1 GHz (which corresponds to a 1 ns clock cycle time). Since the capacitance is also inversely dependent on the distance, making smaller circuits yields larger spurious capacitances, and therefore smaller impedances, the move toward smaller devices is made more difficult. Since electrostatic potentials are only limited by the time in which the electron density rearranges in a molecule (<1 fs) and there are no electron currents, the problems of low impedance between supposedly isolated circuits are eliminated. This point deserves a lot of attention since it is sometimes believed that smaller circuits are faster; however, at microscopic sizes, in the range of present integrated circuits, it is well known that the faster circuits are relatively bigger than the slower ones. This is because the cut-off frequency is larger in larger circuits.

COMPUTATIONAL TECHNIQUES

Matter, in the form we know it, follows the laws of quantum mechanics (QM). Although at the macroscopic level Newton's laws can be applied very precisely, at the atomistic level the only possible way to study accurately the behavior of matter is by using quantum mechanics. The laws of quantum mechanics regarding problems in chemistry have been well understood since the 1930s; however, there is still much controversy regarding their interpretation. Nevertheless the predictions of quantum mechanics have never been in contradiction with the observed behavior of matter.

So far the only limitation, and also the constraint in predictions, is the mathematical complexity of QM equations. However, since the solutions of the quantum-mechanical equations allow us to predict the behavior of matter, efforts to develop the most sophisticated techniques and algorithms are increasing widely and exponentially with such encouraging results that theory now has the ability to correct several of the conclusions drawn from experiments. For example, several techniques are very accurate for predicting bond energies with errors smaller than 1%; even more sophisticated methods have recently claimed tolerances of less than 0.2 kcal/mol,[23] which are better than those from most of the standard experimental methods.

Several levels of theory are available to study matter such that it can be very confusing even to specialists. In order to clarify this situation, let us classify the several techniques into two major categories: the quantum-mechanical and the classical-mechanical. Both can be called first principles, but since we are dealing in most of the cases at the atomistic level, the equivalent term *ab initio* has been reserved for the quantum-mechanical techniques. In addition, both techniques, quantum and classical, can be used with empirical parameters obtained from experiments, or obtained by other *ab initio* calculations on smaller or prototype systems. This adds a problem to the nomenclature since the parameters now are not empirical, but are derived from first-principles calculations on smaller systems. Terms like *molecular modeling* for instance, which have been extensively used to describe a molecular system using classical mechanics techniques with empirical parameters, sometimes are confused with present quantum-mechanical procedures to solve the Schrödinger equation, which is an exact equation of nature and does not require experimental input.

In this section, we describe several quantum mechanical techniques and their extensions. Matter most of the time is in a stationary state that can be studied by the time-independent Schrödinger equation. In addition, most of the systems of interest deal with situations where relativistic effects are not too important, and if they are, they can still be treated as corrections to the nonrelativistic effects. One further simplification is given by the fact that electrons move much faster than nuclei, thereby permitting us to consider the wavefunction for electrons separated from the wavefunction for nuclei. In any case, if corrections due to any of the above effects are necessary, they can always be treated by a perturbational approach. Time-dependent cases are treated using the famous Fermi golden rule or modifications of it, where the time-dependence is considered as a perturbation. For large systems where time-dependent effects involve pressure and temperature, in most cases these effects can be treated by introducing classical-mechanics corrections to the quantum results.

The simplest useful case corresponds to a two-body problem, the hydrogen-like systems, one nucleus and one electron. For this system the Schrödinger equation can be solved exactly and the wavefunction and the energy can be expressed in closed form. The solutions to this system, hydrogenic wavefunctions ($1s$, $2s$, $2p_m$, ...), are the building blocks or basis to construct the solutions of multielectron systems, but are expressed in a functionally different form (Slater-type functions, Gaussian-type functions, etc.). The next case correspond to the helium-like atoms. The helium case is actually the proof of the validity and incredible precision of the Schrödinger equation. Several variational and perturbational techniques have been developed to solve the helium atom.[24] The precision obtained goes beyond the precision of any other theory or method in any area of knowledge. The present value is −2.903724377034119479 Ha (1 Ha = 627.509 kcal/mol) where at least 15 significant figures are correct! This amazing precision is equivalent to predicting a distance between two points: one in New York and the other in Paris with less than 10 Å of error. Notice that chemical accuracy (about 1 kcal/mol) is already surpassed with the fourth decimal figure.

The situation becomes more complicated when we try to solve the Schrödinger equation for molecules or larger atoms. Basis sets have to be of reduced complexity because of the large number of integrals needed to solve the quantum eigenvalue problem and the practical convenience of performing these integrals analytically rather than numerically. There are two well-known routes to solve the electronic, nonrelativistic, time-independent Schrödinger equation,

$$\hat{H}\Psi = E\Psi ,$$

with Hamiltonian partitioned as

$$\hat{H} = \hat{T} + \hat{V}_{ne} + \hat{V}_{ee} ,$$

n-electron wavefunction

$$\Psi = \Psi(\vec{r}_1, \vec{r}_2, \vec{r}_3, \dots \vec{r}_N) ,$$

and Hamiltonian explicitly written as

$$\hat{H} = -\sum_{i=1}^{N} \frac{1}{2}\nabla_i^2 - \sum_{i=1}^{N}\sum_{A=1}^{M} \frac{Z_A}{r_{Ai}} + \sum_{i=1}^{N-1}\sum_{j=i+1}^{N} \frac{1}{r_{ij}} :$$

the standard *ab initio* methods and the density functional theory (DFT) techniques.

The Standard ab Initio Techniques

These techniques start from the Hartree-Fock approximation, and by corrections using several different methods, eventually they will take us to the correct solution of the Schrödinger equation. However, technical problems due to the limitation of present computational resources severely limit the applications of precise levels of theory to relatively small realistic systems. In the Hartree-Fock (HF) approximation, it is assumed that each electron interacts with the mean field created by the other electrons and nuclei. Therefore, the effect due to instantaneous fluctuations from the mean field is ignored. Unfortunately these "small" deviations from the average are really small with respect to the total energy of the system, but "large" with respect to the energy difference between systems and states, and much larger than the desired chemical accuracy. Chemical accuracy has been presently set at 1 kcal/mol. These small fluctuations from the average produce a stabilizing effect that has been widely known as correlation energy. The advantage of the Hartree-Fock approximation is that the wavefunction can be expressed as a determinant made of one-electron molecular orbitals (MO), ϕ_i,

$$\Psi_0 = \Psi_{HF} = |\phi_1\phi_2\phi_3 \cdots \phi_{i-1}\phi_i\phi_{i+1} \cdots \phi_{n-1}\phi_n|$$

which in turn can be expanded as a linear combination of atomic orbitals (LCAO) made of atomic basis sets of the constituting atoms in the molecular systems.

$$\phi_i = \sum_j c_{ij}\varphi_j$$

where the φ_j are the atomic orbitals of the constituting atoms. This basis set, in principle, could be made of the exact hydrogenic solutions of the corresponding hydrogen-like atoms.

The idea behind the HF approximation is equivalent to partition the total Hamiltonian in two parts, one that depends only on one-electron operators, and other that contains the so-called correlation effects

$$H = H_{\text{HF}} + H_{\text{corr}}$$

Since the HF Hamiltonian is a one-electron operator, it can be easily solved using an N-electron wavefunction made of one-electron molecular orbitals.

$$H_{\text{HF}}\Psi_{\text{HF}} = E_0\Psi_{\text{HF}}$$

where Ψ_{HF} is the HF wavefunction and E_0 is the zero-order energy. The HF energy is obtained from

$$E_{\text{HF}} = \langle\Psi_{\text{HF}}|H_{\text{HF}}|\Psi_{\text{HF}}\rangle + \langle\Psi_{\text{HF}}|H_{\text{corr}}|\Psi_{\text{HF}}\rangle$$

Notice that E_0 is not E_{HF}. In perturbation theory terms, E_0 is the zero-order energy and E_{HF} is the sum of the zero- and first-order energy corrections. The difference below is the correlation energy which contains the higher-order contributions to the total energy, or in other words, the many-electron or many-body effects.

$$E_{\text{corr}} = E - E_{\text{HF}}$$

To go beyond the HF approximation, several routes have been devised. These are called correlated methods. The most used and probably the most practical ones are configuration interaction (CI), Møller-Plesset perturbation theory (MP), and coupled cluster theory (CC).

There are several ideal properties that a correlated method should have. It should be applicable to a wide range of systems. It should be variational and well-defined, that is, not *ad hoc* choices. More importantly, it should be size-consistent, that is, it should allow a proper description of dissociation energies. In addition, a good cost-accuracy relation would be highly preferred. No method so far is able to fulfill all the above characteristics.

Configuration Interaction

Basically, the idea is to mix in excited determinants using the variational principle

$$\Psi^{\text{CI}} = \Psi_0 + \sum_{ia} t_i^a \Psi_{i\to a} + \sum_{ijab} t_{ij}^{ab}\Psi_{ij\to ab} + \cdots$$

where the coefficients $t_{i\ldots}^{a\ldots}$ minimize the total energy

$$E = \frac{\langle\Psi^{\mathrm{CI}}|H|\Psi^{\mathrm{CI}}\rangle}{\langle\Psi^{\mathrm{CI}}|\Psi^{\mathrm{CI}}\rangle}$$

When only single excitations are used (CIS), the trial wavefunction contains determinants made by substituting one of the unoccupied orbitals (a) by one of the occupied (i) in the reference determinant, i.e.,

$$\Psi_{i\to a} = |\phi_1\phi_2\phi_3 \cdots \phi_{i-1}\phi_a\phi_{i+1}\cdots \phi_{n-1}\phi_n|$$

Therefore, when double excitations (CID) are used, we have doubly excited determinants in the wavefunction,

$$\Psi_{ij\to ab} = |\phi_1\phi_2\phi_3 \cdots \phi_{i-1}\phi_a\phi_{i+1} \cdots \phi_{j-1}\phi_b\phi_{j+1} \cdots \phi_{n-1}\phi_n|,$$

and this could go up to nth excitations, where n is the total number of occupied states. In general,

$$\Psi^{\mathrm{CI}} = \sum_s t_s\Psi_s$$

where $s = i \to a,\ ij \to ab,\ ijk \to abc, \ldots$. Therefore, the total energy is given by

$$E = \frac{\sum_{su} t_s t_u \langle\Psi_s|H|\Psi_u\rangle}{\sum_s t_s^2},$$

and using $\frac{\partial E}{\partial t_s} = 0$ we end up with the eigenvalue problem

$$\sum_u H_{su}t_u - Et_s = 0 \text{ or } H\vec{t} = E\vec{t}$$

H can be fully diagonalized; usually it is solved iteratively starting with a trial $\vec{t}$ then forming $H\vec{t}$, and then updating to a new $\vec{t}$ until the desired convergence is reached. There are several extrapolation schemes. Technical limitations only allow us to use up to CISD for practical cases. For CISD the matrix elements are given by

$$\langle\Psi_0|H|\Psi_0\rangle = E^{\mathrm{HF}}$$

$$\langle\Psi_0|H|\Psi_{i\to a}\rangle = h_{ia} + \sum_j \langle ij||ab\rangle = F_{ia} = 0$$

$$\langle\Psi_0|H|\Psi_{i\to ab}\rangle = \langle ij||ab\rangle$$

The iterative process for the calculation of these terms scale as O^2V^4 and O^3V^3 in CPU time (O is the number of occupied orbitals and V is the number of virtual ones), and both are in the same order of magnitude as N. It is for this reason that the method formally scales as N^6 for single and double excitations CI, or CISD. When all possible excitations for a given basis are used, the method is called full-CI (FCI), which yields exact results for a given atomic orbital (AO) basis set. If an infinite basis set were used, FCI yields the exact solution of the Schrödinger equation. FCI is used only with very small basis sets on very small systems for reference calculations only and scales as N^N.

The terms in the expansion of the CI wavefunction can be expressed using an operator expansion, for instance, for CISD, defining

$$\hat{T}_1\Psi_0 = \sum_{ia} t_i^a \Psi_{i\to a}$$

and

$$\hat{T}_2\Psi_0 = \sum_{ijab} t_{ij}^{ab} \Psi_{ij\to ab},$$

we can write

$$\Psi^{\mathrm{CI}} = (1 + \hat{T}_1 + \hat{T}_2)\Psi_0$$

and the matrix elements simplify to

$$\langle\Psi_0|H|\hat{T}_2\Psi_0\rangle = E_{\mathrm{corr}}$$

$$\langle\Psi_{i\to a}|\tilde{H}|(\hat{T}_1 + \hat{T}_2)\Psi_0\rangle = t_i^a E_{\mathrm{corr}}$$

$$\langle\Psi_{ij\to ab}|\tilde{H}|(1 + \hat{T}_1 + \hat{T}_2)\Psi_0\rangle = t_{ij}^{ab} E_{\mathrm{corr}}$$

All CI methods are variational procedures and therefore they yield upper bounds to the exact energy. They can be applied to excited states since the interleaving theorem is also valid. However, the major problem for its applications to chemistry is that any truncated Cl is not size-consistent and therefore will not yield a proper description of dissociation energies.

One way out of the problem is to add to the CI wavefunction those terms needed to restore the size-consistency. This new method is called quadratic CI (QCI). One of the most used QCI methods corresponds to the CISD or simply QCISD where the last two matrix elements of the CISD have been modified to

$$\langle\Psi_{i\to a}|\tilde{H}|(\hat{T}_1 + \hat{T}_2 + \hat{T}_1\hat{T}_2)\Psi_0\rangle = t_i^a E_{\mathrm{corr}}$$

$$\langle\Psi_{ij\to ab}|\tilde{H}|\left(1 + \hat{T}_1 + \hat{T}_2 + \frac{1}{2}\hat{T}_2^2\right)\Psi_0\rangle = t_{ij}^{ab} E_{\mathrm{corr}}$$

When triple contributions are added noniteratively, the method is called QCISD(T), which right now is considered one of the most precise *ab initio* methods. Usually the triple-excitations in the wavefunction are needed for a correct description of multiple bonds.

Another avenue can be taken to accelerate the convergence to the correlation energy. The correlation energy can be divided in a qualitative way into dynamic and static correlation. Since static correlation is associated with the multireference nature of the wavefunction, it would be more convenient to have the reference or zero-order wavefunction be composed of more than one determinant. The use of a multireference wavefunction instead of a single-reference determinant in the HF theory has lead to methods like multireference HF or MRHF also called MRSCF. One particular version of this method is the complete active space SCF or CAS-SCF. Usually in most cases of chemistry, the states of interest are made of one dominant configuration and there is little need to recur to these techniques. However, in cases of multibond reactions where bond breaking and bond creation are not synchronous, in homolitic bond fission, or in avoided crossings, there are two or more dominant configurations; therefore, the use of a multireference (or multideterminant) wavefunction is needed. However, other techniques based on density functional theory (DFT) are a much better way out of this problem. The dynamic correlation is obtained mainly by the lowest Cl or other correlated single-reference methods, and the nondynamic by the multireference methods. However, these two methods can be combined in what is called the multireference CI (MRCI), providing a better estimate of the total correlation energy at very extremely high computational costs.

Perturbation Theory

In perturbation theory the Hamiltonian is partitioned in

$$H = H_0 + V$$

where H_0 corresponds to a solvable case. MøllerPlesset theory[25] implies to have H_0 as the H_{HF} so the ground state single determinant and excited determinants can be used, since all of them are eigenfunctions of H_0. Therefore,

$$H_0\Psi_0 = E_0\Psi_0 = \left(\sum_i \varepsilon_i\right)\Psi_0$$

$$H_0\Psi_{i\to a} = (E_0 + \varepsilon_a + \varepsilon_i)\Psi_{i\to a}$$

$$H_0\Psi_{ij\to ab} = (E_0 + \varepsilon_a + \varepsilon_b + \varepsilon_i + \varepsilon_j)\Psi_{ij\to ab}$$

and the energy contributions to the total energy are given by,

$$E^{(0)} = \langle\Psi_0|H|\Psi_0\rangle = \sum_i \varepsilon_i$$

$$E^{(1)} = \langle\Psi_0|V|\Psi_0\rangle = E^{\mathrm{HF}}$$

$$E^{(2)} = E^{\mathrm{HF}} - \sum_s \frac{|\langle\Psi_0|V|\Psi_0\rangle|^2}{E_s - E_0}$$

Also notice that

$$\langle\Psi_0|V|\Psi_{1\to a}\rangle = 0$$

and

$$\langle\Psi_0|V|\Psi_{ijk\to abc}\rangle = 0$$

Therefore s is only made of $ij \to ab$ terms in the second-order energy. The t for triple goes as $O^3V^4 + O^4V^3$, which mean a scaling as N^7. The t for quadruple goes as O^4V^4, although a factorization procedure can avoid these terms. Møller-Plesset expansions show in general a good hierarchy of models. Presently, MP2 is not too difficult to use in large systems. MP4 is good for most systems. The series tend to oscillate when HF is a poor starting wavefunction or when large spin contamination occurs. Since this is an order-by-order perturbation approach, higher orders are often needed. Moreover, this procedure is not exact for two-electron systems. However, the MP procedure is size-extensive, which makes it the most popular method for studying chemical systems. The second-order energy is given by

$$E^{(2)} = E^{\mathrm{HF}} - \frac{1}{4}\sum_{ijab} \frac{|\langle ij||ab\rangle|^2}{\Delta_{ij}^{ab}}$$

where

$$\Delta_{ij}^{ab} = E_{ij \to ab} - E_0 = \varepsilon_a + \varepsilon_b - \varepsilon_i - \varepsilon_j$$

Therefore we can rewrite it as

$$E^{(2)} = E^{\mathrm{HF}} - \frac{1}{4}\sum_{ijab} a_{ij}^{ab} |\langle ij||ab\rangle|^2$$

and the first-order wavefunction is

$$\Psi^{(1)} = \Psi^{\mathrm{HF}} + \frac{1}{4}\sum_{ijab} a_{ij}^{ab} \Psi_{ij \to ab}$$

The terms become more and more complicated as the order of the perturbation is increased. For the third- and fourth-order energy, we have

$$E^{(3)} = E^{(2)} + \frac{1}{4}\sum_{st} \frac{\langle \Psi_0|V|\Psi_s\rangle\langle \Psi_s|\bar{V}|\Psi_t\rangle\langle \Psi_t|V|\Psi_0\rangle}{(E_s - E_0)(E_t - E_0)} = \sum_{st}^{D} a_s \bar{V}_{st}\, a_t = \sum a_{ij}^{ab} u_{ij}^{ab}$$

and

$$E^{(4)} = E^{(3)} + \sum_{stu} \frac{\langle \Psi_0|V|\Psi_s\rangle\langle \Psi_s|\bar{V}|\Psi_t\rangle\langle \Psi_t|\bar{V}|\Psi_u\rangle\langle \Psi_u|V|\Psi_0\rangle}{(E_s - E_0)(E_t - E_0)(E_u - E_0)}$$

$$= E^{(3)} + \sum_{stu} \frac{a_s \bar{V}_{st} V_{tu} a_u}{\Delta_t} = E^{(3)} + \sum \frac{|u_t|^2}{\Delta_t}$$

where s and u subscripts correspond to doubles and t to SDTQ. These methods can also be treated with a reference multideterminant wavefunction MRMP or MR-MBPT (MR-MP2 for instance).

Coupled Cluster Methods

In coupled cluster (CC) theory, the wavefunction is assumed to have the following form

$$\Psi^{CC} = e^{\hat{T}}\Psi_0 = \left(1 + \hat{T} + \frac{1}{2}\hat{T}^2 + \frac{1}{3!}\hat{T}^3 + \ldots\right)\Psi_0$$

where

$$\hat{T} = \hat{T}_1 + \hat{T}_2 + \hat{T}_3 + \ldots + \hat{T}_n$$

CC also solves equations iteratively; each iteration of CCD is like a MP4DQ calculation. It has the advantage that includes doubles to infinite order. However, CCD is not much better than MP4DQ, although CCSD shows a large improvement. For two-electron systems CCSD is identical to CISD. CCSD is correct to fourth order (i.e., like MP4SDQ). These methods are not variational, but they are size-consistent. CCSD scales as N^6 and CCSD(T) scales as N^7. In CCD, the wavefunction is expressed as

$$\Psi^{CCD} = e^{\hat{T}_2}\Psi_0 = \left(1 + \hat{T}_2 + \frac{1}{2}\hat{T}_2^2 + \ldots\right)\Psi_0$$

that is, it includes quadruples, sextuples, etc., excitations up to all n-tuples for even n electrons, and the computational effort is similar to CID. The amplitudes (or coefficients t) cannot be solved variationally; therefore, they project onto lowest-order excitation. For CCD we have the following equations:

$$\langle\Psi_0|H - E|e^{\hat{T}_2}\Psi_0\rangle = 0 \text{ therefore } \langle\Psi_0|H|\hat{T}_2\Psi_0\rangle = E_{corr}, \text{ and}$$

$$\langle\Psi_{ij\rightarrow ab}|H - E|e^{\hat{T}_2}\Psi_0\rangle = 0 \text{ implies}$$

$$\langle\Psi_{ij\rightarrow ab}|H|\left(\hat{T}_2 + \frac{1}{2}\hat{T}_2^2\right)\Psi_0\rangle = t_{ij}^{ab}E_{corr}$$

$$\langle\Psi_{ij\rightarrow ab}|\overline{H}|\hat{T}_2^2\Psi_0\rangle = v_{ij}^{ab}$$

As with other correlated methods, these also can have a multireference zero-order wavefunction (MR-CC) like MR-CCSD, etc., in order to account for a faster convergence of the static correlation energy.

Density Functional Theory (DFT) Methods

The methods in the last section are usually very precise. However, they require extremely large amounts of computational resources. In order to obtain just chemical accuracy, methods that scale as N^7 or higher are required. In addition, the need for larger basis sets also increases with the level of theory. For instance, methods that scale as N^7 usually require polarization *f*- and *g*-functions or even of higher angular momentum. Therefore, standard systems of interest for chemistry are practically intractable with these levels of theory. Parallel to the development of the standard *ab initio* methods when HF was the only alternative, several routes for its simplification were developed. The bottleneck of HF was the calculation of the two-electron four-center integrals that appear in the direct and exchange terms of the electron-electron repulsion energy. Therefore, methods were developed in order to avoid these terms, obtaining in this way an approximation to the HF procedure. The direct terms were no major problem because they are quadratic in the density; therefore, by expressing the density in some basis sets, the four-center integrals are eliminated. The problem was with the exchange terms. There is no straightforward way to eliminate the four-center integrals; therefore functionals of the density were developed that were able to express the exchange energy as a functional of the density. This functional was calculated for a uniform electron gas (also called jellium) by Bloch, who found that the exchange energy for this system was proportional to $\rho^{1/3}$.[26] Later, this expression was also used by Dirac[27] in a well-known paper where the exchange was included in the Thomas-Fermi atom.[28,29] This form of exchange functional with an empirical parameter α was later on used by Slater[30] in order to simplify the HF procedure.

The time-independent, nonrelativistic, Born-Oppenheimer approximated Schrödinger equation,

$$\hat{H}\Psi = E\Psi ,$$

$$(\hat{T} + \hat{\upsilon}_{\text{ext}} + \hat{V}_{\text{ee}})\Psi = E\Psi$$

yields the energy E,

$$E = \int \upsilon_{\text{ext}}(\vec{r}) N \int d\vec{r}_2 d\vec{r}_3 \ldots d\vec{r}_N \left| \Psi(\vec{r}_1, \vec{r}_2, \vec{r}_3, \ldots, \vec{r}_N) \right|^2 d\vec{r} + \langle \Psi | \hat{T} + \hat{V}_{\text{ee}} | \Psi \rangle$$

where the nuclei-electron energy interaction is explicitly calculated using the many-electron wavefunction,

$$V_{\text{ne}} = \int \upsilon_{\text{ext}}(\vec{r}) N \int d\vec{r}_2 d\vec{r}_3 \ldots d\vec{r}_N \left| \Psi(\vec{r}_1, \vec{r}_2, \vec{r}_3, \ldots, \vec{r}_N) \right|^2 d\vec{r} .$$

Because electrons are much faster than nuclei, and the classical energy interaction between a static charge distribution with a set of fixed charges or external potential υ_{ext} is

$$V_{\text{ne}}^{\text{class}} = \int \upsilon_{\text{ext}}(\vec{r}) \rho(\vec{r}) d\vec{r} ,$$

the function

$$N \int d\vec{r}_2 d\vec{r}_3 \ldots d\vec{r}_N \left| \Psi(\vec{r}_1, \vec{r}_2, \vec{r}_3, \ldots, \vec{r}_N) \right|^2$$

has exactly the same meaning as the classical function $\rho(\vec{r})$ and therefore, provided that the Born-Oppenheimer approximation is valid, the function

$$N \int d\vec{r}_2 d\vec{r}_3 \ldots d\vec{r}_N \left| \Psi(\vec{r}_1, \vec{r}_2, \vec{r}_3, \ldots, \vec{r}_N) \right|^2$$

can be regarded as a charge distribution, or more precisely, as the electron density. This is another statement of the Hellmann-Feymnan electrostatic theorem.[31,32] Therefore using

$$E = \langle\Psi|\hat{H}|\Psi\rangle = \int \upsilon_{ext}(\vec{r})\rho(\vec{r})d\vec{r} + \langle\Psi|\hat{T} + \hat{V}_{ee}|\Psi\rangle,$$

and the variational principle, Hohenberg and Kohn,[33] by a trivial *reductio ad absurdum* showed that there is a one-to-one relation between the density function ρ and the external potential υ_{ext}. This implies that the energy, E, and any expectation value are functionals of the electronic density of the ground state, that is, for a given external potential υ

$$E = E_{\upsilon}[\rho]$$

This can be rewritten as

$$E_{\upsilon}[\rho] = \int \upsilon(\vec{r})\rho(\vec{r})d\vec{r} + \langle\Psi|\hat{T} + \hat{V}_{ee}|\Psi\rangle = \int \upsilon(\vec{r})\rho(\vec{r})d\vec{r} + F[\rho]$$

where $F[\rho]$ is a universal functional (it does not depend on the external potential) identical for any system. Together with this theorem, Hohenberg and Kohn also derived a second theorem which provides a variational principle for the energy which established that for any trial density $\tilde{\rho}$, the energy is an upper bound to the exact energy of the ground state, that is,

$$E = \min_{\rho}\left\{\int \upsilon(\vec{r})\rho(\vec{r})d\vec{r} + F[\rho]\right\}$$

For the exact density ρ, we have

$$E = \int \upsilon(\vec{r})\rho(\vec{r})d\vec{r} + F[\rho]$$

There are certain cases in which these theorems have problems associated with the so-called v- and N-representability. These problems were subsequently solved by Levy[34] and the correct ground-state energy can be unambiguously expressed as

$$E = \min_{\rho}\left\{\int \upsilon(\vec{r})\rho(\vec{r})d\vec{r} + \langle\Psi_{\rho}^{min}|\hat{T} + \hat{V}_{ee}|\Psi_{\rho}^{min}\rangle\right\}$$

This expression is what constitutes the Levy's constrained search, with the universal functional

$$F[\tilde{\rho}] = \langle\Psi_{\tilde{\rho}}^{min}|\hat{T} + \hat{V}_{ee}|\Psi_{\tilde{\rho}}^{min}\rangle$$

Notice that Ψ_{ρ}^{min} is the wavefunction that yields $\tilde{\rho}$ and minimizes only $\hat{T} + \hat{V}_{ee}$ and not the total Hamiltonian. A similar universal functional was developed by Percus for the noninteracting system.[35]

Let us now assume a system with Hamiltonian[36]

$$\hat{H}_\lambda = \hat{T} + \hat{\upsilon}_\lambda + \lambda V_{ee}$$

where λ can vary between 0 and 1. In addition, we constrain the $\hat{\upsilon}_\lambda$ such that it always yields the density $\rho(\vec{r})$ of the real system.[37] The Schrödinger equation for this system is

$$\hat{H}_\lambda \Psi_\lambda = E_\lambda \Psi_\lambda$$

where E_λ and Ψ_λ are the corresponding eigenvalue and eigenfuction of the system with the constraint on the density $\rho(\vec{r})$. The case $\lambda = 0$ corresponds to an ideal system of noninteracting electrons having the density of the real system under an effective potential υ_{eff} or υ_s and the case $\lambda = 1$ corresponds to the real system with υ_{ext}. The system of noninteracting electrons is trivially solved since the exact solution is, in most cases, a Slater determinant.

Therefore, the energy E_λ can be written as

$$E_\lambda = \langle\Psi_\lambda|\hat{H}|\Psi_\lambda\rangle = \int \upsilon_\lambda(\vec{r})\rho(\vec{r})d\vec{r} + \langle\Psi_\lambda|\hat{T} + \lambda\hat{V}_{ee}|\Psi_\lambda\rangle$$

Derivation and integration of the above equation with respect to λ yield the difference of the energies of the real and noninteracting systems ($E - E_s$) in the left hand side, and E_s is exactly $V_s + T_s$, that is , the sum of the potential energy plus the kinetic energy of a noninteractive system of electrons with a density identical to the one of the real system. Using the Hellmann Feynman theorem and rearranging terms, we have

$$E(\rho) = \int \upsilon_{ext}(\vec{r})\rho(\vec{r})d\vec{r} + T_s + \int_0^1 \langle\Psi_\lambda|\hat{V}_{ee}|\Psi_\lambda\rangle d\lambda$$

where the integral on λ is still unsolvable. It is convenient to extract the classical electron-electron interaction energy, V_{class}, from the second integral and define the difference as the exchange-correlation energy, because of its close resemblance to their *ab initio* counterparts.

$$\int_0^1 \langle\Psi_\lambda|\hat{V}_{ee}|\Psi_\lambda\rangle d\lambda = \frac{1}{2}\iint d\vec{r}_1 d\vec{r}_2 \frac{\rho(\vec{r}_1)\rho(\vec{r}_2)}{|\vec{r}_1 - \vec{r}_2|} + E_{xc}(\rho)$$

However, it can be easily noted that E_{xc} does not correspond to the $E_c + E_x$ of the standard HF-based definitions. The total energy of the electronic system can be written as

$$E[\rho] = \int \upsilon_{ext}(\vec{r})\rho(\vec{r})d\vec{r} + T_s + V_{class}[\rho] + E_{xc}[\rho]$$

This expression constitutes the basis of the famous Kohn-Sham (K-S) procedure,[38] which gives a practical use of the Hohenberg-Kohn theorem. Up to this point, there are not approximations; not even the mean field approximation has been used, as is the case of the HF procedure. The K-S procedure uses a set of molecular orbitals with the sole purpose of representing the electron density of the system of noninteracting electrons. All this development comes out directly from the Schrödinger equation. All functionals or quantities in the above equation are known, except the E_{xc}. T_s is an unknown functional of the density, but it can be obtained exactly using the wavefunction of the noninteracting system. Therefore, it is not required to have a functional, either in the interactive or noninteractive systems, for its exact calculation.

The only unknown functional is the $E_{xc}[\rho]$. All the others are known or can be easily calculated from the wavefunction of the noninteracting system. Once E_{xc} is known, the total energy E can be minimized with respect to the density ρ, yielding the Kohn-Sham equations that can be solved self-consistently:

$$\left[-\frac{1}{2}\nabla^2 + \upsilon_{\text{eff}} + \upsilon_{\text{class}} + \upsilon_{\text{xc}}\right]\varphi_i = \varepsilon_i\varphi_i$$

where υ_{class} and υ_{xc} are the potentials corresponding to the classical repulsive energy between electrons and to the exchange correlation energy, respectively,

$$\upsilon_{\text{class}}(\vec{r}) = \int d\vec{r}' \frac{\rho(\vec{r}')}{|\vec{r} - \vec{r}'|}$$

and

$$\upsilon_{\text{xc}}(\vec{r}) = \frac{\delta E_{\text{xc}}[\rho(\vec{r})]}{\delta\rho(\vec{r})}$$

The K-S orbitals in this procedure are the solutions for the noninteractive system of electrons that are under the effective potential υ_{eff}

$$\upsilon_{\text{eff}} = \upsilon_{\text{ext}} + \upsilon_{\text{class}} + \upsilon_{\text{ext}}$$

In other words, υ_{eff} is the potential such that, when applied to a noninteractive system, $\lambda = 0$, will yield the same density as the real system, $\lambda = 1$. Therefore,

$$E_s = \sum_{i=1}^{N} \varepsilon_i = T_s + V_s = \sum_{i=1}^{N} \langle\varphi_i| -\tfrac{1}{2}\nabla^2|\varphi_i\rangle + \int \upsilon_{\text{eff}}(\vec{r})\rho(\vec{r})d\vec{r}$$

where E_s, T_s, and V_s are the total, kinetic, and potential energies of the noninteractive system, respectively. The wavefunction of this noninteractive system is exactly the determinant made using the orbitals φ_i,

$$\Psi_s(\vec{r}_1, \vec{r}_2, \ldots, \vec{r}_N) = \det|\varphi_1\varphi_2\ldots\varphi_N|$$

We always have to keep in mind that Ψ_s is not the wavefunction of the real system, neither are the φs related to real electrons, except for fact that they yield the same density through the sum:

$$\rho(\vec{r}) = \sum_{i=1}^{N} \left|\varphi_i(\vec{r})\right|^2$$

The total energy for the real system will be evaluated using those orbitals that yield the minimum energy, which in most of the cases corresponds to those with the lowest energies. The first functionals related to the E_{xc} were not made with DFT in mind, but were oriented to the description of exchange effects. Therefore, as a practical way, the partition,

$$E_{xc} = E_x + E_c$$

is invoked. The exchange energy using the formula for the uniform electron gas is the local density approximation (LDA), written in our notation as

$$E_x^{\text{LDA}}[\rho] = -\int d\vec{r}\frac{3}{4}\left(\frac{3}{\pi}\right)^{\frac{1}{3}}\rho^{\frac{4}{3}}$$

The corresponding potential is

$$\upsilon_x^{\text{LDA}}(\vec{r}) = -\left(\frac{3}{\pi}\rho(\vec{r})\right)^{\frac{1}{3}}$$

The above formulas and the following ones are expressed, in many cases, using a parameter r_s, equivalent to the radius of a sphere with constant charge density ρ and a total charge of one electron, also known as the Wigner-Seits radius,

$$r_s = \left(\frac{4\pi\rho}{3}\right)^{-\frac{1}{3}}$$

Therefore,

$$\upsilon_x^{\text{LDA}} = \left(\frac{3}{2\pi}\right)^{\frac{1}{3}}\frac{1}{r_s}$$

Similarly, this uniform gas of electrons brings us the first approximation for the E_{xc},

$$E_{xc}^{\text{LDA}} = \int \varepsilon_{xc}^{\text{LDA}}[\rho(\vec{r})]\rho(\vec{r})d\vec{r}$$

where $\varepsilon_{xc}^{\text{LDA}}$ is the exchange-correlation energy distribution per unit of volume and it only depends on the density at the point where it is evaluated. The functional for the ideal gas of electrons of uniform density reduces to

$$E[\rho] = T_s + E_{xc}[\rho]$$

which can be solved numerically as precisely as it is needed for any situation using plane waves, and analytically for some limiting cases. Therefore, it will be possible to get the E_{xc}, and subtracting the E_x already known, the E_c can be calculated.

Perdew and Wang (PW92)[39] have proposed the following local correlation functional

$$\varepsilon_c^{\mathrm{PW92}} = -2a\rho(1+\alpha_1 r_s)\ln\left[1+\frac{1}{2a\left(\beta_1 r_s^{\frac{1}{2}}+\beta_2 r_s+\beta_3 r_s^{\frac{3}{2}}+\beta_4 r_s^{2}\right)}\right]$$

where $\alpha = 0.0310907$, $\alpha_1 = 0.21370$, $\beta_1 = 7.5957$, $\beta_2 = 3.5876$, $\beta_3 = 1.6382$, and $\beta_4 = 0.49294$. To correct for the nonuniformity of the electron density, gradients of the density are introduced into the exchange and correlation functionals, creating in this way nonlocal or semilocal functionals. The first gradient approximation was not successful since it did not fulfill many of the requirements of the exchange-correlation functional; it was even worse than the local approximation. To solve this problem of the gradient approximation, generalized gradient approximations (GGA) were designed:

$$E_{xc}^{\mathrm{GGA}}[\rho] = \int d\vec{r}\,\varepsilon_{xc}[\rho, |\nabla\rho|, \nabla^2\rho]$$

As a result of a series of improvements to the gradient approximation, the Langreth-Mehl exchange-correlation functional[40] was developed in 1983. The Perdew-Wang (PW86) for exchange and the Perdew (P86) for correlation functionals were of wider application in the most important DFT programs. Finally the Perdew-Wang (PW91) exchange and correlation[39,41,42] functionals were developed, the first GGAs designed from purely first principles. The exchange functional is

$$\varepsilon_c^{\mathrm{PW91}} = \varepsilon_x^{\mathrm{LDA}}(\rho)\left[\frac{1+sa_1\sinh^{-1}(sa_2)+\left(a_3+a_4e^{-100s^2}\right)s^2}{1+sa_1\sinh^{-1}(sa_2)+a_5s^4}\right]$$

where $a_1 = 0.19645$, $a_2 = 7.7956$, $a_3 = 0.2743$, $a_4 = -0.1508$, and $a_5 = 0.004$, and the correlation functional is

$$\varepsilon_c^{\mathrm{PW91}} = \varepsilon_x^{\mathrm{LDA}}[\rho] + \rho H[\rho, s, t]$$

where

$$H = \frac{\beta^2}{2\alpha}\ln\left[1+\frac{2\alpha}{\beta}\frac{t^2+At^4}{1+At^2+A^2t^4}\right] + C_{c0}[C_c(\rho)-C_{c1}]t^2e^{-100s^2},$$

with

$$A = \frac{2\alpha}{\beta}\left[e^{-2\alpha\varepsilon_C^{\mathrm{LDA}}(\rho)/(\rho\beta^2)} - 1\right]^{-1}$$

and

$$t = \frac{\left(\frac{\pi}{3}\right)^{\frac{1}{6}}}{4}\frac{|\nabla\rho|}{\rho^{\frac{7}{6}}}$$

where $\alpha = 0.09$, $\beta = 0.0667263212$, $C_{c0} = 15.7559$, $C_{c1} = 0.0035521$, and

$$C_c(\rho) = C_1 + \frac{C_2 + C_3 r_s + C_4 r_s^2}{1 + C_5 r_s + C_6 r_s^2 + C_7 r_s^3}$$

the constants, $C_1 = 0.001667$, $C_2 = 0.002568$, $C_3 = 0.023266$, $C_4 = 7.389 \times 10^{-6}$, $C_5 = 8.723$, $C_6 = 0.472$, and $C_7 = 0.07389$.

APPLICATIONS OF THE METHODS AND *AB INITIO* NATURE OF DFT

Specifically, we have used DFT techniques since these are the most powerful tools to deal with a relatively large number of atoms.[22] The use of DFT is fully justified due to the fact that it is an *ab initio* technique able to deal with a broad variety of systems. Several successful applications of DFT have been reported using the hybrid functionals where a portion of the exchange functional is calculated as a fully nonlocal functional of the wavefunction of an auxiliary noninteracting system of electrons. Since this resembles the exchange in the HF procedure (actually in any wavefunction procedure), it is common to refer to this functional, or procedure, as a DFT-HF hybrid. However, we have to consider that the exchange is being calculated using a noninteractive wavefunction whose density, but not its wavefunction, corresponds to the real system. A detailed analysis of its theoretical rigor was recently made.[43] The functional that we used is the B3PW91 with the triple split-valence basis set 6-311G**, which includes polarization functions for all atoms.[39,42,44]

A pure *ab initio* method is one that solves exactly or approximately the Schrödinger equation. This means using only the principles of quantum mechanics and not using empirical or other information extrapolated, interpolated, or transferred from other systems. Whether or not the method is a very rough approximation is not going to eliminate its *ab initio* character. In this sense, we have a spectrum of methods from the very precise to the very inaccurate, forcing us to a trade-off because the cost of the method is usually a direct function of its precision. In general, any practical computational method will be part *ab initio* and part empirical in character, and it is not much of a problem to determine whether a given method can be classified as *ab initio*, empirical, or semiempirical. The selection and use of a basis set is always based on experience. Whether, theoretical or experimental, one can always create or improve a basis set on the basis of its previous performances. The basis set tells the method the spectral region to look for the solutions, but it does not constitute the solution itself. A 100% analytical *ab initio* method, would be limited to the solution of the hydrogen-like atoms. Another 100% *ab initio*, but nevertheless approximated method, would be any of the variational-perturbational methods that solves the helium-like atoms with about 15 figures of precision, and where the basis set has been chosen with the only criteria of the lowest possible energy. The standard *ab initio* or HF-based methods (HF, MP, CC, CI, MRHF, ...) present several degrees of accuracy and also several degrees of small empirical (due to experiment or experience from earlier calculations) corrections and truncations. We call all of them *ab initio*, and there is an implicit understanding that this is where our definition of *ab initio* is settled. Small corrections to those methods, like 10% correction to the ZPE, do not change the *ab initio* character of the methods. There exists the accepted Gaussian2 theory, where corrections, not necessarily small, are made to the energies based on precise experimental results. Methods are being made faster, but not more precise. MP4 was widely available in 1982, and MP5 just appeared in 1992. It will take more time until we have MP7 or MP8 widely available. This is why higher performances of computers are required.

Now to the question whether DFT methods are *ab initio*: First, we derived DFT directly from the Schrödinger equation even without recurring, as in HF methods, to the mean field approximation. The fact that a well-defined but nowadays unsolvable functional shows up can not exclude the *ab initio* character of DFT. As in any of the *ab initio* methods above, approximations are made. A classification could be done whether the functional was derived using entirely first-principles information or whether it was obtained by using experimental or other transferred information. Functionals like the local approximation and gradient approximation are exact functionals up to their order, and they were obtained from purely first-principles arguments. The GGA of Perdew and Wang for exchange-correlation, PW91, is strictly *ab initio* and there is evidence that it performs as well as QCI and MP4 methods. A predecessor to the PW91, the PW86, has been used with systems containing up to transition metals with great success. DFT methods can also be developed using fully numerical techniques, eliminating the use of basis sets. Using this method together with one of the *ab initio* functionals yields a 100% *ab initio* procedure, that is, more *ab initio* than any of the standard ones.

We have presented several convergent synthetic routes to conjugated molecules with various potential digital device applications[45] including a two-terminal molecular wire with a tunnel barrier, a molecular wire with a quantum well to serve as a resonant-tunneling diode (RTD), three-terminal systems with switch-like possibilities, and four-terminal systems that could serve as logical gates without the use of multiple transistor arrays. In each case, the conducting segments are based on π-conjugated *oligo*(phenylene ethynylene)s.[46] The transport barriers are based on methylene (CH2) units since SPM studies showed us that alkyl units pose larger electronic transport barriers (higher resistance) than the p-conjugated moieties in single molecular systems.[47] The ethyl groups impart solubility to the larger molecules so that synthetic manipulations and depositions can be freely executed in solution.[46] All of the final compounds described have termini bearing self-assembling moieties, namely, protected thiols to serve as protected "alligator clips."[48] The precise termini needed for the ultimate arrangement in a molecular-based CPU is, or course, not

TABLE 3. Average errors (kcal/mol) of the theoretical atomization energies for the 32 molecules from the Gaussian-1 basis[52] and error per formal bond, and the formal scaling and number of basis per heavy atom (Li, Be, B, C, N, 0, and F)[a]

Method	Error	Error/bond	Formal Scaling	Basis/heavy-atom
MP4SDTQ/6-311 G(d,p)	15.6	6.2	N^7	18
MP4SDTQ/6-311+G(d,p)	14.6	5.8	N^7	22
MP4SDTQ/6-311 G(2dfp)	8.3	3.3	N^7	30
QCISD(T)/6-311 G(d,p)	16.7	6.7	N^7	18
PW86/DZP	5.3	2.1	N^3	15

NOTE: The molecules in the Gaussian-1 basis are H_2, LiH, BeH, CH, $CH_2(^1A_1)$, $CH_2(^3B_1)$, CH_3, CH_4, NH, NH_2, NH_3, OH, OH_2, FH, Li_2, LiF, C_2H_2, C_2H_4, C_2H_6, CN, HCN, CO, HCO, H_2CO, H_3COH, N_2, N_2H_4, NO, O_2, O_2H_2, F_2, and CO_2. MP4 and QCl errors were calculated from the energies reported in Ref. 52. PW86 errors were calculated from total energies reported in Ref. 53. Experimental atomization energies are reported in Ref. 52 where pointers to the experimental reports are indicated. The DZP basis set is equivalent to a 6-31G(d,p) which also uses 15 basis function for the heavy atoms.

[a]The errors are with respect to precise experimental atomization energies which have been measured with tolerances smaller than 1 kcal/mol.

known, but the capability of modern synthetic methods to construct precise molecular frameworks was exemplified.

Results in TABLE 3 indicate that density functional theory, which is computationally less expensive, is generally more accurate than full fourth-order Møller-Plesset perturbation theory and quadratic configuration interaction. These are the convinc-

TABLE 4. Water atonization energy errors (kcal/mol)

Method	Error[a]	Error/bond
HF/6-311 G(2dfp)	79.9	40.0
MP2/6-311 G(2dfp)	11.7	5.9
MP4/6-311 G* *	16.9	8.5
MP4/6-311+G	12.5	6.3
MP4/6-311 G(2d,fp)	13.2	6.6
QCISD(T)/6-311 G*	17.3	8.5
CCSD(T)/6-311 G*	17.2	8.6
LSD(VWN)/DZVP	32.9	16.5
BLYP/6-31 G*	5.7	2.9
BLYP/6-311G(3df,3pd)	1.2	0.6
PW86/DZVP	4.3	2.2
PW86/TZVP	3.7	1.9
PW91/TZVP	1.9	0.9
BP86/6-31G**	1.9	0.9
B3PW91/6-311G(2df,p)	3.0	1.5

NOTE: The experimental atomization energy of water is based on the heat of formation given in the JANAF tables.[54]

[a]This column shows the errors with respect to the experimental value of 219.3 kcal/mol at 0 K.

ing arguments for the use of DFT methods. The atomization energy is one of the most difficult energies to compute since it implies the rupture of all the bonds on the molecule; therefore, most of the error cancellations in standard calculations are not present. Modern functionals like the PW91 and B3PW91 perform even better than the PW86 shown above, with the additional advantage that the PW91 functional is purely first-principles functional; it has no fitted parameters. TABLE 4 shows these errors for the particular case of water. It can be noticed that the local density approximation is not suitable for precise calculations; however, any of the nonlocal density functionals performs much better than any of the standard *ab initio* methods. The study of nanometer devices depends strongly on the precise determination of energies.

Preliminary studies have permitted us to analyze, using DFT, the effects on real experimental situations performed at our own and other's laboratories. For instance, the modeling of gold[49] or palladium tips[50] to be used as contacts for molecular devices, the study of the effects on conductivity of single molecules,[51] and the rationale for molecular scales systems[45] have all been investigated. These works are the first of their kind. Several *ab initio* methods including density functional theory[22] will be used to determined the viability of several target molecules to be used as logical devices. In addition, these studies will aid in directing us to more viable target molecules for synthesis. This studies are the initial step toward the design of a molecular CPU.

ACKNOWLEDGMENTS

This work is supported by the Defense Advanced Research Projects Agency. We also thank NASA for supercomputer support.

REFERENCES

1. KIRK, W.P. & M.A. REED Eds. 1992. Nanostructures and Mesoscopic Systems. Academic Press: New York.
2. AVIRAM, A., Eds. 1992. Molecular Electronics: Science and Technology. American Institute of Physics. New York.
3. BIRGE, R.R. Ed. 1991. Molecular and Biomolecular Electronics. American Chemical Society. Washington, DC.
4. MONTEMERLO, M.S. *et al.* 1996. *In* Technologies and Designs for Electronic Nanocomputers. Eds. MITRE. McLean., Virginia.
5. BENNETT, C.H. 1995. Quantum information and computation. Physics Today **48:** 24–30.
6. KOUWENHOVEN, L. 1977. Single-molecule transistors. Science **275:** 1896–1897.
7. SERVICE, R.F. 1996. A captivating carbon form. Science **274:** 345–346.
8. TOUR, J.M. 1996. Conjugated macromolecules of precise length and constitution: organic synthesis for the construction of nanoarchitectures. Chem. Rev. **96:** 537–553.
9. BOCKRATH, M. et al. 1977. Single-electron transport in ropes of carbon nanotubes. Science **275:** 1922–1925.
10. CREDI, A. *et al.* 1997. Logic Operations at the molecular level. An XOR gate based on a molecular machine. J. Am. Chem. Soc **119:** 2679–268 1.

11. DOROGI, M. *et al.* 1995. Room-temperature Coulomb blockade from a self-assembled molecular nanostructure. Phys. Rev. B **52:** 9071–9077.
12. ANDRES, R.P. *et al.* 1996. "Coulomb Staircase" at room temperature in a self-assembled molecular nanostructure. Science **272:** 1323–1325.
13. AVIRAM, A.J. 1988. Molecules for memory, logic, and amplification. J. Am. Chem. Soc. **110:** 5687–5692.
14. JOACHIM, C. *et al.* 1995. Electronic transparence of a single C_{60} molecule. Phys. Rev. Lett. **74:** 2102–2105.
15. YAZDANI, A., D.M. EIGLER & N.D. LANG. 1996. Off-resonance conduction through atomic wires. Science **272:** 1921–1924.
16. FISCHER, C.M. *et al.* 1995. Microstructured Gold/Langmuir-Blodgett Film/Gold tunneling junctions. Appl. Phys. Lett. **66:** 3331–3333.
17. GRABERT, H. 1991. Single charge tunneling: A brief introduction. Z. Phys. B **85:** 319–325.
18. JOACHIM, C. & J. K. GIMZEWSKI. 1997. Electromechanic amplifier single molecule. Chem. Phys. Lett. **256:** 353–357.
19. MUJICA, V., M. KEMP & M. A. RATNER. 1994. Electron conduction un molecular wires. ii. Application to scanning tunneling microscopy. J. Chem. Phys. **101:** 6856–6864.
20. TANS, S.J. et al. 1997. Individual Single-wall carbon nanotubes. Nature **386:** 474–477.
21. METZGER, R.M. *et al.* Unimolecular electrical rectification in hexadecylquinolinium tricyanoquinodimethanide. J. Am. Chem. Soc. **119:** 10455–10466.
22. SEMINARIO, J.M., Eds. 1996. Recent Developments and Applications of Modem Density Functional Theory. Elsevier. Amsterdam.
23. MARTIN, J.M.L. 1997. Benchmark *ab initio* calculations of the total atomization energies of the first-row hydrides AH_n (A = Li–F). Chem. Phys. Lett. **273:** 98–106.
24. SANDERS, F.C. 1966. An Introduction to high-precision computational methods for simple atomic and molecular systems. *In* Recent Developments and Applications of Modem Density Functional Theory. Seminario, J.M., Eds.: 99–150. Elsevier. Amsterdam.
25. MØLLER, C. & M.S. PLESSET. 1934. Note on an approximation treatment for many-electron systems. Phys. Rev. **46:** 618–622.
26. BLOCH, F. 1929. Bemerkung zur Elektronentheorie des Ferromagnetismus und der elektrischen Leitfähigkeit. Z. Phyzik **57:** 545–555.
27. DIRAC, P.A.M. 1930. Note on exchange phenomena in the Thomas atom. Proc. Camb, Phil. Soc. **26:** 376–385.
28. THOMAS. 1926. The calculation of atomic fields. Proc. Camb. Phil. Soc. **23:** 542–548.
29. FERMI. 1928. Eine statistische Methode zur Bestimmung einiger Eigenschaften des Atoms und ihre Anwendung auf die Theorie des periodischen Systems der Elemente. Z. Physik **48:** 73–79.
30. SLATER, J. 1974. The Self-Consistent Field for Molecules and Solids: Quantum Theory of Molecules and Solids. Mc Graw-Hill. New York.
31. HELLMANN. 1937. Einfürung in die Quantenchemie. Deuticke. Leipzig.
32. FEYNMAN, R. 1939. Forces in molecules. Phys. Rev. **56:** 340–343.
33. HOHENBERG, P. & W. KOHN. 1964. Inhomogeneous electron gas. Phys. Rev. B **136:** 864–871.
34. LEVY, M. 1979. Universal variational functionals of electron densities, first-order density matrices, and natural spin-orbitals and solution of the v-representability problem. Proc. Nat. Acad. Sci. USA **76:** 6062–6065.

35. PERCUS, J.K. 1978. The role of model systems in the few-body reduction of the N-Fermion problem. Int. J. Quantum Chem. **13:** 89–124.
36. HARRIS, J. & R.0. JONES. 1974. The surface energy of a bounded electron gas. J. Phys. F **4:** 1170–1186.
37. LANGRETH, D. C. & J. P. PERDEW. 1975. The exchange-correlation energy of a metallic surface. Solid State Commun. **17:** 1425–1429.
38. KOHN, W. & L.J. SHAM. 1965. Self-consistent equations including exchange and correlation effects. Phys. Rev. A **140:** 1133–1138.
39. PERDEW, J.P. & Y. WANG. 1992. Accurate and simple analytic representation of the electron gas correlation energy. Phys. Rev. B **45:** 13244-13249.
40. LANGRETH, D.C. & M.J. MEHL. 1983. Beyond the local-density approximation in calculation of ground-state electronic properties. Phys. Rev. B **28:** 1809–1834 [*erratum* in Phys Rev. B **29:** 2310].
41. PERDEW, J.P. 1991. Unified theory of exchange and correlation beyond the local density approximation. *In* Electronic Structure of Solids. Ziesche, P. & H. Eschrig, Eds.: 11–20. Akademie Verlag. Berlin.
42. PERDEW, J.P. *et al.* 1992. Atoms, Molecules, Solids, and Surfaces: Applications of the Generalized Gradient Approximation for Exchange and Correlation. Phys. Rev. B **46:** 6671–6687.
43. GORLING, A. & M. LEVY. 1997. Hybrid schemes combining the Hartree-Fock method and density-functional theory: Underlying formalism and properties of correlation functionals. J. Chem. Phys. **106:** 2675–2680.
44. BECKE, A.D. 1993. A new mixing of Hartree-Fock and local density -functional theories. J. Chem. Phys. **98:** 1372–1377.
45. TOUR, J.M., M. KOSAKI & J.M. SEMINARIO. 1997. Molecular scale electronics: Approaches to nanoscale digital computing. Nature. Submitted for publication.
46. JONES, L., II, S. SCHUMM & J.M. TOUR. 1997. Rapid Solution and Solid Phase Syntheses of Oligo(1,4-phenylene-ethylene)s with Thioester Termini: Molecular Scale Wires with Alligator Clips. Derivation of Iterative Reaction Efficiencies on a Polymer Support. J. Org. Chem. **62:** 1388–1410.
47. BUMM, L.A. *et al.* 1996. Are single molecular wires conducting? Science **271:** 1705–1707.
48. TOUR, J.M. *et al.* 1995. Self-assembled monolayers and multilayers of conjugated thiols, α,ω-dithiols, and thioacetyl-containing adsorbates. Understanding attachments between potential molecular wires and gold surfaces. J. Am. Chem. Soc. **117:** 9529–9534.
49. SEMINARIO, J.M. & J.A TOUR. 1997. Systematic study of the lowest energy states of Au_n (n = 1–4) using DFT. Int. J. Quantum Chem. **65:** 749–758.
50. ZACARIAS, A.G. *et al.* 1997. Lowest energy states of Pd4 using DFT: Nanotips for single molecule electronics. Int. J. Quantum Chem. Submitted for publication.
51. SEMINARIO, J.M., A.G. ZACARIAS & J.M. TOUR. 1998. Theoretical interpretation of conductivity measurements of thiotolane sandwich. A molecular scale electronic controller. J. Am. Chem. Soc. **120:** 3970–3974.
52. Pople, J.A. *et al.* 1989. Gaussian-1 theory: A general procedure for prediction of molecular energies. J. Chem. Phys. **90:** 5622–5629.
53. SEMINARIO, J.M. 1993. Energetics using DFT: Comparisons to precise *ab initio* and experiment. Chem. Phys. Lett. **206:** 547–554.
54. CHASE, A.W. *et al.* 1985. JANAF Thermochemical Tables. J. Phys. Chem. Ref. Data **14:** 1–1856.

Rectification by a Single Molecule of Hexadecylquinolinium Tricyanoquinodimethanide

ROBERT M. METZGER[a] AND MICHAEL P. CAVA

Chemistry Department, University of Alabama, Tuscaloosa, Alabama 35487-0336, USA

ABSTRACT: Asymmetries in the macroscopic and the nanoscopic direct-current electrical conductivity through Langmuir-Blodgett multilayers and monolayers of γ-(*n*-hexadecyl)quinolinum tricyanoquinodimethanide, HDQ-3CNQ (**1**) can be ascribed to a transition from the ground-state zwitterionic state (shown) to a presumably neutral excited-state conformer, thus confirming indubitably rectification of electrical current by a single molecule.

INTRODUCTION

A unimolecular rectifier of electrical current is a molecule through which there is a strongly asymmetric flow of electrons because the electrical conductivity from, say, its left "side" to its right "side" is considerably larger than the conductivity in the reverse direction.[1]

Vacuum diodes that rectify electrical current were discovered in 1904;[2] the first solid-state rectifier was a junction of *p*-doped Ge accosted to *n*-doped Ge.[3] For inorganic semiconductors, rectification will occur either at a sharp "Schottky barrier" between phases with dissimilar work functions,[4] or within a "depletion layer," which can be several µm wide.[3] Junctions of µm-thick layers of "*n*-doped" organic semiconductors with µm-thick layers of "*p*-doped" organic semiconductors are also rectifiers.[5–8] Rectification by Langmuir-Blodgett (LB) multilayers [doped with electron donors (D), topped by insulating LB monolayers, then topped by multilayers "doped" with electron acceptors (A)] was demonstrated by Kuhn and co-workers,[9] and confirmed by other groups.[10,11] Fujihira discovered an electrochemical LB photodiode, or photon-induced rectifier;[12] electrochemical rectification at a monolayer-modified electrode has been reported.[13,14] If one monitors the tunneling current in scanning tunneling microscopy (STM), one performs scanning tunneling spectroscopy (STS). Unsymmetrical STS currents (current (I)–voltage (V), or IV curves, with left-right asymmetry) have been seen through a porphyrin covalently attached to a carboxylated highly oriented pyrolytic graphite (HOPG) surface,[15] through an alkylated hexabenzocoronene deposited on graphite[16] (salt formation?), and through an oligo(phenylethynyl)benzenethiol on Au(111) and Ag(111).[17]

Aviram and Ratner (AR) proposed in 1973[1] unimolecular rectification, i.e., asymmetrical electrical conduction through the molecular orbitals of a single D–σ–A molecule ("through-bond tunneling"), e.g., the proposed "Gedankenmolekül" (**2**);

[a]Address for telecommunication: Phone, 205/348-5952; fax, 205/348-9104; e-mail, rmetzger@ualvm.ua.edu

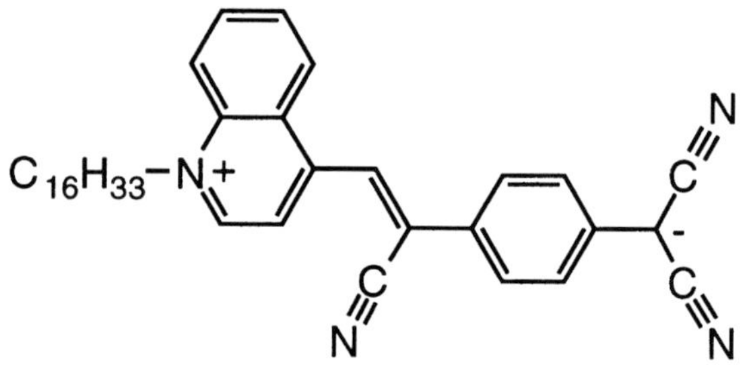

1 , HDQ-3CNQ

SCHEME 1

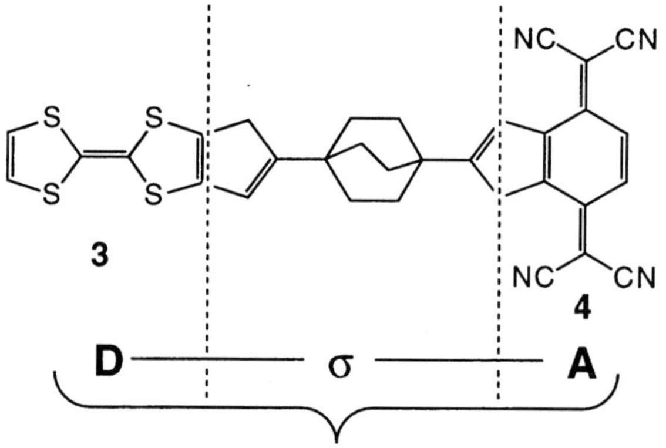

2

SCHEMES 2 – 4

proving its validity has been a major experimental challenge. Here D = “good” one-electron donor with low ionization potential, A = “good” one-electron acceptor with high electron affinity, and σ = insulating saturated covalent “bridge”.[1] This unimolecular rectification would work if the dominant electron flow is “through-bond” rather than “through space,” and if the excited zwitterionic state $D^+–σ–A^-$ is relatively accessible from the ground neutral state D–σ–A; the opposite zwitterion $D^-–σ–A^+$ should be several eV higher in energy, and therefore be relatively inaccessible. Indeed, good organic one-electron donors (e.g., tetrathiafulvalene, TTF, **3**) are poor acceptors, while good acceptors (e.g., tetracyanoquinodimethane, TCNQ, **4**) are very poor one-electron donors:

$$D–\sigma–A \rightarrow D^{+}–\sigma–A^{-} + \approx 1–2 \text{ eV} \quad (1)$$

$$D–\sigma–A \rightarrow D^{-}–\sigma–A^{+} + \approx 6–7 \text{ eV} \quad (2)$$

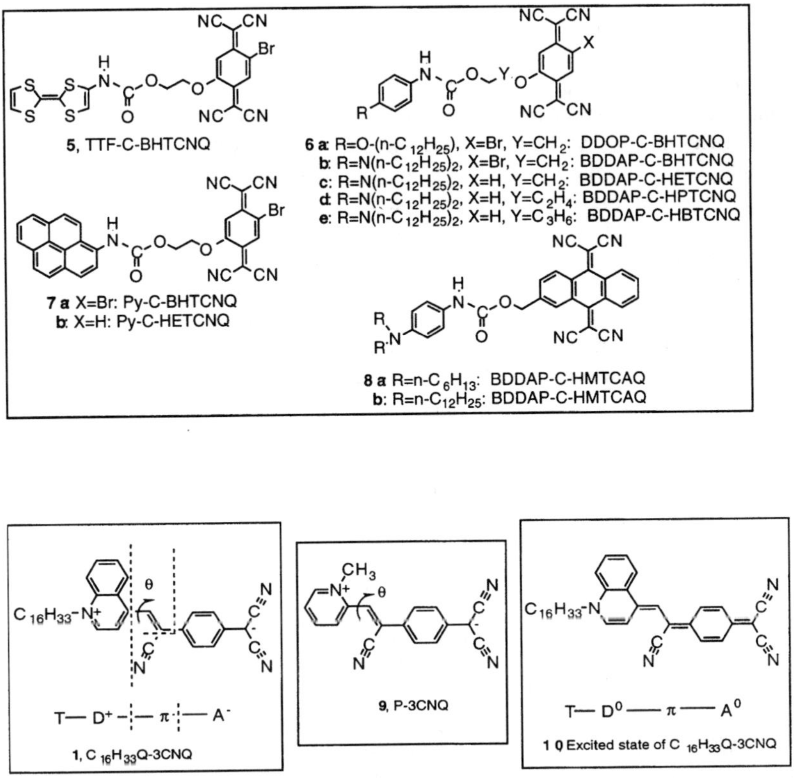

SCHEMES 5 –10

In the 1980s, many D–σ–A molecules (**5–8**) that form LB films were synthesized at the University of Mississippi, but rectification was not found by primitive methods.[18–20]

For one of these molecules, DDOP-C-BHTCNQ, **6a**[21] (D = DDOP = dodecyloxyphenyl, C = carbamate = covalent σ bridge, A = BHTCNQ = bromohydroxyethoxy-tetracyanoquinodimethane), Sambles and co-workers showed IV asymmetries for sandwiches:

Pt | LB multilayer of **6a** | Mg | Ag **(3)**

despite the fact that DDOP is a weak one-electron donor, with high ionization potential.[22] Since a Schottky barrier can form[23] between Mg and TCNQ (i.e., $Mg^{++}(TCNQ^{-})_2$ or $Mg^{++}TCNQ^{--}$ can form); therefore the molecular origin of the IV asymmetries was put into doubt.[23]

Ashwell and co-workers discovered a new class of zwitterionic molecules: a quaternary picolinium or lepidinium halide can react with an alkali salt of TCNQ, and yield ground-state zwitterions. The first example was picolyltricyanoquinodimethane, or picolinium tricyanoquino-dimethanide, P-3CNQ, **9**. Its crystal structure[24] revealed that **9** is a zwitterion, with a $\theta = 30°$ twist[24] between the pyridinium ring and the central six-membered ring of $3CNQ^-$, and a calculated ground-state dipole moment (INDO) of 26 Debyes.[24] These zwitterionic molecules have a very narrow and solvatochromic absorption band; when made amphiphilic by adding long alkyl chains, they yield Z-type LB films that are excellent $\chi^{(2)}$ materials.[25]

An interesting member of this class is Z-β-(*N*-hexadecyl-4-quinolinium)-α-cyano-4-styryldicyanomethanide or γ-(*n*-hexadecyl)quinolinum tricyanoquinodimethanide, HDQ-3CNQ (**1**) which forms a blue acetonitrile solution.[25] That **1** is a zwitterion can be seen by calling it a $T–D^+–\pi–A^-$ molecule, where T = hexadecyl (required for LB films), D^+ = quinolinium y, π = pi-electron bridge, A^- = tricyanoquinodimethanide ($3CNQ^-$). Z-type LB multilayers of **1** have $\chi^{(2)}{}_{zzz} = 180$ pm V^{-1}.[26] The blue absorption band in solution is probably an intramolecular intervalence transfer band (IVT); when several molecules are packed in a multilayer, an intermolecular charge transfer band (ICT) may also occur. For a single crystal of **9**, these two bands were identified by polarized spectroscopy.[27] It is believed[25–27] that the ground state of **1** is very polar, while its first excited state is less polar, of the type T–D–π–A (**10**), and that the efficiency for frequency doubling is due to the transition:

$$T–D^+–\pi–A^- \longrightarrow T–D^-\pi–A \qquad \textbf{(4)}$$

In **1**, the quinolinium and tricyanoquinodimethanide rings may be twisted by some unknown angle θ (which was 30° for **9**[24]). The neutral excited state T–D–π–A might be planar ($\theta = 0°$); if $\theta = 90°$ then IVT = 0, and the blue color should disappear.

When LB multilayers of **1** were sandwiched between a noble metal electrode (Pt or Ag) and a magnesium electrode (protected by an overlayer of Ag), then asymmetric currents through the laterally macroscopic multilayer indicated electrical rectification through the molecule[28]:

Pt | LB multilayer of **1** | Mg | Ag **(5)**

and even in the sandwich:

Pt | LB monolayer of **1** | Mg | Ag **(6)**

Because of doubts about a possible Schottky barrier at the Mg electrode,[23] an insulating monolayer of a fatty acid was inserted between the metal electrodes and 3 monolayers of **1**[29]; I-V asymmetries were seen again, and ascribed to molecular rectification.[29]

This rectification (5) has generated theoretical interest.[30,31] One issue is whether the ground state is "zwitterionic," as in **1**, or "neutral," as in **10**. In the ground state a mixture of neutral and zwitterionic resonance forms may co-exist,[31] and allow for adiabatic intramolecular electron transfer within a monolayer, as predicted by Marcus-Hush theory[32,33]:

$$k_{ET} = \nu_a \{\pi^2\Delta^2 [\, 2h(\pi RT\lambda)^{1/2}]^{-1}\} \exp(-\Delta G^*/RT). \qquad \textbf{(7)}$$

Here k_{ET} is the electron transfer rate, ν_a is the frequency of formation of an activated complex, whose free energy is ΔG^*, λ is the Franck-Condon reorganization energy, Δ is the coupling between the vibrational envelopes of reactant and product at the point of avoided crossing, h is Planck's constant, R is the gas constant, and T is the absolute temperature. The "non-adiabatic" electron transfer factor, given in {braces}, becomes unity for an "adiabatic" transfer.

For a single molecule (e.g., in a monolayer) of **1** Δ is large, and rectification can be understood [31]; for a multilayer of **1** Δ is tiny, and rectification may involve defect states.[31] If the twist angle θ in **1** becomes 90°, then the molecule is a zwitterion.[31] Both **9** and **1** may belong to a class of twisted internal charge transfer (TICT) systems,[31] but such twisting during the electron transfer would make the rectification process too slow for practical use. The energy-minimized "gas-phase" theoretical calculations give for **9** a smaller ground-state dipole moment[31] than calculated for the geometry of **9** in the crystal,[24] unless efforts are made to account approximately for "solvent effect."[31] For the first excited state $\theta \approx 90°$,[31] while for the ground state ("gas phase") $\theta = 9°–11°$.[31]

Following two preliminary studies of **1**[34,35] a detailed new report[36] gives a novel, very convenient and efficient synthesis of **1**, confirms its ground-state zwitterionic structure, confirms electrical rectification by both LB monolayers and multilayers by macroscopic and nanoscopic methods, and identifies incontrovertibly the direction of higher conductivity relative to the orientation of the molecules in the multilayers: **1** is a unimolecular rectifier.[36] Several reviews of these results have been presented elsewhere.[37–40] The published results[36] and other recent results[41–43] are also discussed below.

SYNTHESIS, NMR, SOLUTION VISIBLE-UV SPECTRA, CYCLIC VOLTAMMOGRAM, EPR OF RADICAL ANION, AND DIPOLE MOMENT

The previous syntheses of **1**[45–47] were inefficient. A reproducible synthesis of **1** uses N-cetyl-lepidinium tosylate and a twofold excess of the lithium salt of TCNQ radical anion, in dry DMSO solution with some pyridine added.[36] Presumably, the lepidinium salt first loses $H^\bullet$ to $TCNQ^{\bullet-}$, is converted to a radical cation salt, then couples with free $TCNQ^{\bullet-}$, loses HCN, and leads to the product **1** in 59% yield within 21 hours.[36] The NMR spectrum of the compound corroborates a dipolar ground state.[36]

The visible-ultraviolet spectrum of **1** is solvent-dependent, particularly the long-wavelength maximum in the visible region. There is a single maximum at 711 nm in CH_3CN,[42] which splits into two vibronically split maxima at 830 and 910 nm in CH_2Cl_2,[42] and at 838 and 933 nm in $CHCl_3$.[42] For CH_3CN and CH_2Cl_2 the absorbance is linear with concentration (range 0 to 0.06 mg/mL).[36] The plot of $1/\lambda_{max}$ is linear with the dielectric constant (solvatochromism).[42] The cyclic voltammogram of **1** (FIG. 1) shows a one-electron quasi-reversible reduction ($E_{1/2} = -0.513$ V versus SCE), a second irreversible reduction at $E_{p,r} = -1.16$ V, and an irreversible oxidation at $E_{p,o} = 0.49$ V versus SCE.[36]

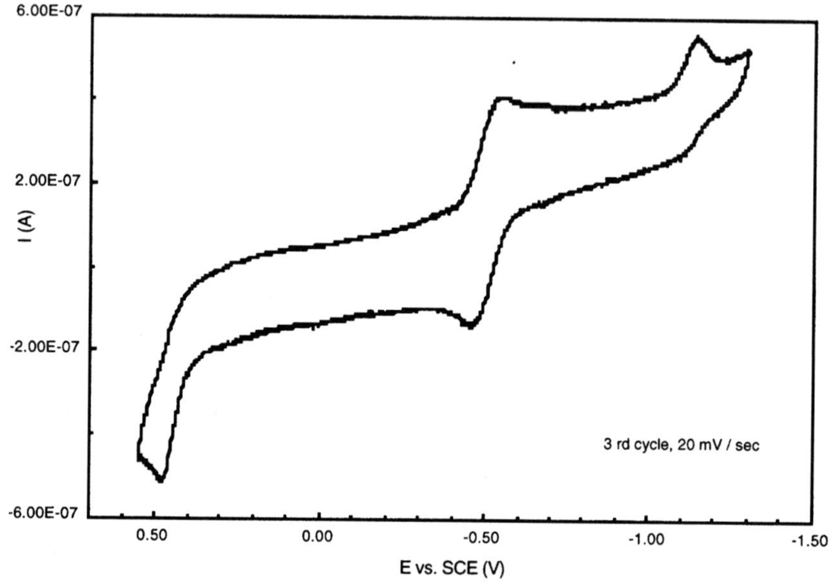

FIGURE 1. Cyclic voltammogram of a 10^{-4} M solution of **1** in CH_2Cl_2 (working electrode=Pt wire, reference electrode=SCE, electrolyte=0.1 M NBu_4ClO_4, N_2 gas, scan rate= 0.020 V/s).[36]

An EPR spectrum of the electrogenerated radical anion of **1** in CH_2Cl_2, at −0.51 V versus SCE, shows a 17-line spectrum, which is an overlay of a broad EPR signal, and a resolved hyperfine spectrum, due to 4 ^{1}H and 2 ^{14}N nuclei, i.e., the spin density is localized on the 3CNQ ring, not including the uncharged CN group.[42] To obtain the dipole moment of **1**, the dielectric constants of dilute solutions of **1** in dichloromethane were measured, as a function of temperature ($-10°C < T < 30°C$), at 1 kHz (coaxial capacitance cell, Hewlett-Packard 4263B LCR meter).[36] Since **1** does not dissolve in nonpolar solvents, corrections were applied to extract the molecular dipole moment from the solvent atmosphere, using the Kirkwood-Fröhlich equation,[44] and corrections for the molar polarization of the solute. FIGURE 2 shows that equivalent results were obtained from both methods.[36] The extrapolation to zero concentration yields a dipole moment of 40 to 45 D (average = 43 ± 8 D) for **1** in CH_2Cl_2.[36] The ground state of **1** is zwitterionic.

THEORETICAL CALCULATIONS

Semiempirical AM1 and PM3 calculations of **1** are confusing, as before.[31] A geometry-optimized PM3 calculation for **1** yields a heat of formation of 6.02 eV, and an angle between the quinolinium ring and the central phenyl ring of 3CNQ of about 30°. The computed dipole moment for the ground state singlet of **1** is only 10.7 D, with the highest occupied molecular orbital (HOMO) at −7.98 eV (MO #106, local-

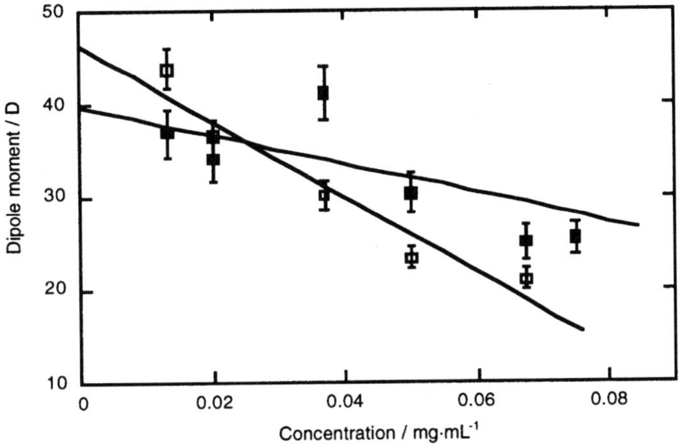

FIGURE 2. Effective dipole moments $g^{1/2}\mu$ of compound **1** in CH_2Cl_2 versus concentration, using the Kirkwood-Fröhlich equation. □ Data obtained directly from the temperature dependence of the dielectric constant of the solution. ■ Data obtained by "subtracting" the contribution of the polar solvent. The solid lines are least-squares fits.[36]

ized on quinolinium, the π bridge, and the 3CNQ end) and the lowest unoccupied molecular orbital (LUMO) at −2.42 eV (MO # 107, localized on $3CNQ^-$). The computed bond lengths are quinonoid, rather than benzenoid. A PM3 calculation for the excited singlet state of **1** provided a heat of formation of 8.469 eV, a twisted geometry ($\theta \approx 90°$), and a high dipole moment of 45.4 D. These results agree with Broo and Zerner.[31] With a shorter alkyl chain (octyl instead of hexadecyl) and AM1,[48] the ground-state dipole moment becomes 10.56 D and a heat of formation becomes 7.61 eV; the first excited singlet state has a smaller dipole moment, 6.36 D, and a heat of formation of 10.83 eV, i.e., 3.22 eV higher.[48]

POCKELS-LANGMUIR MONOLAYERS AT THE AIR-WATER INTERFACE

The Π-A isotherm of **1**, measured under a green safelight, shows a collapse point at 50 $Å^2$ molecule-1 and 34 mN m^{-1} (FIG. 3): this isotherm is moderately temperature dependent.[35,36]

LB FILMS; X-RAY AND UV-VISIBLE SPECTRA OF MULTILAYERS, ELLIPSOMETRY, FTIR, AND XPS SPECTRUM OF A MONOLAYER

Monolayers and multilayers of **1** were transferred to solid substrates at 15 mm min^{-1} at 20 mN m^{-1}: **1** first transfers on the downstroke onto hydrophobic HOPG, and on the upstroke onto a freshly cleaned hydrophilic Au(111) or onto Al. The hydrophobic hexadecyl group is thought to be the end of the molecule closest to

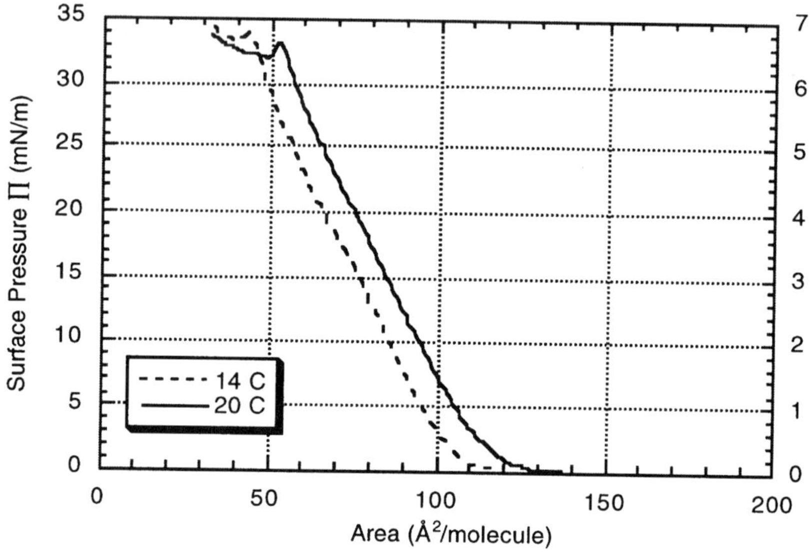

FIGURE 3. Pressure-Area isotherms for **1** at 14°C and 20°C.[35,36]

HOPG, while the hydrophilic 3CNQ$^-$ (A$^-$) end should be closest to Au(111). At 15 mm min^{-1} the LB multilayers (past the first monolayer) are Z-type, while at 45 mm min^{-1}, the LB multilayers are Y-type (centrosymmetric). X-ray diffraction shows that the Y-type films reorder to Z-type some time after transfer, with a thickness of 23 Å per monolayer: for multilayers of **1**, the Z-type ordering is more stable.

Hence, LB multilayers of **1** transferred onto HOPG at 15 mm min^{-1} have orientation:

$$\text{HOPG} \mid \rightarrow \leftarrow \leftarrow \leftarrow \leftarrow \ \ldots \leftarrow \leftarrow \quad \textbf{(8)}$$

Here the arrow → indicates the rough molecular orientation; it points from the hexadecyl tail T and the quinolinium ring D$^+$ (tail of arrow) towards the dicyanomethylene head of A$^-$ (head of arrow). The first monolayer adheres with the hexadecyl tail T next to HOPG (X-type); all other layers have Z-type orientation (tails away from HOPG). The transfer ratios were 50% for the first layer, and from 50% to 100% for the Z-type layers.

The film transfer to Al is Z-type for all layers:

$$\text{Al} \mid \leftarrow \leftarrow \leftarrow \leftarrow \leftarrow \leftarrow \leftarrow \ \text{etc.}; \quad \textbf{(9)}$$

the transfer ratio for the first layer is now over 90%.

A sharp absorption band, with peak at 570 nm, is seen in LB films transferred onto quartz (FIG. 4). This band disappears if the films are protonated, but reappears when exposed to NH_3;[47] it disappears irreversibly if the film, or a solution of **1**, are exposed to bright light and air for a long time.

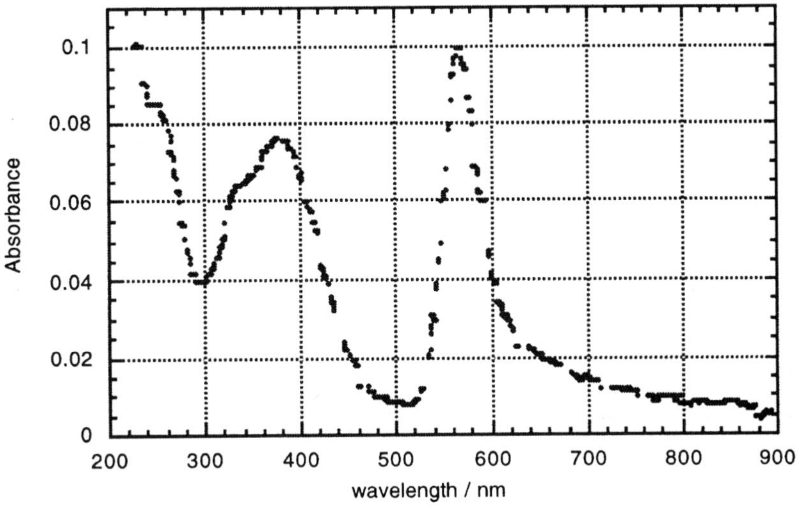

FIGURE 4. Visible spectrum of an 11-layer LB film of **1** on quartz; the narrow absorbance has a peak at 570 nm (2.17 eV).[35,36]

The thickness per monolayer, measured by ellipsometry ($\lambda = 670$ nm) and by surface plasmon resonance, is 22 ± 2 Å;[47] thus the approximate molecular axis may be tilted 48 ± 5° from the normal to the monolayer plane.

A grazing-angle Fourier transform infrared spectrum for a monolayer of **1** on a glass slide covered by Al shows absorbances at 2850 cm^{-1} (symmetric C–H), 2920 cm^{-1} (asymmetric $-CH_2-$), 2139 cm^{-1} (–C≡N) and 2175 cm^{-1} (–C≡N), indicating cyano and methylene groups tilted from the monolayer plane. For a monolayer of **1** on either Au(111) or HOPG, the C–H and $-CH_2-$ signals can be seen clearly; a –C≡N signal at 2175 cm^{-1} is barely visible for the monolayer on Au(111).[48]

An XPS spectrum [Kratos Analytical Axis 165, Al K_α photons (1486.6 eV)] of a multilayer of **1** on Al shows three N 1s bands at 398.8 eV (N^-, dicyanomethylene end), 402.2 eV (N^+, quinolinium ring), and a broad band at 400.5 eV (which is also seen for a monolayer of **1**): **1** degrades somewhat under X-ray irradiation.[42]

MACROSCOPIC CONDUCTIVITY OF LB FILMS, FROM 300K TO 78K

Macroscopic "metal | organic | metal" sandwiches were prepared (Edwards 306A 4-source evaporator, with a liquid nitrogen-cooled substrate holder) and their I-V curves were measured (PAR 270 potentiostat or Hewlett-Packard 3245A universal source and Hewlett-Packard 3457A multimeter controlled by a Gateway 2000 microcomputer).[36] Low-temperature measurements (so far to 77K) were performed in a Janis Varitemp dewar, with computer-controlled dual heaters and thermometers.[43]

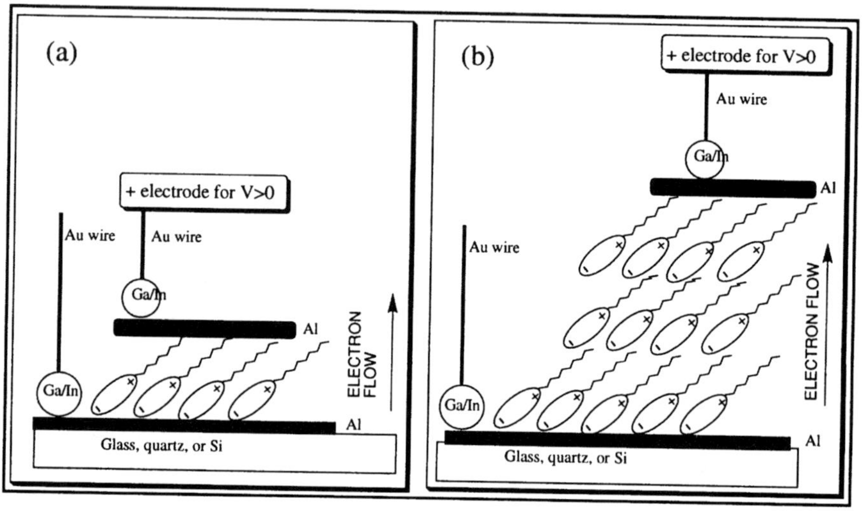

FIGURE 5. Orientation of the LB monolayer **(a)** or multilayer **(b)** on the glass, quartz, or Si substrate; the electrode (+) for positive bias, and the direction of "easy" electron flow for V > 0 are marked. In some cases Ag paste was used instead of the Ga/In eutectic.[36]

The glass, quartz, or silicon substrates were covered by an Al layer (25 mm × 75 mm × 100 nm), then LB monolayer or multilayer films were transferred under a green safelight, then the sample was dried for 2 days in a vacuum desiccator with P_2O_5. Twelve cylindrical "pads" of Al (4 each of areas 2.8, 4.5, and 6.6 mm^2, either 100 nm or 300 nm thick) were finally deposited per microscope slide, atop the organic layer, with the substrate cooled to 77K, to form the sandwich:

$$\text{glass} \mid \text{Al} \mid 1 \text{ LB monolayers of } \mathbf{1} \mid \text{Al} \qquad \mathbf{(10)}$$

as shown in FIGURE 5(a), or the sandwich:

$$\text{glass} \mid \text{Al} \mid 4 \text{ LB monolayers of } \mathbf{1} \mid \text{Al} \qquad \mathbf{(11)}$$

as shown in FIGURE 5(b).

Electrical contact to the Al electrodes was obtained with a Ga/In eutectic, or Ag paste. The electrodes (Al) are symmetrical in Eq. (10), but probably covered by an oxide layer. A very asymmetric I-V curve for the four-monolayer film is shown in FIGURE 6. Reversing the electrode connections reversed the direction of rectification. A similarly asymmetric curve was observed when a Mg layer was put atop the organic layer before an Al overcoat layer was added.[36]

The work with monolayers was more difficult. Of 39 pads checked (20 with areas of 2.8 mm^2, 15 with areas of 4.5 mm^2, 4 with areas of 6.6 mm^2) with two thicknesses of the Al pad (12 with thickness 300 nm, 27 with thickness 100 nm), 17 were electrical short circuits, either because of monolayer defects, or because the Ga/In or Ag paste created new defects during attachment. The 22 "good" junctions had DC resis-

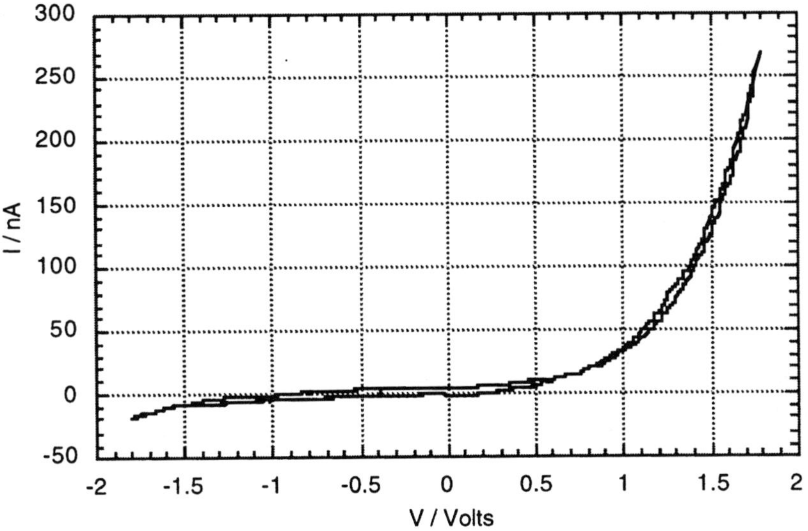

FIGURE 6. Conductivity of a sandwich "Ga/In eutectic | Al (100 nm) | 4 LB monolayers of **1** (Z-type) | Al (100 nm) | Ga/In eutectic." Scan rate 100 mV/s. The counter and reference electrodes (shorted together) are connected to one Al electrode; the working electrode is connected to the other Al electrode. The direction of easier electron flow (I > 0 for V > 0) is from the bottom electrode through the film to the top electrode.[36]

tances from 1 to 400 MΩ; of these, 4 exhibited rectifying behavior. FIGURE 7(a) shows a typical rectification curve for a monolayer. Above a threshold voltage V_t, the monolayer shows larger currents for positive bias than for negative bias. This V_t varies from junction to junction in the range $V_t = 0.8$ to 1.3 V.

The rectification ratio *RR* can be defined as:

$$RR = (\text{current at } V_o)/(\text{current at } -V_o) \tag{12}$$

where V_o is the highest positive bias used. *RR* ranges from 2.4 to 26.4 the first time the film is measured. As the cycle is repeated, *RR* drops steadily, and disappears after 4 to 6 cycles. Similarly, applying a constant positive or negative bias, a slow decrease is seen, over 15–20 minutes, of the current for both polarities. It appears that, under the intense electric fields, the molecular dipoles tend to reorient to minimize energy. For some of the junctions, short circuits form across the monolayer during the measurement.[36]

Since a 4.5 mm^2 Al pad, covering a monolayer of **1** generates a current of 0.4 μA at ≈1.5 V (FIG. 7(a)), the electron transfer rate is estimated as: $k_{ET} = 0.33$ electrons $molecule^{-1}\ s^{-1}$.[36]

In the range −0.5 V to 0.5 V, the resistance follows Ohm's law. In the reverse regime (negative bias), the logarithm of the current is linear with the applied bias (through-space tunneling) from −0.5 V to the highest negative bias applied (−1.5 V

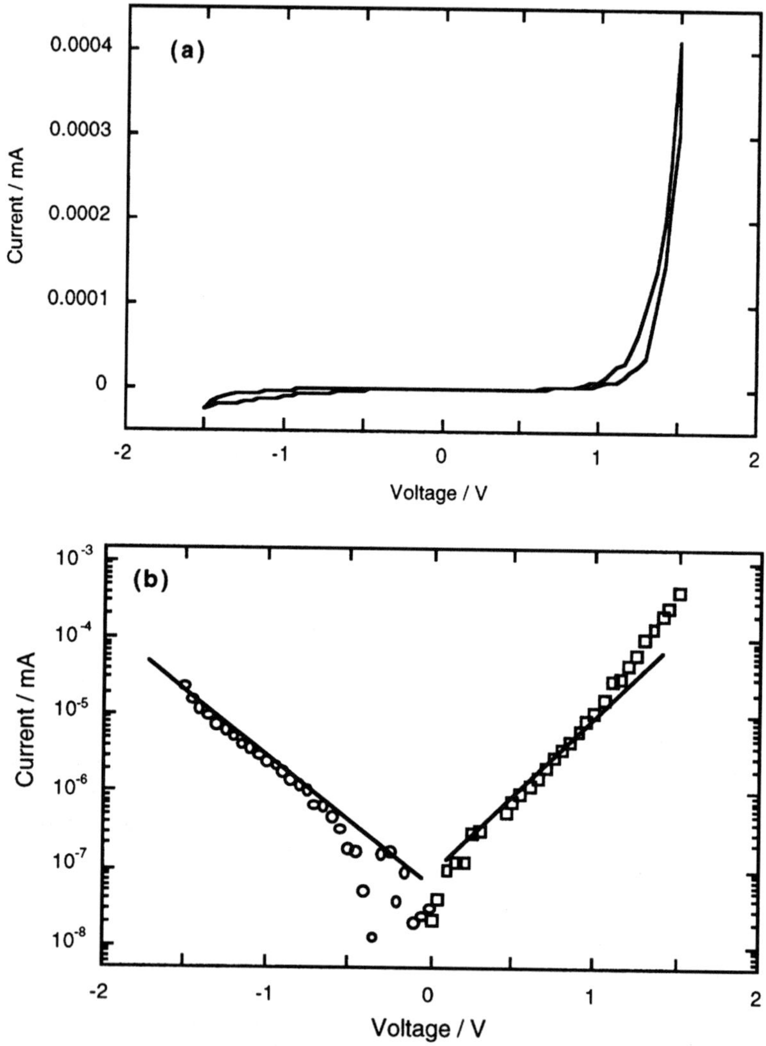

FIGURE 7. Rectification through a single monolayer of **1** sandwiched between Al electrodes (top Al pad area 4.5 mm^2, thickness 100 nm), using Ga/In eutectic and Au wires. The DC voltage is swept at a rate of 10 mV s^{-1}. **(a)** Linear plot of the DC current I versus the DC applied voltage V; **(b)** plot of $\log_{10}$ I versus V.[36]

or −1.8 V). In the forward regime, we again saw $\ln_e(I) \propto V$ for $V > 0.5$, up to a threshold voltage (in the range 0.8 to 1.3 V, sample dependent), above which the current rises severalfold. This deviation from the $\ln_e(I) \propto V$ law may be ascribed to an enhanced through-bond electron transfer.[36] This behavior resembles that observed by

Martin *et al.* for multilayers of **1**.[29] A detailed discussion of the electrical measurements will be presented elsewhere.[41]

The results on both monolayers and multilayers show that the electrons preferentially flow by IVT from 3CNQ (A^-) to quinolinium (D^+), in agreement with the AR Ansatz. For multilayers, electron transfers between successive monolayers from the D^+ of layer N to the A^- of layer N−1 is possible (by tunneling through the alkyl chains), provided that the electric field is high enough to move the electron between the relevant molecular orbitals.

More recently, rectification for a single monolayer of **1** between Al electrodes was measured between 300 K and 78 K.[43] The rectification persists down to 78K, but a temperature dependence of the rectification ratio was not determined, since the molecules tend to "turn" upon repeated cycling of the voltage. During some measurements, "spikes" appear, i.e., filamentary conduction through the monolayer increases, and short circuits also appear: these short circuits often "heal" when the monolayer is brought back to room temperature, but, thereafter, the IV curves deviate from "normal behavior:" part of the monolayer may have "turned around." A nomogram of rectification ratios, measured at different temperatures, at slightly different voltages V_o, and for different samples, is shown in FIGURE 8: no claim is made for the temperature dependence.

NANOSCOPIC STRUCTURE OF LB FILMS

With a Nanoscope II (pre-amplifier modified to allow for 100 times more current than usual) a Langmuir-Schaefer[49] monolayer of **1** deposited on HOPG yielded an image with surprising resolution (FIG. 9). This showed the molecules seen from the

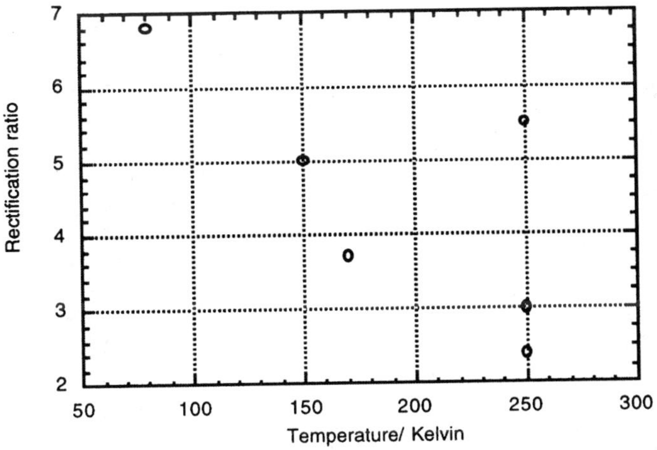

FIGURE 8. Rectification ratios *RR* measured at different temperatures, slightly different voltages V_o, for different samples of Al | monolayer of **1** | Al.[43]

FIGURE 9. STM micrograph of Langmuir-Schaefer monolayer of **1** on HOPG, viewed with a Pt/Ir nanotip (Nanoscope II). Scan size = 2 nm × 2 nm, nominal setpoint currrent = 1.3 nA (actual = 130 nA), bias = 92.5 mV.[34,36]

dicyanomethylene end.[34] The repeat area of 4.5 × 13 Å is of a "light" region (closer to the nanotip) of about 4.5 × 7 Å, plus by a dark region (4.5 × 6 Å). Using a low-current STM (Nanoscope III), an LB monolayer of **1** showed a similar image with lower resolution, with a repeat of 6 Å × 12 Å (FIG. 10).[36] Both images are consistent with the molecules viewed end-on.

NANOSCOPIC CONDUCTIVITY OF LB FILMS

An early effort to detect rectification by scanning tunneling microscopy of LB monolayers of **1** on HOPG showed no reproducible asymmetries in the scanning tunneling spectroscopy (STS) I-V plots.[34]

Recently, multilayers of **1** on HOPG were studied using a Pt/Ir nanotip at room temperature in air by STM and STS.[35,36] Asymmetries were seen in the I-V plot in 15-layer films (FIG. 11);[35] the corresponding STM images were of a superstructure of graphite.[35] In agreement with the I-V characteristics of macroscopic multilayers sandwiched between Al electrodes, the STS current in FIGURE 11 is consistent with electrons flowing by intramolecular intervalence electron transfer (IVT) between LB layers 2 through 15, and with "jumping over" layer 1 (FIG. 12), but, as discussed above, the electrons must somehow jump between adjacent monolayers. The current must also get across the poorly transferred first monolayer (which is oriented the "wrong way," compared with the other 14 layers).

An STS spectrum for a monolayer of **1** shows a slightly higher current at positive bias, in the expected direction for the first (partial) monolayer on HOPG, adhering with the alkyl tail onto the HOPG surface (FIG. 13).

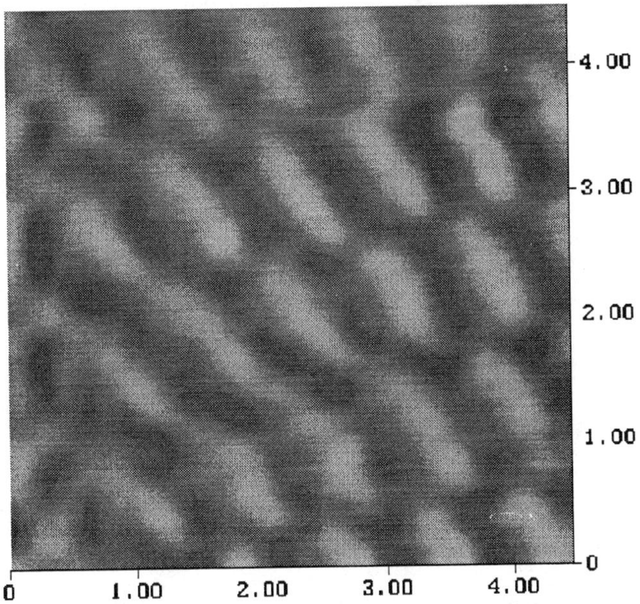

FIGURE 10. STM image of a LB monolayer of **1** on HOPG,with Pt/Ir tip (Nanoscope III). Scan size = 4.5 nm × 4.5 nm, Z-range = 2.3 pA, bias = –316 mV; setpoint current = 3.2 pA.[36]

REVIEW OF THE AR ANSATZ AND ITS MODIFICATION FOR A GROUND-STATE ZWITTERION

The AR mode[1] of unimolecular rectification places a single molecule D–σ–A (D = good electron donor, A = good electron acceptor, σ = covalent linkage between D and A) between two macroscopic metal electrodes M_1 and M_2 with work functions ϕ_1 and ϕ_2, respectively. Under "forward bias" an electron is transferred from M_2 to M_1 in two steps:

$$M_1 \mid D{-}\sigma{-}A \mid M_2 \text{ —— (Step 1, ET)} \longrightarrow M_1^- \mid D^+{-}\sigma{-}A^- \mid M_2^+ \quad \textbf{(13)}$$

$$M_1^- \mid D^+{-}\sigma{-}A^- \mid M_2^+ \text{ —— (Step 2, IVT)} \longrightarrow M_1^- \mid D{-}\sigma{-}A \mid M_2^+ \quad \textbf{(14)}$$

In Eq. (13) one electron hops by elastic tunneling (ET) from the HOMO of the neutral donor molecule, through a chemisorptive barrier, onto the Fermi level of the metal electrode M_1 whose ϕ_1 at the same energy as, i.e., in resonance with, the HOMO. One electron is thus added to the metal conduction band, forming, formally, M_1^- and leaving the donor as the donor radical cation D^+. Similarly, an electron from the Fermi level of the metal electrode M_2 hops through a chemisorptive barrier

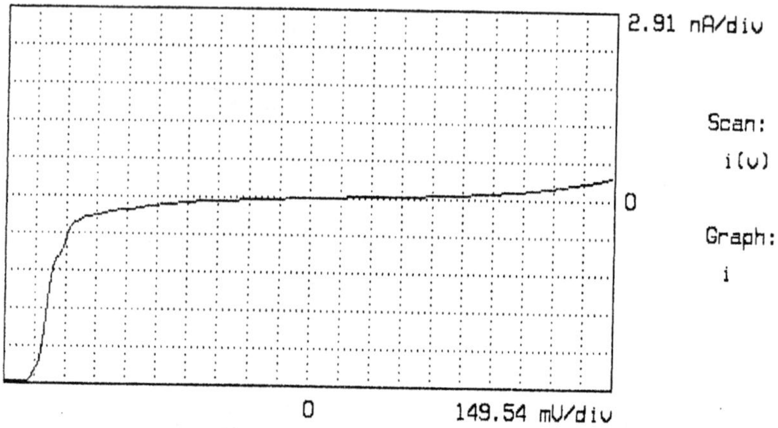

FIGURE 11. STS I-V curve for an LB film of 15 Z-type monolayers of **1** on HOPG, with Pt/Ir tip. The higher current for $V < -1.35$ Volt corresponds to electron flow from HOPG through the molecules to the Pt/Ir tip.[35,36]

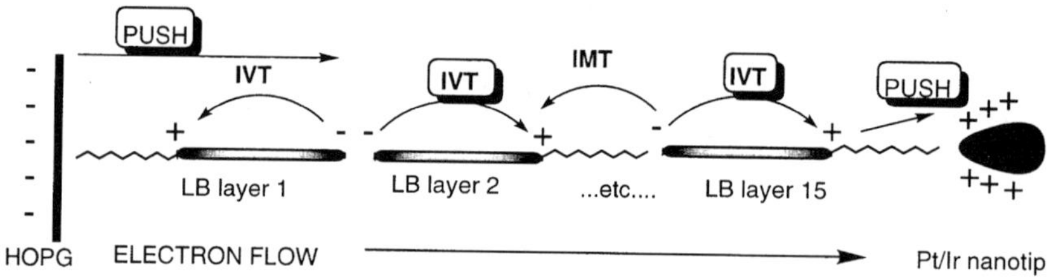

FIGURE 12. Direction of observed STS electron current through 15-layer LB multilayer of **1**.[35,36]

onto the LUMO of the neutral acceptor molecule, which is at resonance with ϕ^2. This forms the acceptor radical anion A^-, and removes one electron from the metal conduction band to form M_2^+. This leaves the molecule as an excited-state zwitterion $D^+–\sigma–A^-$. In Step 2 (Eq. (14)), an intervalence transfer (IVT) occurs: since the LUMO of the neutral acceptor A (A = TCNQ) is about 3 eV higher than the HOMO of the neutral donor D (D = TTF), therefore down-hill inelastic tunneling ensues through the molecule (through-bond tunneling), and the molecule returns to its ground state D–σ–A, but one electron has been moved by steps 1 and 2 from M_2 to M_1. It is guessed that the $D^+–\sigma–A^-$ state lies about 1 to 2 eV above the D–σ–A ground state.

Since good organic one-electron donors are bad acceptors, and good electron acceptors are bad donors, therefore the "reverse bias" process (Step 3, Eq. (15)) would involve an excited state that is estimated to be ≥4 eV higher than the excited state $D^+–\sigma–A^-$, and can be thus neglected.

$$M_1 \mid D–\sigma–A \mid M_2 \text{—— (Step 3, ET) ——XX}\!\longrightarrow M_1^+ \mid D^-–\sigma–A^+ \mid M_2^- \qquad \textbf{(15)}$$

(The only "hermaphroditic" donor that is an equally good acceptor is graphite, for which no rectification would occur.)

If Step 3, Eq. (15) is unlikely, but Steps 1 and 2, Eqs. (13) and (14) are favored, then molecular rectification should occur. A competing "reverse-bias" process is autoionization (Step 4, Eq. (16)), followed by charge migration to the two metal electrodes (Step 5, Eq. (17)):

$$M_1 \mid D\text{–}\sigma\text{–}A \mid M_2 \xrightarrow{\text{(Step 4)}} M_1 \mid D^{+}\text{–}\sigma\text{–}A^{-} \mid M_2 \qquad (16)$$

$$M_1 \mid D^{+}\text{–}\sigma\text{–}A\text{-} \mid M_2 \xrightarrow{\text{(Step 5, IVT, ET)}} M_1^{+} \mid D\text{–}\sigma\text{–}A \mid M_2^{-} \qquad (17)$$

The AR process will work if Step 1, Eq. (13), is much more likely than Step 4, Eq. (16), and if the through-molecule tunneling in Step 2, Eq. (14), is much larger than elastic through-space tunneling. The situation is reviewed diagrammatically in the top half of in FIGURE 13.

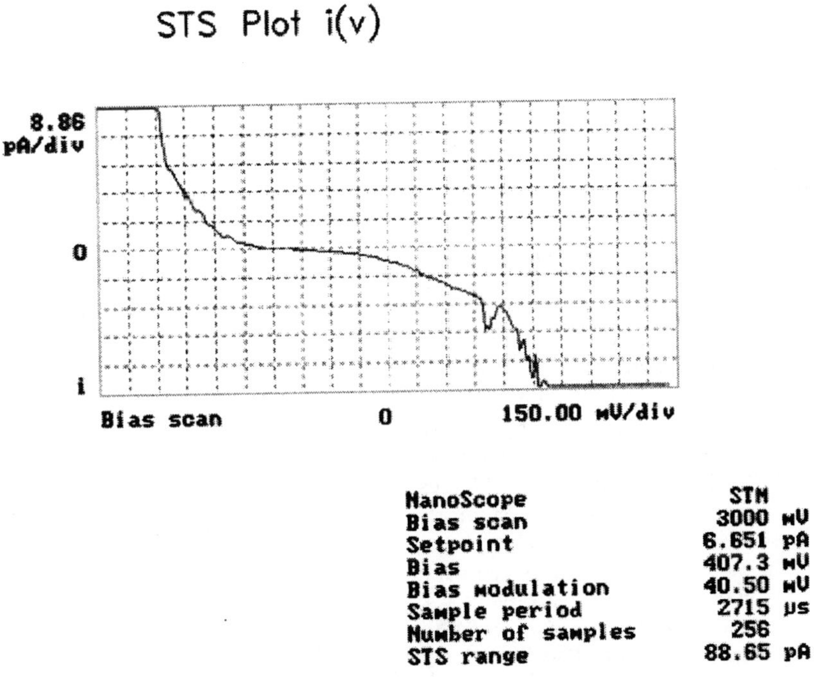

FIGURE 13. STS I-V curve for an LB monolayer of **1** on HOPG, with Pt/Ir tip (Nanoscope III). The higher current for $V \geq 0.5$ Volt corresponds to electron flow from Pt/Ir nanotip through molecules to HOPG.[36]

The AR Ansatz can be adapted to the case when one starts from a ground-state zwitterion $D^+–\pi–A^-$, whose first excited state is the neutral molecule $D^0–\pi–A^0$. Under "forward bias" an electron is transferred from M_2 to M_1 in two steps:

$$M_1 \mid D^+–\pi–A^- \mid M_2 \text{—(Step 6, IVT)} \rightarrow M_1 \mid D^0–\pi–A^0 \mid M_2 \quad \mathbf{(18)}$$

$$M_1 \mid D^0–\pi–A^0 \mid M_2 \text{—(Step 7, ET)} \rightarrow M_1\text{-} \mid D^+–\pi–A^- \mid M_2^+ \quad \mathbf{(19)}$$

The reverse bias process is:

$$M_1 \mid D^+–\pi–A^- \mid M_2 \text{—(Step 8, ET)} \rightarrow M_1^+ \mid D^0–\pi–A^0 \mid M_2^- \quad \mathbf{(20)}$$

$$M_1^+ \mid D^0–\pi–A^0 \mid M_2^- \text{—(Step 9, IVT)} \rightarrow M_1^+ \mid D^+–\pi–A^- \mid M_2^- \quad \mathbf{(21)}$$

Once again, the molecule would rectify (Steps 6, 7, Eqs. (18) and (19)), so that electrons are moved from M_2 to M_1, provided that Step 8, Eq. (20) is less likely than Step 6, Eq. (18).

DISCUSSION

The synthesis of **1** was vastly improved by using a 2:1 molar ratio of TCNQ salt to lepidinium salt. The ground state of the molecule is zwitterionic, as demonstrated by its NMR spectrum and its solution dipole moment $\mu = 43 \pm 8$ D. This value corresponds to an electron-hole pair separated by 9 ± 1 Å, while the calculated distance from the bridgehead C atom in the dicyanomethylene end of **1** to the quinolinium N atom is about 10.5 Å. The monolayer thickness (23 Å by X-ray diffraction, 22 ± 2 Å by ellipsometry and surface plasmon resonance) suggests molecules inclined by 45–55° from the surface normal. The oxidation of **1** in dichloromethane is irreversible; its first reduction is reversible, but **1** is not a powerful electron acceptor.

The STM repeat areas of 4.5 × 13 Å (FIG. 9) or 6 × 12 Å (FIG. 10) are roughly consistent with a calculated molecular cross-section (12.4 Å × 4.5 Å), and also with the observed collapse area of 50 $Å^2$ at 20°C (FIG. 3).[36]

The rectification by a monolayer of **1** on HOPG (FIG. 13), with higher electron flow at positive bias from nanotip to HOPG, is consistent with IVT across a molecule oriented with the dicyanomethylene end closer to the Pt/Ir nanotip (*cf.* LB layer 1 in FIG. 12), but the monolayer adheres poorly to HOPG and is easily moved across the surface by the tip, so other, symmetric STS curves (through-space tunneling) are also seen.[36]

The STS rectification across an LB multilayer (FIG. 11) is very reproducible; the direction of higher electron flow at $V < -1.35$ V corresponds to IVT electron flow *within* layers 2 to 15 of a 15-layer film, as explained in FIGURE 12 (boxed IVT), but the contrary flow across layer 1 (IVT) and the electron flow mechanism between layers (IMT) are not explained.[35,36]

Monolayer rectification between Al electrodes (FIG. 8) can be seen in 10% of the pads examined (one must find regions that are defect-free). Repeated cycling decreases the rectification ratio, as the molecules reorient in the intense electrical fields. Rectification persists down to 78K.

1 is not only a multilayer rectifier, but also a monolayer rectifier,[36] by an improvement of Sambles and co-workers' original technique,[28,29] and in accord with a plausible modification of the Aviram-Ratner Ansatz.[1] The dilemma due to the previous use[28,29] of two metal electrodes with different work functions (Mg and Al) has been eliminated.[36] Although asymmetric STS currents through single halogen adatoms on a STM tip have been seen,[50] it may be claimed that **1** is the "world's smallest electronic device."

CONCLUSIONS

A monolayer of molecule **1** has been shown to rectify by intramolecular tunneling; monolayers and multilayers of **1** rectify both as macroscopic films and on a nanoscopic level.

ACKNOWLEDGMENTS

We are deeply indebted to Profs. Geoffrey J. Ashwell and Ludwig Brehmer, to Drs. M.V. Lakshmikantham, Dominique Vuillaume, Tsuyoshi Kawai, Hiroaki Tachibana, Xiangli Wu, and Hiromi Sakurai; to Messrs. Bo Chen, Ulf Höpfner, and Jeffrey W. Baldwin, and to Miss Christina Hosch. We are most grateful for the support of the National Science Foundation (grant DMR-94-20699), the Department of Energy (grant DE-FC02-91-ER-75678), and the Scientific Affairs Division of the North Atlantic Treaty Organization (grant CRG 960656).

REFERENCES

1. AVIRAM, A. & M.A. RATNER. 1974. Chem. Phys. Lett. **29:** 277–283.
2. FLEMING, J.A. 1904. British Patent **24,850**. Date of application. 16 November.
3. SHOCKLEY, W. 1949. Bell System Techn. J. **28:** 435–489.
4. SCHOTTKY, W. 1942. Z. Phys. **118:** 539–592.
5. MEINHARD, J.E. 1964. J. Appl. Phys **35:** 3059–3060.
6. ANDERSON, T.L., G.C. KOMPLIN & W.J. PIETRO. 1993. J. Phys. Chem. **97:** 6577–6578.
7. PIETRO, W.J. 1994. Adv. Mater. **6:** 239–242.
8. HAMM, S. & H. WACHTEL. 1994. J. Chem. Phys. **103:** 10689–10695.
9. POLYMEROPOULOS, E.E., D. MÖBIUS & H. KUHN. 1980. Thin Solid Films **68:** 173–190.
10. SUGI, M., K. SAKAI, M. SAITO, Y. KAWABATA & S. IIZIMA. 1985. Thin Solid Films **132:** 69–76.
11. FISCHER, C.M., M. BURGHARD, S. ROTH & K.V. KLITZING. 1994. Europhys. Lett. **28:** 129–134.
12. FUJIHIRA, M., K. NISHIYAMA & H. YAMADA. 1986. Thin Solid Films **132:** 77–82.
13. SATO, Y., H. ITOIGAWA & K. UOSAKI. 1993. Bull. Chem. Soc. Jpn. **66:** 1032–1037.
14. ALLEMAN, K.S., K. WEBER & S.E. CREAGER. 1996. J. Phys. Chem. **100:** 17050–17058.

15. POMERANTZ, M., A. AVIRAM, R.A. MCCORKLE, L. LI & A.G. ASCHOTT. 1992. Science **255:** 1115–1118.
16. STABEL, A., P. HERWIG, K. MÜLLEN & J.P. RABE. 1994. Angew. Chem. Intl. Ed. Engl. **34:** 1609–1611.
17. DHIRANI, A., P.-H. LIN, P. GUYOT-SIONNEST, R.W. ZEHNER & L.R. SITA. 1997. J. Chem. Phys. **106:** 5249–5253.
18. METZGER, R.M. & C.A. PANETTA. 1991. New J. Chem. **15:** 209–221.
19. METZGER, R.M. 1994. *In* Molecular and Biomolecular Electronics. RR. Birge, Ed. **240:** 81–129. ACS Adv. Chem. Ser. American Chemical Society. Washington, DC.
20. METZGER, R. M. 1995. Mater. Sci. Engrg. **C3:** 277–285.
21. METZGER, R. M., C.A. PANETTA, N.E. HEIMER, A.M. BHATTI, E. TORRES, G.F. BLACKBURN, S.K. TRIPATHY & L.A. SAMUELSON. 1986. J. Mol. Electronics **2:** 119–124.
22. GEDDES, N.J., J.R. SAMBLES, D.J. JARVIS, W.G. PARKER & D.J. SANDMAN. 1990. Appl. Phys. Lett. **56:** 1916–1918.
23. GEDDES, N.J., J.R. SAMBLES, D.J. JARVIS, W.G. PARKER & D.J. SANDMAN. 1992. J. Appl. Phys. **71:** 756–768.
24. METZGER, R.M., N.E. HEIMER & G.J. ASHWELL. 1984. Mol. Cryst. Liq. Cryst. **107:** 133–149.
25. ASHWELL, G.J., M. SZABLEWSKI & A. P. KUCZYNSKI. 1991. *In* Lower-Dimensional Systems and Molecular Electronics, R.M. Metzger, P. Day & G.C. Papavassiliou, Eds. **B248:** 647–652. NATO ASI. Plenum, New York, NY.
26. ASHWELL, G. J. 1993. *In* Organic Materials for Nonlinear Optics. G.J. Ashwell & D. Bloor, Eds.: 31–39. Royal Soc. of Chem. Cambridge, UK.
27. AKHTAR, S., J. TANAKA, R. M. METZGER & G. J. ASHWELL. 1986. Mol. Cryst. Liq. Cryst. **139:** 353–364.
28. ASHWELL, G.J., J.R. SAMBLES, A.S. MARTIN, W.G. PARKER & M. SZABLEWSKI. 1990. J. Chem. Soc. Chem. Commun. 1374–1376.
29. MARTIN, A.S., J.R. SAMBLES & G.J. ASHWELL. 1993. Phys. Rev. Lett. **70:** 218–221.
30. BROO, A. & M.C. ZERNER. 1996.Chem. Phys. 196: 407–422.
31. BROO, A. & M.C. ZERNER. 1996. Chem. Phys. **196:** 423–426.
32. MARCUS, R.A. 1965. J. Chem. Phys. **43:** 679–701.
33. HUSH, N.S. 1961. Trans. Faraday Soc. **57:** 557–580.
34. WU, X.-L., M. SHAMSUZZOHA, R.M. METZGER & G.J. ASHWELL. 1993. Synth. Metals **55–57:** 3836–3841.
35. METZGER, R.M., H. TACHIBANA, X. WU, U. HÖPFNER, B. CHEN, M.V. LAKSHMIKANTHAM & M.P. CAVA. 1997. Synth. Metals **85:** 1359–1360.
36. METZGER, R.M., B. CHEN, U. HÖPFNER, M.V. LAKSHMIKANTHAM, D. VUILLAUME, T. KAWAI, X. WU, H. TACHIBANA, T.V. HUGHES, H. SAKURAI, J. W. BALDWIN, C. HOSCH, M.P. CAVA, L. BREHMER & G.J. ASHWELL. 1997. J. Am. Chem. Soc. **119:** 10455–10466.
37. METZGER, R.M. 1998. *In* Hyper-Structured Molecules I: Chemistry, Physics & Applications, H. Sasabe, Ed. Gordon & Breach Science Publishers. New York. In press.
38. METZGER, R.M., B. CHEN, D. VUILLAUME, M.V. LAKSHMIKANTHAM, U. HÖPFNER, T. KAWAI, J.W. BALDWIN, X. WU, H. TACHIBANA & M.P. CAVA. 1998. Thin Solid Films. In press.
39. METZGER, R.M. 1998. Adv. Mater. Optics & Electronics. Submitted for publication.
40. METZGER, R.M., B. CHEN, D. VUILLAUME, U. HÖPFNER , J.W. BALDWIN, T. KAWAI, H. TACHIBANA, H. SAKURAI, M. V. LAKSHMIKANTHAM & M.P. CAVA. 1998. *In* Electrical, Optical, and Magnetic Properties of Organic Solid-State Materials. L.Y.

Chiang, L.R. Dalton, A.Y. Jen, J. Reynolds & M. Rubner, Eds. MRS Proceedings. Materials Research Society. Pittsburgh, PA. In press.
41. Vuillaume, D., B. Chen & R. M. Metzger. 1997. Unpublished results.
42. Baldwin, J.W., V. Fidler, S.C. Street, V.I. Konovalov, H. Sakurai, M. P. Cava, L.D. Kispert & R. M. Metzger. 1997. Unpublished results.
43. Chen, B. & R.M. Metzger. 1997. Unpublished results.
44. Böttcher, C.J.F., O.C. van Belle, P. Bordewijk & A. Rip. 1973. Theory of Electric Polarization. **1**: 245–284. Elsevier. Amsterdam, the Netherlands.
45. Ashwell, G.J. 1990. Thin Solid Films **186:** 155–165.
46. Ashwell, G.J., E.J.C. Dawney, A. P. Kuczynski, M. Szablewski, I. M. Sandy, M.R. Bryce, A.M. Grainger & M. Hasan. 1990. J. Chem. Soc. Faraday Trans. **86:** 1117–1121.
47. Ashwell, G.J., G. Jefferies, E.J.C. Dawnay, A. Kuczynski, D.E.J. Lynch, Y. Gongda & D. J. Bucknall, 1985. J. Mater. Chem. **5:** 975–980.
48. Wu, X. 1995. Ph. D. Dissertation. University of Alabama, Tuscaloosa, AL.
49. Langmuir, I. & V. J. Schaefer. 1938. J. Am. Chem. Soc. **60:** 1351–1360.
50. Schott, J. H. & H. S. White. 1994. J. Phys. Chem. **98:** 297–302.

Chemical Approaches of Molecular Switches

J.-P. LAUNAY[a] AND C. COUDRET

Molecular Electronics Group, Center for Material Elaboration and Structural Studies, CNRS, 29 Rue Jeanne Marvig, 31055 Toulouse Cedex, France

ABSTRACT: Molecular switches are defined as molecules existing in two different states, an ON state, allowing a complete intramolecular electron transfer between two redox sites, and an OFF state, for which this transfer is blocked. The efficiency of electron transfer is measured by the V_{ab} electronic coupling parameter, which can be obtained from intervalence transitions, and its various definitions are recalled. Then different switching mechanisms are reviewed: torsion, saturation/unsaturation of double bonds, and quantum interferences. Synthetic strategies such as creation of an unsaturation, condensation to generate a double bond, or connection of two conjugated systems to form a single bond, are discussed and put in perspective.

Switching at the molecular scale is one of the great challenges of molecular electronics. The expression "molecular switch" has been frequently used in the case of molecules able to exist in two different states with markedly different properties,[1] which makes this concept very close to "molecular bistability." In recent papers, Lehn and colleagues have described photochromic systems for which the two isomers have different properties with respect to electron delocalization, thus generating integrated photochromic and electrochromic properties.[2] However, there was no complete electron transfer process in this case.

By contrast, our definition of "molecular switches" is close to the usual definition used in electrical engineering, i.e., the molecular switch must present the following two properties: (1) it must be able to exist in two different states, with the possibility to drive the system in a controlled way from one state to the other (bistability); (2) in one of the states (ON), it must be able to allow a *complete electron transfer* to occur (i.e., a real movement of one unit charge from one part of the molecule to the other), while in the other state (OFF), this electron transfer must be blocked. Note that the term "transfer" designates the situation in which just one electron is moved in a molecular compound isolated in solution. If the molecule, or at least its active part, is embedded between two ultrathin metallic wires acting as electron reservoirs (nanowires), then we may expect a permanent electron flux (a current) to occur in the ON state, and the term "transport" will be used. Examples of devices using single molecules for electron transport are described in this volume.[3,4]

The present paper describes our recent realizations in the field of molecular switches studied in solution, i.e., for electron transfer purposes. They are studied by the technique of intervalence transitions occurring in mixed valence species.[5,6] We consider systems for which the electronic structure is partly localized (the so-called class II compounds in mixed valence terminology),[5,6] so that the potential energy

[a]Address for telecommunication: Phone, 33 (0) 5 62 25 78 32; fax, 33 (0) 5 62 25 79 99; e-mail, launay@cemes.fr

curve presents two minima. The intervalence transition is then a low-energy (typically near 1 eV) electronic transition corresponding to a charge migration between the two redox sites constituting the mixed valence species. The intensity of this transition can be used to characterize quantitatively the degree of electronic coupling between the redox sites. Thus, mixed valence involves electron transfer between chemically equivalent sites, allowing the study of pure electron transfer (i.e., with zero driving force).

DIFFERENT TYPES OF MOLECULAR SWITCHES

When included into a mixed valence complex, a molecular switch can be viewed as a three-component molecule (FIG. 1): (1) two redox centers, usually transition metal complexes; (2) one "effector," which can be triggered by a perturbation external to the molecule; and (3) "connectors" to maintain in a rigid geometry the whole molecule as well as a good electronic coupling between the redox centers. Effector, connectors, and part of the coordination sphere of each metal ion are incorporated in what is usually called the "bridging ligand."

ORBITAL INTERPRETATION OF MOLECULAR SWITCHING

Since molecular switching is based in our definition on a drastic modification of the electronic coupling parameter, a reminder of its definition is necessary. We consider a metal-ligand-metal system with a mixed valence stoichiometry such as Ru^{II}–L–Ru^{III}. We assume that the electronic coupling is small enough so that the system at a given instant exists in one of the localized states, which implies that the electron transfer must be thermally activated. This process is usually described by the well

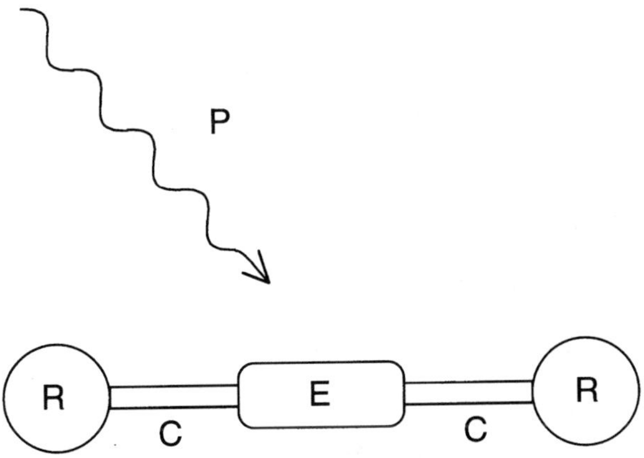

FIGURE 1. A molecular switch as a three-components molecule. R: redox sites; E: effector; P: perturbation; C: connector.

known Marcus-Hush curves,[5,6] and at the crossing point, where the geometrical structure is perfectly symmetrical, the two electronic states Ru_A^{II}—L—Ru_B^{III} and Ru_A^{III}—L—Ru_B^{II} are coupled by some interaction, which will be denoted as V_{ab}. If Ru_A and Ru_B were only separated by free space, V_{ab} would merely be the resonance integral between orbitals located on sites A and B. All difficulties encountered in V_{ab} definitions come ultimately from the mixing of the energy levels of metal and bridging ligand.

The most fundamental definition is based on the dynamic evolution of the system which has been prepared in a non-stationary state such as Ru_A^{II}—L—Ru_B^{III} (i.e., with the extra electron located on site A).[7] If we look at the probability $g_B(t)$ to observe the extra electron on B, this will appear as a periodic function with a major Fourier component. (Note that if there is no ligand between sites A and B, i.e., purely through-space interaction, the time evolution will be a simple sinusoidal function.) The V_{ab} coupling is then given by

$$V_{ab} = h\nu_{ab} / 2$$

where ν_{ab} is the frequency of the major Fourier component, and h is Planck's constant.

Another definition is based on Effective Hamiltonian Theory.[8,9] Calling $|\Psi_A\rangle$ and $|\Psi_B\rangle$ the wavefunctions describing pure localization on A and B, respectively, the off-diagonal matrix element $\langle\Psi_A|H|\Psi_B\rangle$ (where H is the system's Hamiltonian) represents in fact the through-space electronic coupling, which is in our case a small component of the total coupling V_{ab}. To obtain the latter, one has to compute instead $\langle\Psi_A|H_{\text{eff}}|\Psi_B\rangle$, where H_{eff} is the Effective Hamiltonian. H_{eff} is obtained from H by a transformation $U^{-1}HU$ followed by a projection on the subspace defined by $|\Psi_A\rangle$ and $|\Psi_B\rangle$. The U transformation is chosen such as to minimize the distance between eigenvectors and their projections on the $|\Psi_A\rangle$, $|\Psi_B\rangle$ subspace.[8]

Finally, a last possibility is to diagonalize the system Hamiltonian, that is, to solve the *static problem*. A number of eigenvectors will be obtained, among which some of them have the general form:

$$\Psi_1 = c_{1A}|\Psi_A\rangle + c_{1B}|\Psi_B\rangle + \Sigma\, c_{1n}|\Psi_n\rangle$$

$$\Psi_2 = c_{2A}|\Psi_A\rangle + c_{2B}|\Psi_B\rangle + \Sigma\, c_{2n}|\Psi_n\rangle$$

where $|\Psi_n\rangle$ are atomic orbitals belonging to the bridging ligand.

The interesting eigenvectors are the ones for which the weights on A and B are the largest. (Note that if the bridging ligand was not present, these eigenvectors would be, in the limit of zero overlap, the simple symmetry combinations $2^{-1/2}(|\Psi_A\rangle + |\Psi_B\rangle)$ and $2^{-1/2}(|\Psi_A\rangle - |\Psi_B\rangle)$). In the presence of the ligand, we may expect to encounter eigenvectors with for instance $c_{1A} = c_{1B} \approx 2^{-1/2}$ and $c_{2A} = -c_{2B} \approx 2^{-1/2}$, and small c_{1n} and c_{2n} coefficients. Once these relevant (crucial) eigenvectors are identified, the electronic coupling is then simply:

$$V_{ab} = |E_2 - E_1|/2.$$

This last method can be called the "dimer splitting method."[7,9]

As said above, the mother method is the dynamic one. The selection of eigenvectors by their weights can be considered as a filter generating the other two methods, which are equivalent for a symmetrical system (two equivalent sites), i.e., they provide the same V_{ab} value. However, the "dimer splitting" method has the favor of chemists, because it has an immediate visual significance in molecular orbital theory. Thus V_{ab} is determined by the properties of *two orbitals* (not just one, contrary to common belief), and the problem of molecular switching can now be formulated: How to modify as much as possible the splitting between the two crucial orbitals, taking into account, that in most cases both orbitals will be modified by the switching process? The analysis of the shape of the crucial orbitals can give a clue to the possible places where an active group can modify their energies.

Finally, it can be noticed that all computations of V_{ab} require the choice of a quantum chemistry method. The extended Hückel method (EH) appears extremely valuable, although more elaborate (*ab initio*, semi-empirical, self-consistent) methods are increasingly available. This remarkable efficiency of the EH method could be due to the fact that the basic electron transfer process involves only *one electron* moving between a half-filled orbital and a vacant one (or the hole equivalent of this process). It does not depend on spin considerations nor on electron-electron repulsion, two effects which are not explicitly taken into account in EH theory.

DIFFERENT TYPES OF MOLECULAR SWITCHES

Several effects can lead to a large V_{ab} variation and are potentially useful for molecular swithching. Only a few examples, taken from our work, will be mentioned here (a more exhaustive review is given by Ward[1]).

Torsion Process

This is the most obvious possibility, consisting in twisting the molecule around an essentially single bond in order to achieve orbital decoupling in the OFF state. The theoretical analysis of 4,4′-bipyridine[9] or bipyridylbutadiene[8] as bridging ligand showed that internal torsion or isomerization could modify V_{ab} (FIG. 2). But the effect will be modest unless strict orthogonality between aromatic cycles is realized in the OFF state. Experimentally, we have studied a Ru—L—Ru system where L was a 4,4′-bipyridine in which the two cycles were connected by a crown-ether chain, in order to modify the molecular conformation by alkaline ion complexation.[10,11] However, the effect on V_{ab} was found to be very small, probably because the geometrical change was very modest.

An attractive possibility was provided by the so-called TICT molecules (Twisted Internal Charge Transfer), for which a strict orthogonality is realized in an excited charge transfer state.[12] It is possible to prepare TICT-like *bridging* molecules. Some of these molecules exhibit enough coupling to give an intervalence band in the ground (planar) state.[13] But such molecules are *monostable* systems, and no simple way to "lock" the twisted conformation has been found.

Saturation/Unsaturation

Here we transform a double bond, part of a conjugated system, into a single one, thus breaking conjugation. This can be done either by a formal reduction, or by a molecular reorganization, for instance by involving a photochemical reaction, or by a cycloaddition (Diels-Alder) reaction, involving another molecule.

The first case is illustrated by the azopyridine/bipyridylhydrazine system. The 4,4′-azopyridine ligand (azpy) can be used to prepare bimetallic mixed valence com-

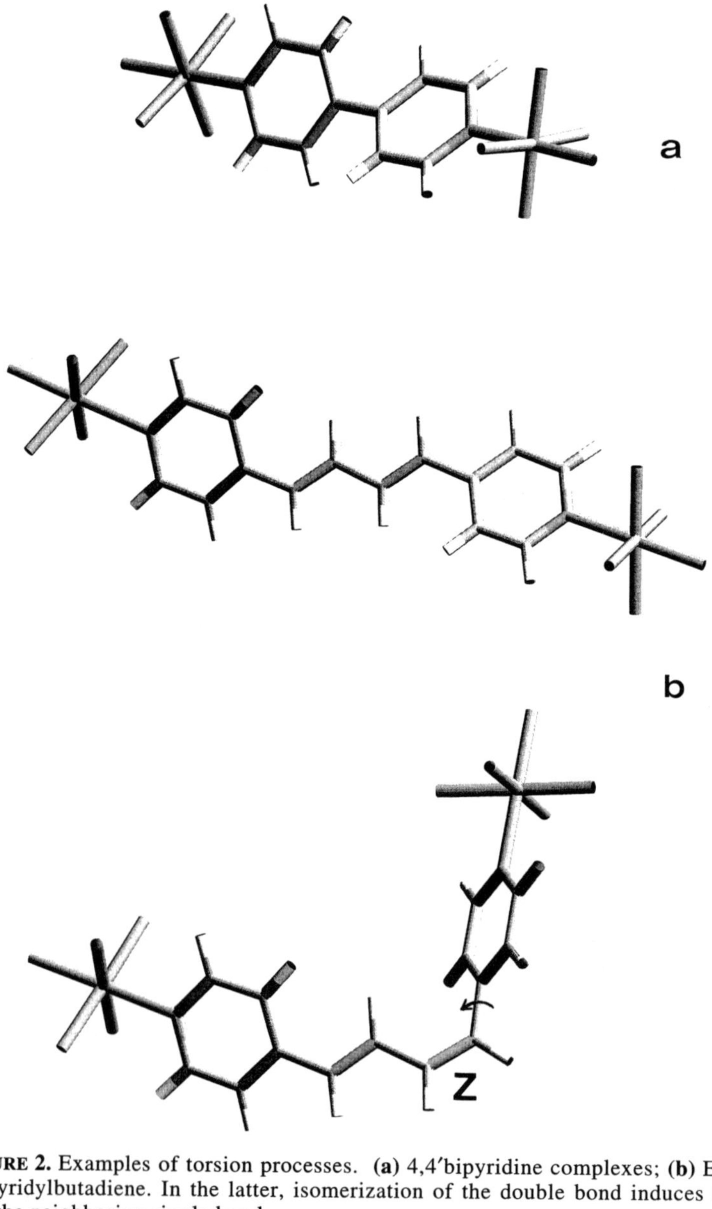

FIGURE 2. Examples of torsion processes. **(a)** 4,4′bipyridine complexes; **(b)** E,E and E,Z bipyridylbutadiene. In the latter, isomerization of the double bond induces a twist around the neighboring single bond.

plexes with rutheniumpentaamine end groups, or oligomeric chains with ruthenium porphyrins.[14] In these systems, the bridging ligand ensures an efficient metal-metal communication thanks to the presence of the azo -N=N- linkage. Upon acidification of Ru(II)-azpy-Ru(II), a pH-induced reaction is observed yielding Ru(III) and the reduced form of the ligand, i.e., 1,2-bis(4-pyridyl)hydrazine (bph). Subsequent redox experiments to generate the mixed valence species bridged by "bph" show that this replacement of the central azo -N=N- linkage by the hydrazo one -NH–NH- cuts indeed the metal-metal interaction.

As an example of the second possibility, we consider the case of the dicyanonorbornadiene/dicyanoquadricyclane (dcnor/dcqc) photoisomerization.[15] The bis(pentaamineruthenium(II)) complex of dcnor can be prepared and photoisomerized in dcqc complex by irradiation at 532 nm in methanol. On the dcnor isomer, partial oxidation generates the mixed valence Ru(II)–Ru(III) species, which exhibits an intervalence band from which an appreciable V_{ab} coupling (0.023 eV) can be computed. Conversely, the dcqc isomer does not present a detectable transition.[15] Interestingly, the general shape of the molecule is retained in the photoisomerization process. The main changes concern bond lengths as a result of the conversion of two double bonds into a cyclobutane ring (FIG. 3). Although this system demonstrates the feasibility of electron transfer interruption, it is not suitable for further development, because the back-reaction could not be observed, despite several attempts with various photosensitizers.

A Diels-Alder reaction is a third possibility to suppress double bonds. We have prepared 2,6-diferrocenylanthracene and studied its reaction with tetracyanoethylene (TCNE).[16] The adduct is easily obtained at room temperature and, from all relevant literature and molecular modeling, it is clear that it must exhibit a "bent" structure, the central 9,10 carbon atoms being sp^3 hybridized (FIG. 4). For 2,6-diferrocenylanthracene, a V_{ab} value of 0.032 eV is obtained, while for the TCNE adduct the value is 0.019 eV. This is somewhat disappointing because we could have expected a complete disappearence of V_{ab} with conjugation breakdown. Upon analysis of the shape of the crucial orbitals, it turns out that in the starting molecule, one of the orbitals has strong weights on carbon atoms 9 and 10, while the other has almost no weight, a situation which should be favorable for a strong effect on V_{ab}. As a consequence, in the adduct, the "sensitive" orbital has moved much more than the other, but unfortunately a crossing occurs, thus restoring some electronic interaction![16] It appears that the complete extinction of the interaction will probably necessitate a fine tuning of the energy levels.

Related to conjugation breakdown are the protonation/deprotonation reactions studied by Haga and colleagues,[17,18] although they occur in aromatic systems for which double and single bonds cannot be clearly distinguished. Bridging ligands containing benzimidazole units have yielded bimetallic ruthenium complexes for which an intervalence band can be switched ON or OFF by either deprotonation or protonation.

Quantum Interferences

While the previous processes are all based on a geometrical change of the molecular structure, the quantum interference switching is mainly based on electronic effects and fully uses the wave nature of electrons. Some geometrical change is

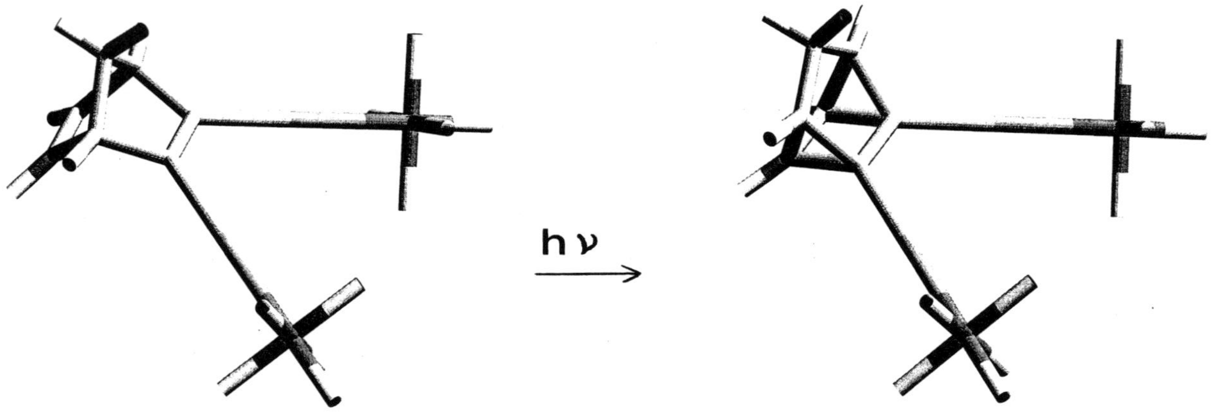

FIGURE 3. Photoisomerization of the dicyanonorbornadiene ruthenium complex into dicyanoquadricyclane complex.

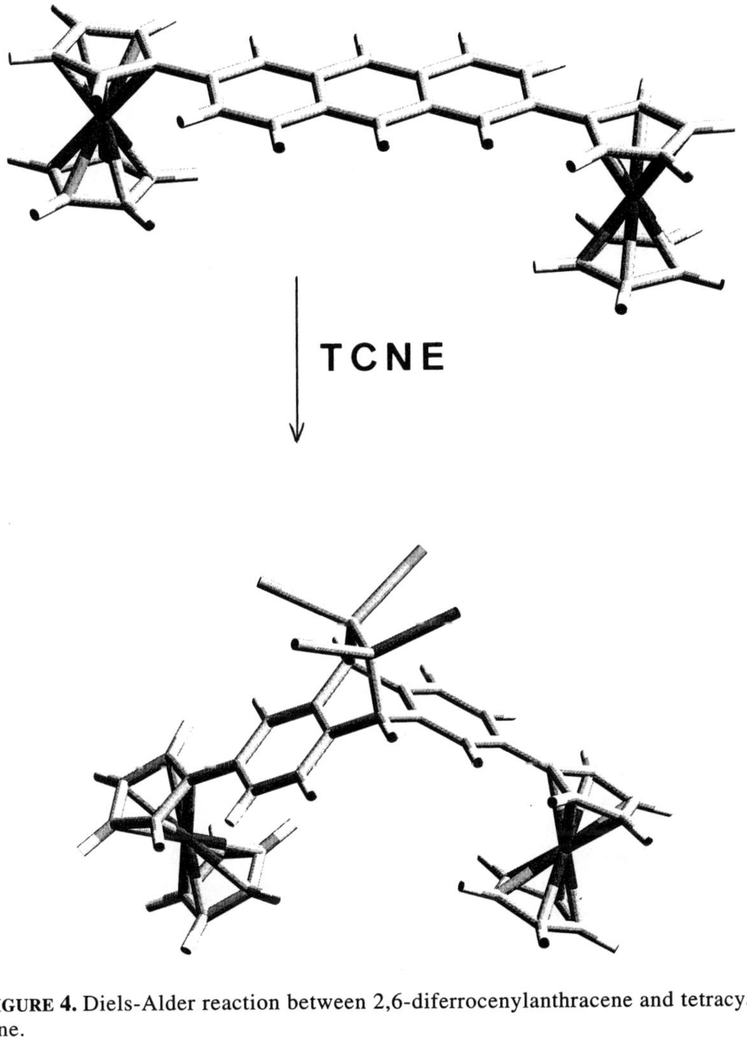

FIGURE 4. Diels-Alder reaction between 2,6-diferrocenylanthracene and tetracyanoethylene.

probably necessary to distinguish the two forms of the molecule, but it does not alone determine the switching effect. We have investigated the properties of substituted 1,3-diferrocenylbenzenes in comparison with the unsubstituted molecule (*meta* isomer) and the *para* isomer (FIG. 5).[19] As already observed by Taube and colleagues for complexes of either 1,3- or 1,4-dicyanobenzene,[20] the electronic coupling is much weaker in the 1,3- than in the 1,4-isomer. A detailed molecular orbital analysis (dimer splitting method) has been performed, considering the complexes as made of a bis(cyclopentadienyl)benzene ligand and two iron atoms. In the *meta* isomer, it is found that the symmetry combinations of metal orbitals interact in almost

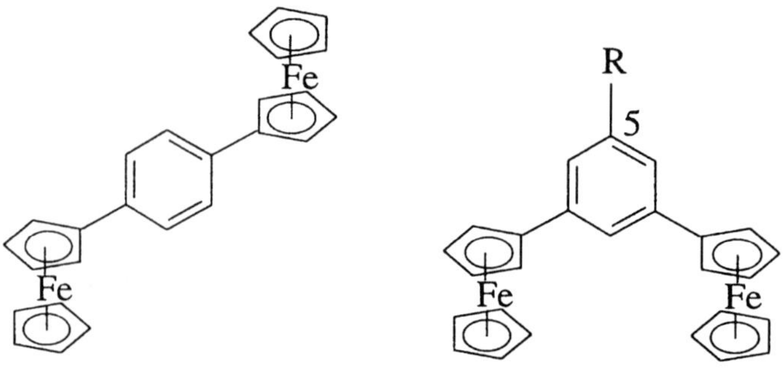

FIGURE 5. Structures of *para* and *meta* diferrocenylbenzenes.

the same way with quasi-degenerate ligand orbitals presenting opposite symmetries, thus resulting in a very small V_{ab}. Additional insight is given by a computation of the dynamical behavior when at time 0, an electron is localized on one ferrocenyl site. The presence of two ligand orbitals with opposite symmetries has a dramatic effect on the time evolution with respect to the one-ligand-orbital case, resulting in a marked decrease in ν_{ab}, the frequency of electronic oscillation.[19] Thus, there is a cancellation of the contributions of two electron transfer paths, each one corresponding to a coupling with a given ligand orbital. (Note that one cannot, in general, associate a coupling path with a given branch of the ligand, i.e., with the topological scheme, contrary to a conventional interferometer).

Molecular switching could be envisaged by grafting some active group in the 5-position of 1,3-diferrocenylbenzenes in order to destroy the delicate balance of this quantum interference effect. Unfortunately, the effect of substitution is found to be extremely small, both experimentally and theoretically.[19] This is due to the insensitivity of the crucial orbitals to substitution. For one of these, the weight in 5-position is strictly zero for symmetry reasons, while for the other it is almost zero by accident. However, using other substituents than ferrocenes could restore this weight, and the principle of quantum interference switching deserves further study.

SYNTHETIC ASPECTS

At least in one of its states (usually the "ON" state), the organic backbone is fully conjugated, that is, made of trigonal or digonal atoms (carbon or nitrogen). Such a situation, identical to the requirements of a molecular wire, implies in most of the cases an even number of atoms and a systematic alternation of single and multiple bonds (FIG. 6). In the following, we will focus our attention on how to build single-stranded molecular wires, keeping in mind that molecular switches are a special class of such conjugated molecules.

Conjugated systems are found in many natural compounds or dyes and are the core of a large variety of light-active materials, such as organic compounds for non

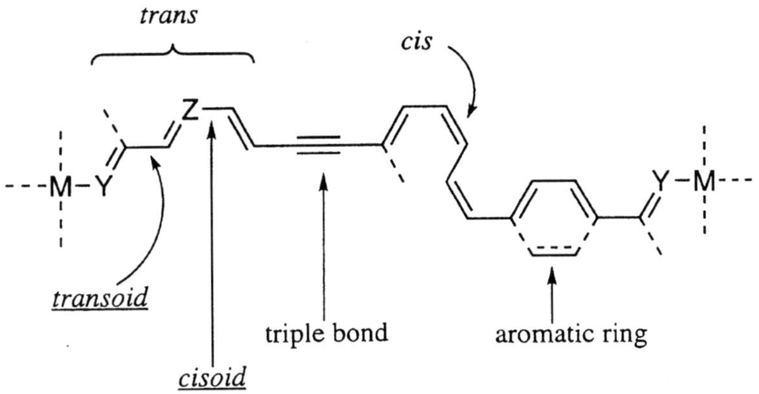

FIGURE 6. General scheme for a single stranded molecular wire, showing the systematic alternance of single and multiple bonds. M: metal center, Y: coordinating atom (Y, Z = C, N). *Italics*: double bond stereochemistry. *Underlined italics*: single bond conformation.

linear optics.[21,22] Three approaches are used to build a polyunsaturated molecule (FIG. 7). In one, the skeleton of the molecule is prepared prior to the creation of the unsaturation (route *a*) while the other routes *b* and *c* imply the simultaneous construction of the sigma and pi system (route *b* via the creation of a multiple bond, route *c* via the creation of a single one).[23] Organic chemistry provides a large variety of unsaturated functional groups. Their selection will mainly rely on synthetic methods to introduce them in a molecule and also on their physicochemical properties, such as stability, solubility, light absorption, and redox properties.

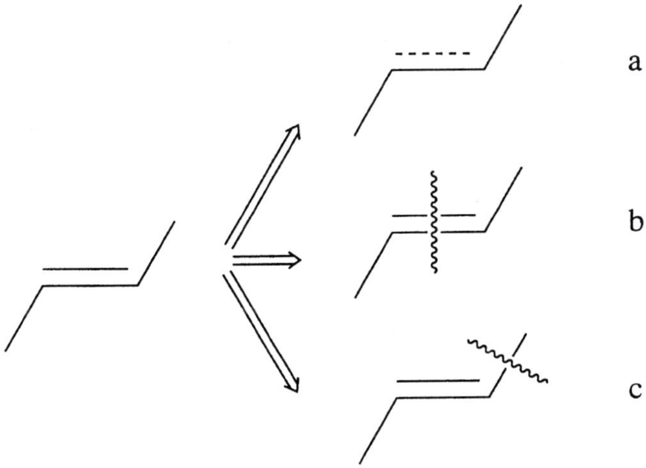

FIGURE 7. Basic retrosyntheses of polyunsaturated compounds. *Route a*: unsaturation made after construction of the molecular backbone. *Route b*: double bond creation. *Route c*: single bond creation.

Due to their high rigidity, large unsaturated molecules are usually poorly soluble. Such a problem can be limited by grafting other groups such as alkyl chains or by using charged metallic centers.[24,25] Let us now give some examples of the various strategies.

TABLE 1. Catalog of various unsaturated groups frequently used for molecular wires

Functional Group	Synthetic Approach	Advantages, *Drawbacks*	Uses
	a, b, c	good V_{ab}, photoisomerizable *stereoisomers, confomers, low stability (O_2) for polyenes*	spacer, effector
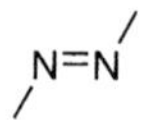	b, c	isomerizable *reducible and requires redox-based preparations, conformers*	spacer, effector
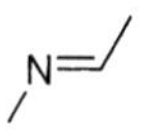	b	easy to prepare *relatively unstable conformers*	spacer
	c	rigidity, no isomers, no conformers *solubility, unstable if more than four triple bonds*	spacer
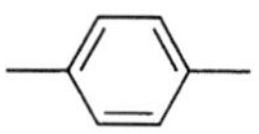	a,c	homoaromatics high stability, rich chemistry *moderate V_{ab} (quantum interferences, biphenyl dihedral angle)*	spacer
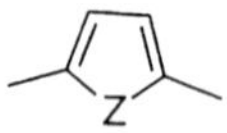		heteroaromatics good V_{ab} *moderate (Z=S) to low (Z=O) stability*	
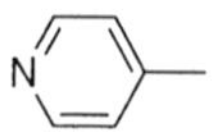	a, c	stable, good coordination site (especially in a chelating ligand 2,2′-bipyridine) *chemistry is difficult*	coordination site
	a,c	easily prepared, no conformers *weak coordinating site, no chelate possible*	coordination site

Route a

Creation of an insaturation can be achieved by two different ways: an elimination reaction or a redox reaction. To the latter category belong the rearomatization reactions classically used in fused polyaromatic chemistry, such as dehydrogenations (oxidation) or quinone reduction.[16,26]

Route b

For a long time condensation reactions have been the best approach for a polyunsaturated compound made of trigonal atoms. Based on a one-pot addition/elimination process of a nucleophilic atom (such as a more or less stabilized carbanion or a nitrogen) on a carbonyl-containing molecule, they are strongly related to *route a* strategy since in some cases the elimination step can be delayed enough to allow the isolation of the addition compound. Condensations are particularly well adapted to the creation of unsymmetrically substituted double bonds (including imines).[21,22,24,27] An elegant approach for highly symmetrical ones is the very convergent reductive ketone dimerization, developed by McMurry (reaction not used in the present studies).[28] Introduction of an azo group between two aromatic rings is traditionally achieved by the well known diazoic copulation, reacting a diazonium salt with an electron rich aromatic compound. An alternative route is merely an arylamine oxidation which provides a quick entry to symmetrical diazo molecules with a great regioselectivity since the newly created bond replaces the former amino group.[14]

Route c

Connection of two conjugated systems by a single bond is not a trivial step. Apart specific cases,[2] a nucleophilic process (i.e., successive addition on an electron deficient fragment followed by an elimination step to restore the conjugation) is almost limited to biaryl syntheses.[29] In fact, in most of the syntheses using such a strategy, this simple bond is obtained by a redox reaction, the best example being the copper-catalyzed diyne syntheses by alkyne coupling.[30] For the preparation of biaryls, it has been longly limited to Wurtz-like or Ullmann coupling of organic halides, to the oxidative dimerization of phenols or the reductive dimerization of pyridines,[10,11] and to the Gomberg-Bachmann arylation.[16,26] It is now the best entry to conjugated systems including those based on an alternation of triple/double bond with aryl groups (poly-tolanes or -stilbenes), thanks to palladium chemistry. This metal catalyzes a formal nucleophilic substitution of an organic halide by an organometallic species. In fact, the key step is a so-called *reductive* elimination, which justifies its belonging to redox reactions. Various organometallics are suitable (mostly organo-boron, -tin, -zinc or (alkynyl)cuprates) and because the reaction medium is usually weakly reducing and nucleophilic, this type of coupling is of a large scope. This methodology has become extremely popular during the past years and a large number of unsaturated molecules has been made.[31–34]

Finally, since they can be used to create the molecular skeleton as well as to destroy or create some of its unsaturations, pericyclic reactions deserve a special place between *routes a* and *c*. Three examples are mentioned in this paper: the [4+2] Diels-Alder reaction, used to make the norbornadiene skeletton,[15] or to destroy the aromaticity of the anthracene inner ring,[16] and the [2+2] photoinduced dimerization converting a diene into a cyclobutane.[15]

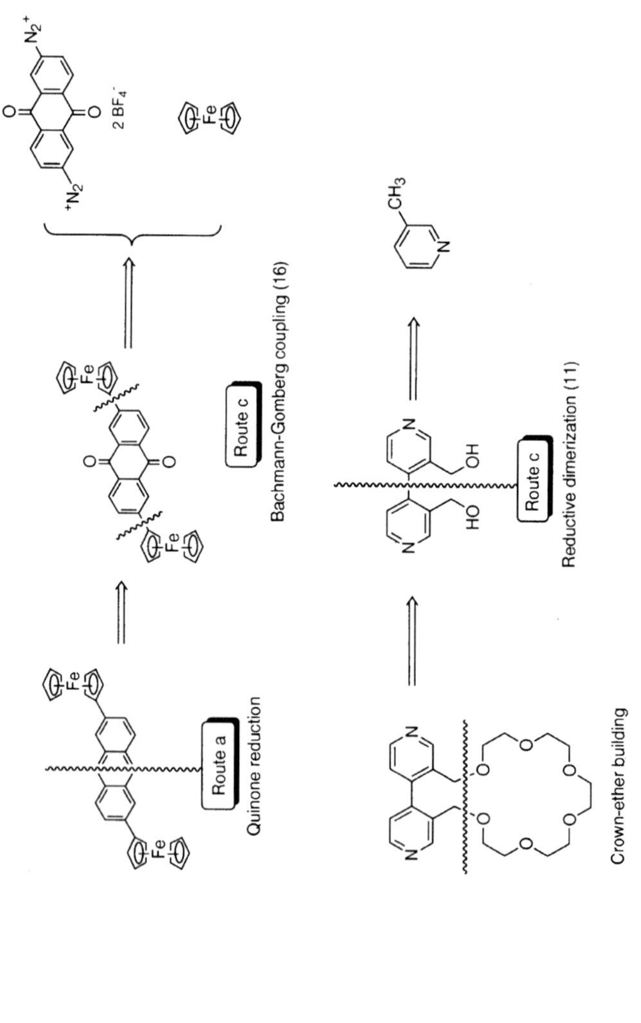

Route a
Quinone reduction
Route c
Bachmann-Gomberg coupling (16)
2 BF4-
Route c
Reductive dimerization (11)
Crown-ether building

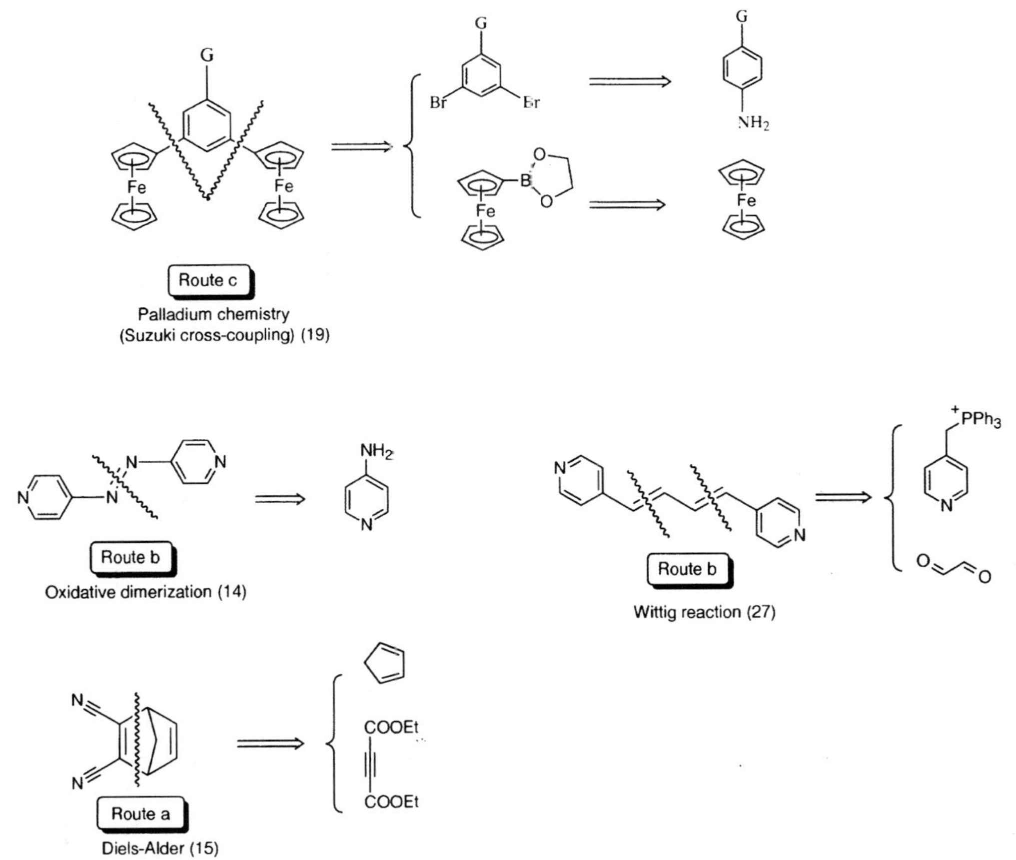

FIGURE 8. Examples of synthetic strategies following the different routes. Bibliographic references are in parentheses.

Generally speaking, the construction of a given molecule will proceed via a stepwise strategy made of a combination of *routes a*, *b*, and *c* and various examples are given in FIGURE 8.

CONCLUSIONS AND PERSPECTIVES

Switching at the molecular scale appears now a realistic possibility, although many problems remain to be solved: complete interruption in the OFF state, full reversibility and stability for a large number of cycles, switching time, inclusion in a useful solid state device, etc. Besides electron transport, such molecules could also play a role in energy transfer when it occurs by the electron exchange (Dexter) mechanism.

ACKNOWLEDGMENT

The financial support of the Human Capital and Mobility program of the European Community (contract CHRX-CT94-0538, "Electron and Energy Transfer in Model Systems and their implication for Molecular Electronics") is gratefully acknowledged.

REFERENCES

1. WARD, M. 1997. Current developments in molecular switches. Chem. Ind. 640–645.
2. KAWAI, S.H., S.L. GILAT, R. PONSINET & J.-M. LEHN. 1995. A dual-mode molecular switching device: bisphenolic diarylethenes with integrated photochromic and electrochromic properties. Chem. Eur. J. **1:** 285–293.
3. JOACHIM, C., H. TANG, C. BERGAUD, H. PINNA & J.K. GIMZEWSKI. 1998. Is there a minimum size and a maximum speed for a nanoscale amplifier? Ann. N.Y. Acad. Sci. This volume.
4. GOURDON, A. & H. TANG. 1998. Molecular landers. Ann. N.Y. Acad. Sci. This volume.
5. CREUTZ, C. 1983. Mixed valence complexes of d^5–d^6 metal centers. Progr. Inorg. Chem. **30:** 1–73
6. CRUTCHLEY, R.J. 1994. Intervalence charge transfer and electron exchange studies of dinuclear ruthenium complexes. Adv. Inorg. Chem. **41:** 273–325.
7. JOACHIM, C., J.-P. LAUNAY & S. WOITELLIER. 1990. Distance dependence of the effective coupling parameters through conjugated ligands of the polyene type. Chem. Phys. **147:** 131–141.
8. JOACHIM, C. & J.-P. LAUNAY. 1986. Bloch effective hamiltonian for the possibility of molecular switching in the ruthenium-bipyridylbutadiene-ruthenium system. Chem. Phys. **109:** 93–99.
9. WOITELLIER, S., J.-P. LAUNAY & C. JOACHIM. 1989. The possibility of molecular switching: Theoretical study of $[(NH_3)_5Ru–4,4'\text{-bipy-}Ru(NH_3)_5]^{5+}$. Chem. Phys. **131:** 481–488.
10. RIBOU, A.-C. 1994. Interaction électronique dans des complexes à valence mixte. Application à la conception de composants moléculaires. Thesis Paul Sabatier University, Toulouse.
11. GOURDON, A. 1992. Molecular switching from conformational change. New J. Chem. **16:** 953–957.

12. RETTIG, W. 1986. Charge separation in excited states of decoupled systems. Angew. Chem. Int. Ed. Engl. **25:** 971–988 and references therein.
13. HATZIDIMITRIOU, A., A. GOURDON, J. DEVILLERS, J.-P. LAUNAY, E. MENA & E. AMOUYAL. 1996. Photoinduced electron transfer in pentaamineruthenium(II) complexes of 1-(4-cyanophenyl)imidazole. Inorg. Chem. **35:** 2212–2219.
14. MARVAUD, V. & J.-P. LAUNAY. 1993. Control of intramolecular electron transfer by protonation: oligomers of ruthenium porphyrins bridged by 4,4′-azopyridine. Inorg. Chem. **32:** 1376–1382.
15. LAINE, P., V. MARVAUD, A. GOURDON, J.-P. LAUNAY, R. ARGAZZI & C.-A. BIGNOZZI. 1996. Electron transfer through norbornadiene and quadricyclane moieties as a model for molecular switching. Inorg. Chem. **35:** 711–714.
16. GONZALO, V., C. COUDRET, C. PATOUX & J.-P. LAUNAY. Modulation of intramolecular electronic coupling in 2,6-diferrocenylanthracene via a Diels-Alder reaction with TCNE. 1998. Manuscript in preparation.
17. HAGA, M., T. ANO, K. KANO & S. YAMABE. 1991. Proton-induced switching of metal-metal interactions in dinuclear ruthenium and osmium complexes bridged by 2,2′-Bis(2-pyridyl)bibenzimidazole. Inorg. Chem. **30:** 3843–3849.
18. HAGA, M., M. MESER ALI, S. KOSEKI, K. FUJIMOTO, A. YOSHIMURA, K. NOZAKI, T. OHNO, K. NAKAJIMA & D. J. STUFKENS. 1996. Proton-induced tuning of electrochemical and photophysical properties in mononuclear and dinuclear ruthenium complexes containing 2,2′-bis(benzimidazol-2-yl)-4,4′-bipyridine. Inorg. Chem. **35:** 3335–3347.
19. PATOUX, C., C. COUDRET, J.-P. LAUNAY, C. JOACHIM & A. GOURDON. 1997. Topological effects on intramolecular electron transfer via quantum interference. Inorg. Chem. **36:** 5037–5049.
20. RICHARDSON, D. E. & H. TAUBE. 1983. Electronic interaction in mixed-valence molecules as mediated by organic bridging groups. J. Am. Chem. Soc. **105:** 40–51.
21. ARRHENIUS, T.S., M. BLANCHARD-DESCE, M. DVOLAITZKY, J.-M. LEHN & J. MALTHETE. 1986. Molecular devices: Caroviologens as an approach to molecular wires. Synthesis and incorporation into vesicles membranes. Proc. Natl. Acad. Sci. USA **83:** 5355–5359.
22. BARZOUKAS, M., A. FORT & M. BLANCHARD-DESCE. 1997. Chain length dependence of the quadratic hyperpolarisability in two series of push-pull polyenes: a two-form model. New J. Chem. **21:** 309–316.
23. MARCH, J. 1985. Advanced Organic Chemistry. Third edition. Wiley Interscience. New York, NY. (For a general survey of organic chemistry)
24. HRADSKY, A., B. BILDSTEIN, N. SCHULER, H. SCHOTTENBERGER, P. JAITNER, K.-H. ONGANIA, K. WURST & J.-P. LAUNAY. 1997. New soluble bis[nona-, octa-, and pentamethylferrocenes] as "molecular wires" with a metal-to-metal distance of up to 40Å. Organometallics **16:** 392–402.
25. BOLGER, J., A. GOURDON, E. ISHOW & J.-P. LAUNAY. 1995. Stepwise Syntheses of mono-and di-nuclear ruthenium tpphz complexes $[(bpy)_2Ru(tpphz)]^{2+}$ and $[(bpy)_2Ru(tpphz)Ru(bpy)_2]^{4+}$ {tpphz = tetrapyrido[3,2–a : 2′, 3′–c : 3″, 2″–h : 2″, 3‴ –j]phenazine}. J. Chem. Soc. Chem. Commun. 1799–1800.
26. ROBERTS, R.M.G. 1990. Synthesis of ferrocenyl derivative of anthracene. J. Organomet. Chem. **388:** 181–186.
27. WOITELLIER, S., J.-P. LAUNAY & C.W. SPANGLER. 1989. Intervalence transfer in pentaamineruthenium complexes of α,ω-dipyridyl polyenes. Inorg. Chem. **28:** 758–762.
28. MCMURRY, J.E. 1989. Carbonyl-coupling reactions using low-valent titanium. Chem. Rev. **89:** 1513–1524.

29. HUTCHINGS, R.H. & A.I. MEYERS. 1996. An oxazoline-mediated synthesis of the pyrrolophenanthridine alkaloids and some novel derivatives. J. Org. Chem. **61:** 1004–1013. SAEDNYA, A. & H. HART. 1996. Two efficient routes to m-terphenyls from 1,3-dichlorobenzenes. Synthesis 1455–1458.
30. EGLINTON, G. & W. MCCRAE. 1963. The Coupling of Acetylenic Compounds. Adv. Org. Chem. **4:** 225–328.
31. STILLE, J.K. 1986. The palladium-catalyzed cross-coupling of organotin reagents with organic electrophiles. Angew. Chem. Int. Ed. Engl. **25:** 508–524.
32. MIYAURA, N. & A. SUZUKI. 1995. Palladium-catalyzed cross-coupling reactions of organoboron compounds. Chem. Rev. **95:** 2457–2483.
33. KNOCHEL, P. & R.S. SINGER. 1993. Preparation and reactions of polyfunctional organozinc reagents in organic synthesis. Chem. Rev. **93:** 2117–2188.
34. XU, Z., M. KAHR, K. L. WALKER, C. L. WILKINS & J. S. MOORE . 1994. Phenylacetylene dendrimers by the divergent, convergent and double-stage convergent methods. J. Am. Chem. Soc. **116:** 4537–4550.

The Electrical Measurement of Molecular Junctions

M. A. REED, C. ZHOU, M. R. DESHPANDE, AND C. J. MULLER

Center for Microelectronic Materials and Structures, Yale University, P.O. Box 208284, New Haven, Connecticut 06520-8284, USA

T. P. BURGIN, L. JONES II, AND J. M. TOUR

Department of Chemistry and Biochemistry, University of South Carolina, Columbia, South Carolina 29208, USA

ABSTRACT: We present the investigation of the electrical transport of metal/(organic molecule or monolayer)/metal junctions. Utilizing a novel mechanically controllable break junction to form a statically stable system, we have self-assembled molecules of benzene-1,4-dithiol onto two facing gold electrodes allowing for direct observation of charge transport through the molecules. Current-voltage I(V) measurements provides a quantitative measure of the conductance of a junction containing a single molecule. We have also created a technique to form well-defined, stable, and reproducible metallic contacts to a self-assembled monolayer of 4-thioacetylbiphenyl with nanoscale area. Electronic transport measurements show a prominent rectifying behavior arising from the asymmetry of the molecular heterostructure. Variable-temperature measurements reveal the dominant transport mechanisms, such as thermionic emission for the Ti-organic system. These techniques demonstrate the capability of electrically characterizing and engineering conductive molecular systems for future potential device applications.

The measurement of charge transport in single organic molecules, and the determination of their conductance, is a long-sought goal. Such measurements are experimentally challenging and intriguing since one can test the validity of transport approximations at the molecular level. A conceptually simple configuration would be to connect a single molecule between metallic contacts. Such a metal-molecule-metal configuration would present the molecular embodiment of a system analogous to a quantum dot,[1–9] with the potential barriers of the semiconductor system replaced by any existing contact barrier of the molecule/metal interface. Previous measurements on atomic and molecular systems have been done by scanning tunneling microscopes (STM),[10–12] and can yield conductivity information.[13–15] Experiments with an evaporated-metal-top contact/molecules/metallic-bottom-contact configuration, which has ten of thousands of parallel active molecules, have also been demonstrated.[16,17] One experiment on an organic system[18] reported evidence for Coulomb charging. We have performed measurements in the configuration of a single molecule between metallic contacts; specifically on benzene-1,4-dithiolate connected between stable proximal metallic gold contacts, at room temperature. This approach complements previous approaches by presenting statically stable contacts, and concurrently restricts the number of active molecule(s) to as few as one.

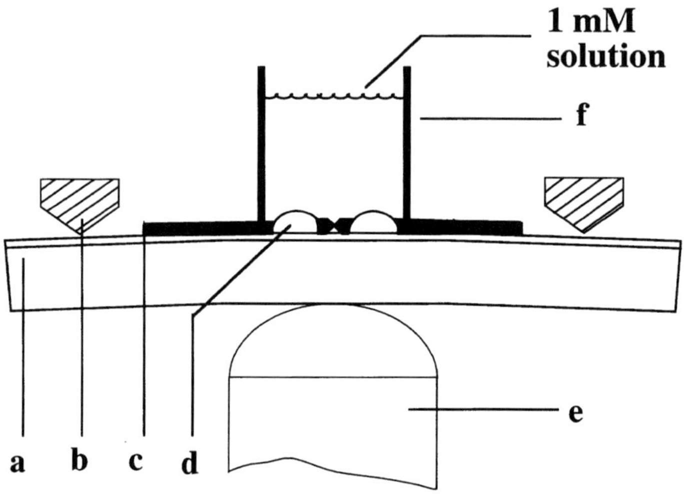

FIGURE 1. A schematic of the MCB junction with (a) the bending beam, (b) the counter supports, (c) the notched gold wire, (d) the glue contacts, (e) the piezoelement, and (f) the glass tube containing the solution.

Experiments were conducted at room temperature using a mechanically controllable break junction (MCB)[19] (FIG. 1). In this approach, a notched metal wire is glued onto a flexible substrate and is fractured by bending the substrate, after which an adjustable tunneling gap can be established. A large reduction factor between the piezo elongation and the electrode separation ensures an inherently stable contact or tunnel junction. The wire contacts are atomically sharp when broken, demonstrated in the conductance quantization as previously reported.[20] In the experiments reported here, benzene-1,4-dithiol was adsorbed from a 1-millimolar (mM) solution in tetrahydrofuran (THF) onto the two facing gold electrodes of the break junction broken in solution under an Ar atmosphere,[21] resulting in formation of a self-assembled monolayer (SAM) on the gold electrodes nearly perpendicular to the surface.[21] The THF solvent was allowed to evaporate in the ambient Ar atmosphere prior to the conductance measurements, and there was no further surface preparation or cleaning. The removal of the THF led to thermal gradients, which disturbed the picometer static dimensional stability of the MCB, requiring the tips to be withdrawn and then returned to measure the electrical properties of the molecule(s) adsorbed on the surfaces (FIG. 2). The configuration of FIGURE 3 is probable, since the displacement of thiols has been shown,[10] and the formation of a disulfide bridge would require oxygen.[21]

As the tips were brought together, current-voltage I(V) and conductance G(V) (= dI/dV) measurements showed characteristic features (FIG. 4A) which proved to be highly reproducible (FIG. 4B). The spacing between the electrodes is ~8 Å, set by the piezo voltage as determined by previous calibration of the spacing-to-piezo voltage conversion factor established by the exponential dependence of the current with distance in the tunneling regime.[22] However, calibration shift due to the solvent

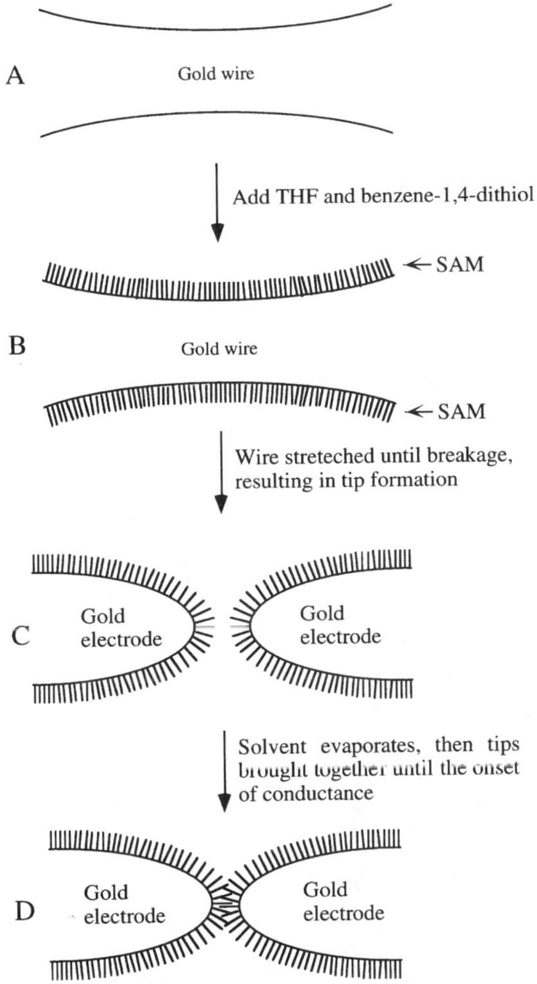

FIGURE 2. Schematic of the measurement process. (A) The gold wire of the break junction, prior to breaking and tip formation; (B) after adding benzene-1,4-dithiol, a SAM forms on the gold wire surface; (C) after mechanically breaking the wire in solution, which produces two opposing gold contacts that are SAM-covered; (D) after the solvent is evaporated, the gold contacts are slowly moved together until the onset of conductance is achieved. Steps C to D (without solution) can be repeated numerous times to test for reprodicibility.

evaporation cannot be eliminated. By comparison, an approximate molecule length of 8.46 Å was calculated using an MM2 force field and measuring to the center of the two gold radii minus the covalent radii of both gold atoms. An apparent gap of ~0.7 V was observed in all cases. The first derivative of I(V) shows two steps in both

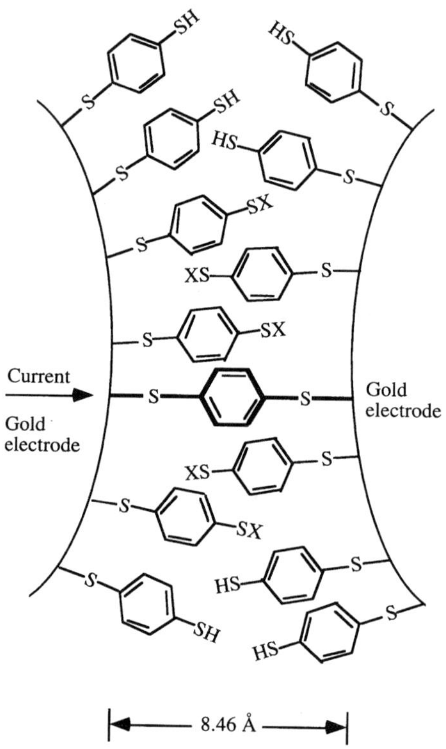

FIGURE 3. A schematic of a benzene-1,4-dithiolate SAM between proximal gold electrodes formed in a MCB. The thiolate is normally H-terminated after deposition; end-groups denoted as X can be either H or Au, the Au potentially arising from a previous contact/retraction event. These molecules remain nearly perpendicular to the Au surface, making other molecule orientations unlikely.[21]

bias directions with the lower step ~22.2 Mohm (0.045 μS) and the higher step ~13.3 Mohm (0.075 μS), possibly indicative of a Coulomb staircase. It is noted that the high Fermi energy of the gold contacts (~2 eV) as compared to the low energies of semiconductor quantum dot systems (<100 meV) precludes the observation of negative differential resistance in the present system,[23] which is often seen in semiconductor systems.[1–7] A control experiment with unevaporated THF solvent alone (that is, without the benzene-1,4-dithiolate) exhibited a resistance of 1-2 Gohm (linear up to 10V independent of electrode spacing, implying ionic conduction through the solvent. When the solvent was evaporated, regular vacuum tunneling with a much higher resistance was observed,[19] with exponential dependence of the current with applied voltage implying the absence of deleterious effects on the MCB due to the solvent. The first step for these three measurements gives resistance values of 22.2, 22.2, and 22.7 Mohm (top to bottom); the next step gives resistance values of 12.5, 13.3, and 14.3 Mohm. This is compared to a resistance of ~9 Mohm[11] and 18 ± 12 Mohm[12] deduced from measurements of an ensemble of similar molecules contacted to a gold nanocrystal, and a calculated resistance of this system of 100 Kohm.[24] A resistance greater than ~22 Mohm was not observed in our measure-

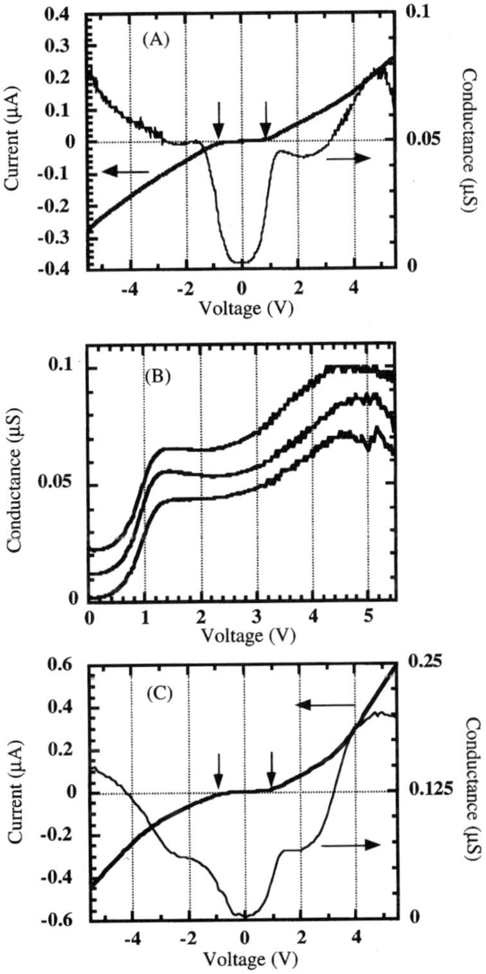

FIGURE 4. **(A)** Typical I(V) characteristics, which illustrate a gap of 0.7 V, and the first derivative G(V), which shows a step-like structure. **(B)** Three independent G(V) measurements, offset for clarity, illustrating reproducibility of the conductance values. The measurements were done using the same MCB, but for different retractions/contacts and thus different contact configurations. Offsets of 0.01 μS for the middle curve and 0.02 μS for the top curve are used for clarity. The first step for these three measurements gives values of 22.2, 22.2, and 22.7 Mohm (*top to bottom*); the next step gives values of 12.5, 13.3, and 14.3 Mohm. The middle curve represents the same data as in **(A)**. **(C)** An I(V) and G(V) measurement illustrating conductance values approximately twice that of the observed minimum conductance values. Resistances of ~14 Mohm for the first step, and 7.1 Mohm (negative bias) and 5 Mohm (positive bias) for the second step were measured.

ments; however, resistances less than this maximum were occasionally observed. FIGURE 4C shows I(V) and G(V) measurements of one singular observation that gave resistances which are approximately half the value of the maximum resistances (using averages, 0.63 and 0.45, respectively). This suggests a configuration of two noninteracting self-assembled molecules in parallel, and substantiates that the threshold resistance of a single molecule is ~22 Mohm, and compares with the previously deduced value of 18 ± 12 Mohm of a similar system.

One interpretation of the observed gap around zero voltage is as a Coulomb gap. Using the "apparent" Coulomb gap, an experimental capacitance of 1.1×10^{-19} F is obtained. Although the charge transport through the molecule is in principle a many-body effect, as a first step one can estimate the capacitance of the aryl group with a crude model. Using a 4.5 Å metallic sphere bound 2.0 Å from proximal metallic planes with intervening vacuum barriers gives a capacitance of 0.4×10^{-19} F, compared to the experimentally derived 1.1×10^{-19} F. However, a definitive demonstration of Coulomb blockade would require a third gate electrode, which is problematic in the present configuration because of the inability of placing a third proximal probe near the molecule. A second interpretation of the observed gap is due to the mismatch between the contact Fermi level and the LUMO (lowest unoccupied molecular orbital). Preliminary calculations using this interpretation give similar characteristics to the experimentally observed data.[25] The reproducibility of the minimum conductance at a consistent value implies that the number of active molecules could be as few as one. A better theoretical understanding of the threshold resistance of this system, either the apparent Coulomb gap derived from the capacitance of a single molecule configuration or the determination of the contact Fermi level-LUMO gap alignment, is needed to compare to the experimental values of ~22 Mohm and ~0.7 V, respectively.

The determination of the effective molecule-metal barrier height in the above experiments requires temperature-dependent measurements, which are not possible in the present configuration because of the temperature sensitivity of the MCB junctions. STM has been used to resolve images of single molecules;[10] however, these experiments could not give the absolute value of the conductivity because of the involvement of a tunneling gap between the STM tip and the molecule. Reifenberger *et al.* has developed a method to deduce the zerobias conductivity of an ensemble of molecular wires;[11,12] nevertheless the bias range applied across the molecules is small and hence no information concerning current-voltage spectroscopy has been derived. Direct contacting to organic thin films has been executed;[16] however, multilayers of the molecular wires and micron scale area of the devices containing a large number of molecules have been used, which complicates the analysis of the transport mechanism of single molecules. Direct contacting to and electrical measurements of a single monolayer of a small number of molecules have not yet been demonstrated.

Here we introduce a novel fabrication technique to directly measure the conduction through a small number of organic molecules. These devices consist of a self-assembled monolayer of conjugated molecular wires sandwiched between top and bottom metallic contacts. This technique guarantees good control over the device area and intrinsic contact stability. Two essential features are included in the process, the first of which is the employment of a nanoscale device area. The area is

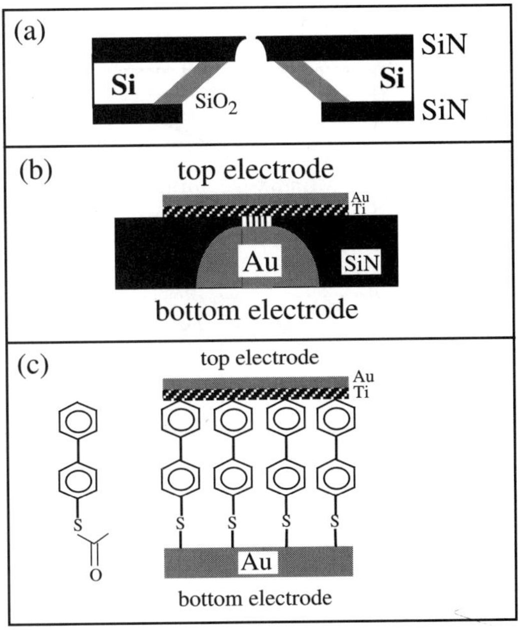

FIGURE 5. Fabrication of the heterostructure. **(a)** Cross section of a silicon wafer showing the bowl-shaped pore etched in suspended SiN membrane with a diameter about 300 A. **(b)** Au-Ti top electrode/self-assembled monolayer/Au bottom electrode sandwich structure in the nanopore. **(c)** 4-thioacetylbiphenyl and detail diagram of the sandwich heterostructure.

made to be smaller than the domain size of the SAM and thus the adsorbed organic layer is highly ordered and mostly defect-free. The second feature is that during the deposition of the top contact, several measures are taken to ensure that the deposited metal atoms accumulate at the SAM surface and do not penetrate into the organic layer. These methods also minimize damage to the SAM during the deposition.

A schematic diagram of the device is shown in FIGURE 5. E-beam lithography followed by plasma etching are used to open a pore in the suspended LPCVD SiN membrane.[26] The pore takes a bowl shape with the opening at the upper edge having diameter of ~30 nm, as shown in FIGURE 5(a). Gold is evaporated from the bottom side to fill the pore. The typical crystallite size for gold is about 50 nm, and thus the upper surface of gold in the nanopore is probably a single crystallite. The samples are then immediately immersed into a 0.3 millimolar solution of the organic molecules for three days under argon atmosphere. The molecules we use are 4-thioacetylbiphenyl, as shown in FIGURE 5(c) with a length about 12 Å. It has been demonstrated that under such conditions the functionalized organic molecules would self-assemble onto the gold surface and form a highly ordered monolayer.[21] After the formation of the SAM, the samples are taken out of the solution and loaded into a high vacuum chamber of 10^{-8} torr to deposit the top electrodes. First 10 Å Ti

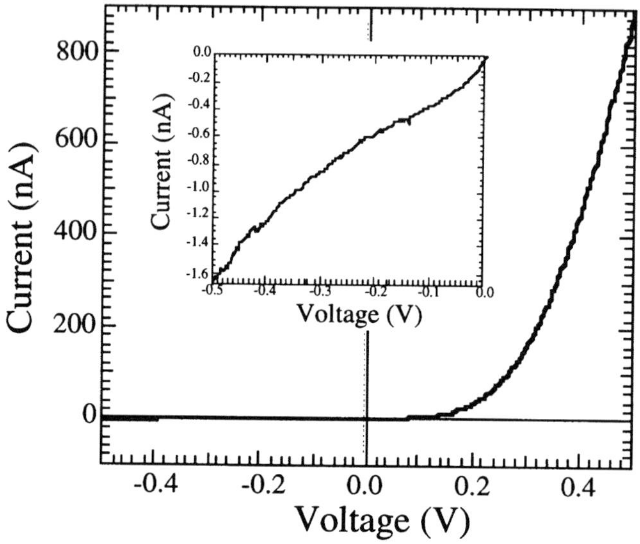

FIGURE 6. Diode-like I-V characteristics at room temperature. *Inset*: magnified view of the negative bias part.

is deposited onto the upper surface of the SAM. This amount of Ti would bond onto the SAM surface and form a continuous film.[27] The samples are then maintained at low temperature while 30 Å Ti and 800 Å Au are deposited onto the upper side of the samples, which are thick enough to make the top electrodes continuous and well conducting. This low-temperature deposition helps to minimize the thermal damage to the SAM. A very low deposition rate (~0.3 Å/s) is also used for the same reason. Detailed schematics of the sandwiched heterostructure are shown in FIGURES 5(b) and (c). The samples are then allowed to warm up to room temperature slowly (~24 hr) and ready for measurements. This technique gives us a yield of about 80% working devices. These devices are stable at room temperature and vary only slightly after thermal cycling to ~50 K.

Two terminal current (I)-voltage (V) characteristics are measured in the temperature range from room temperature down to 57 K. FIGURE 6 shows the I-V curves of the device measured at room temperature. Positive bias corresponds to electrons emitted from the Au/Ti top electrode and collected by the bottom Au electrode. I-V characteristics under negative bias are shown in the inset of FIGURE 6. Prominent rectifying behavior is observed: the current at –1 V bias is about 500 times higher than the current at –1 V bias. While the I-V curve at negative bias is rather linear, the I-V curve at positive bias displays exponential behavior. For different devices, the current magnitude varies slightly due to variation in device area, but the rectifying behavior is always observed. The current is observed to decrease monotonically with temperature under both the positive and negative biases. At 1 V positive bias the current experiences about four orders of magnitude decrease as temperature decreases from room temperature to 57 K, suggesting thermally activated transport mechanism. For thermally activated electron injection from the metal into the SAM, the image force potential causes a lowering of the barrier height and the formula accounting for both thermal activation and image force potential is given by[28,29]:

$$I = Ae^{\frac{a\sqrt{V} - q\Phi}{kT}}, \text{ where } a = \frac{q}{2}\sqrt{\frac{q}{4\pi\varepsilon_i\varepsilon_0 d}} \qquad \textbf{(1)}$$

where A is the effective Richardson constant multiplied by the current injection area, q the electron charge, Φ the thermal emission barrier height, V the bias, k Boltz-

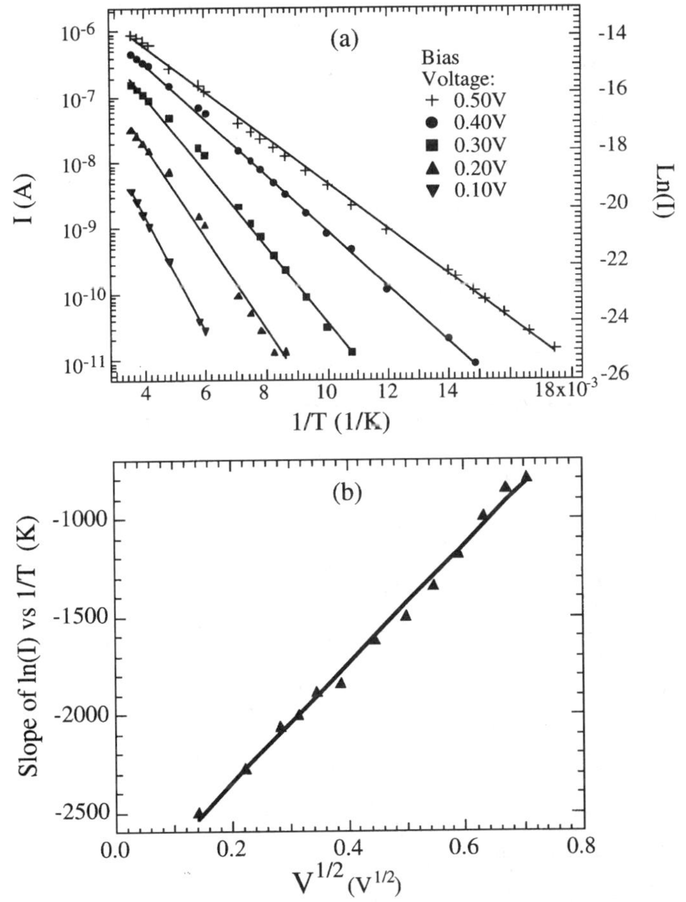

FIGURE 7. **(a)** Plots of I vs. $1/T$ for a series of positive bias voltages. The straight lines are least square fits to the data points for each bias. **(b)** Plot of the slope of the fitted lines in **(a)** vs. $V^{1/2}$. The slope and intercept of the fitted straight line give the value of a and barrier height.

mann's constant, T the temperature, ε_i the relative dielectric constant of the SAM, ε_0 the vacuum dielectric constant, and d the thickness of the molecular dielectric film. FIGURE 7(a) displays the plots of ln(I) versus 1/T for different biases, where the marks represent data points and the lines are least-square fits for each bias. The plots clearly demonstrate that ln(I) is linear with 1/T and the slope of the lines is a function of the bias voltage. The standard thermal emission theory predicts:

$$\text{slope} = \frac{a\sqrt{V} - q\Phi}{k} \qquad \textbf{(2)}$$

One can plot the slope of each line in FIGURE 7(a) versus $V^{1/2}$ and derive the constants a and $q\Phi$, as shown in FIGURE 3(b). The data points are found to lie on a straight line and the least-square fitting yields a value of 0.26 eV for $q\Phi$ and 0.263 $\text{eV}^{1/2}$ for a. Given the film thickness of 10 Å as measured from the ellipsometry data, from formula **(1)**, we can further derive the dielectric constant of the SAM to be 4.2.

The negative bias corresponds to electrons injected from the Au bottom electrode into the conjugated molecules through the thiolates. The current also decreases with temperature; however, detailed analysis reveals that the data fit into the hopping conduction mechanism, which corresponds to electrons hopping from one site to the neighboring ones.[28,30] For different biases, plots of ln(I/V) vs. 1/T fall on one line, as shown in FIGURE 8. The least-square fitting yields a hopping barrier of 0.19 eV. At present it is unclear whether the hopping is related to defects in the SAM, hopping

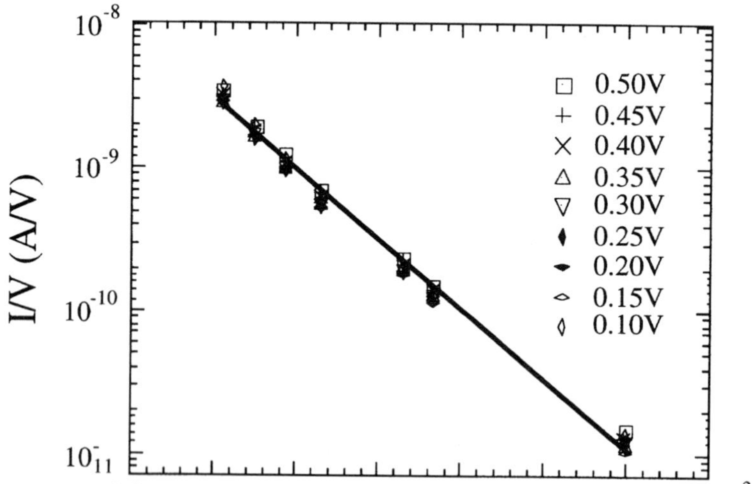

FIGURE 8. Plot of I/V vs. 1/T for different negative biases. The data points collapse onto one line, indicating hopping conduction.

along the molecular wire, or hopping between neighboring molecular wires. Gold is known to have a higher work function than Ti; despite the lack of knowledge about the interface states and the role of sulfur atoms, these results suggest that the barrier for electrons injected from the bottom gold electrode is higher than the barrier for electrons injected from the top Ti electrode. As a result, no prominent thermal emission is observed under negative bias and the current is carried by electrons hopping through the molecular layer instead of electrons thermally activated over the Au/SAM barrier.

CONCLUSION

We have demonstrated new fabrication techniques to study the transport mechanism of organic molecular wires. MCBs and nanoscale fabrication with special metal deposition techniques are employed to provide metallic contacts to the self-assembled monolayer of a small number of conjugated molecular wires. In one case, we have measured the conductivity of a single molecule. In another, because of the asymmetry in the heterostructure, prominent rectifying behavior is observed in the I-V characteristics. These methods can be easily adapted to other molecular wire systems for determination of transport mechanism and band alignment.

ACKNOWLEDGMENTS

We acknowledge DARPA for financial support. We thank M. Rooks and CNF for their help.

REFERENCES

1. REED, M.A. *et al.* 1988. Observation of discrete electronic states in a zero-dimensional semiconductor nanostructures. Phys. Rev. Lett. **60:** 535–538.
2. MEIRAV, U. *et al.* 1990. Single electron charging and periodic conductance resonances in GaAs nanostructures. Phys. Rev. Lett. **65:** 771–774.
3. DELLOW, M. *et al.* 1992. Resonant tunneling through the bound states of a single donor atom in a quantum well. Phys. Rev. Lett. **68:** 1754–1757.
4. KOUWENHOVEN, L.P. *et al.* 1991. Single electron charging effects in semiconductor quantum dots. Z. Phys. B **85:** 367–373.
5. TEWORDT, M. *et al.* 1992. Resonant tunneling in an AlGaAs/GaAs quantum dot as a function of magnetic field. Phys. Rev. B **46:** 3948–3952.
6. SU, B. *et al.* 1992. Single-electron tunneling in nanometer-scale double-barrier heterostructure devices. Phys. Rev. Lett. **46:** 7644–7647.
7. KLEIN, D.L. *et al.* 1996. An approach to electrical studies of single nanocrystals. Appl. Phys. Lett. **68:** 2574–2576.
8. RALPH, D.C. *et al.* 1997. Gate-voltage studies of discrete electronic states in aluminum nanoparticles. Phys. Rev. Lett. **78:** 4087–4090.
9. GRABERT, H. *et al.* 1991. Single Electron Tunneling. Plenum. New York.
10. BUMM, L.A. *et al.* 1996. Are single molecular wires conducting. Science **271:** 1705–1707.
11. DOROGI, M. *et al.* 1995. Room temperature Coulomb blockade from a self-assembled molecular nanostructure. Phys. Rev. B **52:** 9071–9077.

12. ANDRES, R.P. *et al.* 1996. Coulomb staircase at room temperature in a self-assembled molecular nanostructure. Science **272:** 1323–1325.
13. CROMMIE, M.F., C.P. LUTZ & D.M. EIGLER. 1993. Spectroscopy of a single adsorbed atom. Phys. Rev. B **48:** 2851–2854.
14. JOACHIM, C. *et al.* 1995. Electronic transparence of a single C_{60} molecule. Phys. Rev. Lett. **74:** 2102–2105.
15. YAZDANI, A. *et al.* 1996. Off-resonance conduction through atomic wires. Science **272:** 1921–1924.
16. FISCHER, C.M. *et al.* 1995. Microstructured gold/Langmuir-Blodgett film/gold tunneling junctions. Appl. Phys. Lett. **66:** 3331–3333.
17. GRABERT, H. 1991. Single charge tunneling: A brief introduction. Z. Phys. B **85:** 319–325.
18. NEJOH, H. 1991. Incremental charging of a molecule at room temperature using the scanning tunneling microscope. Nature **353:** 640–642.
19. MULLER, C.J. *et al.* 1992. Experimental observation of the transition from weak link to tunnel junction. Physica C **191:** 485–504.
20. MULLER, C.J. *et al.* 1996. Quantization effects in the conductance of metallic contacts at room temperature. Phys. Rev. B **53:** 1022–1025 and references therein.
21. TOUR, J.M. *et al.* 1995. Self-assembled monolayers and multilayers of conjugated thiols, α,ω-dithiol, and thiolacetyl-containing adsorbates. understanding attachments between potential molecular wires and gold surfaces. J. Am. Chem. Soc. **117:** 9529.
22. ZHOU, C. *et al.* 1995. Microfabrication of a mechanically controllable break junction in silicon. Appl. Phys. Lett. **67:** 1160–1162.
23. MUJICA, V. *et al.* 1996. Current-voltage characteristics of molecular wires: Eigenvalue staircase, Coulomb blockade and rectification. J. Chem. Phys. **104:** 7296–7305.
24. SAMANTA, M.P. *et al.* 1996. Electronic conduction through organic molecules. Phys. Rev. B **53:** R7626–R7629.
25. DATTA, S. Unpublished material.
26. RALLS, K.S. *et al.* 1989. Fabrication of thin-film metal nanobridges. Appl. Phys. Lett. **55:** 2459–2461.
27. KONSTADINIDIS, K. *et al.* 1995. An in-situ X-ray photoelectron study of the interaction between vapor-deposited Ti atoms and functional groups at the surfaces of self-assembled monolayers. Surf. Sci. **338:** 300.
28. SZE, S.M. 1981. Physics of Semiconductor Devices, 2nd ed.: 403. Wiley. New York.
29. LAMB, D.R. 1967. Electrical Conduction Mechanisms in Thin Insulating Films: 47. Methuen. London.
30. OKADA, K. *et al.* 1997. Electrical stress-induced variable range hopping conduction in ultrathin silicon dioxide. Appl. Phys. Lett. **70:** 351–353.

Probing Electronic Properties of Conjugated and Saturated Molecules in Self-Assembled Monolayers

P.S. WEISS,[a] L.A. BUMM,[a] T.D. DUNBAR,[a] T.P. BURGIN,[c] J.M. TOUR,[c]
AND D.L. ALLARA[a,b]

Department of Chemistry and [b]Department of Materials Science, The Pennsylvania State University, University Park, Pennsylvania 16802-6300, USA
[c]Department of Chemistry and Biochemistry, University of South Carolina, Columbia, South Carolina 29208, USA

ABSTRACT: We have measured the electron transport properties of single and bundled molecules imbedded in crystalline monolayers of alkanethiols self-assembled on Au{111} surfaces. These measurements have been performed using scanning tunneling microscopy and microwave difference frequency alternating current scanning tunneling microscopy on molecules with saturated and fully conjugated chains. We have learned to separate the contributions of the molecules and of the tunneling junction. We have also found a means to determine the value of the contact conductance due to the choice of chemical connection to the substrate.

INTRODUCTION

Many quantitative measurements of electron transport efficiencies through molecular frameworks have been made.[1–34] These have generally applied optical spectroscopy,[4–9] electrochemistry,[10–15] or direct electronic measurements[16–25] to large ensembles of molecules. The values obtained are thus ensemble averages. Measurements of electron transport through individual molecules are more difficult. Small ensembles of molecules have been measured using break junctions,[17] nanolithographically defined pores,[18] nanometer-scale contacts,[19] and nanometer-scale networks.[20] Carbon nanotubes are particularly amenable to single molecule measurements because of their extreme lengths (microns). Measurements have been made of the molecular electronic properties on single-wall,[21] multi-wall,[22–24] and ropes of single-wall[25] carbon nanotubes.

Scanning tunneling microscopy can be used to probe surface electronic porperties with molecular resolution. We and others have used scanning tunneling microscopy to measure electron transport through individual molecules[26–30] and through atomic assemblies.[31] Atomic force microscopy also can be used when combined with a conducting tip.[24,32]

Scanned probe techniques have the advantage that the measurements are intrinsically localized to nanometer-scale areas. With them we can both spatially select and make direct electronic measurements of the molecules supported on a substrate. If the molecules are sufficiently separated, the scanning tunneling microscope (STM) can be used to select a single molecule for study. In the STM, electrons tunnel through the gap between the tip and the sample. The tip-sample distance is typically adjusted to maintain a constant tunneling current. The tip is then scanned across the surface and the height of the tip is recorded.[35] Such a plot of the tip height versus lateral position is an STM image (constant current STM topograph). Such an image represents a surface of constant tunneling current convolving the electronic and physical structures of both the tip and the sample.[36–37] Unambiguously separating these contributions to the STM topography can only be achieved with the knowledge of one of these components. We do this by preparing well-defined physical structures of known molecular composition.

We are able to measure alkyl chain transconductance for a range of chain lengths and the contribution of the chemical contacts to this conductance. These experiments are performed as pairwise (or higher) combinations of molecular systems so as to eliminate the contributions and perturbations of the measurement systems which here is the STM. We are also able to measure single isolated molecules in a well-defined matrix.[26] This circumvents the difficulty encountered in electrodes defined by nanometer scale contacts[17–19] and networks[20] in that no inference need be made as to the contents of the measurement junction. Rather, the contents are identified and selected directly by the STM.[26,27]

Molecular electronic components and devices[38–44] will require optimized connections, conductivities, and insulation so as to minimize the dissipation of the small amount of current or charge expected to be used. In addition, measurements of the transconductances of contacts and of molecules will give us key insight into the electronic couplings within and between molecules and solids. We describe our initial measurements along these lines and the directions these point for the future. We look at how to quantify the transconductances of a wide range of molecular wire candidates synthesized or proposed.[45–48] In this way we expect to be able to compare directly to the theoretical estimates of Datta,[34] Joachim,[49] Sautet,[32] and others.

BACKGROUND

Electron transport in bulk molecular materials can be characterized by the intensive quantity conductivity. The mechanism of bulk conductivity involves incoherent diffusive intra- and intermolecular electron transport where the electrons thermalize with the matrix and thus is dissipative. When these materials are reduced to their molecular components, what does conductivity mean? At the molecular scale, electron transport through the molecular framework can take place via several different mechanisms. Coherent nonresonant tunneling occurs when the electronic states within the framework are far above the energy of the tunneling electrons. The rate of electron transport is exponentially dependent on the length of the molecular bridge. Coherent resonant tunneling occurs when the energy of the tunneling electrons is resonant with the conduction band of the molecular framework. The rate of electron transport will obey the Landauer formula,[50,51] independent of the length of the molecular bridge. Neither of these mechanisms can be characterized by an *ohmic conductivity* in the usual sense. Therefore we will characterize the electron transport in terms of a transconductance: the ratio of the tunneling current to the bias voltage.

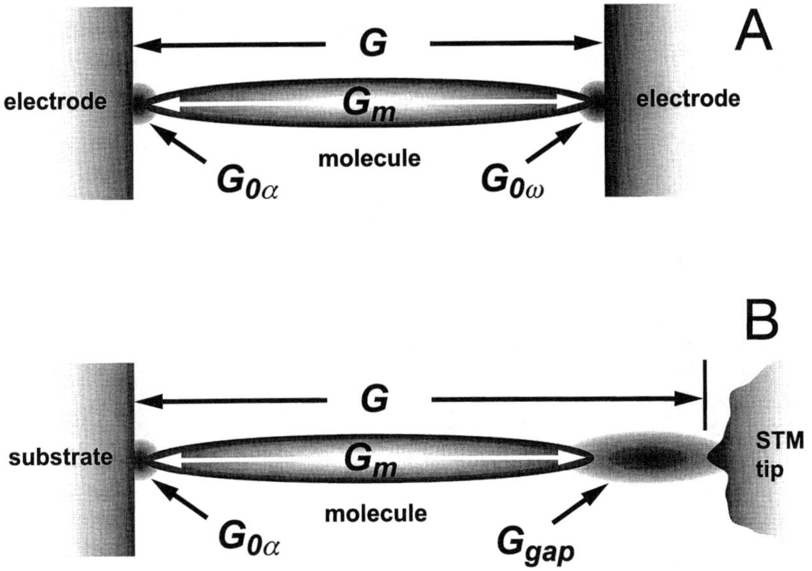

FIGURE 1. Schematic diagram of **(A)** a molecule between two electrodes and **(B)** a molecule attached to one electrode and probed by an STM probe tip.

We are working with molecules in the coherent nonresonant tunneling regime, where the LUMO is far above the energy of the tunneling electrons. Because the transconductance decreases exponentially with distance, this mechanism will only be important up to about 40 Å. Electron tunneling through vacuum also obeys an exponential distance dependence, but decays faster than transport through molecular frameworks. Curiously, in this regime even "insulating" molecules are better conductors than vacuum. The exception can occur for quantum interference in short molecular bridges when electron transfer proceeds through multiple paths each with a different phase. If the interference is destructive the transconductance will be lower than the corresponding vacuum gap.[36] This is observed in STM as contrast reversal—adsorbed molecules appear as depressions in the surface.[36,37]

For a molecule attached to two electrodes as in FIGURE 1A, the transconductance, G, of the molecule can be divided into three contributions, the contact conductance[49] at each end, $G_{0\alpha}$ and $G_{0\omega}$, and that of the molecular framework itself, G_m. Quantum mechanically, these represent the component probabilities for tunneling through the three regions of the barrier — the molecule-electrode and the through-molecule barriers, respectively. Because the transconductances are transport probabilities, the overall transconductance is the product of these components: $G = G_{0\alpha}G_{0\omega}G_m$. We can combine the contact conductances to a single term, G_0, and write G_m as $\exp(-\beta l)$, so that $G = G_0\exp(-\beta l)$, where β is the decay constant and l is the length. The contact conductance is then the transconductance in the limit of $l \rightarrow 0$. For the case of only one end of the molecule in direct contact with the electrode and the other at the variable-size vacuum gap of an STM probe tip, as in FIGURE 1B, $G_{0\omega}$ is then replaced by G_{gap}. The fall-off in the gap transconductance is also exponential, $G_{gap} = G_{0\omega}\exp(-\alpha d)$, where d is the gap width and α is the decay constant. In the limit of $d \rightarrow 0$, $G_{gap} = G_{0\omega}$. For this case the transconductance through the molecule can be written: $G = G_0\exp(-\beta l)\exp(-\alpha d)$. Once again the overall contact conductance G_0 combines the chemical contact at the electrode side and the coupling to the gap on the other.

Connecting the electrodes of a test probe to either end of a molecule is not only technically difficult, but the electronic effect of the contact to the molecule can be very large. The contact conductance depends on the electronic coupling within the contact. Therefore the chemical nature of the interaction of the adsorbate with the surface will have a profound effect. Specifically, the strong interaction present for chemisorption is predicted to yield high transconductance while the weaker interaction present for physisorption is predicted to yield low transconductance.[49,52] An additional bottleneck occurs within the electrode itself as it narrows toward the molecular nanometer-scale contact. Due to the lateral confinement of the electrons the number of states is reduced, resulting in a mismatch between the microscale and nanoscale regions where most of the electrons incident on the contact are reflected.[50,51]

EXPERIMENTAL DETAILS AND METHODS

Oriented thin-film Au{111} substrates were prepared using a procedure developed by Chidsey *et al.*[53] They were prepared by depositing 120 nm of 99.999% pure Au[54] at 0.1 nm/sec onto freshly cleaved muscovite mica[55] *in vacuo* with a base pressure less than 5×10^{-7} torr. The mica substrate had been ramped from room temperature to 340°C at the rate of 32°C/hr and held at the deposition temperature for 18 hours prior to, during, and 1 hour after deposition. The Au films were allowed to cool in vacuum for 18 hours.

The 1-alkanethiols — 1-decanethiol and 1-dodecanethiol — were obtained from Aldrich,[56] purified by vacuum distillation, and stored at −20°C until use. The synthesis of 2′-ethyl-4,4′-di(phenylene-ethynylene)benzothioacetate (PE3-Ac) is described elsewhere.[27,48] We designate these forms via the suffix added to the abbreviation, e.g., PE3-Ac is the thioacetate, PE3-H is the thiol, and PE3 is the thiolate.

The alkanethiolate SAMs were formed by immersing the Au/mica substrates in a 1 mM solution of the alkanethiol in absolute ethanol for 18–24 hr. The SAMs were rinsed sequentially in hexane, acetone, and ethanol and then blown dry with purified N_2.

The two component mosaic structures where prepared using a three step procedure. First a dodecanethiolate (**12**) SAM was prepared as described above. Second, the **12** SAM was immersed in neat 78°C absolute ethanol for 1 hr, then cooled and rinsed in absolute ethanol. Finally the SAM was immersed in a 1 mM solution of decanethiol for 6 hr at room temperature to add decanethiolate (**10**) to the monolayer. The sample then was rinsed and blown dry.[57]

The two-dimensional matrix isolation of the PE3 molecules was a two-step process. First a 1-alkanethiolate SAM, e.g., **10** or **12**, was prepared as already described.

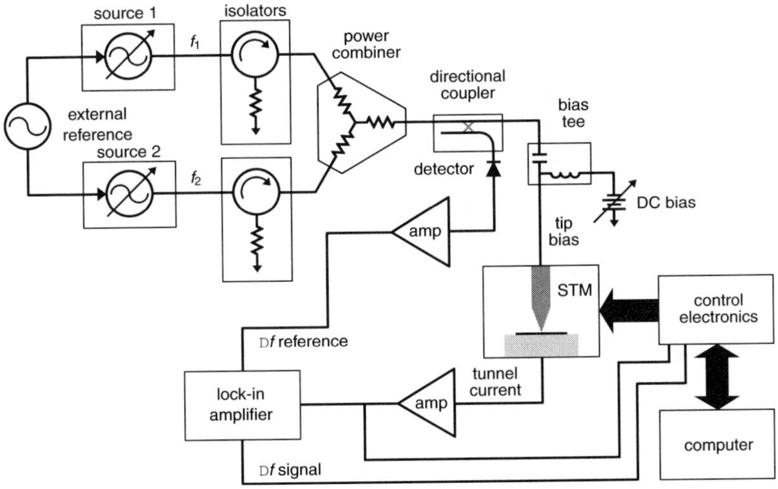

FIGURE 2. Schematic diagram of the ACSTM configured for microwave difference frequency detection. The microwave signals from two HP 83620A synthesized sweepers (source 1 at f_1 and 2 at f_2) are linearly combined using a resistive power combiner. A DC bias voltage is added with a microwave bias tee and applied to the ACSTM tip. The resultant applied bias voltage has frequency components at f_1, f_2, and DC. The difference frequency reference signal for the lock-in amplifier (at $\Delta f = |f_1 - f_2|$) is generated using a broadband Schottky diode (Narda 4507) which is coupled to the microwave signal line via a directional coupler. The tunnel current signal from the transimpedance amplifier is passed to the STM control electronics and to the lock-in amplifier (Stanford Research 850). The bandwidth of the STM feedback is set below Δf so that the servo loop will not respond to the Δf signal. The output of the lock-in amplifier is passed to the control electronics for data acquisition concurrent with DC STM topographic images. In order to achieve a stable difference frequency, the output frequencies of the microwave signal sources are synthesized from a common external 10 MHz reference oscillator. The isolators between each microwave source and the power combiner prevent heterodyne beating and other undesirable effects caused by the interaction of the reflected power with the gain control feedback of the source output amplifiers.

Then, the already formed SAM was immersed in a solution of PE3-H in tetrahydrofuran for 0.5 to 1 hr, rinsed, and blown dry. (*N.B.* due to the susceptibility of PE3-H to air oxidation, it was generated through the *in situ* hydrolysis of the thioacetate form by the addition of 6 µl of 30% NH_4OH per mg of PE3-Ac.[27]) Throughout this discussion we assume that the alkanethiols adsorb to the Au substrate as alkanethiolates.

We use a beetle-style STM[58,59] which has been specially constructed so that tunnel junction bias frequencies from DC to 20 GHz can be applied to the tunnel junction by means of a tunneling tip that is part of and fed by a coaxial transmission line (FIG. 2).[60–71] Any number of bias frequencies in this range can be simultaneously applied to the tunnel junction by means of a bias-tee and a resistive combiner. A simultaneous DC bias is provided by an HP 3245A Universal Source, while two different frequencies are provided by a phase-locked pair of HP 83620A synthesized microwave sources.[72] The tunneling current is detected using an Axon CV4 patch clamp transimpedance amplifier,[73,74] which enables us to acquire STM images at tunnel currents as low as 0.25 pA with a 2.1 kHz bandwidth.[75] Difference frequency signals on the unfiltered output of the transimpedance amplifier (bandwidth DC to 30 kHz) are detected using a Stanford Research model 850 lock-in amplifier[76] at a bandwidth of 100 Hz.

All SAMs were imaged by STM using constant current feedback.[77] Our images were acquired with a tunnel junction transimpedance of 100 GΩ or greater, where the tip is outside the SAM so that the monolayer is not perturbed (high tunnel junction transimpedance corresponds to large tip-sample separations). The STM is enclosed in a controlled atmosphere using a constant N_2 purge. The STM images shown are not filtered.[57,78]

MEASUREMENTS AND DISCUSSION

We can make molecularly resolved electron transport measurements of films in three ways:

(1) Derive the electron transport measurements from STM images of films with known physical structure by comparing to the STM constant current topography.

(2) Map differences in the electronic characteristics of the film using microwave difference frequency ACSTM while simultaneously acquiring a DC STM constant current topograph.

(3) Position the scanning probe directly over the molecular feature of interest and interrogate it by acquiring a series of current-voltage and current-distance curves.

As described below, we have measured the transconductance differences between *n*-alkanethiol chains of different lengths using the first method and contrasted conjugated and saturated chains using the second.

Saturated Chains

We prepare two-component mosaic SAMs where the patches of the mosaic are nearly pure regions of each component and the boundaries between these patches are molecularly sharp. Such well-defined physical structures are ideal for measuring electron transport. An STM constant current topograph is shown in FIGURE 3A of a mosaic SAM composed of **10** and **12**. Domains of **12** are present largely as well-ordered islands — remnants of the treatment in hot ethanol which dramatically reduces the defect density. The islands of **12** are surrounded by **10**, displaying defect densities typical of SAMs formed at room temperature. Interspersed molecules of **12** are observed in the regions of **10** and vice versa. The boundaries between the regions of **10** and **12** are well defined and are molecularly sharp.

A cross-section extracted from this image (the path indicated by the bar in FIG. 3A) is shown in FIGURE 3B. The path was chosen so as to run between two molecular terraces of **12**, passing over a monatomic depression in the Au substrate. The STM topographic height difference between the regions of **10** and **12** on the same Au terrace is 1.0 ± 0.2 Å. Note that the physical height difference between the regions is 2.2 Å.[79]

We can use the STM height difference, Δh_{STM}, of these molecular terraces to determine the difference in tip-film-substrate electron transport through each region. This is possible because the two molecules/regions can be imaged under identical conditions using the same STM tip. The height of a molecule measured by STM, h_{STM}, is not typically the same as the physical film thickness, h_{film} (e.g., from structural models[79]), because the STM tip-sample interaction convolves electronic and physical structure.[36,37,77] Thus, the interpretation of images requires knowledge of the specific sample. At the high tunnel junction transimpedances used for imaging the alkanethiolate molecular lattice, the tip is outside the film.[57] This results in a tunnel junction composed of two distinct layers (FIG. 4): the film, with a thickness and a transconductance determined by its intrinsic properties; and the tip-film gap, with adjustable transconductance determined by its separation, d_{gap}, which is controlled by the scanning tunneling microscope, and the apparent tunneling barrier height, an intrinsic property of the film.[77,85] The constant current STM topograph is a surface of constant tunneling current with height contributions from both the film and the tip-film gap: $h_{STM} = h_{film} + d_{gap}$. Note that we do not directly measure h_{STM} but rather Δh_{STM} between the molecular terraces of **10** and **12**.

For an understanding of how the alkyl chains contribute to the electron tunneling properties of SAMs we can turn to the large body of experimental and theoretical work which has addressed the distance dependence of electron transfer kinetics through σ-bonded hydrocarbon bridges discussed in the previous section. A general feature of these results under nonresonant tunneling conditions is that the transconductance decreases exponentially with increasing length. These measurements show that electrons tunnel through these bridges with a rate, i_t, the tunneling current, which is proportional to $\exp(-\beta l)$, where β ranges from 0.8 to 1.4 $Å^{-1}$ (1.0 to 1.8 per methylene bridge).[2,86] Theoretical calculations also predict that the interactions of the molecule with the electrodes (*viz.*, the tip and the substrate) have very strong effects on the overall transconductance of the substrate-molecule-tip system.[1,3,6,49,52] Hence the chemical nature of the interaction of the adsorbate with the surface is predicted to have a profound effect: chemisorption yields high transconductance while physisorption yields low transconductance.[49,52]

Using this background, we can gain insight into the interpretation of STM topographs by examining our two-layer tunnel junction model in FIGURE 4. The transconductances, G, across each layer are:

$$\text{vacuum gap: } G_{gap} = A\exp(-\alpha d_{gap})$$

and

$$\text{film: } G_{film} = B\exp(-\beta h_{film}),$$

where α and β are the respective decay constants, d_{gap} and h_{film} are the layer thicknesses, and the prefactors, A and B, are the respective contact conductances.[49] Because the transconductance is an electron tunneling probability, as opposed to a dissipative current flow, the transconductance of the composite two layer tunnel junction is the product of the transconductances of the individual layers, $G_t = G_{film}G_{gap}$. A thin film composed of two components (1 and 2) will have a different set of constants associated with each component. The scanning tunneling microscope operating with constant current feedback constrains G_t to be constant over the image, i.e.,

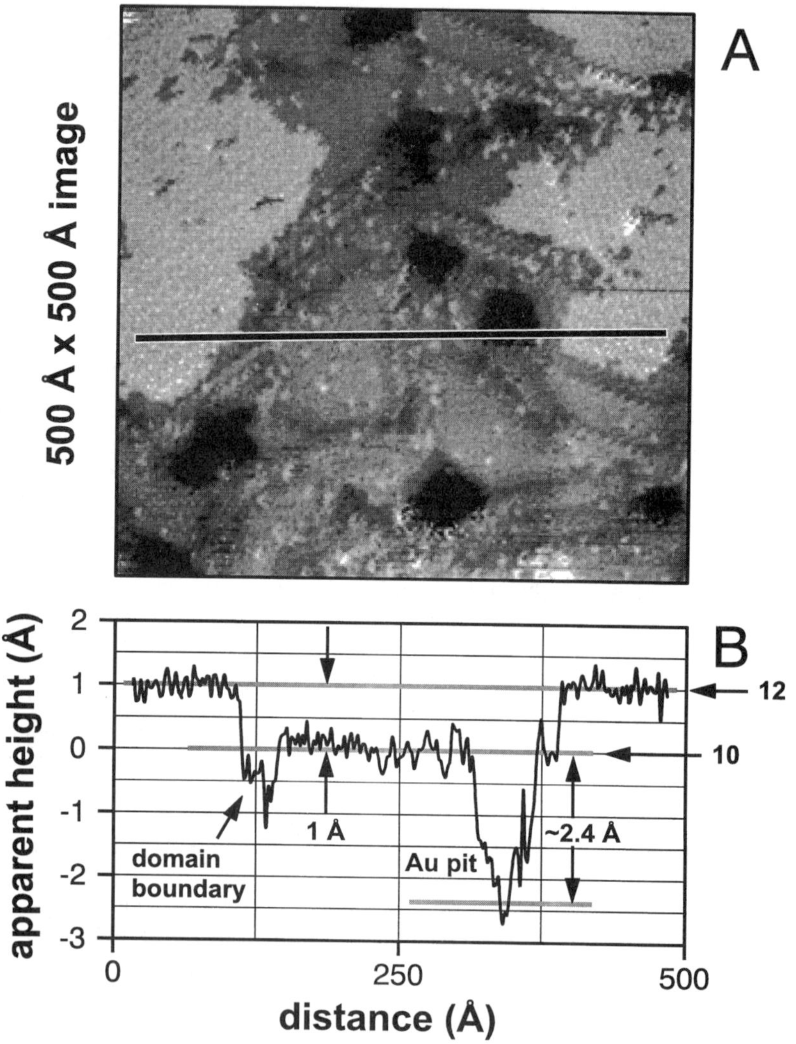

FIGURE 3. (A) STM image of a 500 Å × 500 Å area of the mixed composition mosaic SAM of **10** and **12** showing the alkanethiolate molecular lattice. The topographically higher (lighter) regions are domains of **12**. These are present largely as well-ordered islands (remnants of the treatment in hot ethanol) which dramatically reduces the defect density. The islands of **12** are surrounded by the topographically lower **10**, displaying domain sizes and defect densities typical of SAMs formed at room temperature. These defects are the topographically lower (shown as darker) spots are one atom deep depressions in the underlying Au{111} substrate (Au pits) and the alkanethiolate structural domain boundaries (also darker). Interspersed molecules of **12** are observed in the regions of **10** and vice versa. The boundaries between the regions of **10** and **12** are well defined and are molecularly sharp. The image was recorded with a tip bias of +1.0 V and a tunneling current of 10 pA.

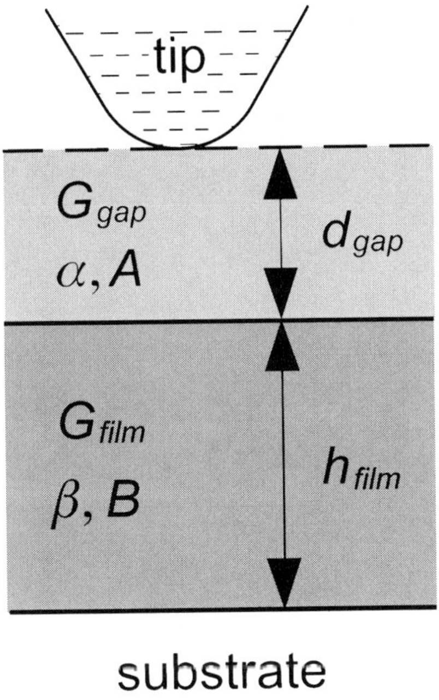

FIGURE 4. The two-layer tunnel junction model is shown. The STM tunnel junction is composed of two distinct layers: the vacuum gap and the film. Each region is characterized by a transconductance, G, which is dependent on the physical thickness of the layer. The film thickness is an intrinsic property while the vacuum gap thickness is controlled by the STM to maintain a constant overall transconductance.

$$G_t = G_{film1}G_{gap1} = G_{film2}G_{gap2} = i_t/e_t = \text{constant},$$

where e_t is the tunnel junction bias voltage. From our model,

$$\Delta h_{STM} \equiv \Delta h_{STM2} - \Delta h_{STM1} = (h_{film2} + d_{gap2}) - (h_{film1} + d_{gap1}) = \Delta h_{film} + \Delta d_{gap}.$$

(B) A topographic cross-section extracted from **A** on the path shown by the black bar. The STM topographic heights are shown for **12**, **10**, and a Au substrate pit. Note that the height difference between the molecular terraces of **12** and **10** is 1.0 ± 0.2 Å, in contrast to the physical height difference of 2.2 Å. The cross-section also passes through a Au substrate pit, 2.4 Å deep.

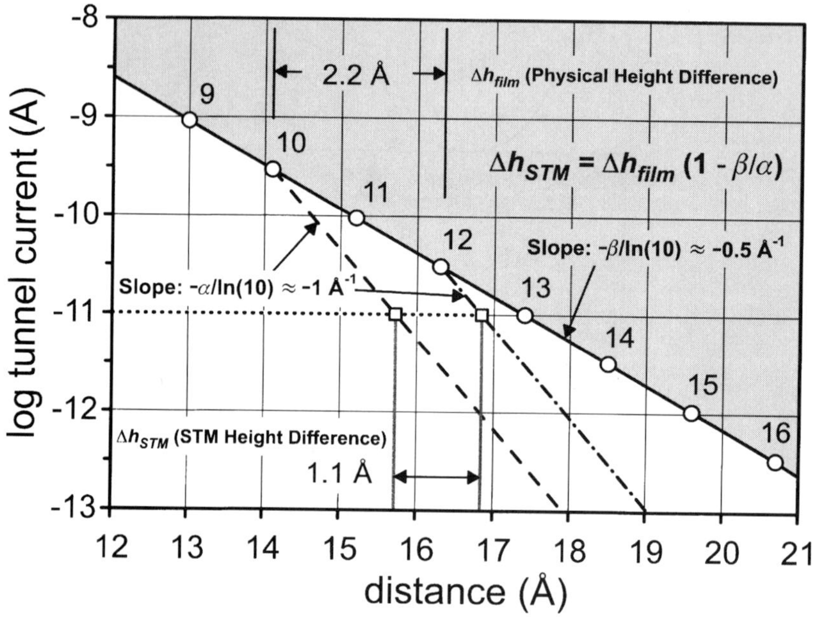

FIGURE 5. Semi-logarithmic plot of the tunneling current versus tip-Au substrate separation for *n*-alkanethiolates on Au{111} at a 1 V tip bias (adapted from Salmeron *et al.*[32]). The open circles represent the hypothetical current-distance point for the tip just in contact with the ends of the alkyl chains with the total number of carbons indicated. These circles fall on a line with a slope $-\beta/\ln(10) = -0.5$ Å^{-1} ($\beta = 1.2$ Å^{-1} or 1.5 per methylene unit). The shaded region above this line corresponds to the condition where the tip has penetrated the alkanethiolate film. The lines intersecting the open circles at alkyl chain lengths of 10 (*dashed*) and 12 (*dot-dashed*) carbon atoms, respectively, represent the tunneling current-distance relationship when the tip is outside the alkanethiolate film (above the ends of the alkanethiolate chains) with a slope of $-\alpha/\ln(10) = -1.0$ Å^{-1}. The tunneling current set point value of 10 pA (*dotted*) used for the data shown in FIGURE 3 positions the probe tip outside the films of both **10** and **12**.[57] The topographic height difference measured by STM is the distance between these two operating points, open squares. The relationship between the STM height difference Δh_{STM} and the physical height difference Δh_{film} is: $\Delta h_{STM} = \Delta h_{film}(1 - \beta/\alpha)$, which for this diagram is $\Delta h_{STM} \approx 0.5\ \Delta h_{film}$.

Applying the simplifying assumptions that the characteristics of the vacuum gap do not change significantly from component 1 to 2 ($\alpha_1 \approx \alpha_2$, $A_1 \approx A_2$), and that the variation in β is also small ($\beta_1 \approx \beta_2$), we get:

$$\Delta h_{STM} = \Delta h_{film}\,(1 - \beta/\alpha) + (1/\alpha)\ln(B_2/B_1) \qquad \textbf{(1)}$$

The first term shows that Δh_{STM} will have a linear dependence on Δh_{film}. The last term is an offset in STM topography due to the differences in the contact conductances of the two thin-film components which can be responsible for a nonzero Δh_{STM} even when $\Delta h_{film} = 0$. Clearly variations in the other constants, e.g., α and β, can also contribute to the Δh_{STM}. Because the molecular components **10** and **12** are members of a homologous series, the 1-alkanethiolates, we can reasonably assume that α, β, A, and B vary very little, if at all, between the different regions of the multicomponent film, where the structures are identical. Therefore, for our experimental system we neglect the last term.

Theoretical estimates of the electronic transmission coefficients for tunneling directly through alkanethiolate molecules have been reported by Sautet and coworkers.[32] If the two-layer tunnel junction model were correct, their transmission coefficients should allow accurate prediction of our observed STM height differences. To test this we have applied Sautet and co-workers' values to the tunneling conditions used to record the data shown in FIGURE 3A. The predicted results are (1) the transconductance of the alkanethiolate chain decreases by an order of magnitude for every two additional methylene units in the chain ($\beta \approx 1.2$ Å^{-1}) and (2) while the tip is outside the SAM, the transconductance of the tunnel junction decreases an order of magnitude for each additional angstrom of tip-film separation ($\alpha \approx 2.3$ Å^{-1}). These results are shown graphically in FIGURE 5 as a semi-logarithmic plot of the tunneling current versus tip-sample separation. The open circles represent the hypothetical condition where the tip is just in contact with the end of the chain for the number of carbon atoms in the molecule as indicated. These points fall on a line with slope $-\beta/\ln(10)$. Points inside the shaded region correspond to the case where the tip has penetrated the SAM. Here we only consider the case where the tip is outside the exposed SAM interface. Outside the film, the current-distance dependence is expressed by the decay constant, α, which is related to the work function of the SAM, ϕ, and the electron mass, m, by $\alpha = (8m\phi)^{1/2}/\hbar$. We assume that the work function of the SAMs is approximately independent of the alkyl chain length.[32] This gives a family of curves each beginning at the open circle corresponding to the number of carbon atoms in the respective alkanethiolate. These are shown for **10** (*dashed*) and **12** (*dot-dashed*). The slopes of these curves can be measured independently by dithering the tip-sample separation and measuring the change in current. This yields a measure of the apparent tunneling barrier height.[77,85]

The scanning tunneling microscope operating point is the tip-sample distance for a given tunneling current, and is unique for each alkanethiolate. Operating points outside the films of **10** and **12** (*open squares*) can be found from the intersection of the 10 pA tunneling current value (*dotted line*) (used to record the data shown in FIGURE 3A) and the respective current-distance curves. The distance between these two operating points is Δh_{STM}. Using our approximation for Eq. (1), the relationship between Δh_{STM} and Δh_{film}, is:

$$\Delta h_{STM} = \Delta h_{film} (1 - \beta/\alpha). \quad \textbf{(2)}$$

In this instance the prediction is that $\Delta h_{STM} \approx 0.5\ \Delta h_{film}$. The plot shows for our STM conditions the apparent height difference would be 1.1 Å. In excellent agreement, we observe 1.0 ± 0.2 Å. This is depicted schematically in FIGURE 6 for the STM tip trajectory over a molecularly sharp **10/12** boundary at constant tunneling current. As the scanning tunneling microscope scans from the region of **10** to the region of **12**, h_{film} increases 2.2 Å while the film transconductance decreases tenfold. To compensate, the STM decreases d_{gap} by 1.1 Å to increase the tip-film gap

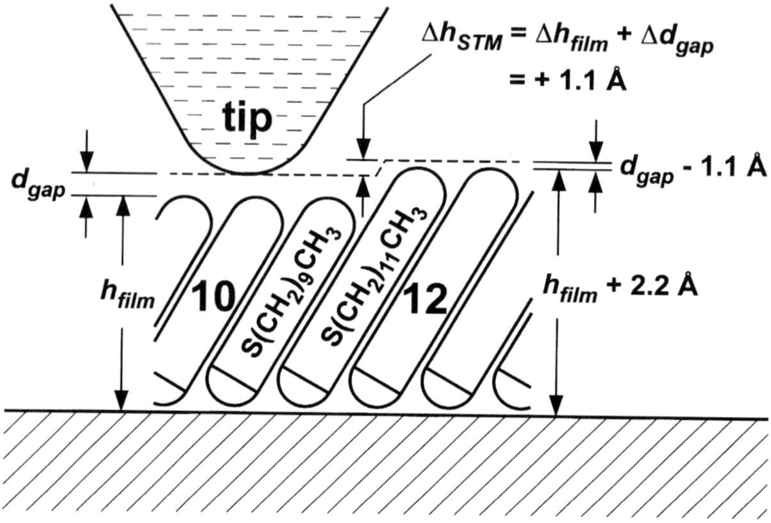

FIGURE 6. A schematic diagram of a molecularly sharp **10/12** boundary showing the STM tip trajectory (*dashed line*) corresponding to the constant tunneling current operating points in FIGURE 3. The STM tunnel junction is composed of two regions, the film and the tip-film gap. The STM topographic height, h_{STM}, is the sum of the physical height, h_{film}, and the tip-film gap, d_{gap}. As the STM scans from the **10** region to the **12** region, h_{film}, increases 2.2 Å and the film transconductance decreases tenfold. To compensate, the STM decreases d_{gap} by 1.1 Å to increase the tip-film gap transconductance by tenfold, thereby maintaining a constant overall tunnel junction transconductance. Consequently $\Delta h_{STM} = \Delta h_{film} + \Delta d_{gap} = +1.1$ Å.

transconductance tenfold, thereby maintaining a constant overall tunnel junction transconductance. Consequently $\Delta h_{STM} = \Delta h_{film} + \Delta d_{gap} = +1.1$ Å.

The diagram presented in FIGURE 5 provides a simple graphical relationship between the STM topographic heights and the electronic and physical structure of the surface. Therefore it is instructive to examine a few special cases. When the conductivity of the film is independent of its thickness ($\beta = 0$), then $\Delta h_{STM} \equiv \Delta h_{film}$. This case corresponds to highly conducting films, e.g., the Au substrate itself, where the STM topographic height difference is identical to the physical height difference. If $\alpha = \beta$ the current-distance dependence of electron tunneling through the film is identical to the current-distance dependence of electron tunneling through the tip-film gap, thus $\Delta h_{STM} \equiv 0$ and no topography due to the film will be observed by STM. For $0 < \beta/\alpha < 1$, the case depicted in FIGURE 5, the STM topographic height difference will be less than the physical height difference, $0 < \Delta h_{STM} < \Delta h_{film}$. The final case is a curious one, for $\beta/\alpha > 1$, the tunneling current decays more rapidly with distance inside the film than it does in the tip-film gap causing the STM topography to appear inverted, $\Delta h_{STM} \propto -\Delta h_{film}$. This instance can arise when the film is immersed in a medium, such as a liquid which has a smaller decay constant than the film or if the apparent tunneling barrier height over the film is very low. Cases where

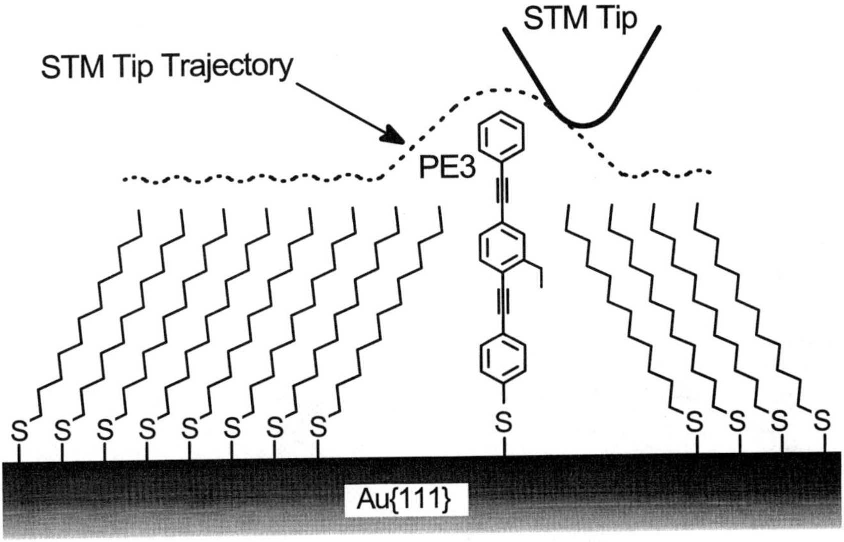

FIGURE 7. A schematic representation of a **12** and PE3 SAM on Au{111}. The height that PE3 extends above the **12** film, if normal to the surface, is 7.3 Å. The trajectory of the STM tip traces out a surface of constant current. The relatively flat **12** layer can be imaged by an atomic-scale asperity on the end of the STM tip with resolution of the molecular lattice. The protruding PE3 molecule "images" the end of the STM tip, resulting in a characteristic feature due largely to the shape of the end of the STM tip for each PE3 molecule, as seen in FIGURES 8 (*inset*) and 9. Note also that the STM tip is closer to the exposed ends of the PE3 molecules than to the exposed ends of the **12** molecules. This is because the apparent tunneling barrier height is higher over the PE3 than over the **12**.

α and/or β are negative correspond to unphysical situations. As can be seen above, the imaging mechanism and results are inexorably linked to the transconductance.

We can construct a plot of experimentally determined current versus distance characteristics for a homologous series of SAMs. As described above, the contact conductance obtained from the intercept of FIGURE 5 contains contributions from the Au-thiolate contact conductance and coupling to the gap. We can measure contact conductance differences between molecules with different head groups, e.g., alkanethiolate-Au versus alkaneselenolate-Au, by comparing the intercepts obtained by measuring these two systems. Similarly, this could be done for the coupling to the gap by comparing the intercepts obtained using different tip materials and/or different exposed functional groups. Note that we require sufficiently low STM junction transconductances that the probe tip remains outside the film. This limiting transconductance is given by the line between the white and shaded regions in FIGURE 5.

Conjugated Chains

Conjugated molecules are predicted to have greater non-resonant tunneling conductivities than saturated molecules, i.e., σ-bonded molecular frameworks. Resonant tunneling may also be observable in some conjugated molecules. Measurement of the electronic properties through the length of an isolated molecule by STM requires that the molecule is standing up on the surface. Isolated molecules will typi-

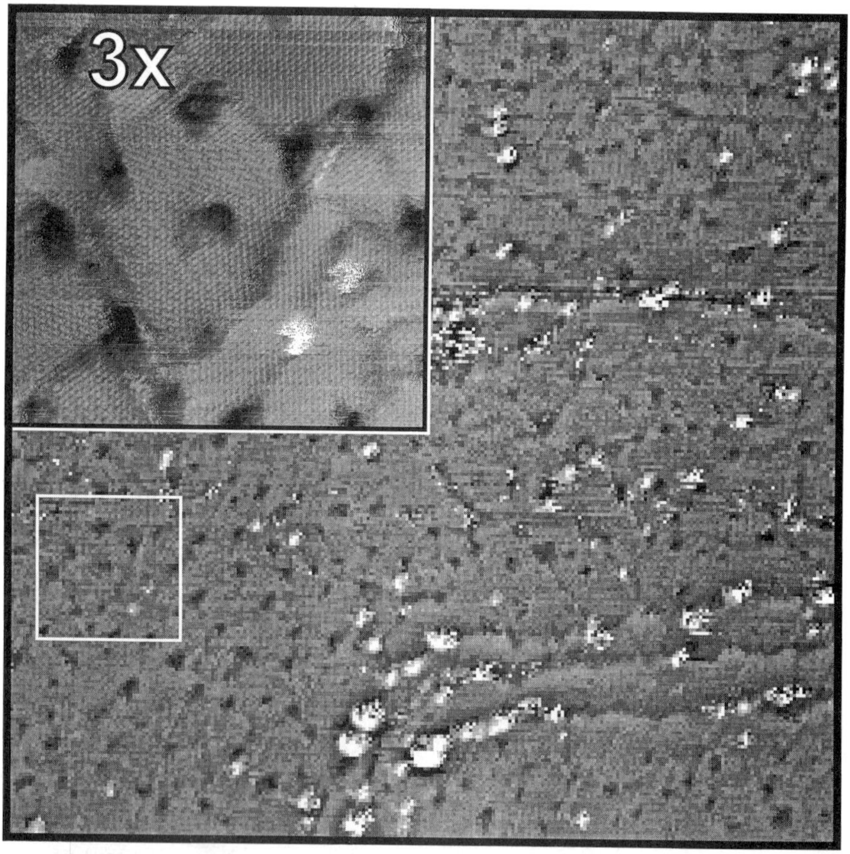

FIGURE 8. STM image of a 2400 Å × 2400 Å area of a **12** SAM modified by exchange. This image has been high-pass spatial filtered to bring out the detail on several consecutive terraces. A 100% **12** SAM was subsequently exposed to a 0.3 mM solution of PE3 in THF for 30 min at room temperature. The PE3, which is 6 Å longer than the thickness of the **12** film, appears as bright features protruding above the surrounding **12** monolayer. Observe that most of the PE3 molecules have exchanged into the monolayer at the Au step edges and may form clusters there. A smaller number of the PE3 molecules have exchanged into the **12** monolayer on the terrace. A 3× higher magnification image (*inset*), taken of the area delineated by the white square, shows that the PE3 on the terraces has exchanged at **12** structural domain boundaries. We have not observed the PE3 within the domains.

cally lie flat on the surface unless some artifice such as a tripodal attachment is used. But even in this case some means must be used to stabilize the geometry during the STM measurement. We have developed a procedure to achieve this using two-dimensional matrix isolation wherein the guest conjugated molecules are imbedded in a host alkanethiolate SAM, shown schematically in FIGURE 7. The host serves both to dilute and to isolate the guest molecules and has a lower conductivity than the guest. The guest molecules are longer than the surrounding matrix film thickness.

In FIGURE 8 the guest molecule PE3 is bound to the Au surface through its terminal sulfur and is embedded in the host **12** SAM. The calculated length of the molecule of PE3 is 21.3 Å, which is 7.3 Å longer than the thickness of the all-trans **12** film. Even if tilted at the 30° angle of the **12** molecules, the PE3 molecule would still protrude 4.0 Å above the **12** film. We assign the (bright) topographic features protruding through the film of **12** as due to the PE3 molecules since these features are not observed in the films of **12** prior to exposure to the solution of PE3-H. While we have not directly observed that these features are single molecules of PE3, we infer that they are from the following observations. (1) Images of these features where the **12** molecular lattice is resolved show that the single-PE3 molecular features are imaged with exactly the same shape, size, and orientation, indicative of features that are much sharper than the STM tip (FIG. 8, *inset*).[26] (2) Larger features (which we assign to clusters of PE3 molecules) are only observed at Au{111} step edges where the **12** SAM is expected to be at lower density and conformationally relaxed[87,88] and the PE3 molecules would be more easily accommodated (FIG. 8). (3) The PE3 molecules are observed to be widely separated on the terraces (FIG. 8) and tend to occur at **12** structural domain boundaries (FIG. 8). Only a small fraction of the **12** structural domain boundaries on the Au{111} terraces are occupied by PE3 features indicating that their insertion into the film is an isolated and improbable event. (4) Direct insertion of clusters is unlikely because neither PE3-H nor PE3-Ac are not known to associate into dimers or larger clusters in solution.

STM topographic images are a convolution of the tip and surface geometry. When features on the surface are much sharper than the tip, such as the PE3 single molecular protrusion ca. 7 Å higher than the **12** film, each molecule is rendered as an image of the tip (FIG. 7).[26,27] This can be seen in the images of PE3 molecules because each appears nearly identical in shape, size, and orientation (FIG. 8, *inset*). An image of the same sample acquired with a different probe tip is shown in FIGURE 9, where the images of the PE3 molecules exhibit a "cross" shape.[27] It is instructive to note that most molecular resolution images are achieved on *atomically flat* surfaces so that sharpness of the tip is determined only by the endmost atomic-scale asperity. Here (FIG. 8, *inset*), with the same tip, we do resolve the molecular lattice on the flat 12 regions.

We have probed the electronic properties of the PE3 molecules using our microwave difference frequency ACSTM.[26,27] In this technique the STM is operated in constant current feedback mode, but in addition to the DC bias applied to the probe tip we simultaneously apply two microwave frequencies. Nonlinear interactions in the tunnel junction create mixing products of the applied bias frequencies, viz. harmonic, sum, difference, and higher combination frequencies. For these experiments we have adjusted the two microwave frequencies, here 5.00000000 GHz and 5.00000500 GHz, so that the difference frequency is 5.00 kHz. We estimate the total

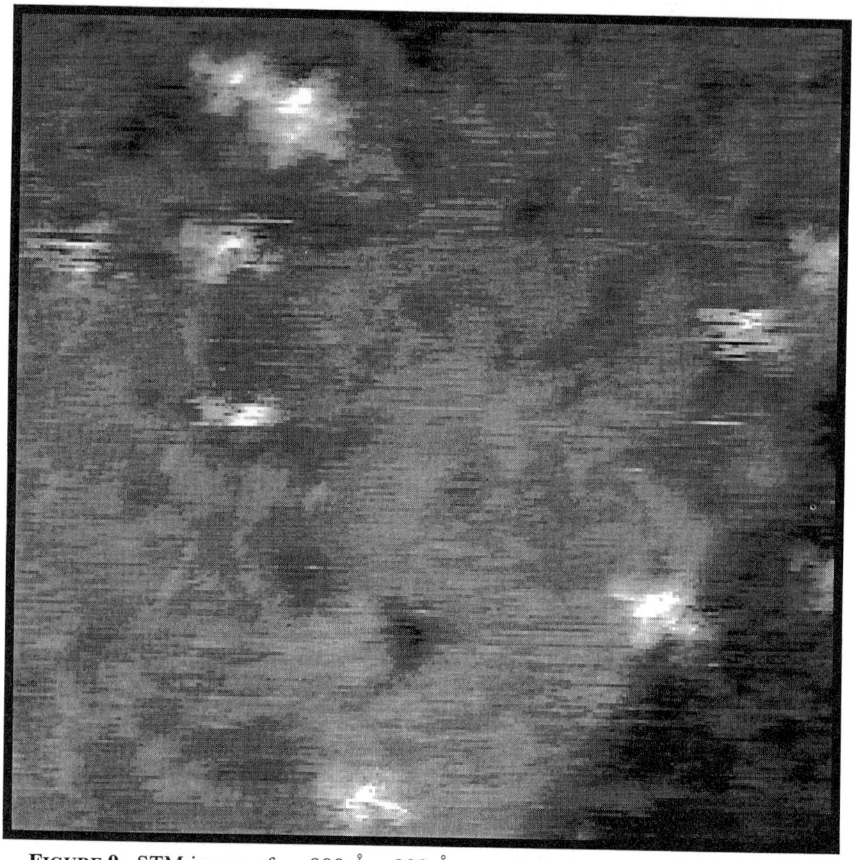

FIGURE 9. STM image of an 800 Å × 800 Å area of the 2D matrix isolated PE3 molecules isolated in a **12** matrix. This image shows that the shape of the PE3 molecules is a characteristic of the STM probe tip. The "cross" shape is the result of imaging PE3 molecules with a particularly shaped tip, where the much sharper molecule produces an image of the probe tip. Note that all of the "cross" features have the identical orientation.

RF power delivered to the tunnel junction to be ~1 mW, which for an open circuited 50 Ω transmission line is 1.3 V peak to peak. This low difference frequency is readily detected by our transimpedance amplifier, but easily rejected by our STM feedback control electronics. This difference frequency is detected within the tunneling current using a lock-in amplifier and is used to create a map of the difference frequency intensity. This is displayed as a 3D surface in FIGURE 10 where the wire frame surface is derived from the constant current DC topograph while the surface shading is derived from the difference frequency signal.

The heuristic explanation of the origin of the MDF contrast lies in the greater polarizability of the extended π system of PE3 compared to the localized α system of

12. The electric field of the AC bias voltage drives the electrons in the molecules at the applied frequencies. The field has a greater effect on the more polarizable molecules, causing their electrons to slosh back and forth along the extended π system. Thus we expect the field-induced polarization of PE3 to be large and of **12** to be small. A field-induced polarization component at the difference frequency is produced by heterodyne beating of the field-induced polarizations of each applied frequency component.

The exact physical mechanism of the MDF signal has not been elucidated, however there are three likely candidates:

(1) Mixing products are generated quasi-statically via the nonlinear current-voltage (*I-V*) characteristic of the tunnel junction due to the modulated bias voltage (DC + RF). For small modulation amplitudes, the difference frequency will be proportional to the d^2I/dV^2 at the DC bias voltage. For large modulation amplitudes, the relationship is more complex.

(2) Field-polarization modulated conductivity is a third-order process analogous to the Kerr effect. In nonlinear optical terms it is four wave mixing (ω_{out}; 0; ω_1; ω_2). The applied AC fields polarize the molecule and the molecular transconductance is dependent on the polarization. Thus the tunnel current flowing through the molecule is modulated by the field-induced polarization at each applied frequency. The difference frequency (and sum frequencies) are generated by the mixing of the two frequency components of the molecular polarization.

(3) Difference frequency generation is a second-order process analogous to rectification. In nonlinear optical terms it is three wave mixing (ω_{out}; ω_1; ω_2). A voltage (or current)[89] is produced in the Au-molecule-tip tunnel junction at the mixing product frequencies (DC, difference, sum, and the second harmonics). Our detector as, configured, is only sensitive to the low frequency 5 kHz difference frequency mixing product. (Any DC component will effectively add to the applied DC bias.)

If (1) were true, the MDF signal could be modeled from the *I-V* characteristic obtained using swept DC measurements. However, at high frequencies, *I-V* curves are often different than those observed at low frequency.[90] This is certainly the case with many semiconductor junctions because the carriers may not reach equilibrium during a cycle.[90] The last two mechanisms can be distinguished by symmetry. The second measures symmetric components while the third measures asymmetric components. (*N.B.*, this symmetry refers to the tip-molecule-substrate tunnel junction, which is practically always highly asymmetric.) We have measured the MDF signal amplitude to be linearly dependent on the amplitude at each applied frequency which is consistent with all three mechanisms.

Despite the low detection frequency scheme used, it is possible to measure a high frequency property of the tunnel junction. This can be understood as an application of Miller's rule.[91] Namely, the *n*th-order susceptibility for difference frequency generation is proportional to the product of the first-order susceptibilities at the frequency of each input and output. Hence, we could map the high frequency response spectrum of the molecules by sweeping the applied frequencies and maintaining their difference constant. Because the relaxation of the electronic polarization of the molecules occurs on a much faster time scale than the 20 GHz upper limit of our source, we would not expect to find a strong frequency dependence. Cursory mea-

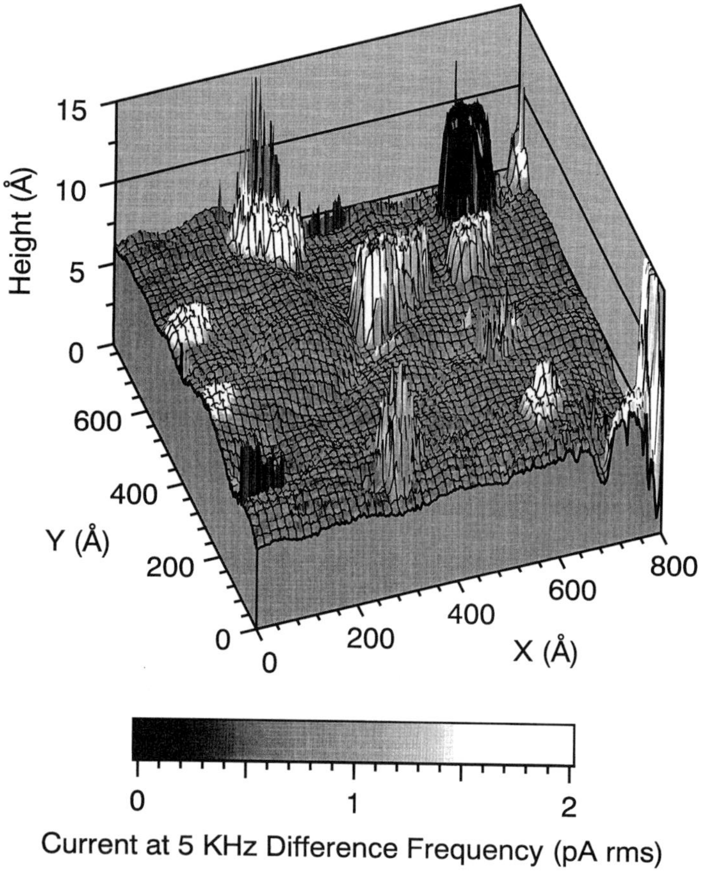

FIGURE 10. A composite ACSTM image showing a 800 Å × 800 Å area of a Au{111} surface covered by **12** and PE3. The wire-frame surface is derived from the constant current topograph and the surface shading from the MDF signal (DC tip bias +1.0 V, DC tunneling current 10 pA, microwave frequencies 5.00000000 GHz and 5.00000500 GHz applied to the ACSTM tip transmission line at ca. 1 mW, detection at 5 kHz).

surements over a range of microwave frequences have confirmed the lack of a strong dependence on applied frequency.[92]

Note that four types of features can be observed in FIGURE 10: (i) topographic maxima that coincide with MDF maxima; (ii) a single topographic maximum that does not coincide with the MDF maxima; (iii) MDF maxima that coincide with small protrusions at 12 film defects; and (iv) a Au{111} substrate atomic step.

The type (i) feature corresponds to our expectation that the PE3 molecules will have a greater MDF signature than the host film of **12**. The type (ii) feature is a rarity. We would assign it to PE3 molecules based on its topographic character, however its MDF signature is actually lower than that of the **12** host film. This may be due to a PE3 molecule in which the bond to the Au substrate is very different leading to a much weaker MDF response. The molecule might not be chemisorbed to the Au substrate, but only held in the host film by van der Waals interactions. It might also be an oxidation product[46] of PE3 or an impurity molecule. The MDF maxima of the type (iii) features we assign to PE3 molecules that are embedded in the **12** SAM. These would be difficult to discern from the host film without the additional contrast from the MDF data channel. We do not observe these features in SAMs without embedded conjugated molecules. We have occasionally found that by scanning the probe tip *through* the matrix, we can liberate the end of the PE3 molecule and a feature of type (iii) on subsequent images appears as a type (i) feature. To do this we scan at gap transconductances which place us in the shaded region of FIGURE 5; in other words with the probe tip in contact with the matrix film. Since the strength of the MDF signal does not depend directly on the topography, we are confident that the MDF channel provides new information about the sample which is different from the DC STM topograph.

We independently map the tunneling barrier height (related to α) for these films while acquiring STM images.[26] This together with the greater MDF signal over the PE3 molecular features lead us to conclude that the embedded PE3 molecules have a higher transconductance than the surrounding **12** matrix.[26] Electrochemical measurements of the electron transfer rate constants of PE3 tethered ferrocene on Au electrodes has confirmed that the transconductance of the PE3 bridge is orders of magnitude higher than the transconductance of similar alkanethiolate tethers. [14]

CONCLUSIONS AND FUTURE DIRECTIONS

We have made initial measurements of the conductances of single and bundled molecules. From the slope and intercept of curves, such as the trace shown in FIGURE 5, we are able to determine the transconductances of chains and the contact conductances of the chemical attachments to various substrates. We can also do such measurements as functions of applied voltage, chain, and attachment identity. These measurements point the way towards the quantification of the conductances of molecular wire candidates as functions of backbone, length, chemical functionality, electrode-molecule connection, substitution, and environment.

We expect such systematic measurements to help advance the field in two ways. First, these measurements will guide the way to coupling electrons more efficiently through molecules and molecular contacts. Second, this will allow a direct, quantitative comparison to theory (such as those elaborated in Refs. 32, 34, and 49) so as to develop a true predictive power for molecular electronics.

ACKNOWLEDGMENTS

The authors would like to thank John Pazik, Mark Ratner, and Mickey Schurr for helpful discussions. The support of the National Science Foundation, the Office of

Naval Research, the Defense Advanced Research Projects Agency, and the John Simon Guggenheim Memorial Foundation are gratefully acknowledged.

REFERENCES

1. BARBARA, P.F., T.J. MEYER & M.A. RATNER. 1996. Contemporary issues in Electron Transfer Research. J. Phys. Chem. **100:** 13148–13168.
2. ARNAUT L.G. & S.J. FORMOSINHO. 1996. Theoretical studies of intramolecular electron transfer reactions: Distance and free energy dependences. J. Photochem. Photobiol. A: Chem. **100:** 15–34.
3. MOSER, C.C., J.M. KESKE, K. WARNCKE, R.S. FARID & J.P. DUTTON. 1992. Nature of biological electron transfer. Nature **355:** 796–802.
4. HOLMLIN, R.E., P.J. DANDLIKER & J. K. BARTON. 1997. Charge transfer through the DNA base stack. Angew. Chem. Int. Ed. Engl. **36:** 2714–2730.
5. GRATZEL, M. 1989. Heterogeneous Photochemical Electron Transfer. CRC Press. Boca Raton, FL. Fox, M.A. & M. Chanon, Eds. 1988. Photoinduced Electron Transfer. Elsevier Science Publishers. Amsterdam, The Netherlands.
6. GRAY, H.B. & J.R. WINKLER. 1996. Electron transfer in proteins. Ann. Rev. Biochem. **65:** 537–561.
7. CLOSS, G.L. & J.R. MILLER. 1988. Intramolecular long-distance electron transfer in organic molecules. Science **240:** 440–447.
8. PADDON-ROW M.N., A.M. OLIVER, J.M. WARMAN, K.J. SMIT, M.P. DE HAAS, H. OEVERING & J.W. VERHOEVEN. 1988. Factors affecting charge separation and recombination in photoexcited rigid donor-insulator-acceptor compounds. J. Phys. Chem. **92:** 6958–6962.
9. PAULSON, B., K. PRAMOD, P. EATON, G. CLOSS & J.R. MILLER. 1993. Long-distance electron transfer through rodlike molecules with cubyl spacers. J. Phys. Chem. **97:** 13042–13045.
10. CHIDSEY, C.E.D. 1991. Free energy and temperature dependence of electron transfer at the metal-electrolyte interface. Science **251:** 919–922.
11. MILLER, C., P. CUENDET & M. GRATZEL. 1991. Adsorbed ω-hydroxy thiol monolayers on gold electrodes: Evidence for electron tunneling to redox species in solution. J. Phys. Chem. **95:** 877–886.
12. FINKLEA, H.0., L. LIU, M.S. RAVENSCROFT & S. PUNTURI. 1996. Multiple electron tunneling paths across self-assembled monolayers of alkanethiols with attached ruthenium(II/III) redox centers. J. Phys. Chem. **100:** 18852–18858.
13. LI, T.T.-T. & M.J. WEAVER. 1984. Intramolecular electron transfer at metal surfaces. 4. Dependence of tunneling probability upon donor-acceptor separation distance. J. Am. Chem. Soc. **106:** 6107–6108.
14. SACHS, S.B., S.P. DUDEK, R.P. HSUNG, L.R. SITA, J.F. SMALLEY, M.D. NEWTON, S.W. FELDBERG & C.E.D. CHIDSEY. 1997. Rates of interfacial electron transfer through π-conjugated spacers. J. Am. Chem. Soc. **119:** 10563–10564.
15. CARTER, M.T, G.K. ROWE, J.N. RICHARDSON, L.M. TENDER, R.H. TERRILL & R.W. MURRAY. 1995. Distance dependence of the low-temperature electron transfer kinetics of (ferrocenylcarboxy)-terminated alkanethiol monolayers. J. Am. Chem. Soc. **117:** 2896–2899.
16. POYMEROPOULOS, E.E. & J. SAGIV. 1978. Electrical conduction through adsorbed monolayers. J. Chem. Phys. **69:** 1836–1847.
17. REED, M.A., C. ZHOU, C.J. MULLER, T.P. BURGIN & J.M. TOUR. 1997. Conductance of a molecular junction. Science. **278:** 252–253.
18. ZHOU, C., M.R. DESHPANDE, M.A. REED, L. JONES II & J.M. TOUR. 1997. Nanoscale metal/self-assembled monolayer/metal heterostructures. Appl. Phys. Lett. **71:** 611–613.

19. ANDRES, R.P., J.I. HENDERSON, R.G. OSIFCHIN, W. MAHONEY, C.P. KUBIAK, R. REIFENBERGER, S. FENG, M. DOROGI & T. BEIN. 1996. "Coulomb staircase" at room temperature in a self-assembled molecular nanostructure. Science. **272:** 1323–1324.
20. ANDRES, R.P., J.D. BIELEFELD, J.I. HENDERSON, D.B. JANES, V.R. KOLAGUNTA, C. P. KUBIAK, W.J. MAHONEY & R.G. OSIFCHIN. 1996. Self-assembly of a two-dimensional superlattice of molecularly linked metal clusters. Science **273:** 1690–1691.
21. TANS, S.J., A. THESS, L.J. GEERLIGS & R.E. SMALLEY. 1997. Individual single-wall carbon nanotubes as quantum wires. Nature **386:** 474–476.
22. LANGER, L., V. BAYOT, E. GRIVEI, J.-P. ISSI, J.P. HEREMANS, C.H. OLK, L. STOCKMAN, C. VAN HAESENDONCK & Y. BRUYNSERAEDE. 1996. Quantum transport in a multiwalled carbon nanotube. Phys. Rev. Lett. **76:** 479–482.
23. EBBESEN, T.W., H.J. LEZEC, H. HIURA, J.W. BENNETT, H.F GHAEMI & T. THIO. 1996. Electrical conductivity of individual carbon nanotubes. Nature **382:** 54–56.
24. DAI, H., E.W. WONG & C.M. LIEBER. 1996. Probing electrical transport in nanomaterials: Conductivity of individual carbon nanontubes. Science **272:** 523–526.
25. THESS, A., R. LEE, P. NIKOLAEV, H. DAI, P. PETIT, J. ROBERT, C. XU, Y.H. LEE, S.G. KIM, A.G. RINZLER, D.T. COLBERT, G.E. SCUSERIA, D. TOMÁNEK, J.E. FISCHER & R.E. SMALLEY 1996. Crystalline ropes of metallic carbon nanotubes. Science **273:** 483–487.
26. BUMM, L.A., J.J. ARNOLD, M.T. CYGAN, I.D. DUNBAR, T.P. BURGIN, L. JONES II, D.L. ALLARA & P.S. WEISS. 1996. Are single molecular wires conducting? Science **271:** 1705–1707.
27. CYGAN, M.T., T.D. DUNBAR, J.J. ARNOLD, L.A. BUMM, N.E. SHEDLOCK, T.P. BURGIN, L. JONES II, D.L. ALLARA, J.M. TOUR & P.S. WEISS. 1998. Insertion, conductivity, and structures of conjugated organic oligomers in self-assembled alkanethiol monolayers on Au{111}. J. Am. Chem. Soc. **120:** 2721–2732.
28. BUMM, L.A., J.J. ARNOLD, T.D. DUNBAR, D.L. ALLARA & P.S. WEISS. Electron transport through organic molecules. Manuscript in preparation.
29. JOACHIM, C., J.K. GIMZEWSKI, R.R. SCHLITTLER & C. CHAVY. 1995. Electronic transparence of a single C_{60} molecule. Phys. Rev. Lett. **74:** 2102–2105.
30. JOACHIM, C. & J.K. GIMZEWSKI. 1997. An electromechanical amplifier using a single molecule. Chem. Phys. Lett. **265:** 353–357.
31. YAZDANI, A., D.M. EIGLER & N.D. LANG. 1996. Off-resonance conduction through atomic wires. Science **272:** 1921–1924.
32. SALMERON, M., G. NEUBAUER, A. FOLCH, M. TOMITORI, D.F. OGLETREE & P. SAUTET. 1993. Viscoelastic and electrical properties of self-assembled monolayers on Au(111) films. Langmuir **9:** 3600–3611.
33. DATTA, S., W. TIAN, S. HONG, R. REIFENBERGER, J.I. HENDERSON & C.P. KUBIAK. 1997. Current-voltage characteristics of self-assembled monolayers by scanning tunneling microscopy. Phys. Rev. Lett. **79:** 2530–2533.
34. SAMANTA M.P., W. TIAN, S. DATTA, J.I. HENDERSON & C.P. KUBIAK. 1996. Electronic conduction through organic molecules. Phys. Rev. B. **53:** R7626–R7629.
35. The height of the tip is measured in the laboratory frame, not relative to the sample surface.
36. SAUTET, P. 1997. Images of adsorbates with the scanning tunneling microscope: Theoretical approaches to the contrast mechanism. Chem. Rev. **97:** 1097–1116.
37. VAN HOVE, M.A., J. CERDA, P. SAUTET, M.-L. BOCQUET & M. SALMERON. 1997. Surface structure determination by STM vs LEED. Prog. Surf. Sci. **54:** 315–329.
38. PETTY, M.C., M.R. BRYCE & D. BLOOR, Eds. 1995. An Introduction to Molecular Electronics. Oxford University Press. New York.

39. AVIRAM, A. Ed. 1992. Molecular Electronics: Science and Technology, Conference Proceeding No. 262. American Institute of Physics. New York.
40. REED, M.A. & W.P. KIRK, Eds. 1989. Nanostructure Physics and Fabrication. Academic Press. New York.
41. REED, M.A. & W.P. KIRK, Eds. 1992. Nanostructures and Mesoscopic Systems. Academic Press. New York.
42. CARTER, F.L., Ed. 1987. Molecular Electronic Devices II. Marcel Dekker. New York.
43. MILLER, J.S. 1990. Molecular electronics? Adv. Mater. **2:** 378–379, 495–497, 601–603.
44. JOACHIM, C. 1991. The conductance of a single molecule. New J. Chem. **15:** 223–229.
45. TOUR, J.M. 1996. Conjugated macromolecules of precise length and constitution. Organic synthesis for the construction of nanoarchitectures. Chem. Rev. **96:** 537–553.
46. TOUR, J.M., L. JONES II, D.L. PEARSON, J.J.S. LAMBA, T.P. BURGIN, G.M. WHITESIDES, D.L. ALLARA, A.N. PARIKH & S.V. ATRE. 1995. Self-assembled monolayers and multilayers of conjugated thiols, α,ω-dithiols, and thioacetyl-containing adsorbates. Understanding attachments between potential molecular wires and gold surfaces. J. Am. Chem. Soc. **117:** 9529–9534.
47. PEARSON, D.L. & J.M. TOUR. 1997. Rapid syntheses of oligo(2,5-thiophene ethynylene)s with thioester termini: Potential molecular scale wires with alligator clips. J. Org. Chem. **62:** 1376–1387.
48. JONES, L., II, J.S. SCHUMM & J.M. TOUR. 1997. Rapid solution and solid phase syntheses of oligo(1,4-phenylene ethynylene)s with thioester termini: Molecular scale wires with alligator clips. Derivation of iterative reaction efficiencies on a polymer support. J. Org. Chem. **62:** 1388–1410.
49. MAGOGA, M. & C. JOACHIM. 1997. Conductance and transparence of long molecular wires. Phys. Rev. B **56:** 4722–4729.
50. LANDAUER, R. 1987. Electrical transport in open and closed systems. Z. Phys. B. **68:** 217–228.
51. LANDAUER, R. 1989. Conductance determined by transmission: Probes and quantized constriction resistance. J. Phys. Condens. Matter. **1:** 8099–8110.
52. JOACHIM, C. & J.F. VINUESA. 1996. Length dependence of the electronic transparence (conductance) of a molecular wire. Europhys. Lett. **33:** 635–640,
53. CHIDSEY, C.E.D., D.N. LOIACONO, T. SLEATOR & S. NAKAHARA. 1988. STM study of the surface morphology of gold on mica. Surf. Sci. **200:** 45–66.
54. REFINING SYSTEMS, P.O. Box 72466, Las Vegas, NV 89170.
55. ASHVILLE-SCHOONMAKER MICA CO., 900 Jefferson Ave., P.O. Box 318, Newport News, VA 23607.
56. ALDRICH, 1001 W. St. Paul Ave., Milwaukee, WI 53233 (http://www.sial.com/).
57. BUMM, L.A., J.J. ARNOLD, L.F CHARLES, T.D. DUNBAR, D.L. ALLARA & P.S. WEISS. Creating molecular terraces with molecularly sharp boundaries in organic monolayers. Manuscript in preparation.
58. BESOCKE, K. 1987. An easily operable scanning tunneling microscope. Surf. Sci. **181:** 145–153.
59. FROHN, J., J.F. WOLF, K. BESOCKE & M. TESKE. 1989. Coarse tip distance adjustment and positioner for a scanning tunneling microscope. Rev. Sci. Instrum. **60:** 1200–1201.
60. KOCHANSKI, G.P. 1989. Nonlinear alternating-current tunneling microscopy. Phys. Rev. Lett. **62:** 2285–2288.

61. MICHEL, B., W. MIZUTANI, R. SCHIERLE, A. JAROSCH, W. KNOP, H. BENEDIKTER, W. BACHTOLD & H. ROHRER. 1992. Scanning surface harmonic microscopy: Scanning probe microscopy based on microwave field-induced harmonic generation. Rev. Sci. Instrum. **63:** 4080–4085.
62. SEIFERT, W., E. GERNER, M. STACHEL & K. DRANSFELD. 1992. Scanning tunneling microscopy at microwave frequencies. Ultramicroscopy **42–44:** 379–387.
63. STRANICK, S.J. & P.S. WEISS. 1993. A versatile microwave frequency-compatible scanning tunneling microscope. Rev. Sci. Instrum. **64:** 1232–1234.
64. BOURGOIN, J.-P., M.B. JOHNSON & B. MICHEL. 1994. Semiconductor characterization with the scanning surface harmonic microscope. Appl. Phys. Lett. **65:** 2045–2047.
65. JOHNSON, M.B., J.-P. BOURGOIN & B. MICHEL. 1994. Doping profiling with scanning surface harmonic microscopy. Microelectronic Engineering. **27:** 539–542.
66. BOURGOIN, J.-P., M.B. JOHNSON & B. MICHEL. 1994. Scanning surface harmonic microscopy: Application to silicon and Langmuir-Blodgett films on silicon. Microsc. Microanal. Microstruct. **5:** 535–543.
67. STRANICK, S.J. & P.S. WEISS. 1994. A tunable microwave frequency alternating current scanning tunneling microscope. Rev. Sci. Instrum. **65:** 918–921.
68. STRANICK, S.J. & P.S. WEISS. 1994. Alternating current scanning tunneling microscopy and nonlinear spectroscopy. J. Phys. Chem. **98:** 1762–1764.
69. STRANICK, S.J., M.M. KAMNA & P.S. WEISS. 1994. A low temperature, ultrahigh vacuum, microwave-frequency-compatible scanning tunneling microscope. Rev. Sci. Instrum. **65:** 3211–3215.
70. STRANICK, S.J., L.A. BUMM, M.M. KAMNA & P.S. WEISS. 1995. Linear and nonlinear spectroscopy with the tunable AC scanning tunneling microscope. *In* Photons and Local Probes. O. Marti & R. Moller, Eds. NATO ASI Series E: Applied Sciences. **300:** 221–233. Kluwer Academic. Dordrecht, The Netherlands.
71. BUMM, L.A. & P.S. WEISS. 1995. Small cavity nonresonant tunable microwave-frequency alternating current scanning tunneling microscope. Rev. Sci. Instrum. **66:** 4140–4145.
72. HEWLETT-PACKARD CO., P.O. Box 4026, Englewood, CO 80155–4026 (http://www.hp.com/).
73. AXON INSTRUMENTS INC., 1101 Chess Dr., Foster City, CA 94404 (http://www.axonet.com/).
74. DUNLOP, D., S. SMITH, C. BUSTAMANTE, J. TAMAYO & R. GARCIA. 1995. A very low current scanning tunneling microscope. Rev. Sci. Instrum. **66:** 4876–4879.
75. We have limited the bandwidth of the CV4 signal used of STM feedback control with a 4 pole Bessel low pass filter. This gives us an excellent signal to noise for current feedback and also the ability to simultaneously use the frequencies above this band for other measurements.
76. STANFORD RESEARCH SYSTEMS, 1290 D Reamwood Ave., Sunnyvale, CA 94089 (http://www.srsys.com/srsys)
77. WEISENDANGER, R. 1994. Scanning Probe Microscopy and Spectroscopy. Cambridge University Press. New York.
78. POIRIER, G.E. 1997. Characterization of organosulfur molecular monolayers on Au(111) using scanning tunneling microscopy. Chem. Rev. **97:** 1117–1127.
79. The difference in film thickness of 2.2 Å was calculated using the following structural factors: C—C bond distance, 1.53 Å;[81] C—C—C bond angle, 112°;[82] and the chain tilt angle, 30°.[80,83,84] This compares to ~2.2–2.4 Å obtained from ellipsometry. Note that chain twist does not effect this value when the lengths differ by an even number of methylene units.

80. DUBOIS, L.H. & R.G. NUZZO. 1992. Synthesis, structure, and properties of model organic surfaces. Annu. Rev. Phys. Chem. **43:** 437–463.
81. ALLEN, F.H., O. KENNARD, D.G. WATSON, L. BRAMMER, A.G. ORPEN & R. TAYLOR. 1987. Tables of bond lengths determined by x-ray and neutron diffraction. Part 1. Bond lengths in organic compounds. J. Chem. Soc. Perkin. Trans. II. **1987:** S1–S19.
82. BURKERT, U. & N.L. ALLINGER. 1982. Molecular Mechanics, ACS monograph. Vol. 177. American Chemical Society. Washington, DC.
83. SHI, J., B. HONG, A.N. PARIKH, R.W. COLLINS & D.L. ALLARA. 1995. Optical characterization of electronic transitions arising from the Au/S interface of self-assembled *n*-alkanethiolate monolayers. Chem. Phys. Lett. **246:** 90–94.
84. LAIBINIS, P.E., G.M. WHITESIDES, D.L. ALLARA, Y.-T. TAO, A.N. PARIKH & R.G. NUZZO. 1991. Comparison of the structures and wetting properties of self-assembled monolayers of *n*-alkanethiols on the coinage metal surfaces, Cu, Ag, Au. J. Am. Chem. Soc. **113:** 7152–7167.
85. OLESEN, L., M. BRANDBYGE, M.R. SØRENSEN, K.W. JACOBSEN, E. LÆGSGAARD, I. STENSGAARD & F. BESENBACHER. 1996. Apparent barrier height in scanning tunneling microscopy revisited. Phys. Rev. Lett. **76:** 1485–1488.
86. HSU, C.-P. & R.A. MARCUS. 1997. A sequential formula for electronic coupling in long range bridge-assisted electron transfer: Formulation of theory and application to alkanethiol monolayers. J. Chem. Phys. **106:** 584–598.
87. STRANICK, S.J., A.N. PARIKH, D.L. ALLARA & P.S. WEISS. 1994. A new mechanism for surface diffusion: Motion of a substrate-adsorbate complex. J. Phys. Chem. **98:** 11135–11142.
88. KROM K.R. 1997. Structural effects of alkanethiol monolayers on the gold surface studied by molecular dynamics. Ph.D. Thesis. The Pennsylvania State University. University Park, PA.
89. The actual figure of merit is the power generated at the difference frequency. Assigning the output as a voltage or as a current is purely arbitrary, usually chosen based by experimental convenience and/or by the system impedance.
90. SZE, S.M. 1981. Physics of Semiconductor Devices. John Wiley & Sons. New York, NY.
91. MILLER, R.C. 1964. Optical second harmonic generation in piezoelectric crystals. Appl. Phys. Lett. **5:** 17–19.
92. It may be possible to observe the effects of dielectric relaxation of the host **12** matrix on the field-induced polarization of PE3.

Transport in (10,10) Carbon Nanotubes with Defects

M.P. ANANTRAM,[a] JIE HAN, AND T. R. GOVINDAN[b]

NASA Ames Research Center, Mail Stop T27-A, Moffett Field, California 94045-1000, USA

[b]Applied Research Laboratory, P.O. Box 30, State College, Pennsylvania 16804-0030, USA

ABSTRACT: In this paper, we numerically calculate the conductance characteristics of a uniform (10,10) carbon nanotube in the presence of an applied bias. It is shown that for long tubes, the conduction mechanism involves only the four nondegenerate sub-bands when the applied bias drops slowly across the tube. The focus of our work is to calculate the transmission coefficient and low bias conductance of carbon nanotubes with defects. Our calculations show that (1) a weak uniform disorder in the tube does not cause a significant change in the zero bias conductance and (2) substitutional defects which are strong scatterers, can cause a gap in the transmission coefficient versus energy at the middle of the band resulting in a significant change in the low bias conductance

INTRODUCTION

Carbon nanotubes (CNT) are hollow cylindrical surfaces of carbon atoms with the lattice structure of a rolled up graphite sheet. They can presently be fabricated as both multi-walled or single-walled tubes, with the multi-wall tubes typically containing concentric cylinders of single-wall tubes.[1] Experimentally, CNT have demonstrated use as STM tips[2] and room temperature electron field emitters.[3] The possibility of making wires and devices led to transport experiments on single-wall nanotubes and their ropes, which measured the low bias conductance as a function of temperature, gate voltage, and magnetic field,[4–6] and the I-V characteristics.[7] The experimental activity in manufacturing CNT coupled with the prospect of some useful applications has generated considerable interest in modeling various properties of these tubes. Calculations have shown that the Young's modulus is comparable to 1 TPa; tubes with different diameters and chirality can have properties varying from that of semiconductors to metals; the low bias conductance can be as large as $4e^2/h$ (molecular wires) and junctions between semiconducting and metallic tubes can be formed.

Defect-free zig-zag and armchair tubes have two propagating modes at the center of band (Fermi energy for infinitely long tubes) with a maximum conductance of $4e^2/h$. To measure a low bias conductance that is comparable to $4e^2/h$ requires both good contacts to the large electron reservoirs where voltage is applied and absence of defects. Both factors lead to reflection of electrons in the CNT causing a decrease in conductance. This paper calculates the I-V characteristics of a uniform CNT in

[a]Address for correspondence: anant@nas.nasa.gov

the presence of an applied bias and examines the influence of defects on the transmission coefficient and low bias conductance. The two types of defects we have considered are the *weak uniform disorder* and *substitutional defects*.

MODEL

The electronic properties are modeled using a single π electron per spin per atom in a sp^2 bonded network.[1,8,9] The corresponding single particle Hamiltonian is,

$$H = \sum_i \varepsilon_i c_i^\dagger c_i + \sum_{i,j} t_{ij} c_i^\dagger c_j(i), \tag{1}$$

where, the summation is over all lattice sites of the tube. $\{c_i^\dagger, c_i\}$ represent the creation and annihilation operators at site i and $j(i)$ represents the three nearest neighbors of a carbon atom in the uniform tube. ε_i is the on-site potential, which was chosen to be zero for an uniform tube and the hopping parameter between neighboring carbon atoms t_{ij} was assumed to be −3.1 eV.[8]

The current, across the device was determined by assuming the contacts to be in local equilibrium with all unique chemical potential assigned to them. We calculate the current in the phase coherent limit, where the effect of phonons and dephasing is neglected. In this limit, the current (I) is related to the Fermi-Dirac factors in the contacts ($f_{\text{source}}(E)$ and $f_{\text{drain}}(E)$) and the transmission coefficient $T(E)$ between the contacts, by the Landaue-Buttiker formula,[11]

$$I = \int dE\, T(E)(f_{\text{source}}(E) - f_{\text{drain}}(E)), \tag{2}$$

where E is the energy coordinate. The temperature enters the calculations only through the Fermi factors of the contacts. That is, we assume that the lattice is at zero temperature and the electrons are not. In computing the transmission coefficient, the spatial variation of the electrostatic potential and the modeling of the connection between the CNT and contacts must be considered. Calculating the contribution of the electrostatic potential through the wire requires a self-consistent solution of the equation for charge density and Poisson's equation. With the objective to obtain broad features of the I-V characteristics, we have calculated the current for different assumed forms of the applied potential. The electrostatic potential in the tube was assumed to have the form,

$$V(x) = V_{app}/2 + Ae^{+x/L_{sc}} + Be^{-x/L_{sc}} \tag{3}$$

where

$$A = -\frac{1}{2}\frac{1 + e^{-L/L_{sc}}}{e^{L/L_{sc}} - e^{-L/L_{sc}}} \text{ and } B = +\frac{1}{2}\frac{1 + e^{+L/L_{sc}}}{e^{+L/L_{sc}} - e^{-L/L_{sc}}}$$

where L is the length of the wire and L_{sc} is a parameter that can be varied to yield different forms for the potential drop. For values of $L_{sc} > L$, $V(x)$ varies linearly across the wire and for values of $L_{sc} < L$, the potential drops abruptly close to the ends of the tube, with the V potential in the bulk of the wire being $V_{app}/2$. Assuming the presence of an adiabatic connection[12] between the contacts and the device, the transmission coefficient was calculated using semi-infinite CNT leads as boundary conditions.

In the case of uniformly distributed weak disorder, small and random variations in the location of the atoms occur due to their interaction with a nonuniform surrounding. As a result, both the on-site potential and hopping parameters acquire a random component. Substitutional defects, on the other hand, correspond to lattice locations onto which electrons cannot hop easily. Our calculations are based on modeling these defects by exclusively varying the on-site potential. The modified on-site potential was $\varepsilon_i^0 + \delta\varepsilon_i$. In a uniformly distributed weak disorder, $\delta\varepsilon_i$ was chosen randomly from the range $\pm\varepsilon_{random}$ at every lattice point. The degree of disorder is proportional to ε_{random}. $\delta\varepsilon_i$ was set to a large number at lattice sites chosen randomly in the case of the substitutional defect.

RESULTS

Defect-Free Wire

For a uniform (10, 10) arm chair tube having twenty atoms per unit cell, FIGURE 1 shows the transmission coefficient and k vector versus energy along the tube axis.[1] Of the forty quasi one-dimensional bands, eighteen are twofold degenerate (*solid lines*) and four are nondegenerate (*dashed lines*). The transmission coefficient increases (decreases) every time a sub-band opens (closes) The change in transmission coefficient by units of two corresponds to the opening and closing of the twofold degenerate sub-bands. Also, note that while the nondegenerate sub-bands meet at E = 0 eV, the degenerate bands have a large band gap.

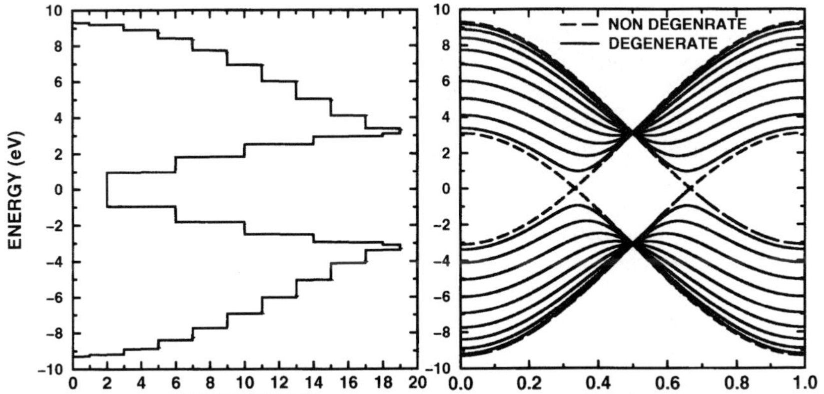

FIGURE 1. **(a)** Transmission and **(b)** k vector versus energy of a (10,10) tube. The transmission increases/decreases when a sub-band in **(b)** opens/closes, The *dashed* and *solid lines* are non-degenerate and degenerate sub-bands respectively.

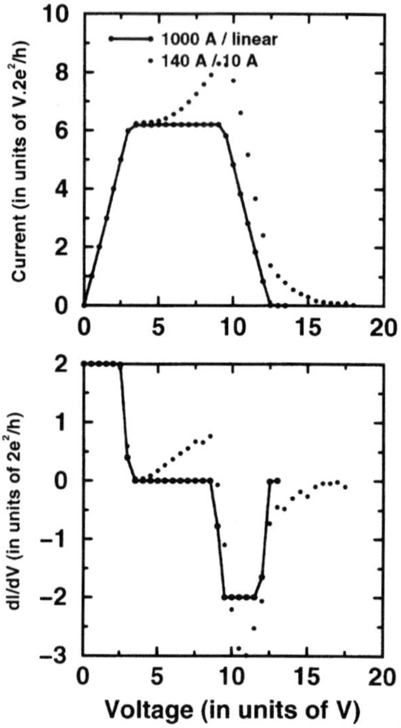

FIGURE 2. The current-versus-voltage curve for a tube of length 1000 Å and L_{sc} = 2000 Å, which corresponds to a linear drop in the potential across the tube (*solid line*). The dots correspond to a wire of 140 Å and L_{sc} = 10 Å. The lower curve is dI/dV.

In the presence of a finite applied bias, we have calculated the current versus voltage for different values of length (L) and L_{sc}. For a 1000 Å wire with L_{sc} = 2000 Å (which corresponds to a linear drop in the potential across the tube), the current-versus-voltage characteristics are shown in FIGURE 2, which can be interpreted as follows. In the presence of a slowly varying potential across the tube the degenerate bands (*solid lines* in FIG. 1b) do not contribute to transport because of the finite band gap between the conduction and valence bands. Further, for a slow variation of the applied potential, band-to-band tunneling is small. Electrons incident from the source contact in the two positive velocity states (source to the left and drain to the right) of the nondegenerate bands gain velocity and k vector as they move to the right and are injected into the drain for applied voltages less than 3.1V. This holds good for electrons in the energy regime of −3.1 eV to 0, which is taken to be the Fermi energy in the source end. The transmission of two states per energy from these quasi one-dimensional bands leads to the conductance of $4e^2/h$ (spin degeneracy included) in a straightforward manner from Eq. (2). In contrast, for voltages larger than 3.1V, the k vector increases to values where Bragg reflection of incident positive velocity states at some energies occurs (for example, close to the Fermi energy of the source). However, this happens in such a way that the current remains fixed until about 9.3V, when the band edge in the drain end of the tube and the Fermi energy of the source are comparable. Further increase in the applied voltage decreases the current. The overall feature of the decrease in current with increase in voltage can be understood as follows. Near the source end, the π electron band structure of the CNT extends

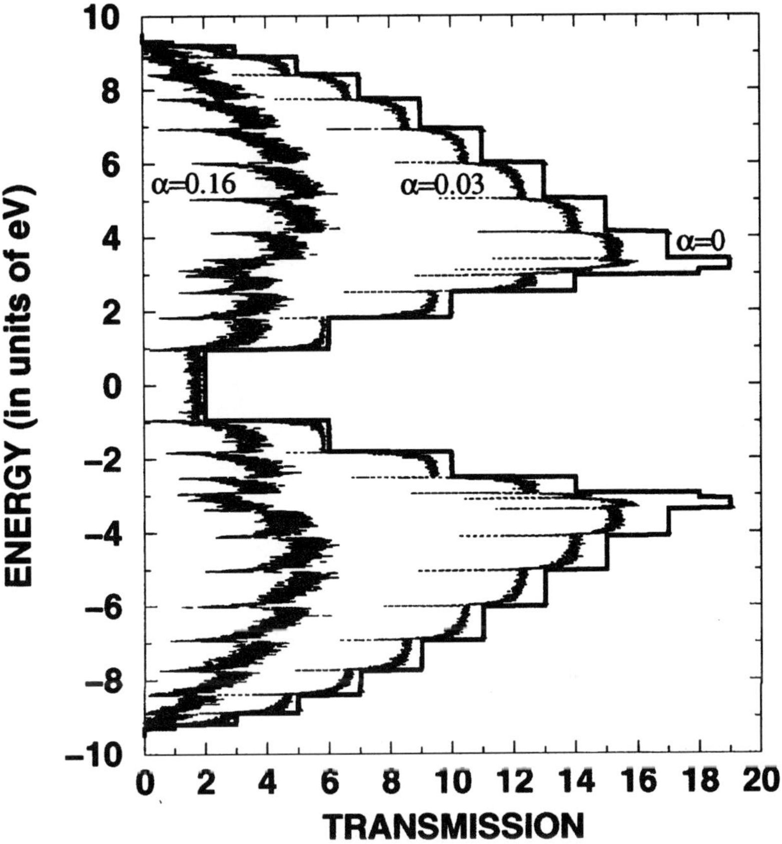

FIGURE 3. The transmission coefficient versus energy in the presence of disorder for a 1000 Å long tube. The disorder increases with increase in $\alpha = \varepsilon_{random}/3.1$ (see text).

from $-9.3eV + eV_{app}$ to $+9.3eV + eV_{app}$ and near the drain end, the band structure extends from $-9.3eV$ to $+9.3eV$. For current flow at an energy, the Fermi factor at this energy at the source and drain should be different and also an incident electron from the source (drain) at this energy should be able to scatter into a state at the drain (source) end at the same energy. The overlap between the bands at the source and drain ends decrease with increasing applied voltage, and for a large enough applied voltage the overlap becomes so small that the current begins to decrease before becoming zero. Note that this is an idealized picture that has neglected the effects of phonons and the other bands that are present in a more sophisticated band structure calculation.

The semiclassical discussion in the previous paragraph is not fully correct when the voltage drops across a short tube or if the screening length is very small because for large values of electric field in the tube, tunneling events between sub-bands, which were negligible at smaller fields, begin to contribute to the current.[13] An example of this is shown using a tube of length 140 Å and L_{sc} = 10 Å (FIG. 2). The deviation in the conductance at the larger applied voltages was consistent with increasing significance of tunneling events at greater electric fields.

Disordered Wires

FIGURE 3 illustrates the case of weak disorder distributed randomly in a tube of length 1000 Å. It is a plot of the transmission coefficient versus energy for various strengths of disorder. Our main observations are: (1) this type of disorder does not change the transmission coefficient significantly at energies near the center of the band; (2) the transmission coefficient versus energy develops sharp resonances; and (3) the transmission coefficient changes by large amounts near the opening and closing of sub-bands.

Weak disorder in the on-site potential introduces randomly distributed quasi-bound states along the length of the tube.[13] Tunneling across the tube via these quasi-bound states causes the resonances in FIG. 3. By increasing the degree of disorder,

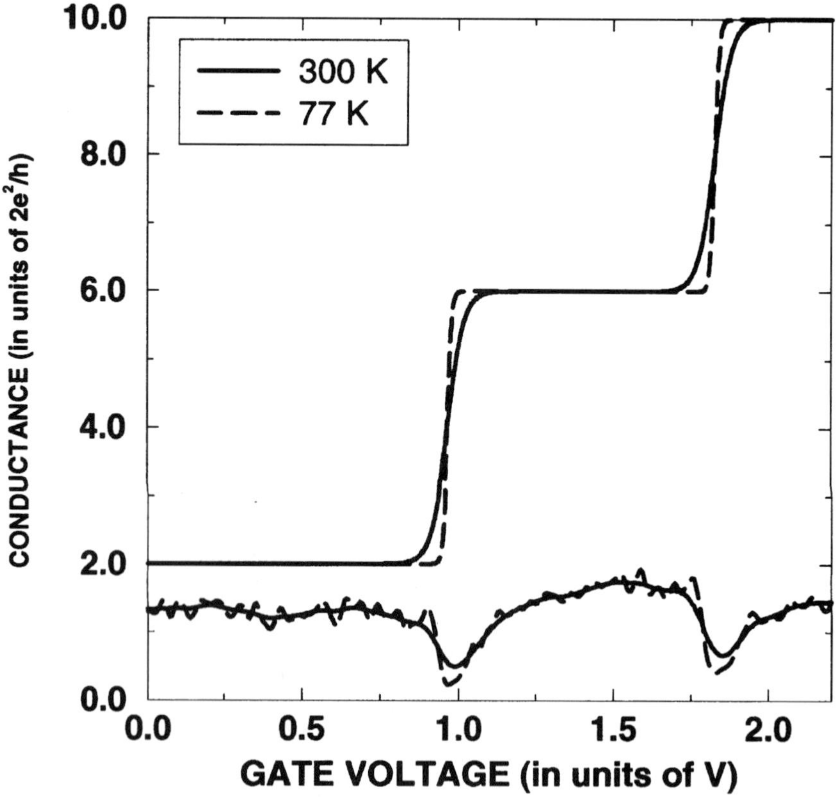

FIGURE 4. Small bias conductance versus gate voltage for the case shown in FIGURE 3. The conductance does not change significantly close to zero gate voltage.

the quasi one-dimensional states localize more strongly in position, leading to an overall decrease in the transmission coefficient. The decrease in the transmission coefficients at sub-band openings and closings in FIG. 3 was in accordance with the explanation that an infinitely long structure with a single sub-band localizes at the band edges, where the velocity is small.[13] A particle incident in one of the nondegenerate sub-bands (*dashed lines* of FIG. 2) at energies close to the second sub-band opening (~0.9 eV) scatters to states at the band edge of the second sub-band. This leads to reflection of the incident state as these low velocity states in the second sub-band are localized. As a result, there is a (111) in the transmission coefficient. The low bias conductance versus the gate voltage calculated using Eq. (2) indicated that the CNT behaves as a good wire, in the presence of weak uniform disorder (FIG. 4) when the Fermi energy was close to the band center. A much larger decrease in the conductance was realized when the Fermi energy was close to a sub-band opening or closing.

In the presence of a single substitutional defect, it was observed by Chico *et al.*[9] that the transmission coefficient decreased at all energies with a dip in the center of the band. The transmission coefficient versus energy characteristics was computed in the presence of many such defects to see if the conductance of $4e^2/h$ was affected significantly. Results are presented for a tube of 140 Å in length with ten defects sprinkled randomly along the tube. The most prominent feature in the transmission versus energy plot (FIG. 5) is that a gap opens up in the middle of the band. A few narrow resonances traverse the entire tube. Our simulations show that increasing the

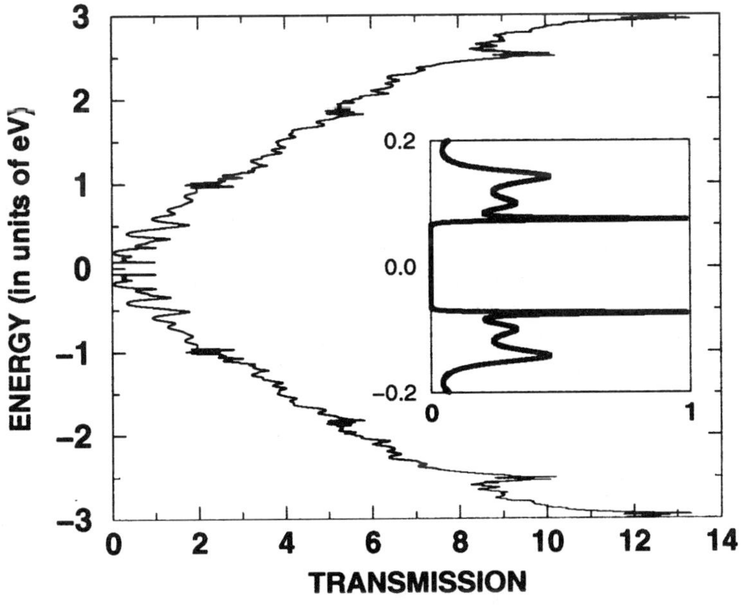

FIGURE 5. Transmission versus energy for a 140 Å long tube in the presence of substitutional defects. A gap opens in the transmission at the middle of the band (*inset*).

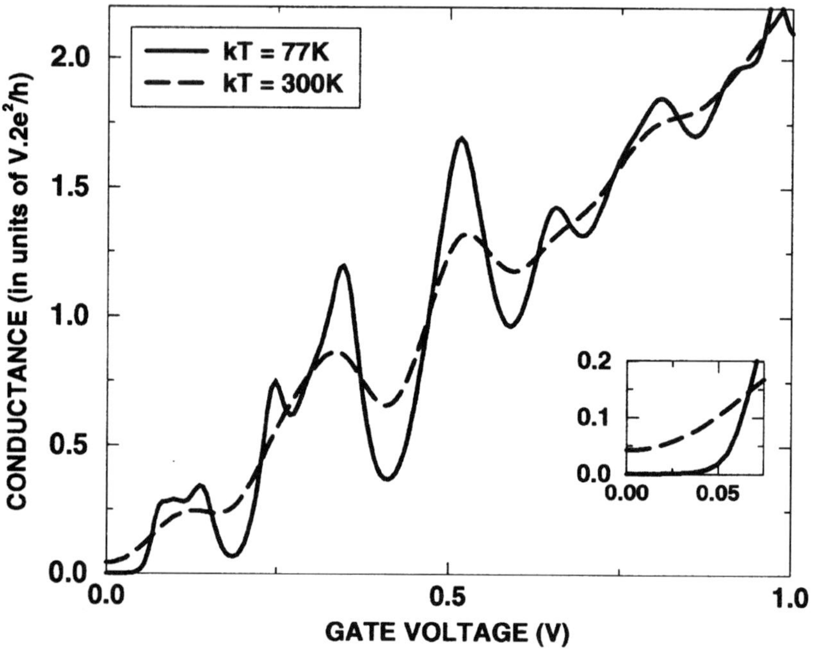

FIGURE 6. Small bias conductance versus gate voltage corresponding to the case in FIGURE 5. Note the significant drop in conductance close to zero gate voltage and the corresponding temperature dependence (*inset*).

number of defects removed these resonances in the transmission gap and further broadened it. The second feature was the stepwise increase in the transmission coefficient (FIG. 1) diminished leading to a comparatively linear behavior. Thirdly, similar to the previous case of disorder discussed, the transmission coefficient developed sharp resonances at energies corresponding to opening and closing of sub-bands. The conductance was calculated from Eq. (2). While the conductance was almost temperature independent in the defect-free case, in the presence of these defects, the conductance was strongly temperature dependent (FIG. 6). The small bias conductance close to zero bias was about forty times smaller than the defect-free value of $4e^2/h$ at a temperature of 300 K. Simulations were performed with different configurations of the defects and results obtained were qualitatively similar.

SUMMARY

The conductance properties of uniform tubes with and without defects were numerically calculated. Typical lengths of the CNT in these calculations were 140 Å and 1000 Å. In the defect-free case, the current-versus-voltage characteristics were computed for two different forms of the potential drop across the tube. We observed that if the electric field in the tube was small (which is the case for the longer tube), transport can be understood just in terms of the four nondegenerate bands. Some preliminary results for shorter tubes where deviations in the current-versus-voltage characteristics occurred were also presented. The second part of the paper attempts to study the influence of many defects in finitely long tubes. The transmission coefficient versus energy and the conductance characteristics were calculated in the presence of two models of defects, namely *weak uniform disorder* and *substitutional defects*. In the first case, we showed that the low bias conductance was not significantly affected for values of the Fermi energy close to the band center and that a sharp decrease in the conductance was observed when the Fermi energy matched that of the opening or closing of sub-bands. On the contrary, substitutional defects led to a strong change in the small bias conductance for values of the Fermi energy at the band center. The calculations were performed in the phase-coherent limit and by assuming perfect connection to the contacts.

REFERENCES

1. DRESSELHAUS, M.S., G. DRESSELHAUS & P. C. EKLUND. 1996. Science of Fullerenes and Carbon Nanobibes. Ch. 19. Academic Press. New York.
2. DAI, H., J.H. HAFNER, A.G. RINZLER, D.T. COLBERT & R.E. SMALLEY. 1996. Nanotubes as nanoprobes in scanning probe microscopy. Nature. **384:** 147–151; CARROLL, D.L., P. REDLICH, P.M. AJAYAN, J.C. CLIARLIER, X. BLASE, A. DE VITA AND R. CAR. 1997. Electronic structure and localized states at carbon nanotube tips. Phys. Rev. Lett. **78:** 2811–2814.
3. DER HEER, W.A., A. CHATELAIN & D. UGARTE. 1997. A carbon nanotube field-emission electron source. Science. **270:** 1179–1180.
4. TATIS, S.J., M. DEVORET, H. DAL, A. THESS, R.E. SMALLEY, L.J. GEERLIGS & C. DEKKER. 1997. Individual single-wall carbon nanotubes as quantum wires. Nature. **386:** 474–477.
5. COBDEN, D.H., M. BOCKRATH, N. CLIOPRA, A. ZETTL, P.L. MCEUEN, A. RINZLER, A. THESS & R.E. STRIALLEY. 1997. Transport spectroscopy of single-walled carbon nanotubes. Preprint.
6. BOCKRATH, M., D.H. COBDEN, P.L. MCELIEII, N.C. CHOPRA, A. ZETTL, A. THESS & R.E. SMALLEY. Single-electron transport in ropes of carbon nanotubes. Science. **275:** 1922–1925.
7. COLLINS, P.G., A. ZETTL, IL BANDO, A. THESS & R.E. SMALLEY. 1997. Nanotube nanodevice. Science **278:** 100–103.
8. CHARLIER, J.-C., T.W. EBBESEN & PH. LAMBIN. 1996. Structural and electron properties of pentagon-heptagon pair defects in carbon nanotubes. Phys. Rev. B **53:** 11108–11113.
9. CHICO, L., L.X. BENEDICT, S.G. LOUIE & M.L. COHEN. 1996. Quantum conductance of carbon nanotubes with defects. Phys. Rev. Lett. **54:** 2600–2606.
10. LOUIE, S.G. 1996. Proceedings of the Robert A. Welch Foundation. 40th Conference on Chemical Research, Chemistry on the Nanometer Scale. Houston, Texas. October 21–22, 1996.
11. DATTA, S. 1995. Electronic Transport in Mesoscopic Systems, Cambridge University Press, Cambridge, U.K.
12. TEKMAN, E. & S. CIRACI. 1991. Theoretical study of transport through a quantum point contact. Phys. Rev. B. **43:** 7145–7169.
13. ZIMAN, J. 1979. Models of Disorder: The Theoretical Physics of Homogeneously Disordered Systems. Cambridge University Press. Cambridge, U.K.

Carbon Nanotubes: Molecular Electronic Components

DEEPAK SRIVASTAVA,[a] SUBHASH SAINI, AND MADHU MENON[b]

[a]*NASA Ames Research Center, MRJ, Mail Stop T27-Al, Moffett Field, California 9403-51000*

[b]*Department of Physics and Astronomy, University of Kentucky, Lexington, Kentucky 40506*

ABSTRACT: The carbon nanotube junctions have recently emerged as excellent candidates for use as the building blocks in the formation of nanoscale molecular electronic networks. While the simple joint of two dissimilar tubes can be generated by the introduction of a pair of heptagon-pentagon defects in an otherwise perfect hexagonal graphene sheet, more complex joints require other mechanisms. In this work we explore structural characteristics of complex 3-point junctions of carbon nanotubes using a generalized tight-binding molecular-dynamics scheme. The study of pi-electron local densities of states (LDOS) of these junctions reveal many interesting features, most prominent among them being the defect-induced states in the gap.

INTRODUCTION

The interest in pure carbon clusters received a tremendous boost by the discovery of the C_{60} molecule.[1] Following this discovery, quasi one-dimensional structures made of carbon atoms, called carbon nanotubes were experimentally observed by Iijima.[2] The carbon nanotubes consist of rolled-up graphene sheet with various chiralities. The electronic structure of these tubes can be either metallic or semiconducting, depending on both the diameter and chirality which can be uniquely determined by the chiral vector (n , m), where n and m are integers. [3–7]

The possibility of connecting nanotubes of different diameter and chirality has generated considerable interest recently.[8–12] This is because of the possibility of the junctions being the building blocks of nanoscale electronic devices. The simplest way to connect two dissimilar nanotubes is found to be via the introduction of a pair of heptagon and pentagon in an otherwise perfect hexagonal graphene sheet.[6] The resulting structure still contains threefold coordination for all carbon atoms forming the junction.

In reality nanotubes have finite lengths and, in most cases, tend to be closed with fullerene caps. The closure introduces a small gap in the electronic structures of these tubes. Also, formation of tube junction results in considerable local strain which is relieved by the relaxation of the atoms. The relaxation is also expected to alter the electronic structure and local density of states (LDOS) of the atoms forming the junction and their neighbors. Although several theoretical models have been used in the study of nanotube heterojunctions, most of them tend to ignore the effects of relaxation altogether.

In this work we investigate the effects of full symmetry unconstrained relaxation of the nanotube heterojunctions on the structural and electronic properties using the generalized tight-binding molecular dynamic (GTBMD) scheme of Menon and Subbaswamy.[7] The GTBMD makes explicit use of the non-orthogonality of the orbitals in treating interactions in covalent systems and has been found to be very reliable in obtaining good agreement with experimental and local density approximation (LDA) results for the structural and vibrational properties of fullerenes and nanotubes.[7,13] Additionally, GTBMD has been applied earlier to obtain equilibrium geometries for small carbon clusters,[14] in good agreement with *ab initio*[15] results for the lowest energy structures of carbon clusters of size up to $N = 10$ (for which *ab initio* results are available).

THREE-POINT JUNCTIONS

T-Junctions

Unlike the simple nanotube junctions, the three-point junctions of single-walled carbon nanotubes could be used as building blocks of nanoscale tunnel junctions in a 2D network of nanoelectronic devices. As a prototype of such junctions we study a T-junction formed by fusing two nanotubes of different diameter and chirality perpendicular to each other.[17]

The T-junctions provide a challenge to the conventional rules applicable to tube bends. This is because unlike the knee joint where one can clearly define the opposite sides of the joint as either the front or the behind, both sides are topologically equivalent. As a result, we can expect a net excess of heptagons over the pentagons at the junction. Furthermore, whereas the bend angles at a two-point tube junction depend on the tube parameters (diameter and chirality) of both the component tubes, no such dependence exists for the T-junction where the angle remains fixed at 90°.

We explore an alternative route to the formation of T-junctions that is not constrained by the usual heptagon-pentagon defect pair considerations. In particular, we examine two metal-semiconductor-metal T-Junctions, namely the (5,5)-(10,0)-(5,5) (FIG. 1a) and (9,0)-(10,0)-(9,0) (FIG. 1b) junctions. We designate the former by the symbol **T1** and the latter by the symbol **T2**. The structure **T1** is composed of 314 atoms while **T2** contains 304 atoms. The numbers chosen are sufficiently large enough to avoid the effects of the dangling π bonds at the edges on the junction. The (10,0) tube is semiconducting, and the (9,0) and (5,5) tubes are semi-metals. The (5,5), (10,0), and (9,0) tubes have armchair, zig-zag, and zig-zag configurations, respectively. In the armchair (zig-zag) configuration the tubes have C—C bonds perpendicular (parallel) to their axis. As seen in FIGURE 1a, in going across the junction in **T1** from the (10,0) side to the (5,5) side, the orientation of the C—C bonds remains unchanged. Crossing the junction in **T2**, on the other hand, involves rotation of 90° for the C—C bonds. Interestingly, **T1** contains six heptagons and no pentagons at the junction (FIG. 1a). **T2** contains eight heptagons and two pentagons at the junction (FIG. 1b). The presence of the additional two heptagon-pentagon pairs in **T2** is as-

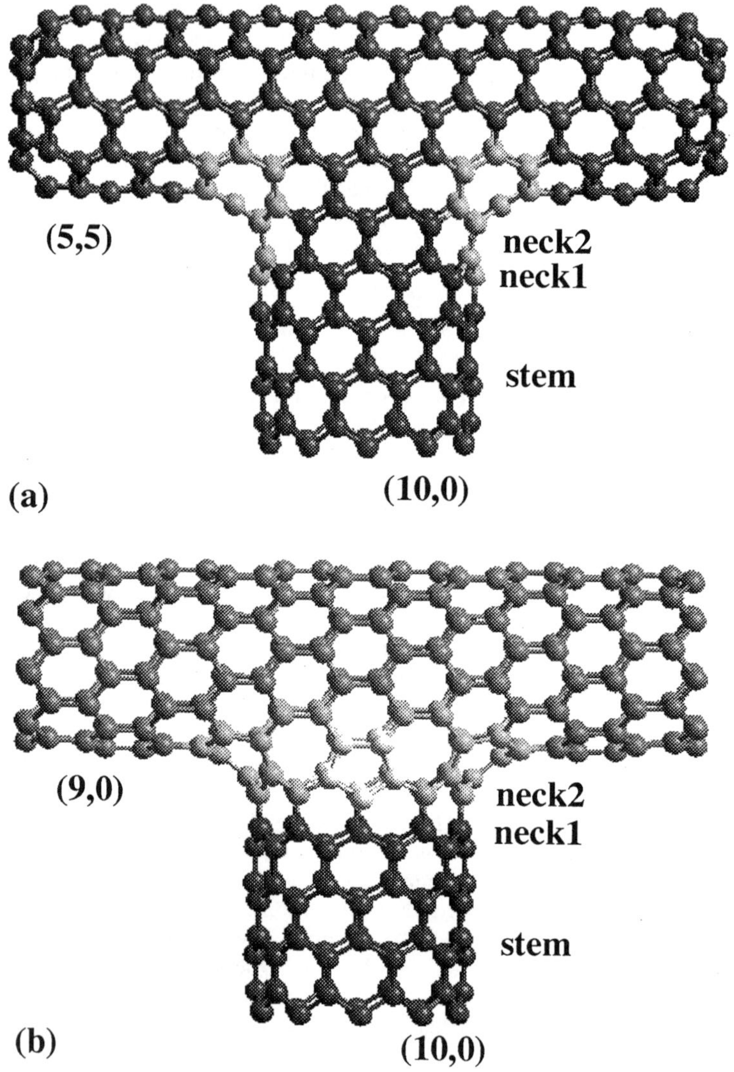

FIGURE 1. **(a)** Fully relaxed (5,5)-(10,0)-(5,5) tube (**T1**). The structure contains six heptagons and no pentagons. **(b)** Fully relaxed (9,0)-(10,0)-(9,0) tube (**T2**). The structure contains eight heptagons and two pentagons.

sociated with the change in orientation of the C–C bonds in going across the junction.

The geometries shown in FIGURE 1 are the GTBMD optimized structures for **T1** and **T2**, respectively. The starting configuration of **T1** has twofold coordinated at-

TABLE 1. Bond length analysis for the T-junction in **T1** and **T2**. The GTBMD scheme has been used to relax both clusters. The bond lengths are averaged over the pentagons, heptagons, and the hexagons

Tube	Average Bond Length in Pentagons (Å)	Average Bond Length in Heptagons (Å)	Average Bond Length in Hexagons (Å)
T1		1.427	1.419
T2	1.43	1.419	1.418

oms at the ends of the armchair portion of the tubule within strong bonding interactions of each other. The GTBMD relaxation results in the closure seen in FIGURE la on relaxation with threefold coordination for all atoms at the armchair ends. No such closure results at the zig-zag ends, however, as the twofold coordinated atoms are sufficiently far from each other. TABLE 1 summarizes the average bond lengths in the pentagons, heptagons and hexagons at the junctions of **T1** and **T2**. In **T1**, the average bond length in the heptagons is found to be longer than the one for the hexagons. In **T2**, however, the average bond length in the heptagons is nearly equivalent to that in the hexagons. Also, as seen in TABLE 1, the pentagons in **T2** have larger bond length on the average when compared to heptagons or hexagons. Both the pentagons in **T2** are almost planar, with the heptagons bearing the brunt of the curvature with slight assistance from hexagons that share edges with the heptagons.

The Fermi levels of the (5,5) nanotube lie within the gap of the (10,0) semiconducting tube. The T-junction forms microscopic tunnel junction, made up entirely of carbon atoms, through which electrons can cross by quantum mechanical tunneling. The tunneling current can be controlled by an application of a potential difference that raises the chemical potential of one side with respect to the other. Furthermore, either n-type or p-type doping of the semi-conducting portion of the T-junction should yield Schottky barrier type devices.

We investigate the local density of states (LDOS) using the tight-binding π-band approximation. Only nearest-neighbor interactions are considered with $Vpp\pi$ = 2.66 eV.[16] FIGURE 2 shows the LDOS for the relaxed (5,5)-(10,0)-(5,5) structure at various cross sections indicated in FIGURE 1. LDOS for panel labeled stem, for example, is for the cross section containing 20 atoms zig-zagging along the circumference of the semi-conducting (10,0) portion. As can be seen in FIGURE 1, in going from the semi-conducting side into the junction, localized states begin to appear in the gap. Detailed study indicates their origin to be the heptagonal defects present in the neck region. These defect states may pin the Fermi level of the system.

Y-Junctions

The Y-junctions pose a different kind of challenge when three-point junctions are formed. As found in the case of T-junctions, the pentagon-heptagon defect pair rule, is also found to be not applicable in the formation of the Y-junctions. We, therefore, explore the formation of large angle bends through octagon-pentagon defect pairs and apply it to study Y-junctions which provide another route to the formation of 3-point junction of dissimilar tubes.

We show that the Y-junctions can be made with the incorporation of pentagons and octagons with no assistance from heptagons, and that incorporation of octagons

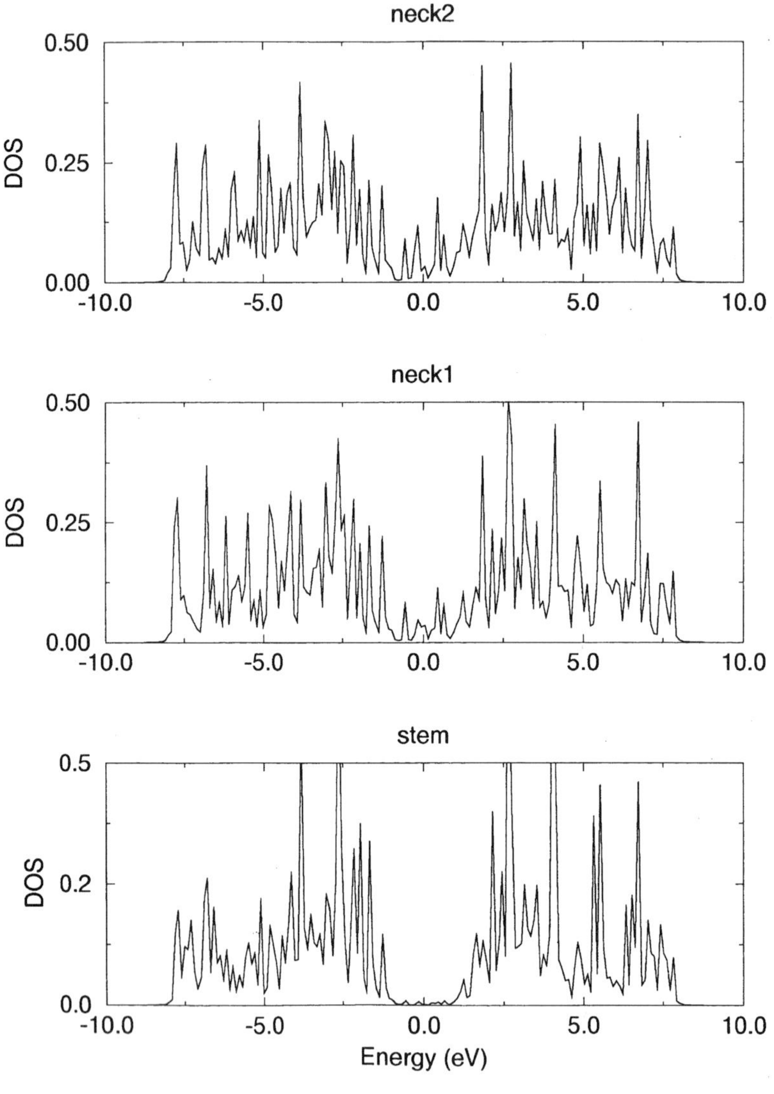

FIGURE 2. LDOS for the relaxed (5,5)-(10,0)-(5,5) structure at various cross sections indicated in FIGURE 1. showing the increase in the defect induced localized states in the gap as the junction is approached.

in place of heptagons gives greater flexibility in the Y-junction formation involving similar or dissimilar nanotubes. Additionally, we find that the number of octagonal defects equal the number of pentagonal defects in the junction region, regardless of whether or not there is a change in chirality when crossing the junction. In particular,

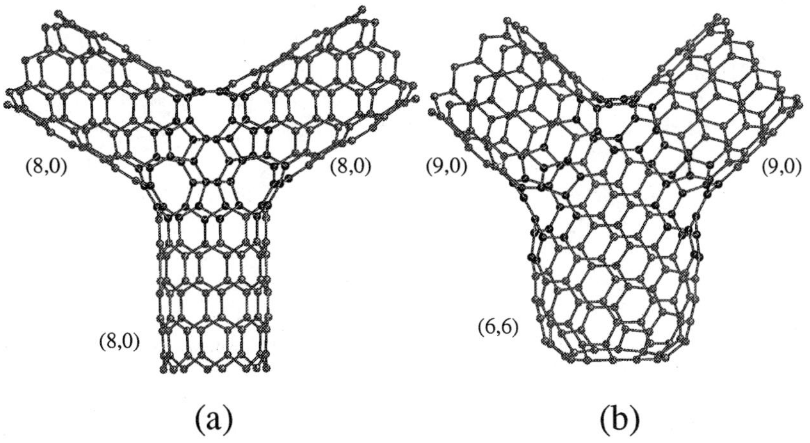

FIGURE 3. **(a)** Fully relaxed Y-junction (I) with (8,0) tubes forming all three arms. **(b)** Relaxed Y-junction (II) with two (9,0) and one (6,6) arm. Both these structures contain six octagons and six pentagons and no heptagons. The carbon atoms forming octagons and pentagons are shown as dark circles.

we examine three distinct Y-junctions made up of nanotubes in arm-chair as well as zigzag configurations. The three different joints investigated are: Y-junction formed by (I) three (8,0) nanotubes, (II) two (9,0) and one (6,6) nanotubes, and (III) two (9,0) and one (8,0) nanotubes. These systems are selected because they encompass structures with radically differing electronic properties.

All these structures were fully optimized without any symmetry constraints using the GTBMD scheme. The three (8,0) tubes forming the arms of structure (I) are each semiconducting and have zig-zag configurations. The structure is shown in FIGURE 3a and contains 252 atoms. The carbon atoms forming octagons and pentagons are shown as dark circles. The number of atoms chosen are sufficiently large enough to avoid the effects of the dangling π bonds at the edges on the junction. On relaxation (I) was found to be stable with six octagons and six pentagons and, interestingly, no heptagons. The brunt of the curvature is borne out by octagons and the hexagons at the junction while the pentagons are almost planar. Also, as seen in FIGURE 3a, all the pentagons are clustered near the center of the junction. The atomic arrangement in (I) is radically different from the case of simple tube bends where both pentagons and heptagons participate in the tube bend. TABLE 2 summarizes the average bond lengths in the pentagons, octagons, and hexagons at the junctions of (I).

The two arms of structure (II) (FIG. 3b) consist of metallic (9,0) tubes that have zig-zag configurations. The third (6,6) arm is also metallic, but with arm-chair configuration. This junction also contains six octagons and six pentagons. This structure contains 258 atoms. Again, the carbon atoms forming octagons and pentagons are shown as dark circles. The octagons are boat shaped as in (I) and bear the brunt of the curvature. Two of the pentagons are located on the (6,6) side and contribute to the curvature while the remaining four pentagons are almost planar. Also, the pentagons are farther out from the center than in (I). This is due to the change in diameter and chirality in going from (9,0) to the (6,6) arm. The GTBMD relaxation results

TABLE 2. Bond length analysis for the Y-junctions in (I), (II), and (III). The GTBMD scheme has been used to relax these clusters. The bond lengths are averaged over the pentagons, octagons, and the hexagons

Tube	Average Bond Length in Pentagons (Å)	Average Bond Length in Octagons (Å)	Average Bond Length in Hexagons (Å)
(I)	1.438	1.418	1.418
(11)	1.435	1.434	1.414
(111)	1.440	1.422	1.417

in the closure seen in FIGURE 3b with threefold coordination for all atoms at the armchair ends. The formation of squares at the armchair ends is plausible considering the fact that the system can lower its energy by forming 3-fold coordination rather than keeping the dangling bonds.

All three arms of structure (III) have zig-zag configurations, but the (9,0) arms are metallic, while the (8,0) arm is semi-conducting. Thus, this structure forms

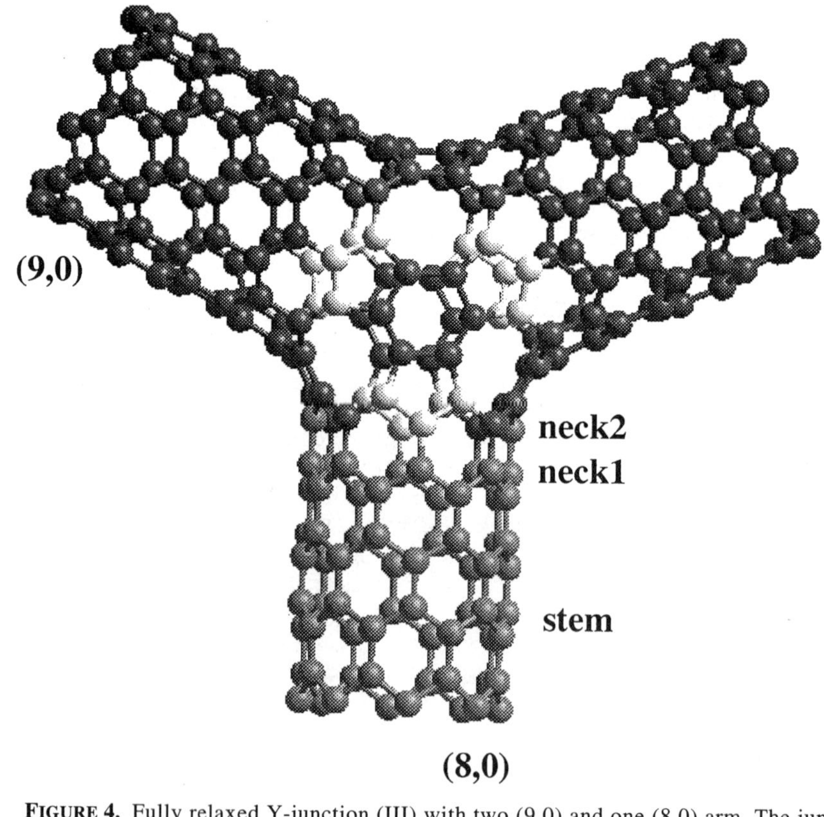

FIGURE 4. Fully relaxed Y-junction (III) with two (9,0) and one (8,0) arm. The junction is formed by six octagons and six pentagons.

nanoscale metal- semi conductor-metal tunnel junction. The relaxed structure is shown in FIGURE 4 and contains 272 atoms. It also contains six octagons, six pentagons, and no heptagons. As in (1), the brunt of the curvature is borne out by octagons and the hexagons at the junction while the pentagons are almost planar. Also, as in (I), all the pentagons are clustered near the center of the junction since no change in chirality results in crossing the junction.

TABLE 2 summarizes the average bond lengths in the pentagons, octagons, and hexagons at the junctions of (I), (II), and (III). The average bond length in the octa-

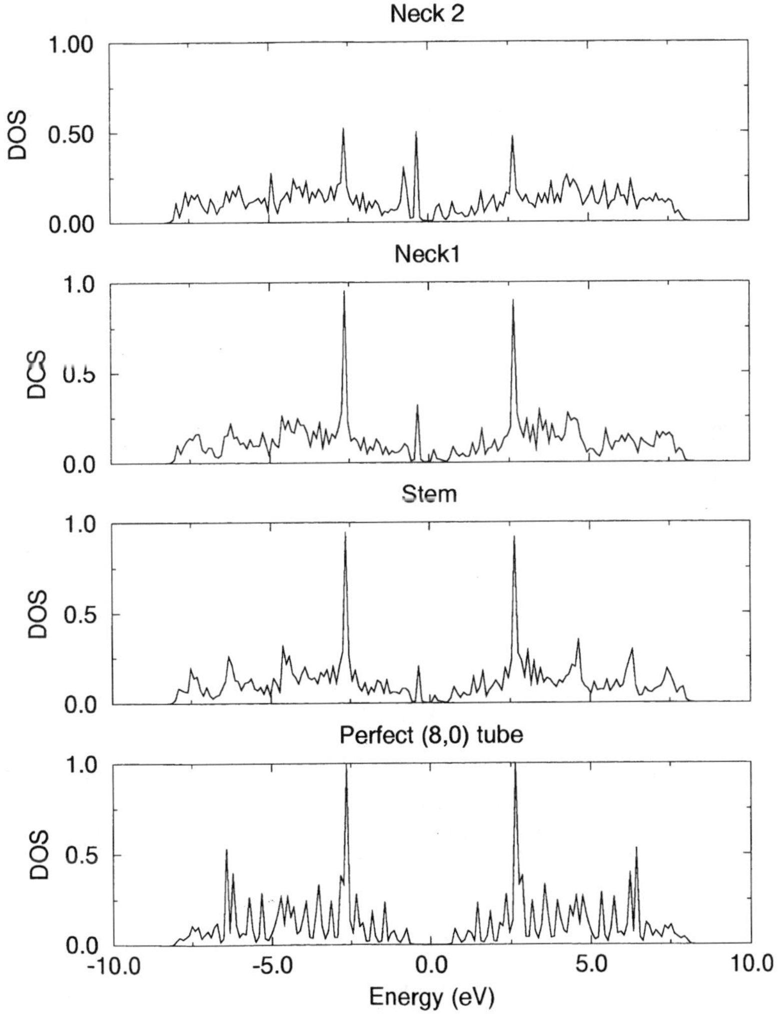

FIGURE 5. LDOS for the relaxed Y-junction (111) at various cross-sections indicated in FIGURE 4. The contribution in the gap is due to presence of octagon-pentagon defects at the junction.

gons is found to be longer than the one for the hexagons, but shorter than those in the pentagons. In all junctions the octagons bear the brunt of the curvature with slight assistance from hexagons that share edges with the octagons.

We investigate the local density of states (LDOS) using the same tight-binding π-band approximation used to study the T-junctions. FIGURE 5 shows the LDOS for the relaxed Y-junction (III) at various cross sections indicated in FIGURE 4. LDOS for panel labeled stem, for example, is for the cross section containing 16 atoms zigzagging along the circumference of the semi-conducting (8,0) portion. As can be seen in FIGURE 5, in going from the semi-conducting side into the junction, localized states begin to appear in the gap. Detailed study indicates their origin to be the octagonal and pentagonal defects present in the neck region. These defect states may pin the Fermi level of the system.

In light of the present work, serious considerations must be given to the possibility of existence of octagonal defects in complex junctions. The octagonal defects may give greater flexibility for such junction formation involving similar or dissimilar nanotubes. The search for larger angle bends or Y-junction formation in the existing experimental data on single and multi-wall nanotubes may point towards the possibility of octagonal defects in the nanotube-based pure carbon structures.

SUMMARY

The implications of these findings are intriguing. Both T-junctions and Y-junctions defy the conventional arguments made in favor of equal number of heptagon-pentagon defect pairs for the stability of dissimilar tube joints. We have shown that T-junction joints can be made without the incorporation of pentagons. Our results also show that Y-junctions can be created with an equal number of pentagons and octagons and, surprisingly, no heptagons. The presence of large localized states in the gap that can pin the Fermi level has interesting implications for complex device modeling. Furthermore, the T-junctions can be used as "universal joints" for forming a 2D network of tubes in which conduction pathways can be controlled. If produced, these junctions could be the prototypes of nanoscale tunnel devices.

ACKNOWLEDGMENTS

Part of this research (MM) was supported by National Science Foundation grant OSR 94-52895, and by the University of Kentucky Center for Computational Sciences. GTBMD relaxation and electronic structure computations were performed at NAS computational facility (DS) at NASA Ames Research Center.

REFERENCES

1. KROTO, H.W., J.R. HEATH, S.C. O'BRIEN, R.F. CURL, & R.E. SMALLEY. 1985. Nature (London) **318:** 162.
2. IIJIMA, S. 1991. Nature **354:** 56.
3. MINTMIRE, J.W., B. L DUNLAP & C.T. WHITE. 1992. Phys. Rev. Lett. **68:** 631.

4. SAITO, R., M. FUJITA, G. DRESSELHAUS & M.S. DRESSELHAUS. 1992. Phys. Rev. **B46:** 1804.
5. HAMADA, N., S. SAWADA & A. OSHIYAMA. 1992. Phys. Rev. Lett. **68:** 1579.
6. DUNLAP, B.I. 1994. Phys. Rev. **B49:** 5643.
7. MENON, M., E. RICHTER & K.R. SUBBASWAMY. 1996. J. Chem. Phys. **104:** 5875.
8. CHICO, L., V.H. CRESPI, L.X. BENEDICT, S.G. LOUIE & M.L. COHEN. 1996. Phys. Rev. Lett. **76:** 971.
9. CHARLIER, J.C., T.W. EBBESEN & PH. LAMBIN. 1996. Phys. Rev. **B53:** 11108.
10. SAITO,R., G. DRESSELHAUS & M.S. DRESSELHAUS. 1996. Phys. Rev. **B53:** 2044.
11. IIJIMA, S., T. LCHIHASHI & Y. ANDO. 1992. Nature **356:** 776.
12. LAMBIN, PH., A. FONSECA, J. VIGNERON, J.B. NAGY & A.A. LUCAS. 1995. Chem. Phys. Lett. **245:** 85.
13. RAO, A.M., E. RICHTER, S. BANDOW, B. CHASE, P. C. EKLUND, K. A. WILLIAMS, S. FANG, K. R. SUBBASWAMY, M. MENON, A. THESS, R. E. SMALLEY, G. DRESSELHAUS & M.S. DRESSELHOUSE. 1997. Science **275:** 187.
14. MENON, M., K.R. SUBBASWAMY & M. SAWTARIE. 1993. Phys. Rev. **B48:** 8398. (1993).
15. RAGHAVACHARI, K. & J.S. BINKLEY. 1987. J. Chem. Phys. **87:** 2191.
16. BLASE, X., L.X. BENEDICT, E.L. SHIRLEY & S.G. LOUIE. 1994. Phys. Rev. Lett. **72:** 1878.
17. MENON, M. & D. SRIVASTAVA. 1997. Phys. Rev. Lett. **79:** 4453.
18. KROTO, H.W. & D.RM. WALTON. 1993. Chem. Phys. Lett. **214:** 353.
19. MENON, M. & D. SRIVASTAVA. Phys. Rev. Lett. Submitted for publication.

A Self-Assembled Single-Electron Tunneling Device

S.H.M. PERSSON,[a,c,d] L. OLOFSSON,[a] L. HEDBERG,[b] D. SUTHERLAND,[b] AND E. OLSSON[a]

[a]*Department of Microelectronics and Nanoscience,* [b]*Department of Applied Physics, and* [c]*Department of Microwave Technology*
Chalmers University of Technology and
University of Göteborg, S-412 96 Göteborg, Sweden

ABSTRACT: Single-electron tunneling devices were made by self-assembly of colloidal ligand-stabilized gold clusters in the small gap between two gold electrodes, which were covered with a self-assembled monolayer of 1,8-octanedithiol. With this method we control the size of the active part of the device which is smaller than the traditional lithographic resolution, with chemical methods and use self-positioning of the particle by weak forces between the substrate and the particle. The current voltage characteristic of the devices exhibit a Coulomb staircase and gate effect at 4.2 K, and a prominent Coulomb blockade at room temperature.

When the size of electronic devices decreases down to the nanometer range and molecular dimensions, it will be possible to observe and utilize single-electron tunneling[1] and quantum effects in electronics at room temperature. However, there is no lithographic tool to make such small features. The minimum size that is possible to make with electron beam lithography and most techniques that use scanning probes is limited to approximately 10 nm. At this limit, variations in size corresponding to atomic layers are unavoidable, but nevertheless critical to device performance. Size control is essential for uniformity and reproducibility and it can be accomplished by using chemical synthesis for the most critical parts, which could be metallic or semiconducting clusters with a very narrow size distribution, or well-defined organic molecules. Chemical synthesis allow nanoscale structures to be built atom by atom, that is, from the bottom, and the interfacing to electrical leads, which is less sensitive to size variations, can be made by conventional lithographic techniques.

Aviram and Ratner[2] proposed the molecular rectifier in 1974, but it was not until recently that molecules have been contacted at the molecular-scale and rectification shown,[3] although rectifying behavior have been observed in macroscopic scale devices.[4] There are several problems related to miniaturization of electronics to the nanometer scale, and the cost for processing can rise rapidly if new methods are not developed. One problem is the alignment of several nanometer-scaled features on the same chip which requires extremely small temperature gradients and implies that the choice of process should be a self-aligned, self-limited, and self-assembled one. A promising route is to use conducting molecules with reactive end groups that can

[d]Address for telecommunication: e-mail, persson@ep.chalmers.se

form self-assembled monolayers (SAM). Hence, a well-defined molecule may be attached on specific metallic surfaces between two metal electrodes.[5,6]

The recent reports on the creation of nanocrystals of metals and semiconductors that can be treated and manipulated as macromolecules make a new approach possible for manufacturing nanoscale transistors. Klein *et al.*[7] have made a single-electron transistor with CdSe clusters trapped between two electrodes and they change the number of electrons on the cluster by applying a voltage to the underlying substrate. The CdSe clusters are chemisorbed on a self-assembled monolayer (SAM) of 1,6-hexanedithiol which covers the gold electrodes. Bezryadin *et al.*[8] used electrostatic forces to trap palladium particles between platinum electrodes, and they observe charging effects in their device. Sato *et al.*[9] have used a self-assembly method to make a multiple junction device, with single-electron transistor behavior at 4.2 K.

We have used a similar method as Klein *et al.*[7] to capture gold clusters in a electrode gap that is smaller than 10 nm. A Coulomb staircase was observed at 4.2 K and a Coulomb blockade was seen at room temperature. It was possible to modulate the current voltage characteristics of the device by applying a voltage to a gate electrode. The gold electrodes covered with 1,8-octanedithiol and the clusters were stabilized by a organic surfactant of 1-dodekanethiol.[10,11] In our case the ligand shell prevents the cluster from an immediate covalent bonding to the electrode. The cluster can move on the surface and stop where attractive forces between it and the substrate are strongest, which is at steps and in the gap between the electrodes. It is possible that covalent bonding can stabilize the position of the cluster by replacing alkanethiol molecule on the cluster by a dithiol molecule already bonded to the electrode surface.

LITHOGRAPHY AND SYNTHESIS OF COLLOIDAL GOLD

The lithograpic patterning of the devices (seen in FIGURE 1) was made in three steps. The first is photolithographic to define the contact pads on an oxidized silicon wafer by deposition of 20 nm NiCr and 80 nm Au. The second is a lift-off process with an electron beam patterned two-layer resist to define a gate electrode. This step was not included on all samples. The bottom resist layer is a 140-nm-thick copolymer and the top layer is 60-nm-thick PMMA. The gate electrode is made by evaporation of four layers: NiCr (40 Å), Au (100 Å), Al (100 Å), and Al_xO_y (20 Å). The aluminium is oxidized in the evaporation chamber before AlxOy is evaporated. The native and evaporated oxides form a barrier a few nanometers thick between the gate and the gap electrodes. The NiCr layer is chosen for good adhesion, the gold layer for good conductivity, the aluminium because it is possible to make a good quality oxide in the evaporation chamber, and the evaporated oxide to make the barrier thicker. The top electrodes are made of 40 + 40 Å NiCr and 100 + 100 Å Au, evaporated with the substrate tilted at two different angles. The angles are adjusted to control the gap between two electrodes. There is a third lateral electrode with a 30- to 40-nm distance from the gap electrodes. FIGURE 1 shows a schematic drawing and one scanning electron micrograph of a device. The distance between the gap can be made smaller than 10 nm, but it is very hard to resolve and determine the actual distance with the used scanning electron microscope.

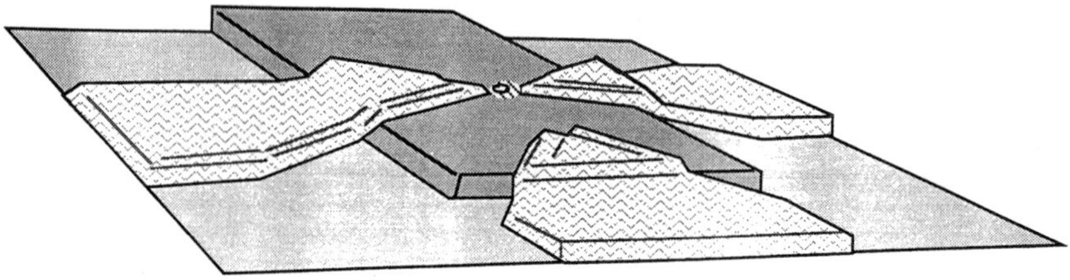

FIGURE 1a. Schematic drawing of the device geometry with three lateral electrodes and a gate electrode beneath.

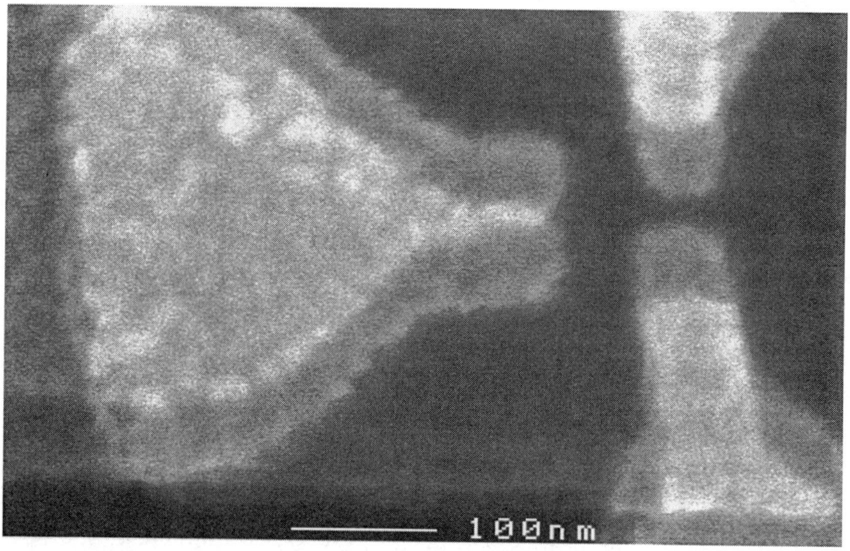

FIGURE 1b. Scanning electron microscope image showing a device with a 10-nm gap.

The chemical synthesis of the gold particles follows the scheme used by Burst *et al.*,[10] where tetrachloroaurate was reduced in a liquid two-phase system in the presence of dodecanethiol. The size of the particles was controlled by the molar ratio between gold and alkylthiol molecules.[11] The alkanethiol molecules cover the whole surface of the particles and prevent agglomeration and make these particles stable and easy to handle. The particles are thermodynamically stabilized by the surfactant molecules and can be dried and then again dissolved in polar solvents. Whetten *et al.*[12] showed that it is possible to make monodisperse clusters by aggregation of large particles and separation by filtering in several steps. We have not performed any purification of our particles, but it is important that clusters with equal size can be made. A molar ratio of 3.5 : 1 between Au and thiol was used in synthesis and that

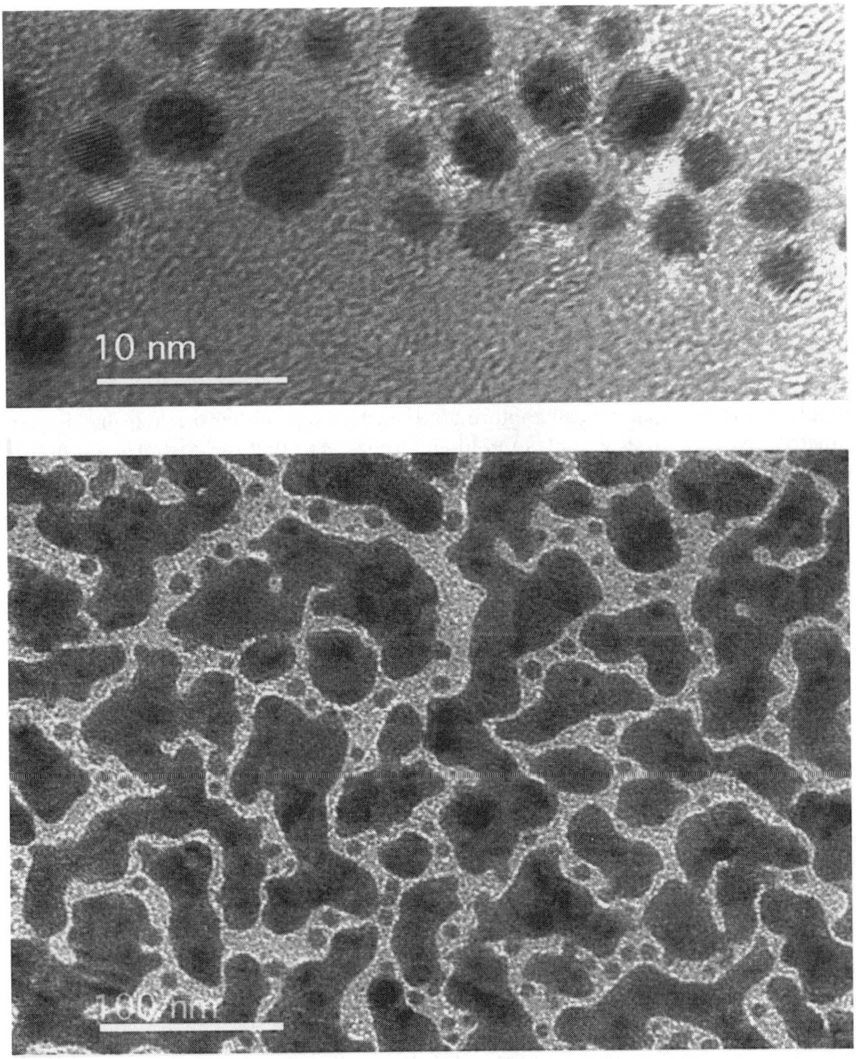

FIGURE 2. **(a, *top*)** Transmission electron micrograph showing gold clusters, with sizes in the range 2–5 nm. **(b, *bottom*)** TEM image of clusters on an evaporated granular gold film covered with 1,8-octanedithiol SAM. The gaps between the grains has the same width as the electrode gap. The clusters that are adsorbed can be seen in these gaps.

should result in a mean particle diameter of 5 nm.[11] We have observed sizes in a range from 2 to 6 nm with the transmission electron microscope. FIGURE 2a shows a TEM-image of such particles on a graphite film.

The substrate was, after the angled evaporation of the gap electrodes, immersed in acetone for lift-off and rinsed in isopropanol and blown dry with nitrogen, placed in a container with 1,8-octanedithiol dissolved in hexane for 1 to 24 hr, where a SAM is formed on the gold surfaces. It was thereafter rinsed in hexane and directly put into a hexane solution of gold particles for 12–24 hours. Terrill *et al.*[13] have studied alkanethiol stabilized particles on flat surfaces with STM and found that particles stick to steps and edges on the surface. The particles are mobile on the surface without electrostatic forces[14] and move to a position where the attractive forces are strongest. Experiments with large particles have shown that passive surface microstructures can trap and induce controlled drift of particles.[15] We assume that van der Waals forces attract particles to the surface and that the interaction is stronger in the gap where the particle is surrounded by the alkanedithiol-covered electrodes, compared to a flat surface. Experiments with adsorption of colloidal gold on 1,8-octanedithiol SAM and on 1-octanethiol SAM showed that equal mass was adsorbed on the surface, but a clean gold surface adsorbed less particles. The mass increase was measured with a quartz microbalance during the adsorption process. It is therefore possible that we can capture gold clusters with either methyl- or thiol-terminated SAM, but not with clean gold electrodes, and that the particles are not directly chemisorbed on the surface. It is possible that clusters are chemisorbed on dithiol SAM and more strongly bound after a longer time, but we have not observed any evidence for this.

FIGURE 2b shows particles adsorbed on a evaporated granular gold film. The gold grains are covered with 1,8-octanedithiol and there exist several gaps that have the same width as the electrode gap. We observe that there are quite few particles adsorbed and that they are positioned in the grooves and in the gaps between the grains. The mobility of the particle and the stronger attractive force in the gap give a higher probability for capturing a particle in the gap, that is, the alkylthiol shell prevents the particles from being stuck anywhere on the surface and keep them mobile until they

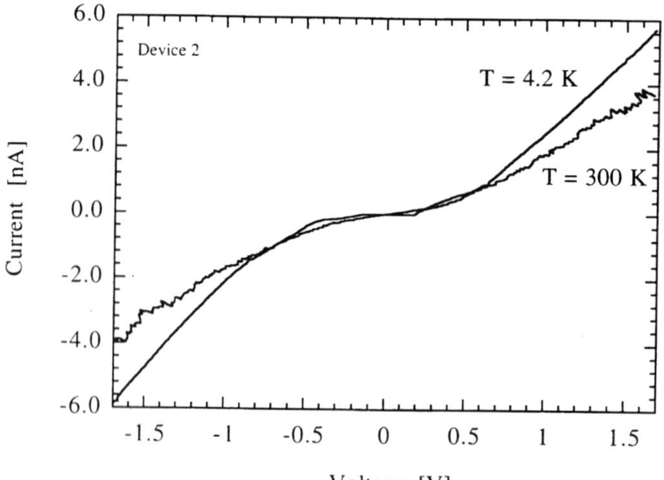

FIGURE 3. Coulomb blockade was observed at room temperature, and when the sample was immersed in liquid helium the Coulomb staircase becomes visible.

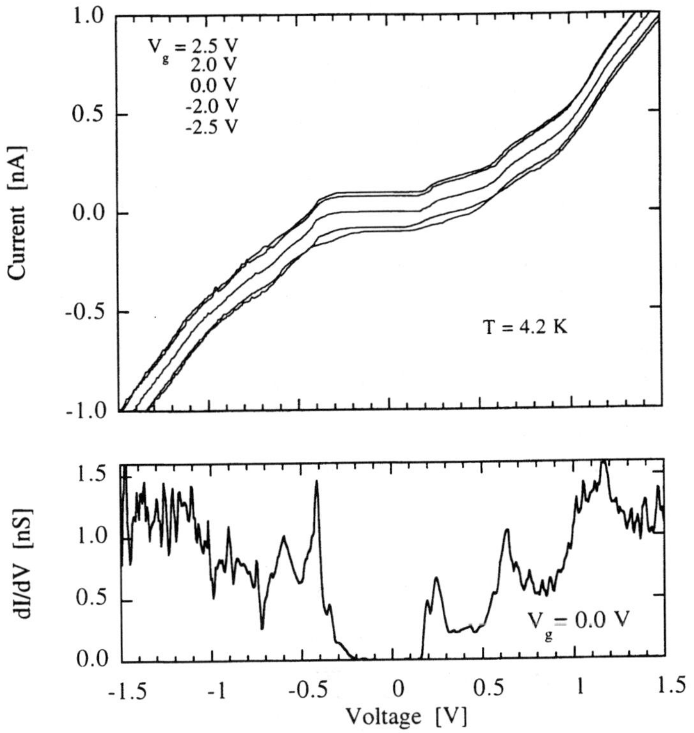

FIGURE 4. **(a, *top*)** IV-curves with different voltages applied to the lateral gate. The blockade voltage decreases, but it is not a complete suppression. The steps of the staircase are shifted. The traces are offset proportional to the applied gate voltage. **(b, *bottom*)** The differential conductance for the $V_g = 0$ V trace.

are trapped. A chemisorption would result in a electrode surface that is more or less completely covered with particles. Electrons would then have larger probability for tunneling in several different current paths, through many adsorbed particles between the electrodes.

ELECTRICAL CHARACTERIZATION

The current voltage characteristics was recorded by applying a symmetric voltage over two (typically 100 MΩ) bias resistors and measuring the current and the voltage over the sample. The IV-characteristics at room temperature and at 4.2 K are shown in FIGURE 3. The Coulomb gap and the staircase are characteristic of tunneling through a small double junction system and it can be modelled analytically[16] or by Monte Carlo simulations. We can therefore assume that we have captured one or a few particles in the gap when we observe the Coulomb staircase. FIGURE 4 shows the IV curve for one device with different gate voltages applied to the lateral gate. We can see that the steps are shifted and the Coulomb gap is decreased. The device was destroyed when higher gate voltages were applied, and we were not able to observe periodic modulation nor finite zero-bias conductance. FIGURE 4b shows the dI/dV versus bias voltage, and from this we can see peaks with a distance of 0.4 V. The IV characteristic of a double junction system can be reproduced by Monte Carlo simulations, and the capacitances and resistances of the junctions can be estimated. We made Monte Carlo simulations of a double junction system with the software package SIMON,[17] which takes account of co-tunneling events. A simulated IV-curve is displayed in FIGURE 5, together with experimental IV curves. The equivalent scheme of the system is shown in the inset. The tunnel junction capacitances are 0.18 aF and 0.3 aF for the simulation. These values are comparable with the self-capacitance of a sphere $C_0 = 4\pi\varepsilon_0\varepsilon_r r = 0.4$ aF for a particle with radius $r = 2$ nm and dielectric constant $\varepsilon_r = 2$. These values are reasonable and indicate that the particle diameter should be in the range of 2–4 nm. The tunneling current is higher in the experimental curves, which can be explained by a suppression of the effective tunnel barrier by the applied voltage.[8] The blockade voltage is for all devices around 0.2 volts, and equally large Coulomb blockade has been observed experimentally by

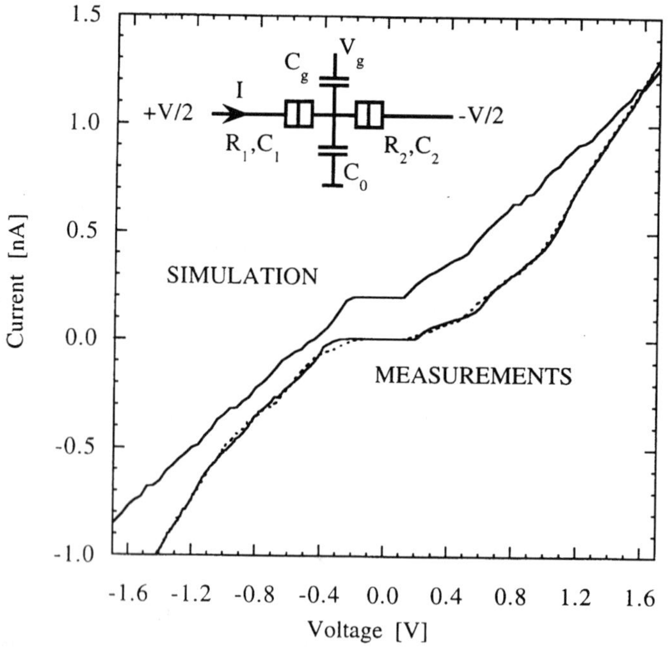

FIGURE 5. Monte Carlo-simulated IV curves showing the Coulomb staircase with reasonable agreement with the experimental IV curves. The parameters used for the simulation are: $C_1 = 0.3$ aF, $C_2 = 0.18$ aF, $R_1 = 1.05$ GΩ, $R_2 = 0.35$ GΩ, $C_0 = 0.2$ aF, $C_g = 0.016$ aF, $V_g = -1.5$ V, and T = 4.2 K.

tunnelling to a gold cluster from an scanning tunneling microscope tip.[18] The measured diameter of the cluster in their experiments was 1.2 nm.

We have observed Coulomb blockade in a self-assembled device at room temperature. The blockade voltage is for the devices around 0.2 V. The current voltage characteristics are modulated at 4.2 K as a function of applied voltage to the lateral gate electrode. We have used 1-dodecanethiol-stabilized gold clusters that arrange themselves in the right position in the gap between two electrodes, and eventually they may chemisorb in the gap because of covalent bonding to the dithiol SAM on the electrode surfaces. With this method it can be possible to realize a single-electron tunneling transistor operating at room temperature.

ACKNOWLEDGMENTS

This work has been supported by the Swedish Research Council for Engineering Sciences (TFR) and the Foundation for Strategic Research (SSF). We used the Swedish Nanometer Laboratory for lithographic processing.

REFERENCES

1. See references *in* GRABERT, H. & M.H. DEVORET, Eds. 1992. Single Charge Tunneling. NATO ASI Series B, Vol. **294**. Plenum, New York.
2. AVIRAM, A. & M.A. RATNER. 1974. Molecular rectifiers. Chem. Phys. Lett. **29:** 277.
3. ZHOU, C., M.R. DESPHANDE, M.A. REED, L. JONES II & J.M. TOUR. 1997. Nanoscale metal/self-assembled monolayer/metal Heterostructures. Appl. Phys. Lett. **71:** 611–613.
4. FISCHER, C.M., M. BURGHARD, S. ROTH & K. VON KLITZING. 1994. Organic quantum wells: Molecular rectification and single electron tunnelling. Europhys. Lett. **28:** 129–134.
5. REED, M.A. 1995. Sub-nanoscale electronic systems and devices. United States patent no. 5475341.
6. TOUR, J.M., *et al.* 1995. Self-assembled monolayers and multilayers of conjugated thiols, a,w-dithols and thioacetyl-containing adsorbates. Understanding attachments between potential molecular wires and gold surfaces. J. Am. Chem. Soc. **117:** 9529–9534.
7. KLEIN, D.L., P.L. MCEUEN, J.E. BOWEN-KATARI, R. ROTH & A.P. ALIVISATOS. 1996. An approach to electrical studies of single nanocrystals. Appl. Phys. Lett. **68:** 2574–2576.; D.L. KLEIN, R. ROTH, A.K.L. LIM, A.P. ALIVISATOS & P.L. MCEUEN. 1997. A single-electron transistor made from a cadmium selenide nanocrystal. Nature. **389:** 699–701.
8. BEZRYADIN, A., C. DEKKER & G. SCHMID. 1997. Electrostatic trapping of single conducting nanoparticles between nanoelectrodes. Appl. Phys. Lett. **71:** 1273–1275.
9. T. SATO, H. AHMED, D. BROWN & B.F.G. JOHNSON. 1997. Single electron transistor using a molecularly linked gold colloidal particle chain. J. Appl. Phys. **82:** 696–701.
10. M. BRUST, M. WALKER, D. BETHELL, D.J. SCHRIFFIN & R. WHYMAN. 1994. Synthesis of thiol-derivatised nanoparticles in a two-phse liquid-liquid system. J. Chem. Soc. Chem. Commun. 801–802.
11. LEFF, D.V., P.C. OHARA, J.R. HEATH & W.M. GELBART. 1995. Thermdynamic control of gold nanocrystal size: theory and experiment. 1995. J. Phys. Chem. **99:** 7036–7041.

12. WHETTEN, R.L., *et al.* 1996. Nanocrystal gold molecules. Adv. Mater. **8:** 428–433.
13. TERRILL, R.H., *et al.* 1995. Monolayers in three dimensions: NMR, SAXS, thermal, and electron hopping studies of alkanethiol stabilized gold clusters. J. Am. Chem. Soc. **117:** 12537–12548.
14. SATO, T., D. BROWN & B.F.G. JOHNSON. 1997. Nucleation and growth of nano-gold colloidal lattices. Chem Commun. 1007–1008.
15. DINSMORE, A.D., A.G. YODH & D.J. PINE. 1996. Entropic control of particle motion using passive surface microstructures. Nature. **383:** 239–242.
16. AMMAN, M., R. WILKINS, E. BEN-JACOB, P.D. MAKER, & R.C. JACLEVIC. 1991. Analytic solution for the current-voltage characteristics of two mesoscopic tunnel junctions coupled in series. Phys. Rev. B **43:** 1146–1149.
17. The commercial software package "SIMON" for simulations of single electronics was used. Information can be found on the WWW: http//:members.magnet.at/cats-meov
18. DOROGI, M., J. GOMEZ, R OSIFCHIN, R.P. ANDRES & R. REIFENBERGER. 1995. Room-temperature Coulomb blockade from a self-assembled molecular nanostructure. Phys. Rev. B **52:** 9071–9077.

Recent Advances in Molecular Scale Electronics

JAMES M. TOUR,[a,b] WILLIAM A. REINERTH,[b] LEROY JONES II,[b] TIMOTHY P. BURGIN,[b] CHONG-WU ZHOU,[c] C.J. MULLER,[c] M.R. DESHPANDE,[c] AND MARK A. REED[c]

[b]*Department of Chemistry and Biochemistry, University of South Carolina, 730 South Main Street, Columbia, South Carolina 29208, USA*

[c]*Department of Electrical Engineering, Yale University, New Haven, Connecticut 06520, USA*

ABSTRACT: This paper describes four significant breakthroughs in the syntheses and testing of molecular scale electronic devices. The 16-mer of *oligo*(2-dodecylphenylene ethynylene) was prepared on Merrifield's resin using the iterative divergent/convergent approach which significantly streamlines the preparation of this molecular scale wire. The formation of self-assembled monolayers (SAMs) and multilayers on gold surfaces of rigid rod conjugated oligomers that have thiol, α,ω-dithiol, thioacetyl, or α,ω-dithioacetyl end groups have been studied. The direct observation of charge transport through molecules of benzene-1,4-dithiol, which have been self-assembled onto two facing gold electrodes, has been achieved. Finally, we are reporting initial studies into what effect varying the molecular alligator clip has on the molecular scale wire's conductivity.

INTRODUCTION

Future computational systems will likely consist of logic devices that are ultra dense, ultra fast, and molecule-sized.[1–23] The slow step in existing computational architectures is not usually the switching time, but the time it takes for an electron to travel between devices. By using molecular scale electronic interconnects,[d,24] the transmit times could be minimized, resulting in computational systems that operate at far greater speeds than is presently attainable from conventional patterned architectural arrays.[1] There is another technical advantage that might also be gained from molecular scale devices. A powerful computational system presently utilizes about 10^{10} silicon-based devices. If devices were to be based upon single molecules,[21] using routine chemical syntheses, one could prepare approximately 10^{23} devices in a single reaction flask. Of course, the task of addressing large arrays of ordered molecular scale devices is presently unattainable; however, the potential is certainly enough to maintain current and future interests.

[a]Address for telecommunication: Phone, 803/777-9517; fax, 803/777-9521.

[d]"Molecular electronics" is a poorly defined term since some authors refer to it as any molecular-based system, such as a film or a liquid crystalline array. Other authors, including us, have preferred to reserve the term "molecular electronics" for single-molecule tasks, such as single molecule-based devices or single molecular wires. Because of this confusion, we have chosen here to follow the Petty *et al.* terminology by using two subcategories, namely "molecular materials for electronics" for bulk applications and "molecular scale electronics" for single molecule applications.

Though it is well-documented that bulk conjugated organic materials can be semiconducting or even conducting when doped,[25] we only recently determined how thiol-ended rigid rod conjugated molecules orient themselves on gold surfaces,[26,27] and how we could record electronic conduction through single undoped conjugated molecules that are end-bound onto a metal probe surface.[28] We have previously described the syntheses of soluble *oligo*(3-ethyl-2,5-thiophene-ethynylene)s and *oligo*(2-alkyl-1,4-phenylene-ethynylene)s, potential molecular scale wires, by a rapid, solution phase iterative divergent/convergent doubling approach,[29,32] as well as the syntheses and attachments of protected thiol moieties to one or both ends of the molecular scale wires. These thiols serve as molecular scale alligator clips for adhesion of the molecular scale wires to the gold probes.[26–28]

Herein we report four significant breakthroughs that could have dramatic implications in the syntheses and testing of molecular scale electronic devices. First, we have completed the synthesis of the dodecyl-containing 16-mer on Merrifield's resin (chloromethyl polystyrene) using the iterative divergent/convergent approach, a method which significantly streamlines the preparation.[33–36] Secondly, we have studied the formation of self-assembled monolayers (SAMs) and multilayers on gold surfaces of rigid rod conjugated oligomers that have thiol, α,ω-dithiol, thioacetyl, or α,ω-dithioacetyl end groups. Thirdly, molecules of benzene-1,4-dithiol have been self-assembled onto the two facing gold electrodes of a mechanically controllable break junction (MCB) to form a statically stable gold-sulfur-aryl-sulfur-gold system, which allows for direct observation of charge transport through the molecules. Finally, we are examining what effect varying the molecular alligator clip has on the molecular scale wire's conductivity.

RESULTS AND DISCUSSION

Solid Phase Oligomer Synthesis

The monomers needed for the polymer supported synthesis were prepared as depicted in SCHEME I. The silylated triazene was divided into two portions; the first portion was desilylated to form **1**, the anchor unit to be attached to the polymer support, while the second portion was iodinated[33–36] to form the iodoarene **2**. Attachment of **1** to the polymer support resin, and the oligomer syntheses on the polymer support are depicted in SCHEME II. This polymer support approach greatly simplifies the isolation and purification of oligomers **5**, **8**, **11**, and **14**.

Self-Assembled Monolayer Formation of Rigid-Rod Conjugated Oligomers

As a prelude to the utilization of molecular scale wires in molecular scale electronic devices, it is necessary to understand the molecular ordering on metal surfaces. These sulfur-terminated conjugated oligomers form self-assembled monolayers (SAMs) on gold surfaces by attachment of the thiol end groups which serve as molecular scale alligator clips.[26] The SAMs were analyzed using ellipsometry, X-ray photoelectron spectroscopy (XPS), and infrared external reflectance spectroscopy. The thiol moieties usually dominate adsorption on the gold sites; interactions with the conjugated π-systems are weaker. Rigid rod α,ω-dithiols form assemblies in which one thiol group binds to the surface while the second thiol moiety projects upward at the exposed surface of the SAM. *In situ* deprotection of the thiol moieties by deacylation of thioacetyl groups using NH_4OH permits formation of SAMs without having to isolate the oxidatively unstable free thiols. Moreover, direct adsorption, without exogenous base, of the thioacetyl-terminated oligomers can be accomplished to generate gold surface-bound thiolates. However, in the non-base-promoted adsorptions, higher concentrations of the thioacetyl groups, relative to that of thiol groups, are required to achieve monolayer coverage in a given interval. A thiol-terminated phenylene ethynylene system was shown to have a tilt angle of the long molecular axis of <20° from the normal to the substrate surface. These aromatic α,ω-dithiol-derived monolayers provide the basis for studies leading to the design of molecular wires capable of bridging proximal gold surfaces.

SCHEME I

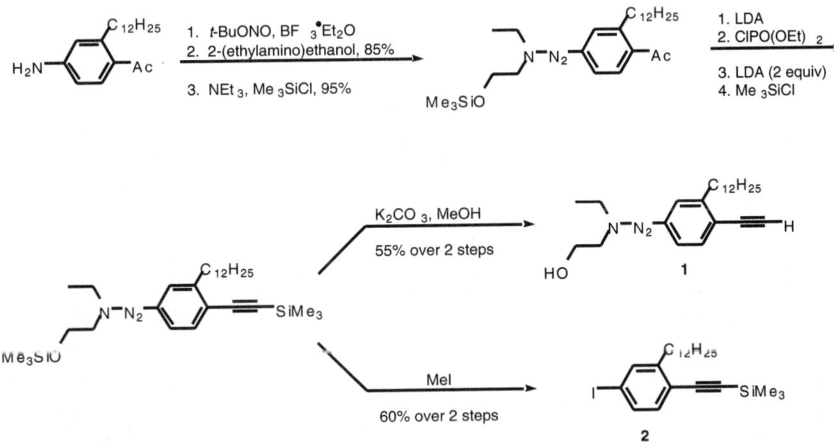

Conductance of a Molecular Junction

Charge transport in and the measurement of the conductance of single organic molecules is an intriguing, experimentally challenging, and long sought goal. These measurements have been performed on benzene-1,4-dithiolate connected between stable proximal metallic gold contacts using a mechanically controllable break junction (MCB).[37] The metal-molecule-metal configuration presents the molecular embodiment of a system analogous to a quantum dot, with the potential barriers replaced by the contact barrier of the gold-thiolate end groups (FIG. 1). The results show a highly reproducible apparent Coulomb gap at about 0.7 V at room temperature. This study provides a direct, quantitative measurement of the molecular conductance of a junction containing a single molecule, a fundamental step in the emerging field of molecular scale electronics.

Variation of a Molecular Alligator Clip

By measuring the conductivity of a series of molecules that are systematically altered, the contribution of each component of a molecular scale electronic device can

SCHEME II

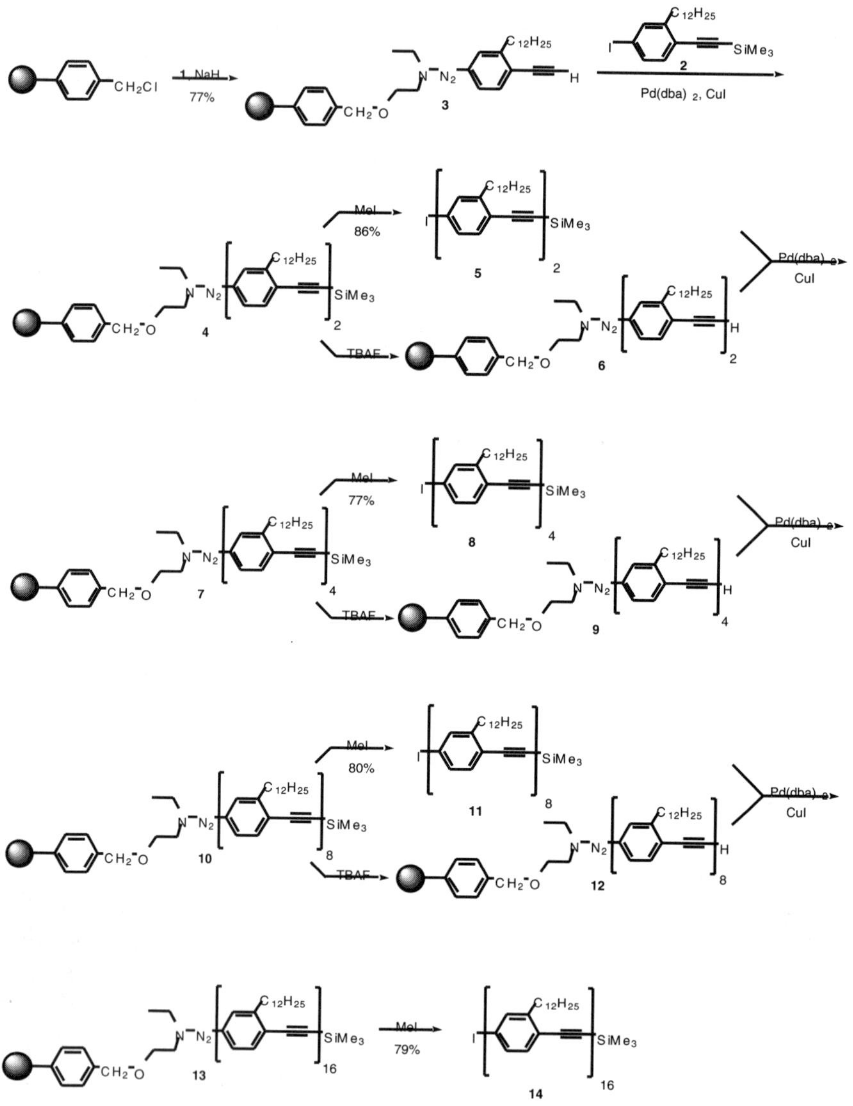

be determined. For example, the simplest experiment of this type would be to measure the conductivity of identical arrays of molecular wires in which only the wires' attachment to the gold surface is altered. Thus, by varying the molecular alligator clip and examining the molecules' conductivity, the contribution of the alligator clip to the resistance of the molecular wire can be determined. Our initial work has focused on varying the alligator clip by proceeding down the chalcogens from S to Se to Te.

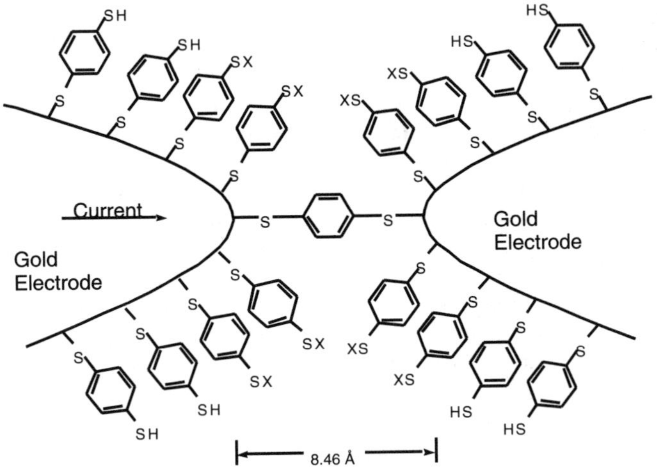

FIGURE 1. The desired placement of benzene-1,4-dithiolate between gold electrodes in a MCB. (X = H, Au).

We first desired a stable, yet easily removable, protecting group for selenium that would allow us to use the same *in situ* deprotection prototcol we had employed for the thioacetates. An investigation of several protecting groups led us to conclude that, as with sulfur, the acetyl group was the protecting group of choice.[38] Recently, molecular arrays of 4,4′-biphenyl thioacetate (**15**), 4,4′-biphenyl selenoacetate (**16**), and bis(4,4′-biphenyl)ditelluride (**17**) have been subjected to conductivity examination using a heterostructure nanopore system (FIG. 2).

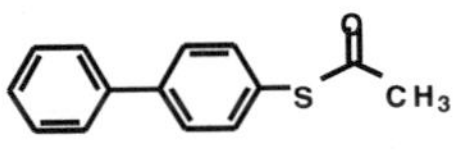

15

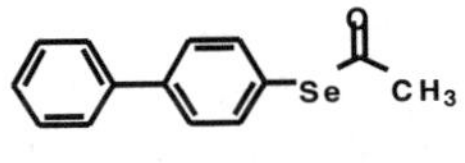

16

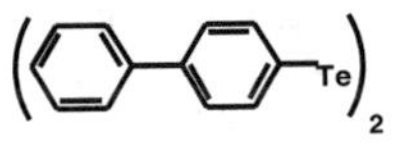

17

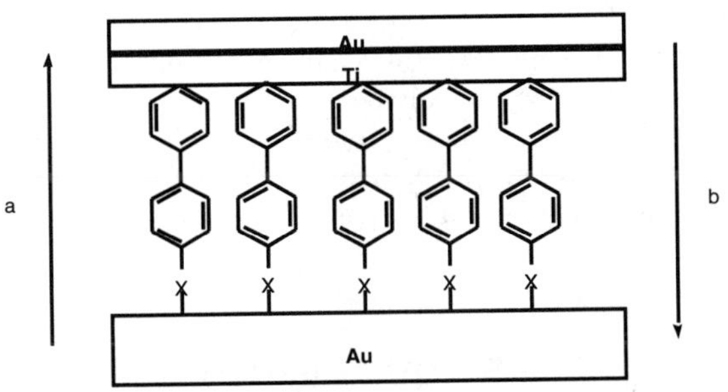

FIGURE 2. Schematic of nanopore experiment (X = S, Se, Te). Arrows indicate direction of electron flow for (a) reverse bias and (b) forward bias.

We have previously examined the electronic properties of **15** using this heterostructure nanopore system (X = S in FIG. 2).[39] Desiring to examine the degree to which the molecule's attachment to the gold surface determines conductivity, we subjected **16** (X = Se) and **17** (X = Te) to examination. The barrier heights that were obtained for each of the chalcogen-terminated molecular wires are listed in TABLE 1.

It appears that the selenium alligator clip creates a higher barrier for electron transport regardless of the direction of electron flow.[40] This is puzzling in light of the greater metallic character of Se versus S which we reasonably assumed would yield a lower barrier for Se. However, tellurium does not exhibit the same trend for barrier height. The Te-terminated molecular wire displays a barrier to conduction comparable to its sulfur congener. This non-periodic trend is puzzling and we are not yet able to fully explain it. However, recent calculations on the HOMO-LUMO gap of chalcogen-terminated aromatics bonded to Au shows a similar trend when the chalcogen bridges three Au atoms on the surface.[41]

TABLE 1. Barrier heights for chalcogen-terminated biphenyl SAMs

X	Forward bias (eV)	Reverse bias (eV)
S	0.256	0.19
Se	0.33	0.25
Te	0.215	0.21

SUMMARY

In summary, we have made four significant breakthroughs in the syntheses and testing of molecular scale electronic devices. First, we have completed the synthesis of the dodecyl-containing 16-mer on Merrifield's resin (chloromethyl polystyrene) using the iterative divergent/convergent approach; a method which significantly streamlines the preparation. Secondly, we have studied the formation of self-assembled monolayers (SAMs) and multilayers on gold surfaces of rigid rod conjugated oligomers that have thiol, α,ω-dithiol, thioacetyl, or α,ω-dithioacetyl end groups. Thirdly, molecules of benzene-1,4-dithiol have been self-assembled onto the two facing gold electrodes of a mechanically controllable break junction (MCB) to form a statically stable gold-sulfur-aryl-sulfur-gold system, which allows for direct observation of charge transport through the molecules. Finally, we have begun investigating what effect varying the molecular alligator clip has on the molecular scale wire's conductivity.[16]

ACKNOWLEDGMENT

We thank the Defense Advanced Research Projects Agency for support of this work.

REFERENCES

1. AVIRAM, A., Ed. 1992. Molecular Electronics: Science and Technology. *In* Conference Proceedings No. 262. American Institute of Physics. New York, NY.
2. BIRGE, R.R., Ed. 1991. Molecular and Biomolecular Electronics. *In* Advances in Chemistry Series 240. American Chemical Society.
3. KIRK, W.P. & M.A. REED, Eds. 1992. Nanostructures and Mesoscopic Systems. Academic Press. New York, NY. In press.
4. MUJICA, V. *et al.* 1996. Current-voltage characteristics of molecular wires: Eigenvalue staircase, Coulomb blockade, and rectification. J. Phys. Chem. **104:** 7296–7305.
5. JOACHIM, C. & J.F. VINUESA. 1996. Length dependence of the electronic transparence (conductance) of a molecular wire. Europhys. Lett. **33:** 635–640.
6. GROSSHENNY, V., A. HARRIMAN & R. ZIESSEL. 1995. Electronic energy transfer across ethynyl-bridged ru^{ii}/os^{ii} terpyridyl complexes. Angew. Chem. Int. Ed. Engl. **34:** 1100–1102.
7. LANGLER, L. *et al.* 1995. Electrical measurements on submicronic synthetic conductors: carbon nanotubes. Synth. Methods. **70:** 1393–1396.
8. PASCUAL, J.I. *et al.* 1995. Properties of metallic nanowires: from conductance quantization to localization. Science **267:** 1793–1795.
9. PURCELL, S.T. *et al.* 1994. Characterization of the conductivity of organic thiols by field emission microscopy and field emission spectroscopy. J. Am. Chem. Soc. **116:** 11985–11989.
10. MARTIN, C.R. 1994. Nanomaterials: A membrane-based synthetic approach. Science **266:** 1961–1966.
11. SETH, J. *et al.* 1994. Investigation of electronic communication in multi-porphyrin light-harvesting arrays. J. Am. Chem. Soc. **116:** 10578–10592.
12. GUST, D. 1994. Molecular wires and girders. Nature **372:** 133–134.
13. WU, C. & T. BEIN. 1994. Conducting carbon wires in ordered, nanometer-sized channels. Science **266**: 1013-1015.
14. WAGNER, R.W. *et al.* 1996. Molecular optoelectronic gates. J. Am. Chem. Soc. **118**: 3996-3997.
15. BRIGELLETTI, F. *et al.* 1994. Rigid rod-like dinuclear ru(ii)/os(ii) terpyridine-type complexes. Electrochemical behavior, absorption spectra, luminescence properties, and electronic energy transfer through phenylene bridges. J. Am. Chem. Soc. **116:** 7692–7699.
16. WU, C. & T. BEIN. 1994. Conducting polyaniline filaments in a mesoporous channel host. Science **264:** 1757–1759.
17. MOERNER, W.E. 1994. Examining nanoenvironments in solids on the scale of a single, isolated impurity molecule. Science **265:** 46–53.
18. SESSLER J.L., V.L. CAPUANO & A. HARRIMAN. 1993. Electronic energy migration and trapping in quinone-substituted, phenyl-linked dimeric and trimeric porphyrins. J. Am. Chem. Soc. **115:** 4618–4628.
19. FARAZDEL, A. *et al.* 1990. Electric field induced intramolecular electron transfer in spiro π-electron systems and their suitability as molecular electronic devices. A theoretical study. J. Am. Chem. Soc. **112:** 4206–4214.
20. TOUR, J.M., R. WU & J.S. SCHUMM. 1991. Extended orthogonally fused conducting oligomers for molecular electronic devices. J. Am. Chem. Soc. **113:** 7064–7066.
21. DAI, H., E.W. WONG & C.M. LIEBER. 1996. Probing electrical transport in nanomaterials: Conductivity of individual carbon nanotubes. Science **272:** 523–526.
22. WU, R. *et al.* 1996. Convergent synthetic routes to orthogonally fused conjugated oligomers directed toward molecular scale electronic device applications. J. Org. Chem. **61:** 6906–6921.
23. WARD, M.D. 1996. Current developments in molecular wires. Chem. Ind. 568–573.

24. PETTY, M.C. *et al.*, Eds. 1995. Introduction to Molecular Electronics. Oxford University Press. New York, NY. In press.
25. SKOTHEIM, T.A., Ed. 1986. Handbook of Conducting Polymers. Marcel Dekker. New York, NY.
26. TOUR, J.M. *et al.* 1995. Self-assembled monolayers and multilayers of conjugated thiols, α,ω-dithiols, and thioacetyl-containing adsorbates. Understanding attachments between potential molecular wires and gold surfaces. J. Am. Chem. Soc. **117:** 9529–9534.
27. DHIRANI, A. *et al.* 1996. Self-assembly of conjugated molecular rods: A high resolution STM study. J. Am. Chem. Soc. **118:** 3319–3320.
28. BUMM, L.A. *et al.* 1996. Are single molecular wires conducting? Science **271:** 1705–1707.
29. SCHUMM, J.S., D.L. PEARSON & J.M. TOUR. 1994. Iterative divergent/convergent approah to linear conjugated oligomers by successive doubling of the molecular length: A rapid route to a 128 Å-long potential molecular wire. Angew. Chem. Int. Ed. Engl. **33:** 1360–1363.
30. PEARSON, D.L., J.S. SCHUMM & J.M. TOUR. 1994. Iterative divergent/convergent approah to linear conjugated oligomers by a doubling of molecular length at each iteration. A rapid route to potential molecular wires. macromolecules **27:** 2348–2350.
31. PEARSON, D.L. & J.M. TOUR. 1997. Rapid syntheses of *oligo*(2,5-thiophene ethynylene)s with thioester termini: Potential molecular scale wires with alligator clips. J. Org. Chem. **62:** 1376–1387.
32. JONES, L., II, J. S. SCHUMM & J. M. TOUR. 1997. Rapid solution and solid phase syntheses of *oligo*(1,4-phenylene ethynylene)s with thioester termini: Molecular scale wires with alligator clips. Derivatization of iterative reaction efficiences on a polymer support. J. Org. Chem. **62:** 1388–1410.
33. YOUNG, J.K., J.C. NELSON & J.S. MOORE. 1994. Synthesis of sequence specific phenylacetylene oligomers on an insoluble solid support. J. Am. Chem. Soc. **116:** 10841–10842.
34. BHARATHI, P. *et al.* 1995. Improvements in the synthesis of phenylacetylene monodendrons including a solid-phase convergent method. Macromolecules **28:** 5955–5963.
35. HODGE, P. & D.C. SHERRINGTON, Eds. 1980. Polymer-Supported Reactions in Organic Synthesis. Wiley. New York, NY. In press.
36. NELSON, J.C., J.K. YOUNG & J.S. MOORE. 1996. Solid-phase synthesis of phenylacetylene oligomers utilizing a novel 3-propyl-3-(benzyl-supported) Triazene Linkage. J. Org. Chem. **61:** 8160–8168.
37. REED, M.A. *et al.* 1997. Conductance of a molecular junction. Science **278:** 252–254.
38. REINERTH, W.A. & J.M. TOUR. 1997. Protecting groups for organoselenium compounds. J. Org. Chem. **63**. 2397–2400.
39. ZHOU, C. *et al.* 1997. Nanoscale metal/self-assembled monolayer/metal heterostructures. Appl. Phys. Lett. **71:** 611–613.
40. ZHOU, C. *et al.* 1997. Variation of a molecular alligator clip. Manuscript in preparation.
41. SEMINARIO, J.M. & J.M. TOUR. Unpublished results.

Macromolecular and Supramolecular Architectures for Molecular Electronics

K. MÜLLEN[a] AND J.P. RABE[b]

[a]*Max-Planck-Institut für Polymerforschung, Ackermannweg 10, D-55128 Mainz, Germany*

[b]*Institut für Physik der Humboldt-Universität zu Berlin, Invalidenstrasse 110, D-10115 Berlin, Germany*

ABSTRACT: The benzene ring is introduced as a modular building block into carbon-rich 1D-, 2D- and 3D-polymers. Remarkable synthetic products are functionalized poly(para-phenyleneethynylene)s as rigid rods, giant polycyclic aromatic hydrocarbons as molecular models for graphite and polyphenylene dendrimers as shape-persistent nanoparticles. Supramolecular ordering of these unprecedented molecular structures is a key prerequisite when elucidating their role as media for charge carrier transport in molecular electronics. Current-potential curves recorded for single disc-type molecules and the drastic increase of charge mobility as a result of supramolecular order in π-stacked columnar arrays are outstanding physical results.

INTRODUCTION

The visualization and manipulation of single molecules[1] have become major themes of scientific research. Related to this is the concept of molecular electronics, which focuses on charge carrier transport through single molecules.[2,3] However, the use of single molecules as active components of electronic devices raises a number of fundamental problems. One of them is the reliable fabrication of solid state electronic circuitry, which operates reproducibly on the nanometer scale. While single junctions have been made,[4] it remains an issue how to fabricate circuits. Regarding suitable molecules, it appears that extended π-conjugation is a prerequisite for efficient charge or energy transport. While materials made from conjugated polymers exhibit usually low charge carrier mobilities, this may be different for single molecules without chemical defects and very high shape persistence, since one may expect there coherent transport through single quantum states. Finally, interfacing molecules to electronic circuitry is a key issue. It can be achieved, e.g., via the tip of a scanning tunneling microscope or, technologically more relevant, via suitable nanoelectrodes on an insulating substrate. For interfacing via the STM tip, the molecules may be immobilized in a supramolecular pattern on a surface. When attempting to bridge the gap between nanoelectrodes on a substrate, a molecular wire can consist of a single molecule or of well-defined aggregates of molecules.[5] Further, charge transport can occur along the backbone of a conjugated organic molecule or polymer — with the length of the chain being a critical question — or through a columnar array of chain- or disc-type molecules.

It is obvious that both approaches for interfacing require a strong impact from synthetic chemistry. What one needs is a rationale design of molecular and supramolecular structures by which one can easily bridge different length scales and create a contact between molecules — or molecular assemblies - and the outside world. The present approach toward molecular electronics therefore rests on a modular concept. It uses the benzene ring as a regular building block which is incorporated into polymeric structures of different dimensionality: (1) conjugated chains, (2) disc-type graphite subunits, and (3) 3D-polyphenylenes as shape persistent nanoparticles. The resulting molecules are carefully processed to yield supramolecular motifs, such as columnar mesophases, monomolecular adsorbate layers, or single nanostructures, like nanowires on surfaces. Weak intermolecular forces such as π–π-stacking or alkyl chain packing serve as structure-forming principles.

POLY(PARA-PHENYLENEETHYNYLENE)S

Rigid-rod type conjugated polymers are ideal candidates for molecular wires. We have recently introduced a great number of polyphenylenes with ladder and step-ladder structures possessing useful electronic and optoelectronic properties, whereby structural homogeneity and processability are important ingredients of the synthetic approach.[6–8]

Herein we describe the poly(para-phenyleneethynylene)s (**3**) as attractive alternative to polyphenylene. Polymers such as **3a** or **3b** have found interest due to their strongly anisotropic electronic properties, their high photoluminescence quantum yields,[9] and their electroluminescence in the blue-green wavelength domain.[10] Our synthetic concept had to fulfill several requirements: perfect uninterrupted π-conjugation; chain-lengths between a few nanometers and about 100 nanometers; solubility and solution processability; and chemical functionalization at both ends, in particular introduction of thiol-groups for establishing contact with both electrodes.

SCHEME 1

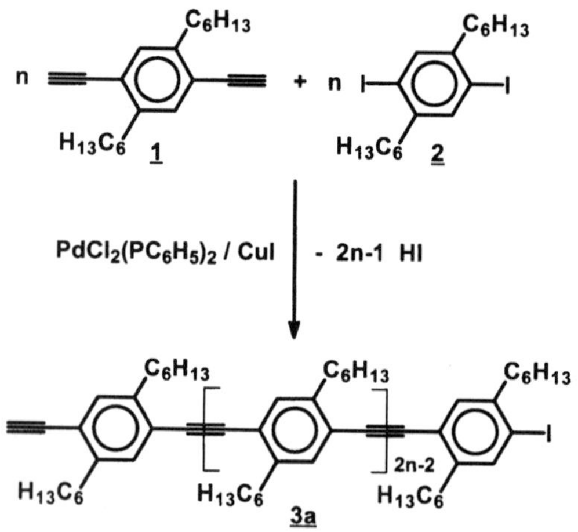

These requirements could be met by two approaches: a modification of the known aryl-ethynyl coupling according to Heck,[11] Hagihara and Sonogashira[12] using building blocks such as aryl dihalides **2** and diethynyl derivatives **1**,[13] and an unprecedented synthesis of poly(phenyleneethynylene)s (**3**) using the metathetic polymerization of 1,4-dipropynylbenzene derivatives **4** under extrusion of butyne.[14] The particular advantage of this method is that the propynyl end groups possess a "living character" and can be functionalized at a given chain length by introducing functionalized propynyl reagents.

SCHEME 2

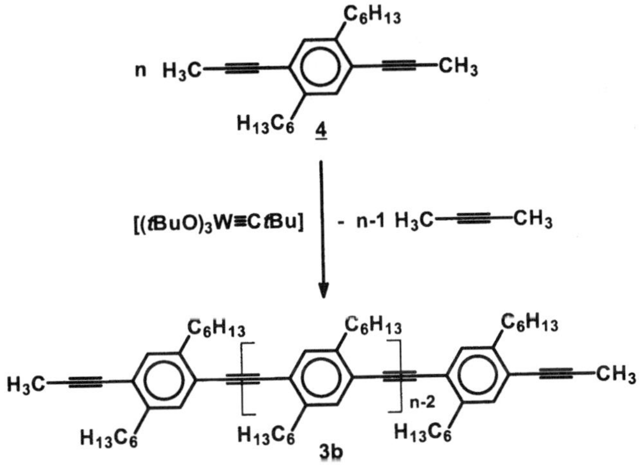

Well defined anisotropic supramolecular nanostructures of **3** and of related monodisperse oligomers can be produced by physisorption from solution on both mica and graphite surfaces. Their visualization with scanning force microscopy and scanning tunneling microscopy, respectively, suggests important conclusions.

On mica, one detects ribbon structures (FIG. 1) with the conjugated backbones parallel to the substrate and perpendicular to the long ribbon axes. Thereby, the distribution of the ribbon widths corresponds to the distribution of molecular weights determined independently.[5] Accordingly, polymer chains with similar molecular weights tend to phase-segregate into ribbon sections with more or less homogeneous widths. The ribbons are typically two molecules thick with their alkyl chains oriented perpendicular to the substrate. It is interesting to note that the supramolecular architecture on the surfaces is similar to the previously studied *bulk* phases of poly(diacetylene)[15,16] and poly(para-phenyleneethynylene) systems.[17] Since conjugated molecules are self-assembled on a non-conductive substrate and since the end groups can be deprotected to yield thiol-functions at the ends, the ribbons can be used to interconnect gold nanoelectrodes. While measurements of conduction through these ribbons will be reported elsewhere, it should be emphasized that the present approach toward nanowires differs from those taken in the literature.[13] The

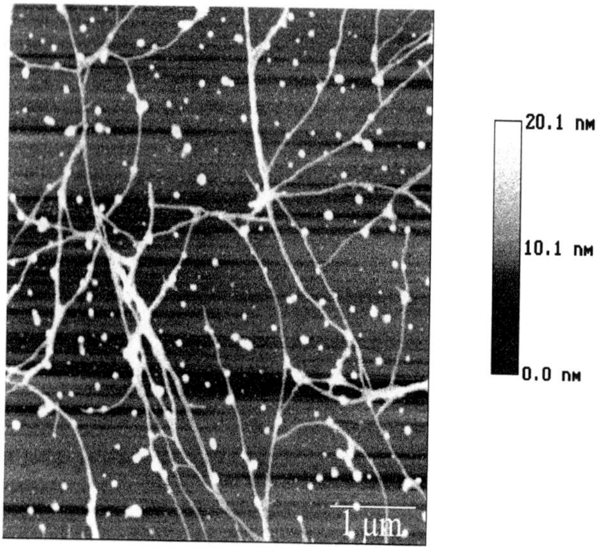

FIGURE 1. Scanning force microscopy image of poly(para-phenyleneethynylene) on mica, which typically exhibits long ribbons. The ribbon widths correspond to the length of single extended polymer chains.[5]

latter use relatively short conjugated oligomers or even monomers, and the emphasis rests on the fabrication of the nano-gap. The emphasis of the present concept is on the synthesis of polymer structures and their self-assembly, thus yielding supramolecular structures at different length scales.

POLYCYCLIC AROMATIC HYDROCARBONS AS GRAPHITE SUBUNITS

Conjugated, all-carbon based molecular structures of spherical and tubular form have attracted interest as molecular wires.[18] Another class of carbon-based structures comprises large discs made up from polycyclic aromatic hydrocarbons (PAHs). While PAHs have long been known, only our recent work has made available extremely large representatives of this class of compounds.[19] Key aspects of our synthesis are the systematic control of size and shape and, in spite of the very large, rigid structures, their processability. This made it possible to organize giant disc structures into tightly packed monomolecular adsorbate layers on surfaces and into π-stacked columnar arrays.

Unlike the common synthesis of large PAHs, which usually gives rise to a wide variety of products, we apply a very smooth and selective intramolecular fusion reaction at ambient temperatures in which soluble precursor oligophenylenes are cyclodehydrogenated to yield well-defined PAHs.[20,21] The design of the precursor oligophenylenes is crucial, because the mode of coupling of the phenylene units must allow transition from a 3D- to a 2D-structure upon dehydrogenation. Suitable oligophenylenes are made by (1) intramolecular Diels-Alder cycloaddition of phenylenevinylene derivatives,[20,21] (2) cyclotrimerization of aryleneethynylene precursors,[2,22] and (3) intermolecular Diels-Alder cycloaddition of phenyleneethynylenes with tetraphenylcyclopentadienones.[19,23]

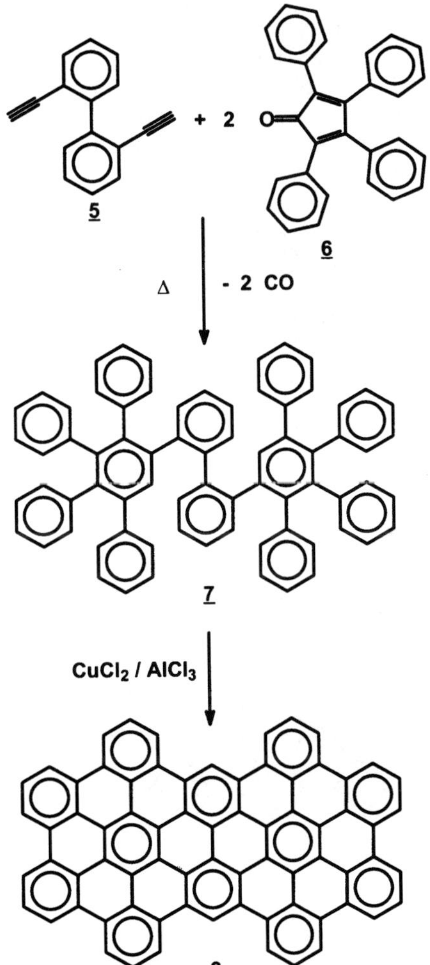

SCHEME 3

A typical example is the reaction of 2,2′-diethynylbiphenyl (**5**) with tetraphenylcyclopentadienone (**6**) to yield the hexiphenylene system **7**. The subsequent step is the dehydrogenation of the oligophenylene compounds, which can be accomplished by reaction with, e.g., copper dichloride and aluminium trichloride in solvents such as methylene chloride, carbon disulfide, or tetrachloroethane. If the reaction conditions are carefully optimized to avoid, e.g., intermolecular oxidative coupling or chlorination, the yields of the dehydrogenation of the intramolecular fusion reactions are generally quite high. The solubility of the planar products is extremely low so that for their purification one is restricted to special techniques such as sublimation under ultra-high vacuum.[21,25,26] An alternative approach rests on the synthesis of alkyl-substituted PAHs which can be achieved by introducing e.g. alkyl substituted tetraphenylcyclopentadienones, into the Diels-Alder reaction. Next to improving the solubility of the final products, n-alkyl substituents play an important role in introducing phase-forming behavior of the disc-type compounds (see below).[2,22,24] A major concern of these syntheses is a variation of size and shape and, in particular, the construction of homologous series of molecules in a nanometer domain as shown in SCHEME 4.

SCHEME 4

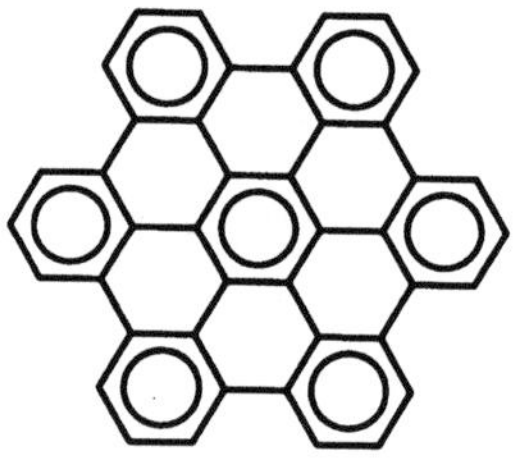

9

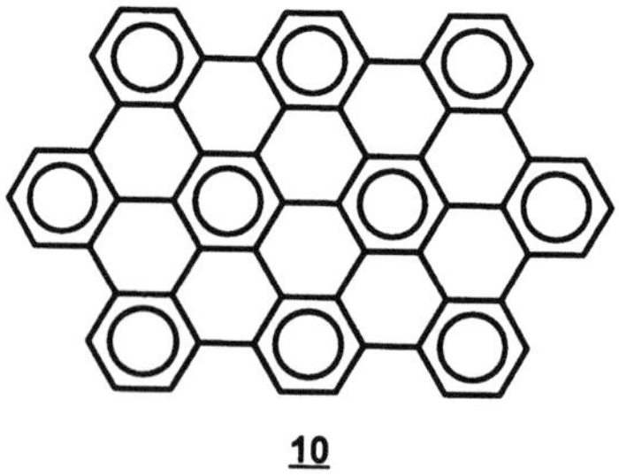

10

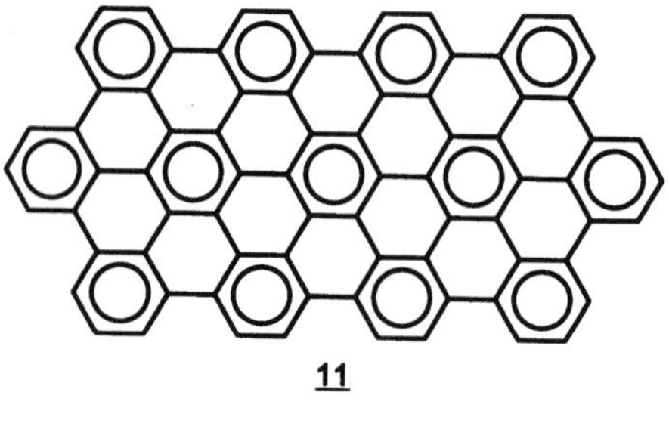

11

In overlooking the disc-shaped molecules which serve as a nanoscale oligomeric version of an inifinite, two-dimensional graphite sheet, their supramolecular ordering in two and three dimensions to cover even higher length scales becomes important. When, e.g., the hexa-perihexabenzocoronene (**9**) carries six peripheral alkyl substituents, it self-assembles into one-dimensional columnar aggregates. A discotic mesophase is formed which persists over an extremely broad temperature range.[2,22] Electronic charge transport in such columnar nanowires is predicted to be mediated by the non-covalent π–π-interaction between the extended discs of neighboring macrocycles. The size of the discs would therefore suggest that charge-transport along the axis should be particularly rapid. The charge-transport properties have been investigated by using a pulse-radiolysis resolved microwave conductivity technique. The charge carrier mobility determined for hexaalkyl substituted hexa-perihexabenzocoronene amounts to 0.13×10^{-4} m^2/Vs.[24] This is the largest value ever determined for a discotic organic material. It should be added that the mobility is constant over a wide range of temperatures. It is the combination of these properties, high charge carrier mobility, liquid cristallinity, and good thermal stability which qualifies such materials as vectorial charge-transport layers in, e.g., xerography, electrophotography, or molecular electronic devices, in which columnar aggregates would play the role of nanowires.

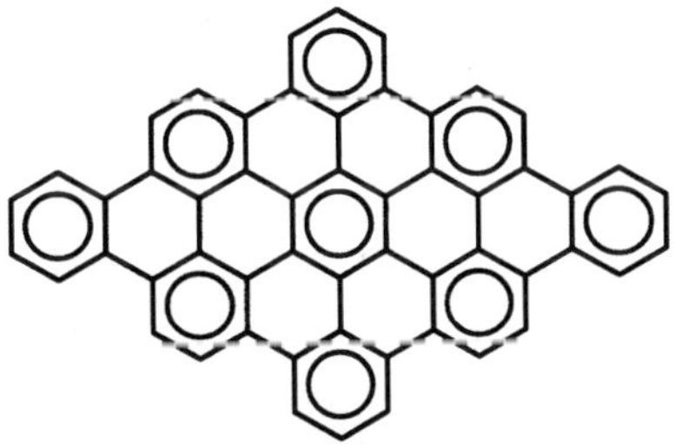

12

SCHEME 5

As concerns the organization of large PAHs into monomolecular adsorbate layers, deposition of, e.g., the hexa-peri-hexabenzocoronene (**9**) or $C_{54}H_{22}$ arene **12** is revealed by low-energy electron diffraction and scanning tunneling microscopy to form densely packed 2D-crystals on substrates, such as highly oriented pyrolytic graphite (HOPG) or MoS_2.[21,25,26] We have carefully investigated the balance of molecule-molecule interaction energy and molecule-substrate interaction energy in the formation of stable ordered surface structures finding, e.g., for $C_{54}H_{22}$ **12** a very high binding energy of the monolayer molecules on the surface.[26]

An easier approach toward the formation of monomolecular adsorbate layers is physisorption from solution of solubilized PAHs onto, e.g., HOPG surfaces. It can be demonstrated by scanning tunneling microscopy (STM) that hexaalkyl substituted derivatives of hexa-peri-hexabenzocoronene (**9**) do form regular monomolecular patterns on substrate surfaces. The key outcome, however, is that as a result of the large size of the discs, STM can be combined with reliable scanning tunneling *spec-*

troscopy. It is possible to record separate current-potential curves for the aromatic and aliphatic domains of single discs.[2] The curve of the aromatic part, unlike that of the aliphatic region shows a strong asymmetry and a diode-like behavior. This example of measuring an electronic function for a single molecule, although intriguing, raises a series of further questions as to the role of the molecular energy levels and the Fermi levels of the electrodes, and to the actual location of the disc between tip and surface. Along such investigations it is, of course, important, to have available discs of different size and shape. We have also made derivatives of **9** carrying chemical functions. Accordingly it is now possible to correlate currents measured for single molecules with bulk properties, such as the oxidation potentials.

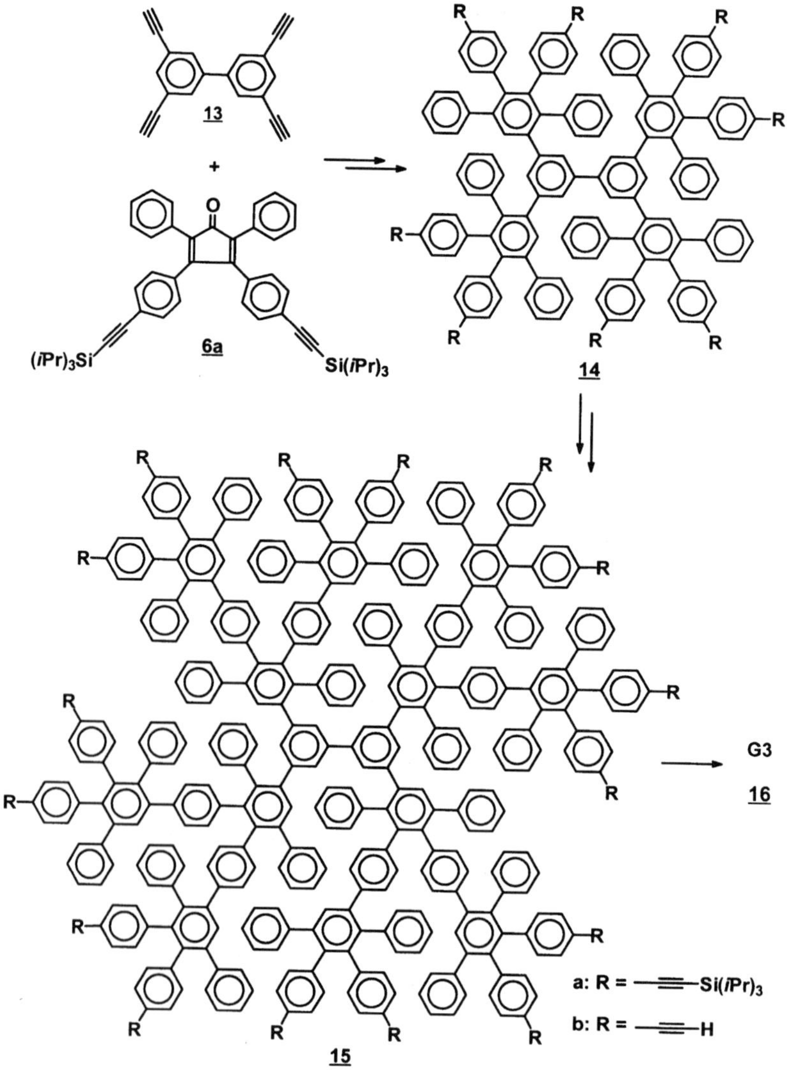

SCHEME 6

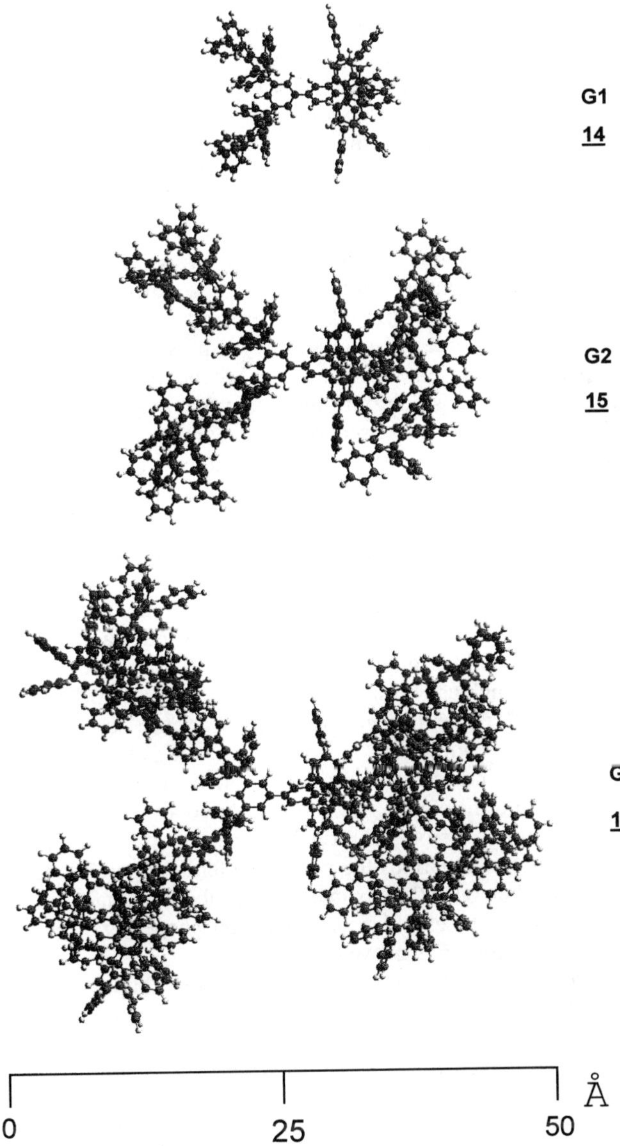

SCHEME 7

POLYPHENYLENE DENDRIMERS AS SHAPE-PERSISTENT NANOPARTICLES

The emphasis has so far been on polymers with extended π-conjugation and on columns of π-stacked discs, both structures constituting media for one-dimensional charge transport. As an alternative, we have been considering nanoparticles from tightly packed, but electronically decoupled benzene rings. Toward that end we have designed a new class of dendritic and hyperbranched polyphenylenes which comprise pentaphenyl benzene repeat units.[23] The significant torsion about the inter-ring single bonds ensures that no extended π-conjugation is established.

The synthetic concept refers to the polyphenylenes which in the above section have been used as precursors for graphite sheets, with the key reagent being a diethynyl-substituted tetraphenylcyclopentadienone. This cyclopentadienone derivative contains a diene and two dienophile functions for Diels-Alder reactions and can thus be regarded as an A_2B-type building block.[23] Not surprisingly, therefore, when this ethynyl substituted compound is allowed to react at elevated temperatures hyperbranched polyphenylenes are formed.[27] These hyperbranched polyphenylenes, while being readily accessable, have a high polydispersity and contain structural irregularities depending on whether or not ethynyl dienophile functions have escaped the Diels-Alder reaction. It is therefore important that the Diels-Alder reactions of the A_2B-type synthons can also be used to obtain monodisperse and structurally perfect dendrimers.[23,26] The key aspect is the attachment of bulky triisopropyl silyl groups to the ethynyl dienophiles, which then do not enter Diels-Alder reactions. This simple trick allows the clean reaction of the ethynyl protected cyclopentadienone **6a** as diene with reactive ethynyl dienophiles, which are attached to a suitable core.

Thus, 3,3′,5,5′-tetraethynylbiphenyl (**13**) gives rise to the first generation **14** of a benzene-based dendrimer upon reaction with the A_2B-unit **6a**. The deprotection of the ethinyl groups by desilylation with ammonium fluoride produces eight reactive ethynyl dienophiles, which can be subjected to another Diels-Alder reaction with **6a** under formation of G2 **15**, and repetition of the sequence can create well-defined dendritic polyphenylenes with up to 142 phenylene units already in G3 **16**. The resulting polyphenylene dendrimers are nicely soluble in organic solvents (the solubility increases upon going to higher generations), have a high thermal stability, and can readily be reprecipitated from solution. What is more important is that according to molecular dynamics calculations the polyphenylenes do not significantly change their size and shape when one allows some rotation about the inter-ring bonds.[23] Accordingly, the polyphenylenes exist as stable and shape-persistent nanoparticles with diameters as indicated in SCHEME 7.

This structural feature is important since the overall shape of the nanoparticles can be modified by the choice of the core and the branching reagent.[29] Using tetraethynylbiphenyl,[13] the tetrahedral shape persists even up to higher generations. In contrast, a propeller shaped dendrimer results when 1,3,5-triethynylbenzene (**17**) serves as core.

Replacing the A_2B-unit by an A_4B-synthon, namely the silyl protected tetraethynyltetraphenylcyclopentadienone **6b**, the Diels-Alder reaction produces an extreme crowding of phenyl rings. Thus, when reacted with the biphenyl core **13**, generation G1 is smoothly formed, however, the 16-fold adduct of G2 is only produced in minor amounts with the 14-fold cycloaddition product being the major component. On the other hand, when G1 is reacted with unsubstituted tetraphenylcyclopentadienone one can readily produce the structurally perfect and clean generation G2 **19** whereby a comparison of the overall shape of G2 **15** and G2 **19** convincingly documents the

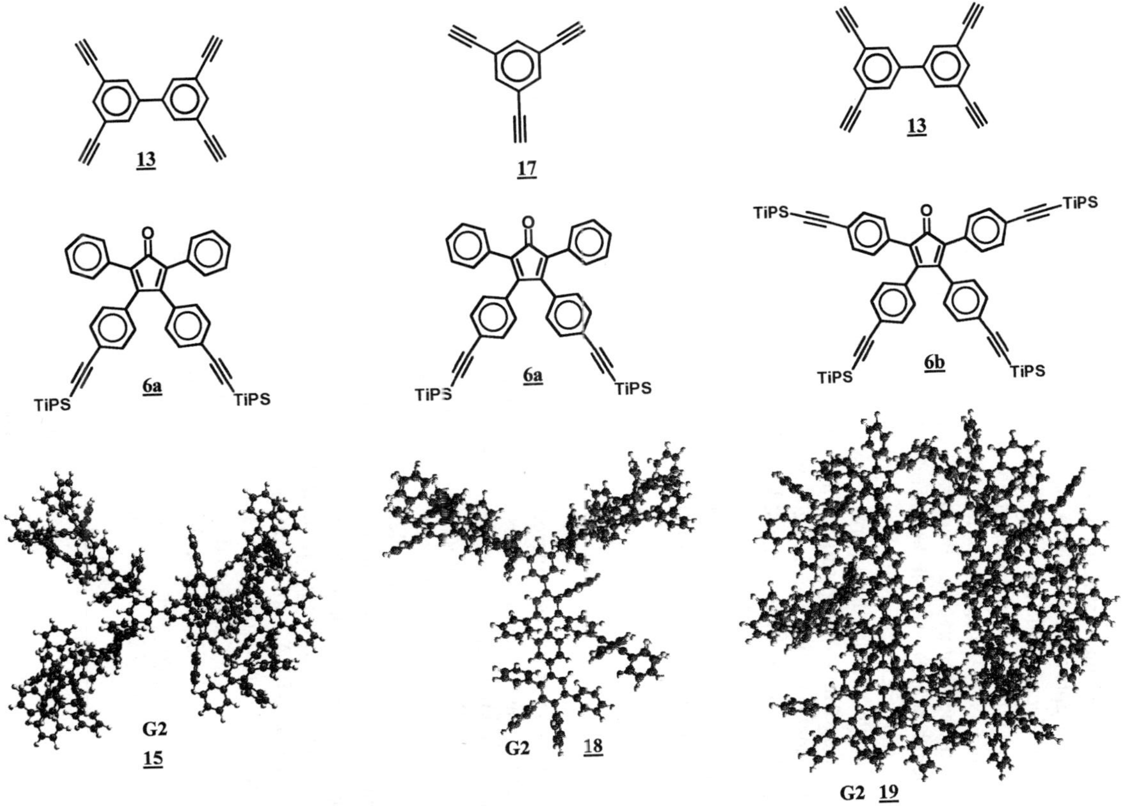
13
17
13
TiPS
TiPS
6a
TiPS
TiPS
6a
TiPS
TiPS
6b
TiPS
TiPS
G2
15
G2
18
G2 19

SCHEME 8

power of the present design principle: While both structures possess about the same diameter, the particles comprise 62 and 102 phenyl rings, respectively. Accordingly, G2 **19** consists of a more tightly packed polyphenylene.

Having available chemically stable nanoparticles of well-defined size and shape, it is now important to attach a variety of chemical functions. This can be achieved by different approaches. On the one hand, one can subject the dendritic polyphenylenes to electrophilic aromatic substitution reactions thus yielding functionalized derivatives in a non-selective, but nevertheless straightforward, fashion. A more defined approach is the application of a suitably functionalized tetraphenylcyclopentadienone in the final cycloaddition.[29] Some typical examples may be mentioned: the polyphenylene particles can be decorated at their surface with protected thiol-functions to study their self-assembly at gold surfaces; fluorophores can be introduced to form strongly emitting nanoparticles whose photoluminescence is studied by single molecule spectroscopy; pyridyl units can be built in during the final cycloaddition under formation of water-soluble nanoparticles, which, after coupling with, e.g., oligopeptides, can be used in the study of recognition processes; and amino or thiol substituted dendrimers can be applied as spacers in the formation of semiconductors, e.g., cadmium sulfide or cadmium selenide nanocrystals, and energy transfer between the latter can be investigated, again, by single molecule spectroscopy.

CONCLUSION AND OUTLOOK

The present article is centered around the design, synthesis, and supramolecular ordering of nanoscale molecules for molecular electronics. While the emphasis has been on the chemical aspects, the corresponding physical aspects are briefly outlined. For a more detailed description the reader is recommended to consult the relevant literature to which reference is given. This text is meant to stress the synthesis driven approaches toward molecular electronics. A good case can be made when looking at hexaalkylhexabenzocoronene, the first molecule for which scanning tunneling spectroscopy was performed at submolecular resolution. In previous studies we had formed and imaged regular 2D-crystals by physisorption of various alkyl-substituted PAHs. A consideration of the conflicting requirements defined by the quest for reproducibility of the spectroscopy and spatial resolution of the resulting images had strongly suggested a systematic increase of the size of the disc-type objects. Accordingly, our approach toward the molecular construction of carbonaceous 1D-, 2D- and 3D-polymers can well be regarded as an alternative to research directed toward spherical or tubular carbon allotropes.

REFERENCES

1. JUNG, T.A., , R.R. SCHLITTER, J.K. GIMZEWSKI, H. TANG & C. JOACHIM. 1996. Controlled room-temperature positioning of individual molecules: Molecular flexure and motion. Science **271:** 181–184.
2. STABEL, A., P. HERWIG, K. MÜLLEN & J.P. RABE. 1995. Diodelike current-voltage curves for a single molecule — tunneling spectroscopy with submolecular resolution of an alkylated, peri-condensed hexabenzocoronene. Angew. Chem. Int. Ed. Engl. **34:** 609–611.

3. TANS, S.J., M.H. DEVERET, H.J. DAI, A. THESS, R.E. SMALLEY, L.J. GEERLINGS & C. DEKKER. 1997. Individual single-wall carbon nanotubes as quantum wires. Nature **386:** 474–477.
4. ZHOU, C., R.M. DESHPANDE, M.A. REED, L. JONES II & J.M. TOUR. 1997. Nanoscale metal/self-assembled monolayer/metal heterostructures. App. Phys. Lett. **71:** 611–613.
5. SAMORI, P., V. FRANCKE, T. MANGEL, K. MÜLLEN & J.P. RABE. 1998. Phenylene-Ethynylene assemblies for a Potential Molecular Nanowire: A SFM Study. Optical Materials **9:** 390–393.
6. REHAHN, M., A.-D. SCHLÜTER & G. WEGNER. 1990. Soluble poly(para-phenylene)s: Variation of the length and the density of the solubilizing side chains, Makromol. Chem. **191:** 1991–2003.
7. KREYENSCHMIDT, M., F. UCKERT & K. MÜLLEN. 1995. A new soluble poly(p-phenylene) with tetrahydropyrene repeating units. Macromolecules **28:** 4577–4582.
8. SCHLÜTER, A.-D. 1991. Ladder polymers: The new generation. Adv. Mater. **3:** 282–291.
9. WEDER, C. & M.S. WRIGHTON. 1996. Efficient solid-state photoluminescence in new poly(2,5-dialkoxy-p-phenyleneethynylene)s. Macromolecules **29:** 5157–5165.
10. TADA, K., M. ONODA, M. HIROHATA, T. KAWAI & K. YOSHINO. 1996. Blue-green electroluminescence in copolymer based poly(1,4-phenylene-ethynylene). Jpn. J. Appl. Phys. **35:** L251–L253.
11. DIECK, H.A. & F.R. HECK. 1975. Synthesis of aryl- and vinyl-substituted acetylene derivatives by the use of nickel and palladium complexes. J. Organomet. Chem. **93:** 253–257.
12. TAKAHASHI, S., Y. KOROYAMA, K. SONOGASHIRA & N. HAGIHARA. 1980. A Convenient synthesis of ethynylarenes and diethynylarenes. Synthesis: 627–630.
13. MANGEL, T. 1997. Übergangsmetallkatalysierte Kupplungsreaktionen zur Darstellung von endfunktionalisierten konjugierten Oligo- und Polymeren, Ph.D. thesis, Johannes Gutenberg-Universität, Mainz.
14. WEISS, K., A. MICHEL, E.-M. AUTH, U.H.F. BUNZ, T. MANGEL & K. MÜLLEN. 1997. Acyclic diyne metathesis (ADIMET), an efficient route to poly(phenylene)ethynylenes (PPEs) and nonconjugated polyalkynyles of high molecular weight, Angew. Chem. Int. Ed. Engl. **36:** 506–509.
15. WANG, W., G. LIESER & G. WEGNER. 1993. Lyotropic liquid crystals of a soluble polydiacetylene, A comparative investigation by means of optical and electron microscopy. Liquid Crystals **15:** 1–24.
16. Huggins, K.E., S. Son & S.I. Stupp. 1997. Two-dimensional supramolecular assemblies of a polydiacetylene. 1. Synthesis, structure, and third order nonlinear optical properties. Macromolecules **30:** 5305–5312.
17. OFER, D., T.M. SWAGER & M.S. WRIGHTON. 1995. Solid-state ordering and potential dependence of conductivity in poly(2,5-dialkoxy-p-phenyleneethynylene. Chem. Mater. 7: 418–425.
18. VAN DE CRAATS, A.M., J.M. WARMAN, M.P. DE HAAS, D. ADAM, J. SIMMERER, D. HAARER & P. SCHUHMACHER. 1996. The mobility of charge carriers in all four phases of the columnar discotic material hexakis(hexylthio)triphenylene: Combined TOF and RP-TRMC Results. Adv. Mater. **8:** 823–826.
19. IYER, V.S., M. WEHMEIER, J.D. BRAND, M.A.KEEGSTRA & K. MÜLLEN. 1997. From hexa-peri-hexabenzocoronene to "Superacenes." Angew. Chem. Int. Ed. Engl. **36:** 1604–1607.
20. MÜLLER, M, H. MAUERMANN-DÜLL, M. WAGNER, V. ENKELMANN & K. MÜLLEN. 1995. A cycloaddition-cyclodehydrogenation route from stilbenoids to extended aromatic hydrocarbons. Angew. Chem. Int. Ed. Engl. **34:** 1583–1586.

21. MÜLLER, M., J. PETERSEN, R. STROHMAIER, C. GÜNTHER, N. KARL & K. MÜLLEN. 1996. Polybenzoid C_{54} hydrocarbons: Synthesis and structural characterization in vapor-deposited ordered monolayers. Angew. Chem. Int. Ed. Engl. **35:** 886–888.
22. HERWIG, P., C.W. KAYSER, K. MÜLLEN & H.W. SPIESS. 1996. Columnar mesophases of alkylated hexa-peri-hexabenzocoronenes with remarkably large phase widths. Adv. Mater. **8:** 510– 513.
23. MORGENROTH, F., E. REUTHER & K. MÜLLEN. 1997. polyphenylene dendrimers: from three-dimensional to two-dimensional structures. Angew. Chem. Int. Ed. Engl. **36:** 631–634.
24. VAN DE CRAATS, A.M., J.M. WARMAN, K. MÜLLEN, Y. GEERTS & J.D. BRAND. 1998. Rapid charge transport along self-assembling graphitic nanowires. Adv. Mater. **10:** 36–38.
25. ZIMMERMANN, U. & N. KARL. 1992. Epitaxial growth of coronene and hexa-peri-benzocoronene on MoS2 (0001) and graphite (0001): A LEED study of molecular size effects. Surface Sci. **268:** 296–306.
26. GÜNTHER, C., N. KARL, R. STROHMAIER, W. EISENMENGER, M. MÜLLER, K. MÜLLEN & N. SATO. 1998. Structure and binding of ordered monolayers of the rhombus-shaped polycondensed aromatic molecule $C_{54}H_{22}$ on graphite, MoS_2 and GeS, studied by LEED, STM and TDS. Surface Sci. Submitted for publication.
27. MORGENROTH, F. & K. MÜLLEN. 1997. Dendritic and hyperbranched polyphenylenes via a simple Diels-Alder route. Tetrahedron **53:** 15349–15366.
28. MORGENROTH, F., C. KÜBEL & K. MÜLLEN 1997. Nanosized polyphenylene dendrimers based upon pentaphenylbenzene units, J. Mater. Chem. **7:** 1207–1211.
29. BERRESHEIM, A.J., F. MORGENROTH, U.-M. WIESLER & K. MÜLLEN. 1998. Dendritic, hyperbranched, and cross-linked polyphenylenes - From benzene to shape-persistent nanoparticles. Polym. Prep. Submitted for publication.

Molecular Landers

ANDRÉ GOURDON[a] AND HAO TANG

Molecular Electronics Group, CEMES, CNRS UP8011, BP 4347, 29 Rue J. Marvig, 31055 Toulouse Cedex, France

ABSTRACT: Molecular landers are molecules designed for the study of electron transport in a tip-molecule-surface experimental set-up. They comprise a fully aromatic and rigid main board linked to several spacers (legs). When the molecular landers are adsorbed on a metallic surface, the spacers maintain the main board at such a distance from the surface that the aromatic part is electronically decoupled from the surface. One end of this board extends beyond the last spacer and can be chimisorbed by docking along a diatomic step. Therefore in this conformation, the current can flow between the STM tip at one position of the molecular board and the surface at the other end, and only through the aromatic part.

Molecular systems able to ensure fast long-distance electron transfer or electron transport are an essential requisite for future applications in molecular electronics. In order to design and to synthesize the best molecules for this purpose, it is very important to understand the parameters controlling this phenomenon. In the case of intramolecular electron transfer, the distance-dependence of the electron-transfer rate, which is itself directly related to the distance-dependence of the electronic coupling parameter, has been widely studied in 5–25-Å-long mixed valence systems[1] and is now rather well understood.[2] This is not the case for long distance electron transport. Although theoretical investigations of the conductance and transparence of long molecular wires between two metallic electrodes have been recently published,[3] the experimental study of electron-transport decay with the molecule length and that with different types of molecules still has to be carried out. For electron transfer, the experimental values of this decrease of metal-metal coupling through a molecule are obtained by synthesizing a series of metallic dimers in which the two metal centers are coupled by bridging ligands of various lengths. The spectroscopic studies of the intervalence bands then provide the coupling parameter for each metal-metal distance. In this case, the molecule geometry and, in particular, the metal-metal distance are well known, and the environmental (solvent) differences are averaged over a very large number of molecules. However, this type of solution study is currently limited to ca. 30 Å by detection threshold of the low-intensity intervalence band.

This type of approach is clearly not possible with the electrode-molecule-electrode experimental set up shown in FIGURE 1. It would require, for each type of molecular wire, first the fabrication of numerous chips with different interelectrode gap lengths and then the connection of these electrodes by molecular wires. Furthermore, at this scale and on one single molecule, there is an absolute necessity at least to observe, to measure, and to extract by calculation the exact physical geometry and

[a]Address for telecommunication: Phone, 33 5 6225 78 59; fax, 33 5 62 25 79 99; e-mail, gourdon @cemes.fr

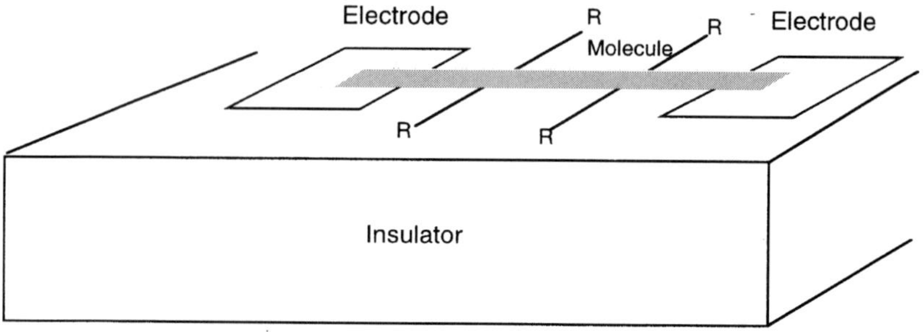

FIGURE 1. Electrode-molecule-electrode experiment.

in particular the geometry of the molecule-electrode contact. Indeed, it has been shown that the molecule-electrode contact resistance is very dependent, by several orders of magnitude, on the electronic coupling (that is, the geometry) between the molecule σ and π orbitals and the electrode atoms.[4] Therefore, the net value of the conductance of a molecular wire in function of its length will be obtained only after extraction by calculation of any geometrical difference for each experiment. Any blind experiment would lead to erratic and meaningless results, even without taking into account all possible artifacts at this scale (e.g., "dust," lithography residues, resins, metal clusters). This experiment has been recently carried out using a gapless single-wall nanotube.[5] But these insoluble "metallic" molecular wires are difficult to handle and to functionalize for more complex devices such as molecular switches. For "nonmetallic" molecular wires, the very low current intensity expected for the interelectrode gap currently made by nanolithography remains a very crucial problem. For instance, for a pad-to-pad distance of 8 nm and a bias voltage of 0.1 V, the

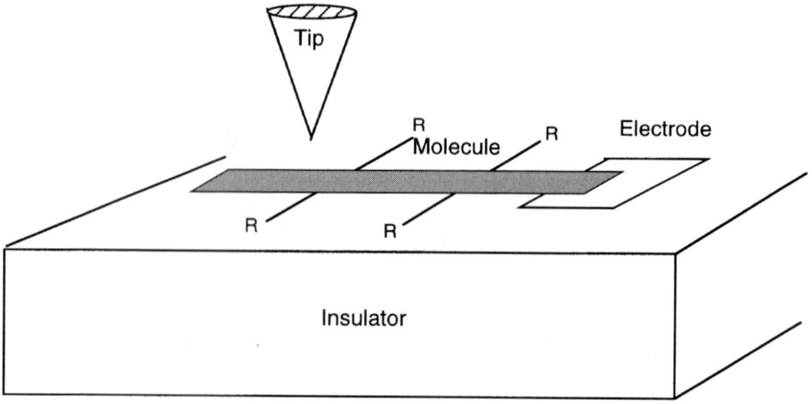

FIGURE 2. Tip-molecule-electrode experiment.

calculation[3] shows that the expected current intensity would be at the maximum ca. 10^{-14}–10^{-15} A, at least for the best molecular wires synthesized so far. On the other hand, with current nanolithography technology, nanoelectrodes with an inter-electrode gap smaller than 5 nm remain difficult to make in a reproducible way.[6]

However, the measurement of electron-tunneling current through a molecular wire in function of its lengths could be studied in another type of experimental setting, such as the tip-molecule-electrode one shown in FIGURE 2. The molecular wire is chemically bound to the electrode and lies on the insulator in which the metal pad is buried. First, the exact image of the geometry is obtained in AFM mode. Switching to the STM mode then allows the measurement of the current flowing through the molecule between the tip and the electrode at any point of the molecular wire directly giving its conductance. This experimental set-up has several important advantages:

(*1*) the molecules can be easily and selectively bound at only one end from solutions to the electrode by a chemical link such as thiol-gold or isonitrile-platinum;

(*2*) substituents R such as alkyl chains, necessary for improving solubility and chemical processing, do not deeply modify the current intensity;

(*3*) mixtures of oligomers with a size distribution, much easier to synthesize than monodisperse long molecules could be used;

(*4*) a buried small pad can be more easily, reproducibly, and less expensively made than a circuit with two electrodes separated by a 5–10-nm gap. On each pad, the measurement can be done on several molecules;

(*5*) a very large number of experiments could then be performed at low cost.

Although we are currently working along these lines, this experiment is hampered by the poor planarity of insulators such as SiO_2. The surfaces prepared by controlled oxidations of silicon crystals, are too bumpy, impeding an accurate AFM imaging of relatively narrow and soft molecules.

In order to overcome this problem, we are currently working on a slightly different experimental set-up as shown in FIGURE 3. The experiment is done under ultra-high vacuum using a clean metallic surface. The molecule has a rigid main board,

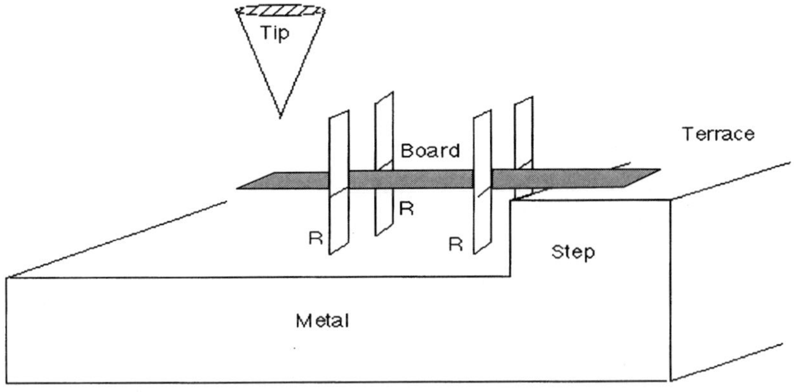

FIGURE 3. Tip-lander-surface experiment.

which is maintained at least 4–6 Å above the surface by several spacers (legs) R in order to prevent a direct electron flow between the tip and surface through and perpendicularly to the board plane. The main board (wire) has an aromatic platform which projects beyond the last spacer for chemisorption with an another terrace of

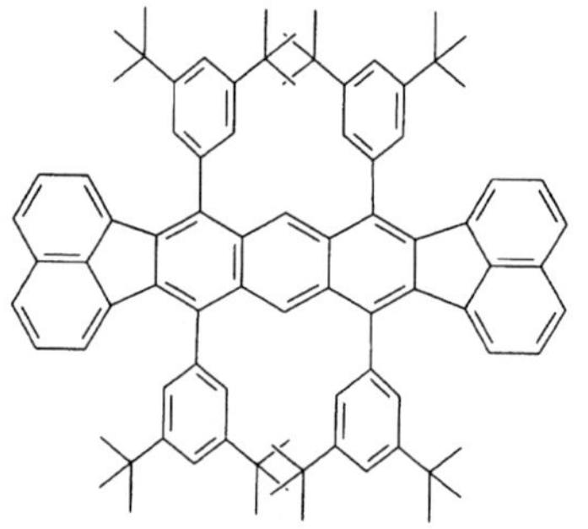

1

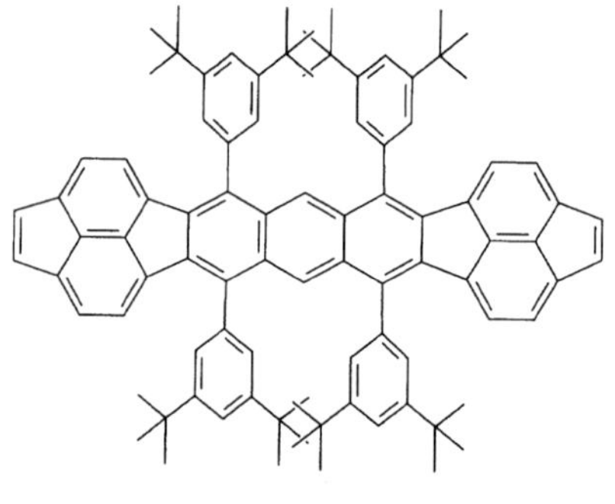

2

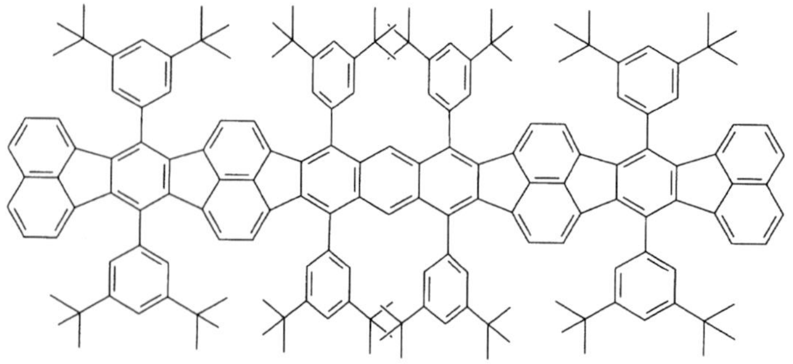

3

FIGURE 4. Molecular landers.

the metal surface. The similarity of these molecular ladders with landing crafts has led us to coin the name *molecular landers*. The measurement of the current intensity versus length is done by moving the tip along the molecular board. The spacers must show small electronic coupling with the main board and with the surface to minimize electron leaks. They must also allow some movement of the molecule on the surface. Although the parameters controlling this mobility are not fully understood,[7] there seems to be a compromise between rigidity, which favors the mobility, and softness (slackness), which sticks the molecule on the surface, and physisorption and chimisorption. And, finally, the spacers must also maintain some solubility to the molecule, at least for purification and spectroscopic studies.

However, in contrast with the previous case where most molecules can be adsorbed from dilute solutions, the type of molecules and functionalities that can be used in this experiment is limited by conditions inherent in ultra-high vacuum conditions. For instance, the molecule cannot be adsorbed from solutions but must be sublimated onto a hot clean metallic surface. Fragile molecules can undergo thermal degradation during heating or catalytic decomposition on the reactive metallic surface, which limits the type of chemical functions.

For these experiments, we have synthesized several types of molecular landers, shown in FIGURE 4. These landers comprise a polyaromatic molecular board ended by a fluoranthene group for chimisorption on the terrace, and four to eight 3,5-bis-*tert*butylphenyl spacers. These spacers are maintained in a conformation so that the phenyl rings are nearly perpendicular to the main board plane by steric crowding, as shown in FIGURES 5 and 6. The *tert*butyl groups increase the board-surface distance, allow some rotation of the "legs" without reducing this distance too much, reduce the phenyl-surface electronic coupling, allow easy displacement of the molecule on the surface, and finally increase the molecule solubility.

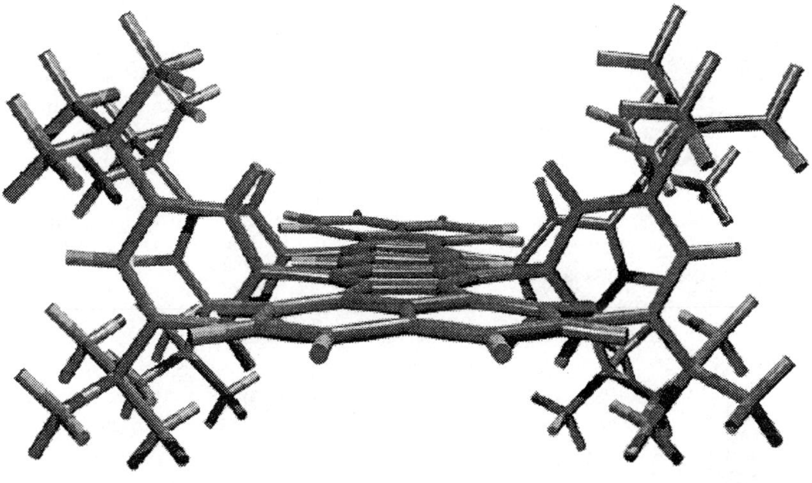

FIGURE 5. View of **1** in one of its minimum energy conformations (in vacuum) showing the planarity of the main board.

FIGURE 6. View of **1** showing the nearly perpendicular conformation of the bis*tert*butylphenyl substituents.

In order to appraise the feasibility of this experiment, a theoretical study of the interactions between tip-lander-surface was carried out for the lander **1**. As a first step, its molecular geometry in a vacuum was optimized. Two minimum energy conformations were found (ΔE = 0.2 Kcal) with parallelar (FIG. 6) or anti-parallelar spacers. The molecule was then drawn near to a Cu (100) surface and the equilibrium position was reached at a board-surface distance of 6.8 Å (FIG. 7). At that distance a docked molecule is at van der Waals contact with a diatomic copper step (3.62 Å). The tip-surface distances at constant current I = 0.02 nA for a bias voltage of 50 mV were calculated for scans along the axis of a lander on a flat surface and along the axis of a lander docked on a double step (FIG. 8) using the ESQC method.[8] Two main results can be stressed:

(*1*) For **1** on a flat surface (*dashed line*), the calculation shows two main bumps **B** and **C** corresponding to the tunneling current through the two pairs of vertical spacers. In this geometry, the fluoranthene ends appears as two smaller peaks **A** and **D**, which shows that there a small current leakage from this end through the spacers.

(*2*) For **1** in interaction with a terrace (*solid line*), the free fluoranthene end is seen as a higher bump **A** , which indicates that a current is flowing from this end to the terrace. For a tip at 2.8 Å above the free end of the molecular board, the current is 0.103 nA, whereas it was 0.013 nA under the same conditions for the previous case. Furthermore, the tip-surface distance increases steadily from this end to the step, which is equivalent to an increase of tunneling current at constant height along the molecular axis. Extraction of the contribution from the spacers in order to get the net conductance of the main board alone is currently in progress.

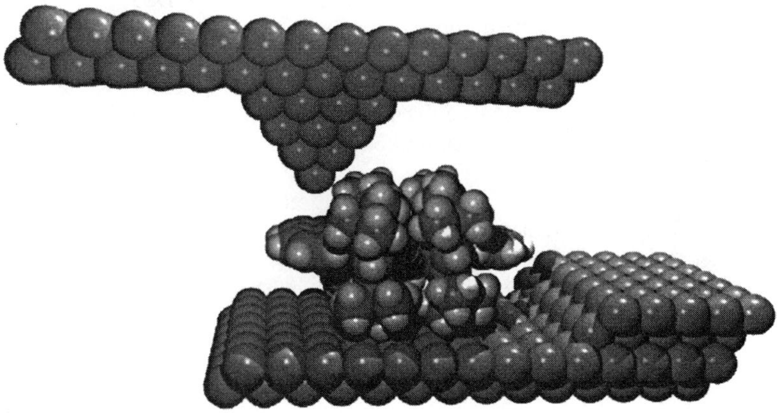

FIGURE 7. Geometry used for the calculation of the tunnel current.

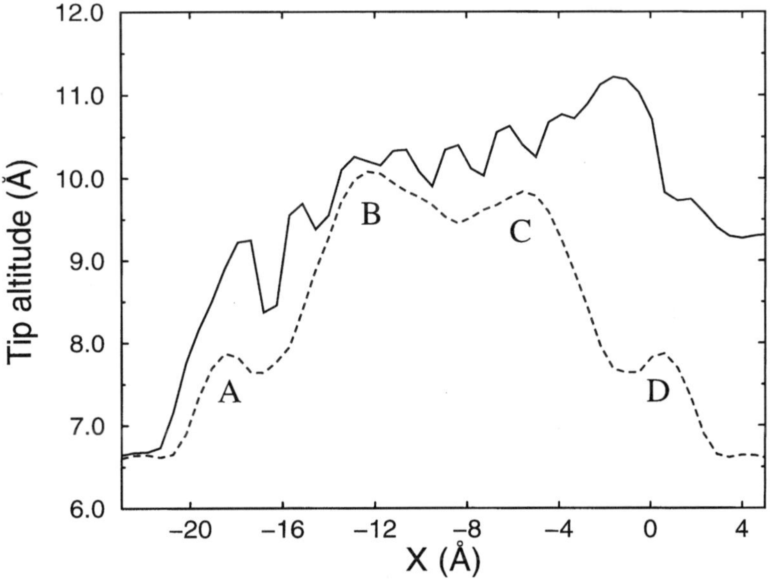

FIGURE 8. Calculated tip-surface distance for **1** on a flat Cu (100) surface (*dashed line*) and for 1 docked on a Cu diatomic step (constant current 0.02 nA, $\Delta E = 50$ mV). The scan origin goes through the edge of the step.

SYNTHESES

The outline of the syntheses of **1** is presented in SCHEME 1: Bromethyl-3,5-bis-*tert*butyl benzene obtained by bromination of commercial 3,5-bis*tert*butyltoluene using NBS in CCl_4. Carbonylating coupling by iron pentacarbonyl and a base in phase-transfer conditions[9] then gives the 1,3-diaryl cetone **4**. Double Knoevenagel condensation with acenaphtenequinone gives the acecyclone **5** in good yields.[10] Reaction of two molecules of this diene **5** with benzadiyne, prepared *in situ* by oxidation of 1,5-diamino-1,5-dihydrobenzo[1,2–d:4,5-d′]bistriazole by lead tetra-acetate,[11] yields the target product **1**. The synthesis of **2** is very similar, the condensation being done with 1,2-pyracycloquinone instead of acenaphtenequinone.

In contrast with similar molecular boards recently described by Shluter *et al.*,[12] **2** cannot be used as a diene for subsequent reaction with a dienophile as the acecyclone **5**. This reaction provides complex mixtures of byproducts and only low yields of the target molecule **3**. The difference stems mainly in the lower solubility of **2** and in more electron-rich character of **5** due to the two 3,5-bis*tert*butylphenyl substituents when compared with α,ω-dodecanyl, which precludes the retro Diels-Alder.

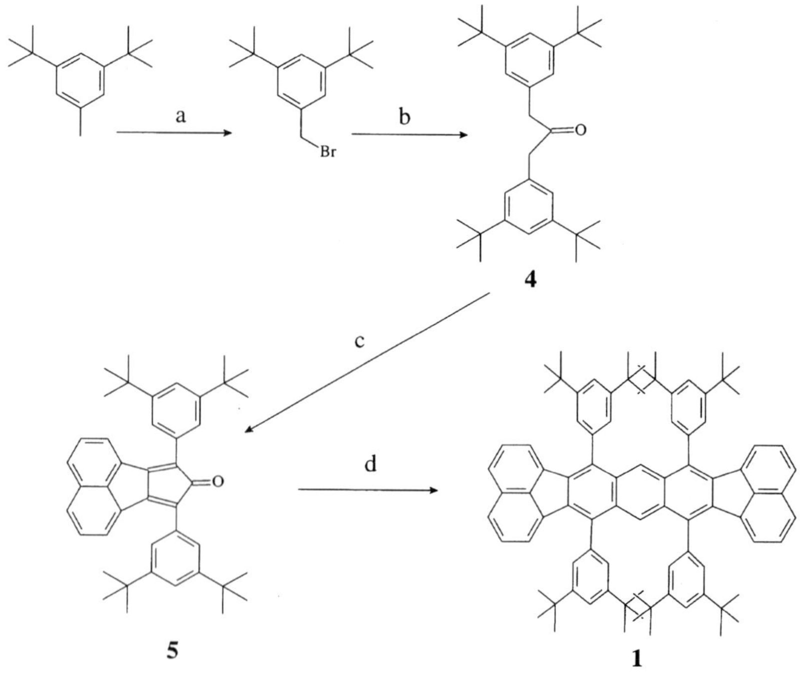

SCHEME 1. **(a)** NBS, dibenzoylperoxyde/CCl_4 rflx, 1 h, yield 82%. **(b)** $Fe(CO)_5$, $[N(CH_3)_2(C_{16}H_{33})_2]Br$ /NaOH 30%, CH_2Cl_2 RT, 3 h, yield 41%. **(c)** acenaphtenequinone, KOH/MeOH, RT, 8 h, yield 86%. **(d)** 1,5-diamino-1,5-dihydrobenzo [1,2–d:4,5-d′]bistriazole, LTA THF, 0.5 h yield 33%.

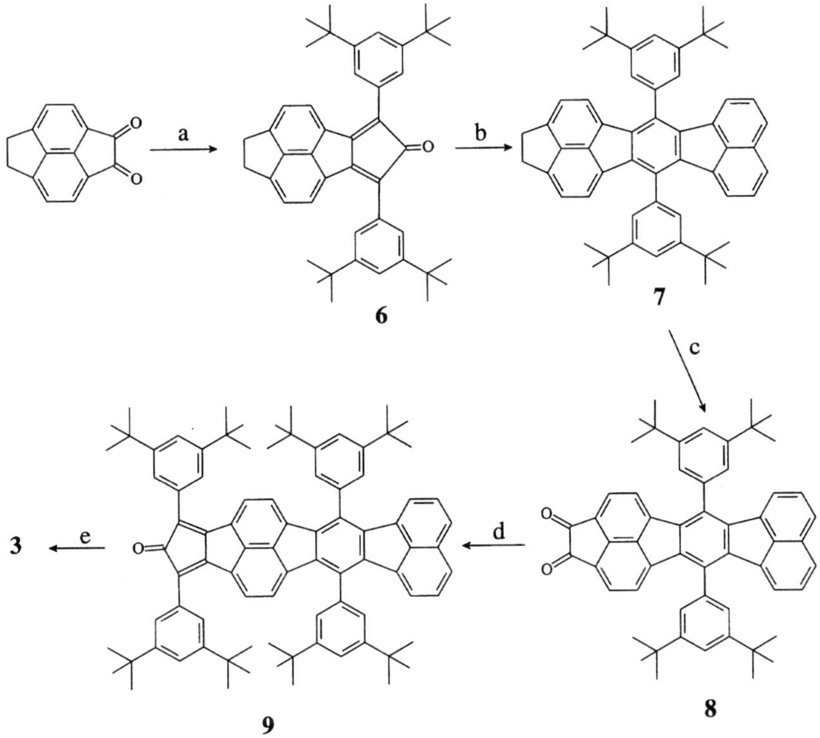

SCHEME 2. (**a**) 4, KOH / MeOH RT, 5 h, yield: 98 %. (**b**) acenaphtylene / toluene rflx, 3 d , yield 73 %. (**c**) benzeneseleninic anhydride/chlorobenzene rflx, 1 d, yield 47%. (**d**) 4, KOH / MeOH RT, 10 h, yield: 77%. (**e**) 1,5-diamino-1,5-dihydrobenzo[1,2-d:4,5-d']bis-triazole, LTA/ THF RT, 0.5 h, yield 27 %.

The synthesis of **3** is shown in SCHEME 2: 1,2-diketopyracene, prepared from acenaphtene by Friedel-Craft reaction,[13] from **4** gives the acecyclone **6**. Diels-Adler with acenaphtylene yielded **7**. Oxidation of **7** by seleninic anhydride[14] afforded the diketone **8**, which condenses again with **4** to give the cyclopentadieneone **9**. Two molecules of **9** can then react with benzadiyne to yield the ladder **3**. These landers with an anthracene core are moderately (**1**) to poorly (**3**) soluble giving from yellow (**1**) to pink-red (**3**) highly luminescent solutions. They can be sublimated under ultra-high vacuum on warm copper or silver surfaces.

PRELIMINARY STM RESULTS

The STM studies of these landers are currently in progress.[15] The first results on **1** show that at low surface coverage, this lander is very mobile, with a 2D gas or liquid behavior at the STM scanning time scale. The molecules can be stopped by a voltage pulse, allowing their imaging. Futhermore, the molecules spontaneously dock along the metal steps by chimisorption of the fluoranthene end. In this conformation, the other fluoranthene end can be imaged, which shows an electron transfer from the tip to the metal step through the molecular board. At higher coverage the landers crystallize in well-ordered 2D crystals, showing different conformations for the spacers with respect to the main board and with the metallic surface.

CONCLUSIONS AND PERSPECTIVES

In this paper we have introduced the concept of molecular landers. These molecules comprise a fully aromatic and rigid main board linked to several spacers (legs). When the molecular landers are adsorbed on a metallic surface, the spacers maintain the main board at such a distance from the surface that the aromatic part is electronically decoupled from the surface. One end of this board extends beyond the last spacer and can be chimisorbed by docking along a diatomic step. Therefore in this conformation, the current can flow between the STM tip at one position of the molecular board and the surface at the other end, and only through the aromatic part. Current work involves simulations of the STM images in order to extract the exact conformation of the molecule on the surface and with respect to the metallic steps, which will give an evaluation of the conductance of these boards in function of their length.

Depending on their mobility and on the surface coverage, the adsorbed molecules can be considered as in a 2D gas or liquid or in a 2D crystal with little intermolecular interactions. One can expect that the chemical and photochemical properties of these adsorbed molecules will be very similar to those of molecules in vacuum with the avantages of the near-field techniques for imaging, manipulation, and excitation.

ACKNOWLEDGMENTS

We thank M. Magoga and C. Joachim, without whom this study would not have been started, for fruitful discussions and calculations. This study was done in collaboration with J. Gimgewski, V. Langlais and R.R. Schittler (IBM Zurich), we are grateful for their enthusiasm and efficiency. Financial support by the CNRS, Ultimatech (France), and the European Esprit IV Nanowires is gratefully acknowledged.

REFERENCES

1. BELEY, M., S. CHODOROWSKI-KIMMES, J.-P. COLLIN, P. LAINE, J.-P. LAUNAY & J.-P. SAUVAGE. 1994. Pronounced electronic coupling in rigidly connected N,C,N-coordinated diruthenium complexes over a distance of up to 20 Å. Angew. Chem. Int. Ed. Engl. **33:** 1775–1778; RIBOU, A.-C., C. LAUNAY, M.L. SACHTLEBE, H. LI & C.W. SPANGLER. 1996. Intervalence electron transfer in mixed valence diferrocenylpolyenes. Decay law of the metal-metal coupling with distance. Inorg. Chem. **35:** 3535–3740.
2. JOACHIM, C., J.-P. LAUNAY & S. WOITELLIER. 1990. Distance dependence of the effective coupling parameters through conjugated ligands of the polyene type. Chem. Phys. **147:** 131–141.

3. MAGOGA, M. & C. JOACHIM. 1997. Conductance and transparence of long molecular wires. Phys. Rev. B. **56:** 4722–4729; MAGOGA, M. & C. JOACHIM. 1998. Minimal attenuation for tunneling through a molecular wire. Phys. Rev. B. **57:** 1820–1823.
4. JOACHIM, C. & J.F. VINUESSA. 1996. Length dependence of the electronic transparence (conductance) of a molecular wire. Europhys. Lett. **33:** 635–640.
5. TANS, S.J., DEVORET, M.H., H. DAI, A. THESS, R.E. SMALLEY, L.J. GEERLIGS & C. DEKKER. 1997. Individual single-wall nanotubes as quantum wires. Nature **386:** 474–477.
6. ROUSSET, V., C. JOACHIM, S. ITOUA, B. ROUSSET & N. FABRE. 1995. Fabrication of buried co-planar metal-insulator-metal nanojunctions with a gap lower than 10 nm. J. Phys. III **5:** 1983–1989; ROUSSET, V., C. JOACHIM, T. ONDARCUHU & B. ROUSSET. 1998. I-V characteristics of coplanar metal-semiconductor-metal nanojunctions. J. Phys. III. EPJ A. In press.
7. JUNG, T.A., R.R. SCHLITTER, J.K. GIMZEWSKI, H. TANG & C. JOACHIM. 1996. Controlled room-temperature positioning of individual molecules: Molecular flexure and motion. Science **271:** 181–184.
8. CHAVY, C., C. JOACHIM & A. ALTIBELLI. 1993. Interpretation of STM images: C60 on the gold (110) surface. Chem. Phys. Letters. **214:** 569–575.
9. TANGUY, G., B. WEIBERGER & H. DES ABBAYES. 1984. Iron pentacarbonyl mediated coupling of benzyl halides to ketones under mild phase transfer conditions. Tetrahedron Lett. **25:** 5529–5532; TANGUY, G., B. WEIBERGER & H. DES ABBAYES. 1983. Catalytic phase transfer carbonylation of benzyl halides with iron pentacarbonyl. Tetrahedron Lett. **24:** 4005–4008.
10. OGLIARUSO, M.A., M.G. ROMANELLI & E.I. BECKER. 1965. Chemistry of cyclopentadienone. Chem. Rev. **65:** 261–367.
11. HART, A. & D. OK. 1986. Synthesis of 1,5-diamino-1,5-dihydrobenzo[1,2-d:4,5-d′]bistriazolc (DABT) and its usc as a 1,4-bcnzadiync cquivalcnt. **51:** 979–986.
12. SCHLICKE, B. & A.-D. SCHLUTER. 1996. Repetitive synthesis of soluble oligofluoranthene and their electrochemical analysis. Synlett.: 425–426;LOFFLER, M. & A.-D. SCHLUTER. 1994. 7,18-9,16-di(α,ω-dodecadiyl)-dipyracyclo[1′,2′-b,i]anthracene: a novel diels-alder-bisdienophile. Synlett. 75–77; SCHLICKE, B., A.-D. SCHLUTER, P. HAUSER & J. HEINZE. 1997. Polycyclic aromatic hydrocarbons in the nanometer range. Angew. Chem. Int. Ed. Engl. **36:** 1996–1998.
13. STILLE, J.K., G.K. NOREN & L. GREEN. 1970. Hydrocarbon ladder aromatics from Diels-Alder reaction. J. Polymer. Sci. **A8:** 2245–2254; CLAYTON, M.D., Z. MARCINOV & P.W. RABIDEAU. 1996. Improved synthesis of a C30H12 buckybowl via benzylic oxidation with benzeneseleninic anhydride. J. Org. Chem. **61:** 6052–6054.
14. BARTON, D.H.R., R.A.H.F. HUI, D.J. LESTER & S.V. LEY. 1979. Preparation of aldehydes and ketones by oxidation of benzylic hydrocarbons with benzeneseleninic anhydride. **35:** 3331–3334.
15. J.J. GIMGEWSKI, V. LANGLAIS & R.R. SCHITTLER, IBM Zurich. Work in progress.

Self-Assembly, Manipulation, and Discrimination of DNA Molecules by Scanning Tunneling Microscope (STM) on Solid Surfaces

TOMOJI KAWAI[a]

ISIR-Sanken, Osaka University, Ibaraki, Osaka 567, Japan

ABSTRACT: DNA base molecules form novel low-dimensional superstructures on Cu(111) surfaces through hydrogen bonding among molecules and through their epitaxial growth on the surface. The chemical inertness of the Cu(111) substrate allows the molecules to diffuse over the surface to self-assemble spontaneously into their own unique structures. Molecular orbital calculations explain the mechanism of the observed self-assemblies induced by base-base hydrogen bonding. High-resolution molecular imaging and manipulation of DNA base molecules and DNA oligomers have also been successfully carried out on $SrTiO_3(100)$ and Cu(111) surfaces under ultra-high vacuum conditions by STM.

DNA molecules are fascinating for the application to molecular electronics. They can be used for molecular-level information storage as shown in FIGURE 1. When DNA base molecules (such as adenine, thymine, cytosine, and guanine) are discriminated by high-resolution imaging of scanning tunneling microscope (STM), we can use these bases for the ultimate unit of information storage at the molecular level. Furthermore, the self-organization of these molecules to form various novel superstructures can be used for inter-connections between molecular devices and for optical devices. The imaging and manipulation of single and double-stranded DNA itself are much more fascinating from the viewpoint of the new direction of molecular electronics. When one of the DNA bases in a DNA chain is replaced by another base, it can be duplicated by PCR (polymerase chain reaction) and products can be made on the macro scale. In this paper, we will show formation of two-dimensional superstructures of DNA base molecules through hydrogen bonding, and also show high-resolution imaging and manipulation of DNA molecules by STM.

FORMATION OF TWO-DIMENSIONAL SUPERSTRUCTURE OF DNA BASE MOLECULES THROUGH HYDROGEN BOND

In DNA, hydrogen bonds play an important role in duplication of the chain and for information transfer between the DNA molecules. We show here that the DNA base molecules form novel superstructures through hydrogen bonding on Cu(111) surfaces. This phenomenon can be used for molecular wires and molecular connections. When DNA molecules, such as adenine, are randomly dispersed on Cu(111) surface at room temperature, they form a two-dimensional superstructure after dif-

[a]Address for telecommunicatoin: Phone, 81/6-879-8445; fax: 81/6-875-2440; e-mail: kawai @sanken.osaka-u.ac.jp

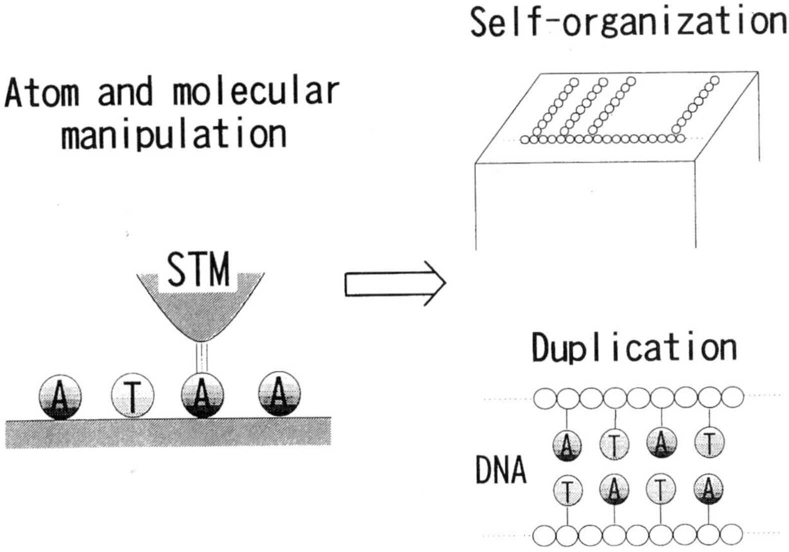

FIGURE 1. Schemes of molecular imaging and manipulation by STM. Self-organization and duplication toward molecular electronics are also shown.

fusion, collision, and bond formation (FIG. 2). As a surface science, this phenomenon is interesting from the viewpoint of the generation of an ordered state from a disordered state. STM images of the self-assembly for the four kinds of DNA bases (adenine, thymine, cytosine, and guanine) have been reported previously.[1] Adenine forms one-dimensional chain structures which then make up two-dimensional networks. Thymine forms two-dimensional islands and guanine forms two-dimensional square lattices. Cytosine, most interestingly, forms zigzag one-dimensional chains toward six different directions reflecting the crystal symmetry of Cu(111) surface. Why are these different structures formed on the Cu(111) surfaces? Essentially, these structures are the superstructures formed through hydrogen bonding as shown in FIGURE 2, in the case of adenine molecules. The molecules migrate on the surfaces and bind to each other through hydrogen bonds of the molecule. In the STM images of FIGURE 3(a), the migration of adenine molecules at room temperature are imaged as white noise by STM.[2] Only the chain structures after self-assembly are imaged as one-dimensional chains, and imaging is hindered because of thermal fluctuations. When the temperature is lowered to 70K, all the motion of the molecules is frozen out.[2] Thus, not only chain structure but also single molecules, dimers, and clusters of adenine can be clearly seen in FIGURE 3(b). The real time resolved growth process is further imaged in FIGURE 4. Cytosine molecules are deposited on Cu(111) surface at room temperature, and these molecules migrate on the surface imaged — white noise at room temperature. When these moving molecules reach the top of the chain, it grows at a rate of one molecule/minute.[3] In this manner, the molecular chains grow in time to form chain structures toward sixfold symmetric directions un-

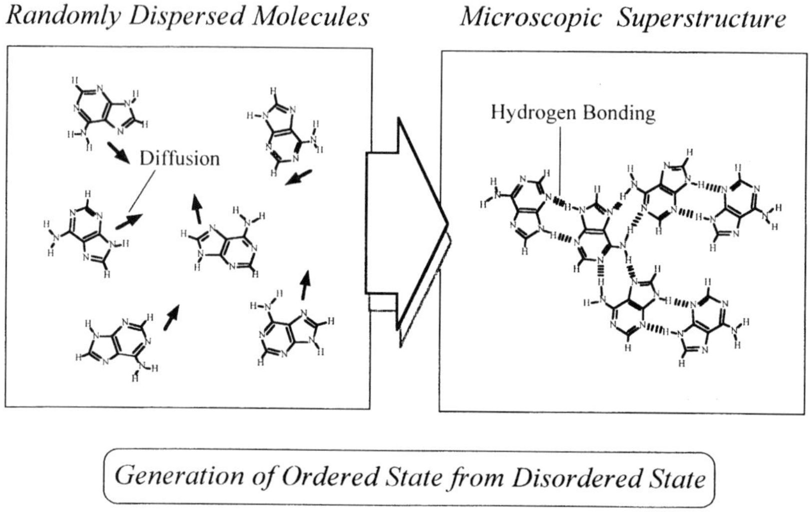

FIGURE 2. Adenine molecules are dispersed randomly on a surface to migrate and form two-dimensional superstructure through hydrogen bond between molecules.

der these conditions. The detailed molecular level mechanisms are shown in FIGURE 5.[2] When adenine molecules encounter each other, they form dimers. For the stability of dimers, the AA I type (FIG. 5b) is the most stable one. Accordingly, the AA I structure is supposed to form first. The other structures such as AA II, III, IV, and V are less stable. Among these, AA II and AAIII cannot be formed by the combination of AA I because these combinations are impossible after the formation of AA I. Thus, the combination of AA IV and AA I can be achieved to form one-dimensional chains, and hexagonal structures through AA V and AA I as shown in FIGURE 5a. In this manner, the detailed microscopic mechanisms have been elucidated.

These superstructures and also the primary structures are dependent on the kind of molecules. For uracil, the RNA base molecule, trimers are the most stable structures actually observed in low-temperature STM at 70K. The trimers further get together to form snow-like crystal structures.[4] The main controlling factor for self-assembly is the molecule-molecule interactions through hydrogen bonding. This is around 20 kJ/mol · hydrogen bond. The second factor is the molecule-surface interaction, especially the activation energy for surface imaging of base molecules. The third factor is thermal energy. At room temperature, thermal energy is 2.5 kJ/mol, and therefore, this amount of energy enhances the surface migration of base molecules on the Cu(111) surface. At low temperature, for instance 70K, the thermal energy is 0.6 kJ/mol. Accordingly, molecules are frozen on the surfaces and observed by STM. At room temperature, only the molecules self-assembled through hydrogen bonding can be imaged by STM. These superstructures, one-dimensional and two-dimensional ones, may be used as quantum wires and also for the inter-connections between molecular level devices.

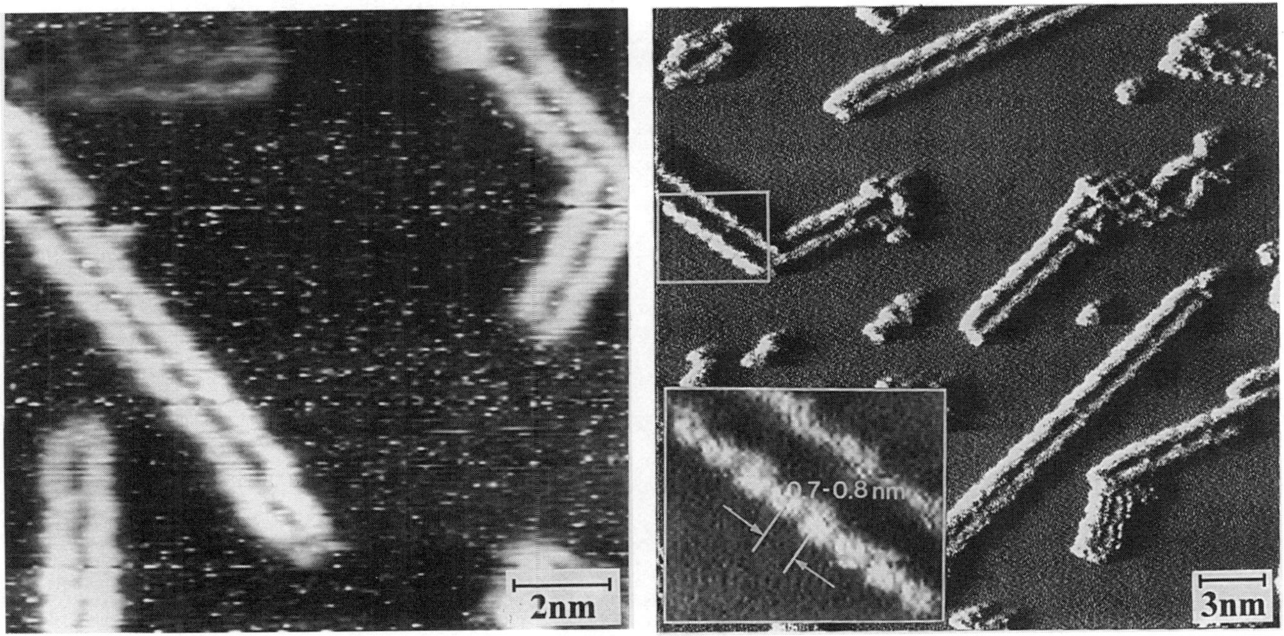

FIGURE 3. STM images of self-assembled adenine molecules with low surface coverage at room temperature (**a**) and at 70K (**b**).

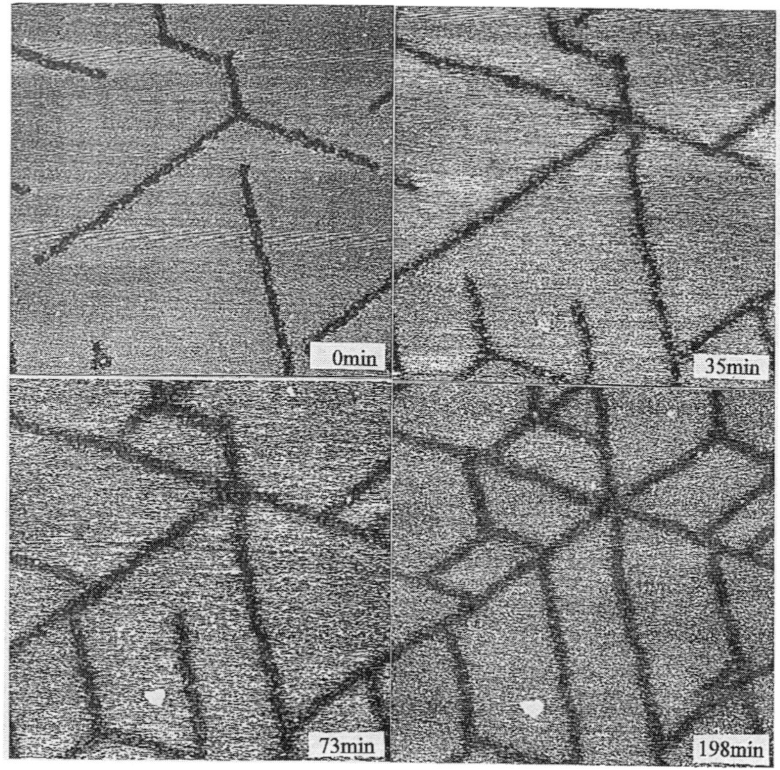

FIGURE 4. Time-dependent STM images of self-organization of cytosine molecules at room temperature. Molecular chains are imaged as dark stripes (Scan area: 1100 Å × 1100 Å).

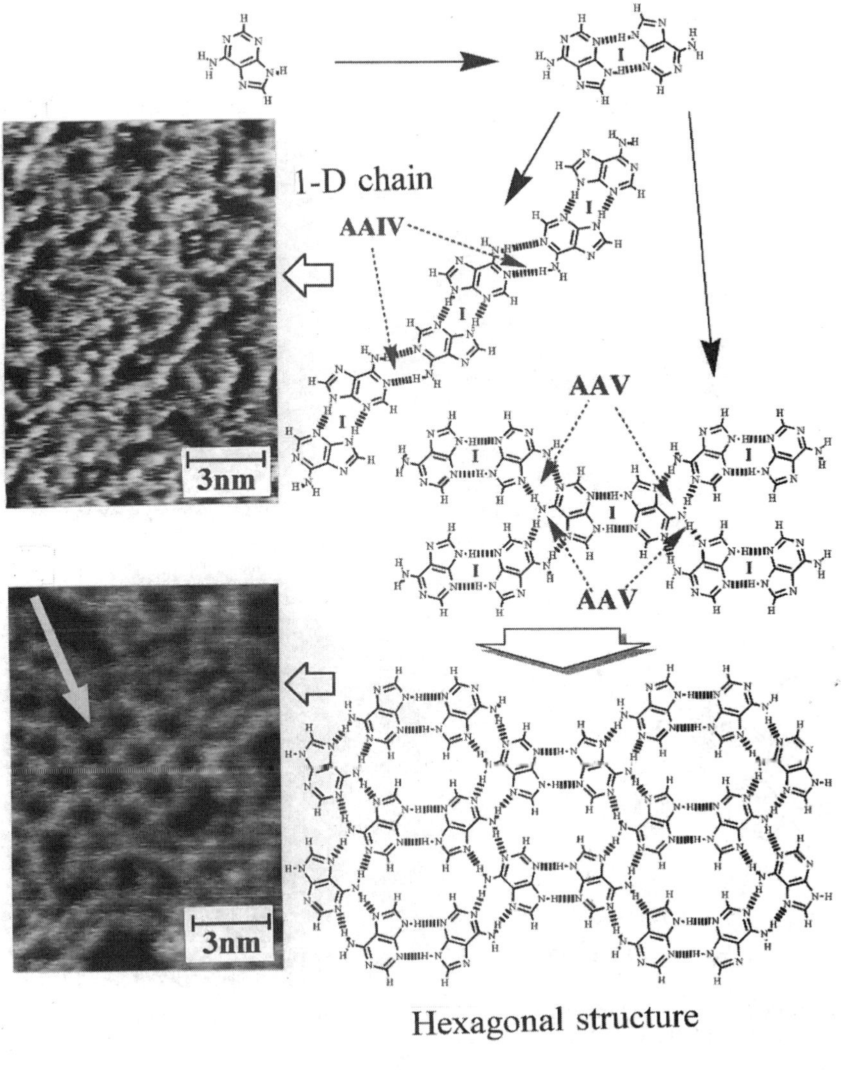

FIGURE 5. (a) Self-assembled superstructures of adenine molecules with high deposition rate and low deposition rate. Microscopic models for the self-organization presented at the right side. Adenine molecules are dimerized to form AA I and 5 combined and through AA IV and AA V to form one-dimensional chains and hexagonal structures, respectively.

IMAGING AND MANIPULATION OF DNA MOLECULES

In FIGURE 6, co-deposited thymine and adenine on the $SrTiO_3(100)$ $\sqrt{5} \times \sqrt{5}$ surface are imaged by STM.[5,6] Adenine molecules are imaged as bright, tall, and large

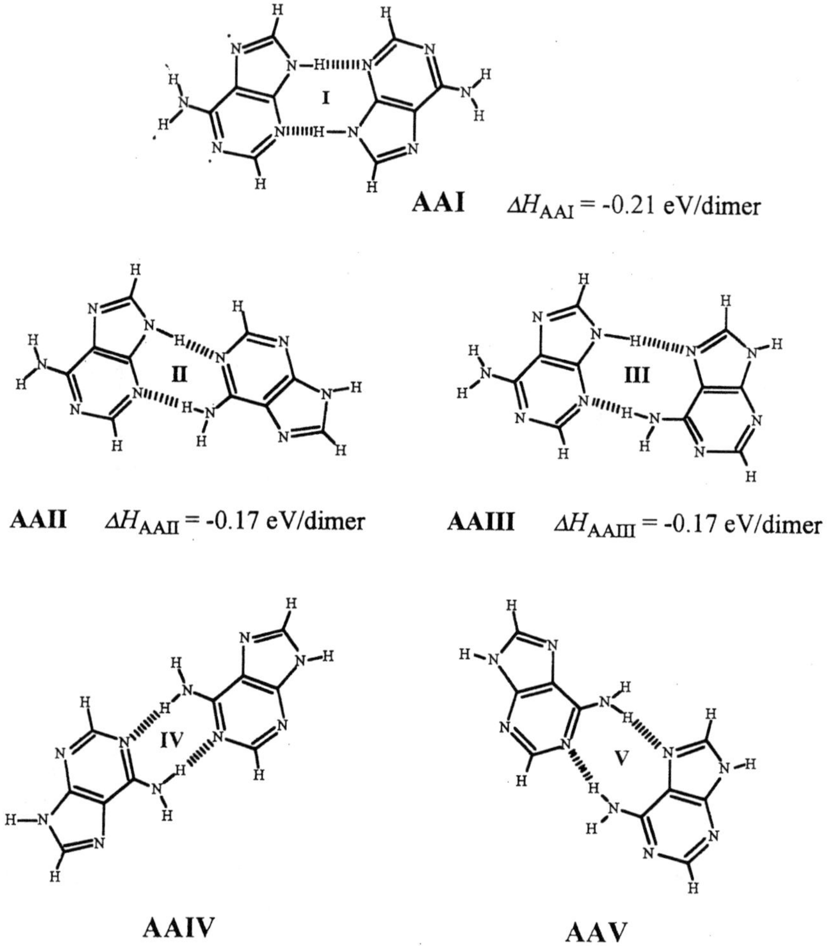

AAI $\Delta H_{AAI} = -0.21$ eV/dimer

AAII $\Delta H_{AAII} = -0.17$ eV/dimer

AAIII $\Delta H_{AAIII} = -0.17$ eV/dimer

AAIV $\Delta H_{AAIV} = -0.14$ eV/dimer

AAV $\Delta H_{AAV} = -0.13$ eV/dimer

FIGURE 5. (b) Structure and stability of five kinds of adenine dimers. Calculation has been made by MOPAC-PM3.

dots, while thymine molecules are imaged as dim, low, and small dots. In this manner, adenine molecules and thymine molecules can be discriminated by high-resolution STM even though the size of adenine and thymine are small enough, 5 and 3 Å, respectively.[7] For molecular electronics, only two digits may be necessary, that is, discrimination of adenine and thymine is useful. The present result that adenine and thymine are discriminated is quite encouraging for molecular-level information storage and reading at molecular levels.[8,9]

The STM imaging and discrimination can be applied not only to DNA base molecules, but also to DNA itself having chain structure. Since many bio-molecules are

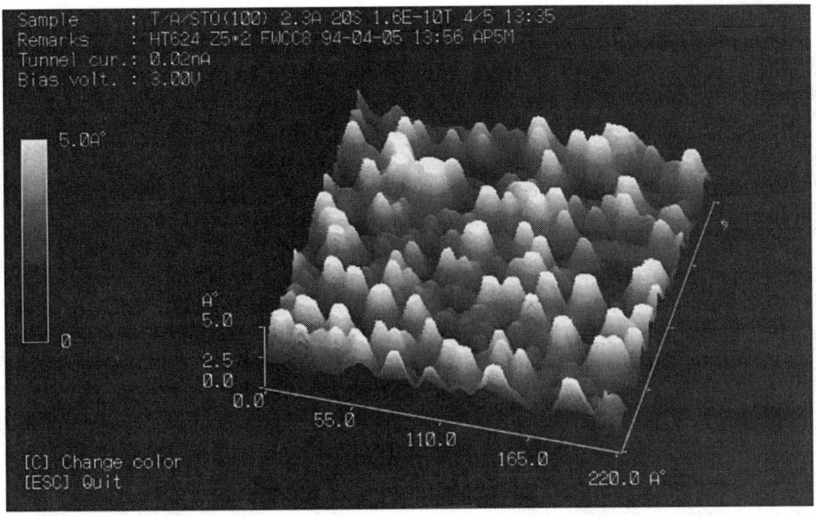

FIGURE 6. Thymine/adenine/ $SrTiO_3(100)$ $\sqrt{5} \times \sqrt{5}$; Vs = –3 V, It = 20 pA, 100 × 90 × 8 Å STM images of co-deposited adenine and thymine molecules on the reduced $SrTiO_3(100)$ surface. Bright dots correspond to adenine molecules and dim dots correspond to thymine molecules. Adenine and thymine molecules are clearly distinguished in this image.

decomposed by thermal evaporation, we have developed a new sample preparation technique called pulse injection.[10] DNA oligomer aqueous solution has been prepared and this solution is pulse injected into vacuum on atomically clean Cu(111) surface. Water is vaporized and pumped out, and only DNA is present on the Cu(111) surfaces. DNA oligomers containing five adenine molecules are clearly imaged as shown (FIGURE 7) with no ambiguity on the atomically flat Cu(111) surfaces. So far, many reports have been made in which DNA is imaged with high-resolution STM.[11] However, most images were artifacts because the measurements were done in air on graphite surfaces.[12] In the present experiment, we have prepared atomically clean flat surface in ultra-high vacuum and developed a new method for depositing DNA-related molecules. Thus, we have been able to obtain reproducible high-resolution DNA images.

The DNA molecules have been manipulated on solid surfaces by the tip of the STM. In the manipulation procedure, we found that two modes exist according to the tip/surface distance and bias voltage as shown in FIGURE 8. Even when tip/surface distances are large, molecules move because of the large electric field by the increase of bias voltage.[13] When the tip comes close to the surface, within 0.5 nm, only the molecules just below the tip move even though the bias voltage is extremely small (~10 mV). This latter procedure should be called single molecule manipulation by STM. Taking advantage of this single molecule manipulation, we have manipulated single-stranded DNA containing five adenine molecules.[10] The distance between STM tip and the surface is changed from 1 nm for imaging to 0.2 nm for

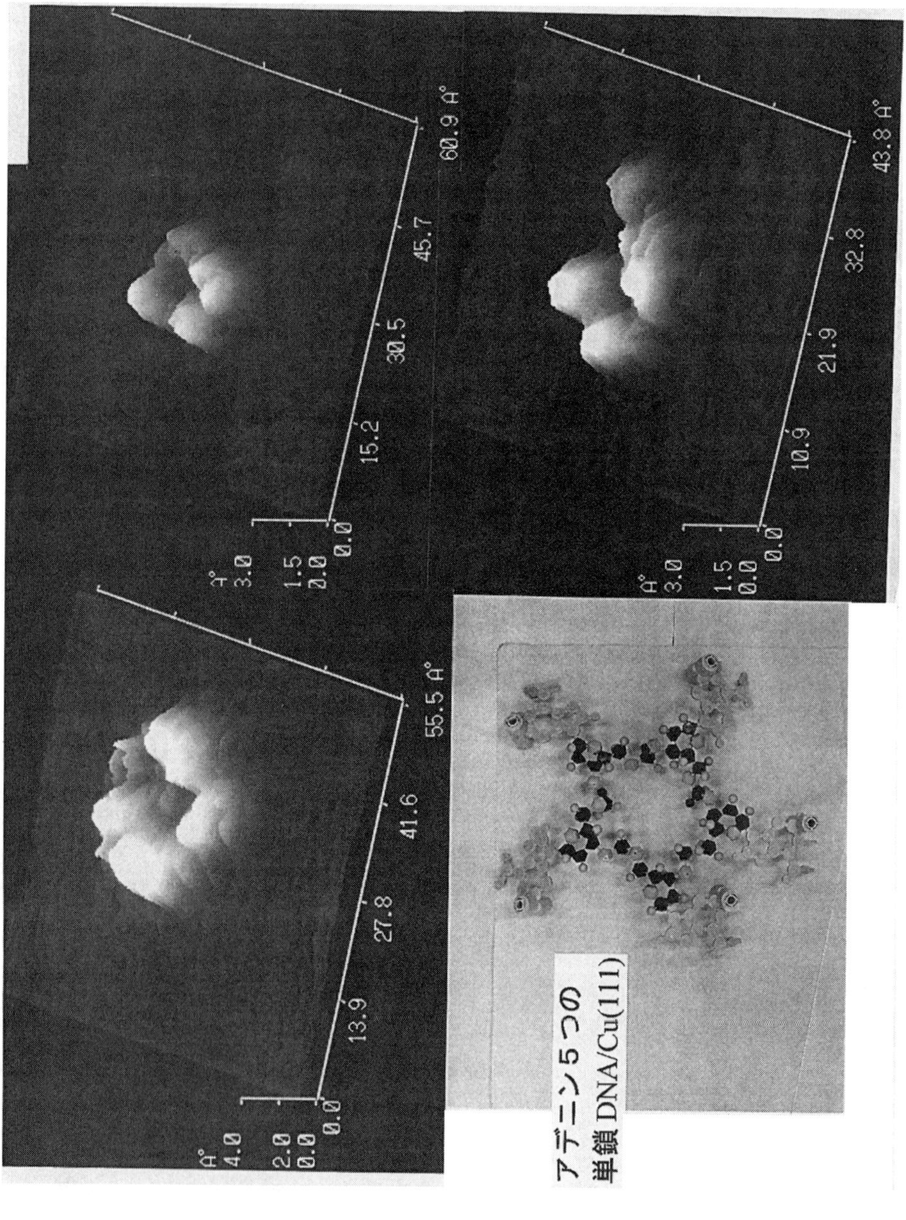
Å 4.0 2.0 0.0
0.0 13.9 27.8 41.6 55.5 Å
Å 3.0 1.5 0.0
0.0 15.2 30.5 45.7 60.9 Å
Å 3.0 1.5 0.0
0.0 10.9 21.9 32.8 43.8 Å
アデニン5つの
単鎖 DNA/Cu(111)

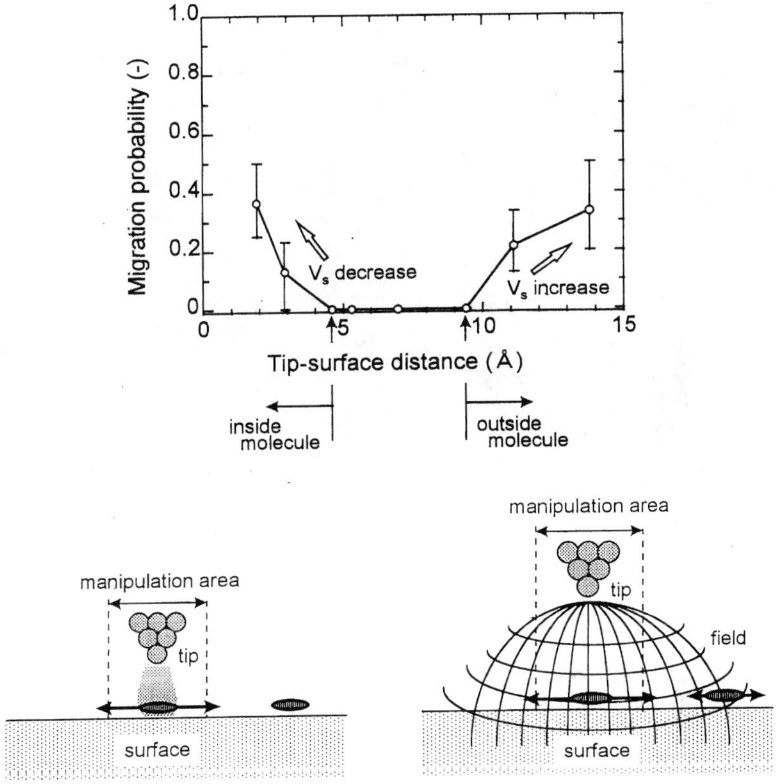

FIGURE 8. Migration probability as a function of tip bias voltage. At large tip-surface distance, molecules outside the scanning area are manipulated by the increase of bias voltage (Vs), while only the molecule just below the tip is manipulated with low bias voltage within the tip-surface distance of 0.5 nm.

FIGURE 7. High-resolution STM images obtained after the injection of a DNA sample having five adenines on a clean Cu(111) surface. In this model, a hydrophilic chain of the sugar and phosphate is located to form a circle inside, while hydrophobic adenines are located outside the circle. The diameter of this model is ~3 nm, in good agreement with the observed value of ~3 nm.

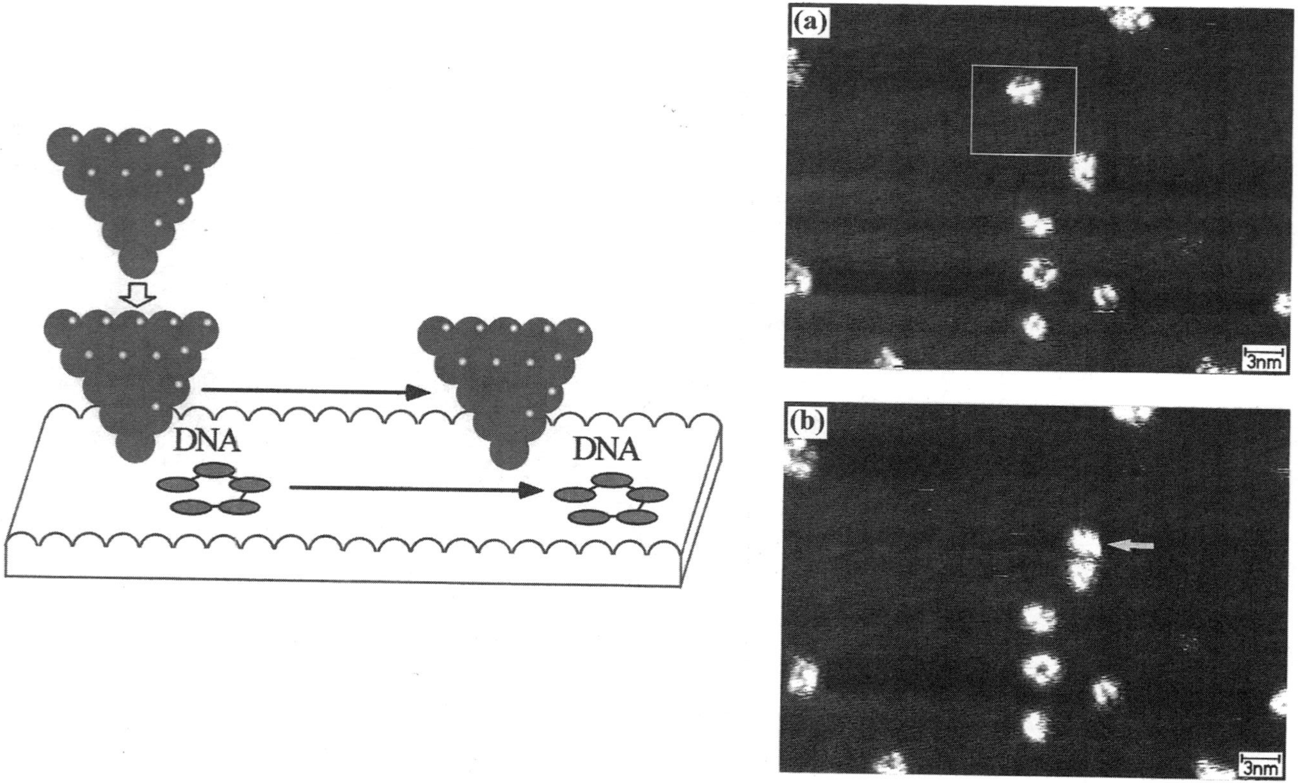

FIGURE 9. A series of STM images of an identical region of the surface, taken **(a)** before and **(b)** after the operation of manipulation of single-stranded DNA. Both images were taken at Vs = −2 V and It = 10 pA. Manipulation was performed by scanning at Vs = −1 mV and It = 10 pA on a region indicated with a white open square in **(a)**.

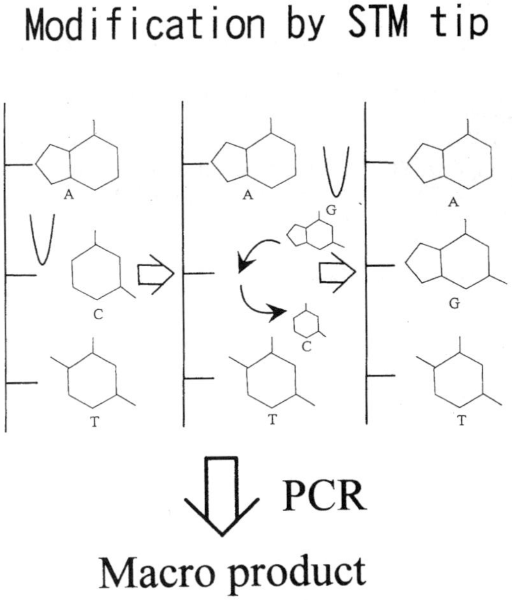

FIGURE 10. Schemes for future DNA manipulation by STM. One of the bases is changed by the tip of STM and the DNA is duplicated by PCR method.

manipulation. Then, the tip comes close to the DNA, and DNA is pushed along the surface as shown in FIGURE 9. In this way, two DNA oligomers are intentionally brought together by the STM tip. For molecular manipulation, it is essential that DNA molecules are arranged as arrays on surfaces. We have found that the $TiO_2(110)$ surface yields a row structure with one unit cell steps.[14] When DNA chains are deposited on these surfaces, DNA chains are aligned one-dimensionally along steps. This technique may be used for the application of DNA base sequencing and also for manipulation. As to the future of DNA manipulation by STM, we propose here the final goal as shown in FIGURE 10. In this figure, we show that one base molecule in DNA is replaced by another base using the STM tip, and the newly formed DNA chains are duplicated by PCR method to produce macro amounts of products. In this way, we hope that molecular-level imaging and manipulation and also self-assembly will provide a new route for molecular electronics.

REFERENCES

1. TANAKA, H., T. NAKAGAWA & T. KAWAI. 1996. Surface Sci. **364:** L575.
2. FURUKAWA, M., H. TANAKA & T. KAWAI. 1997. Surface Sci. **392:** L33.
3. KAWAI, T., H.TANAKA & T. NAKAGAWA. 1997. Surface Sci. **386:** 124.
4. TANAKA, H., T. NAKAGAWAZ & T. KAWAI. 1996. Surface Sci. **364:** L575.
5. TANAKA, H. & T. KAWAI. 1996. Jpn. J. Appl. Phys. **35:** 3759.
6. KIM, J.T., T. KAWAL, J. YOSHINOBU & M. KAWAI. 1996. Surface Sci. **20:** 50.

7. TANAKA, H. & T. KAWAI. 1995. Mat. Sci. Eng. C **3:** 143; 1995. J. Vac. Sci. Tech. B **13:** 1411.
8. TANAKA, H., J. YOSHINOBU, M. KAWAI & T. KAWAJ. 1996. Jpn. J. Appl. Phys. **35:** L244; 1996. Surface Sci. **360:** 50.
9. KASAYA, M., H. TABATA & T. KAWAI. 1995. Surface Sci. **342:** 215; Surface Sci. 1996. **357–358:** 195.
10. TANAKA, H. & T. KAWAI. 1997. J. Vac. Sci. Technol. B **15:** 602.
11. DRISCOLL, R.J., M.G. YOUNGQUIST & J.D. BALDESCHWIELER. 1990. Nature **346:** 29.
12. CLEMMER, C.R. & T.P. BEEBE, JR. 1991. Science **251:** 640.
13. AKIYAINA, R., T. MATSUMOTO, H. TANAKA & T. KAWAI. 1997. Jpn. J. Appl. Phys. **36:** 388.
14. NAKAMURA, T. *et al.* 1998. Manuscript in preparation.

Is There A Minimum Size and a Maximum Speed for a Nanoscale Amplifier?

C. JOACHIM,[a] C. BERGAUD,[b] H. PINNA,[a] H. TANG,[a] AND J.K. GIMZEWSKI[c]

[a]*CEMES-CNRS 29, Rue J. Marvig, B.P. 4347,31055 Toulouse Cedex, France*
[b]*LAAS-CNRS,7, Avenue du Colonel Roche,31077 Toulouse Cedex, France*
[c]*IBM Research Division, Zurich Research Laboratory, 4, Saumerstrass, Ruchlikon, Switzerland*

ABSTRACT: The C_{60} electromechanical amplifier is a working 3-terminal device based on the modulation of the tunnel-transport regime through a C_{60} molecule by a reversible deformation of its cage. We present the advantages of the planarization of this triode such as a surface extension lower than 0.05 μm^2 on the wafer and a large gain ($G_d > 40$). High-speed operation (bandwidth greater than 1 GHz) and low energy dissipation in the attojoules range are discussed together with the technological difficulties to fabricate such a planar molecular triode.

I. INTRODUCTION

A nanoscale molecular triode is a single molecule whose intrinsic elastic conductance (transparence) measured between two electrodes (the source and the drain) depends on the state of a third electrode traditionally called the grid. An amplification results if a small change of the grid state produces a large change in the source-drain current. The first experimental prototype of a nanoscale triode[1] was a molecular amplifier based on the modulation of a C_{60} molecule conductance by the compression of its cage using the tip of a scanning tunneling microscope (STM). The source-drain bias voltage was chosen such that the I-V characteristic of the device remains linear in the range of the grid operation, constraining the amplifier to work in a tunneling transport regime.[2]

This first prototype offers a unique opportunity to experiment on a real device at the nanoscale, introducing new guidelines to improve its gain and bandwidth or to explore the limit of a molecule used as an amplifier. Obviously, in a future planar version of the prototype, the grid must also be turned molecular.

In Section II, we show that the C_{60} amplifier has similarities to all 3-terminal devices with its equivalent electrical diagram. This diagram is used in Section III to optimize the working set point of the amplifier for a better gain and a better bandwidth. Its planar version is discussed in Section IV, analyzing each step of the process required for this planarization. In conclusion, the best C_{60} amplifier performances due to the planarization are discussed.

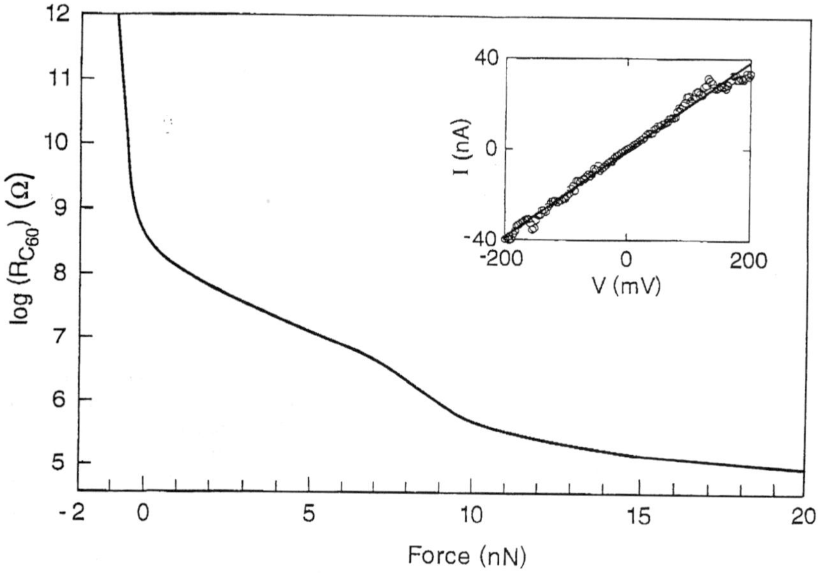

FIGURE 1. Calculated variations of the "STM tip apex-C_{60}-surface" junction resistance as a function of force used to deform the C_{60} cage. For low forces, the resistance increases rapidly because the tip apex is not in electrical contact with the top of the C_{60} molecule (bias voltage is 50 mV).

II. THE C_{60} AMPLIFIER AS A STANDARD 3-TERMINAL DEVICE

The first basic concept of the electromechanical amplifier is the sensitivity of a conjugated molecule to compression[3] (or deformation). The main consequence of a compression is the progressive decrease of the molecule heighest occupied (homo)–lowest unoccupied (lumo) separation which, in turn, changes its elastic resistance[3] (transparence) (FIG. 1). Although this works for saturated and unsaturated molecules, the conductance of an unsaturated molecule is several orders of magnitude higher than for a saturated one.[4] The other basic concept is the device voltage range of operation. To avoid the electronic resonances of the molecule, the bias voltage must be lower than its homo-lumo gap.[2] This enables the device to work in a tunneling transport regime where the I-V characteristic is linear, while also ensuring a better stability than operation via tunneling resonance through the molecule.[5]

In the first experimental prototype, a C_{60} molecule was chosen since its molecular shape ensures good mechanical stability of the molecule under tip apex compression.[1] Here, the grid was rudimentary being attached to the drain (FIG. 2a). The cage compression increases the local density of states near the Fermi level in the tunnel junction between the drain and the source. In that case, the transfer of an electron between the drain and the source is facilitated, that is, the "transfer resistance" through the junction decreases. Notice that in a field-effect transitor, the carrier density is controlled in the junction and not the density of states. Furthermore, the trans-

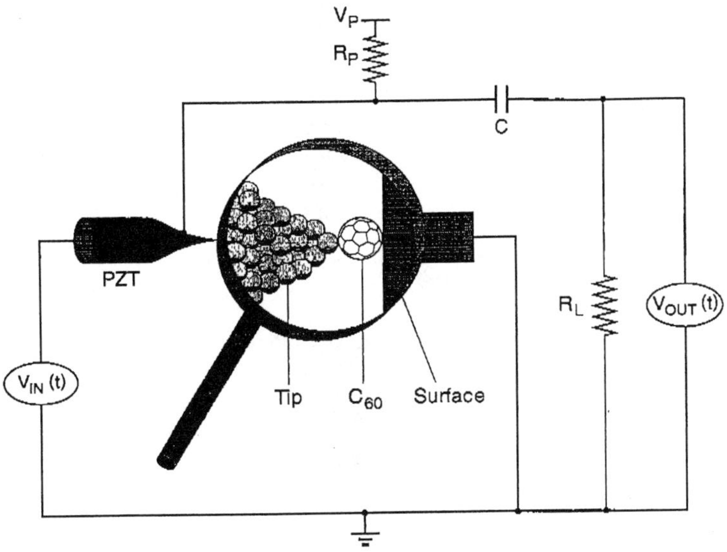

FIGURE 2a. Electrical circuit and connections of the first prototype of a C_{60} amplifier. $V_{in}(t)$ is applied independently on the z-piezo bias voltage compared to the bias voltage V_p which is applied via R_p to the tip apex. The amplifier is loaded by the filtering capacitance C and the load resistance R_L.

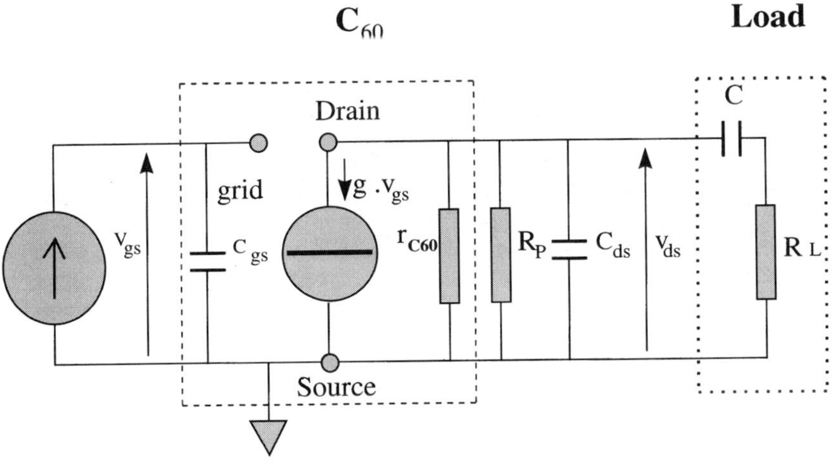

FIGURE 2b. The equivalent dynamic electrical circuit with Cds the drain-source stray capacitance and g the transconductance of the C_{60} molecule.

port regime is Bolzmanian or ballistic for smaller solid-state devices. An equivalent dynamic electrical circuit diagram of the C_{60} amplifier can be easily established, as presented in FIGURE 2b. From this circuit, the frequency-dependent dynamical gain $G_d(\omega)$ of the amplifier is given by:

$$G_d(\omega) = g\frac{R_{C_{60}} \cdot R_p}{R_{C_{60}} + R_p + i\omega C_{ds}R_pR_{C_{60}}} \quad (1)$$

and the electronic cut-off frequency of this gain by:

$$f_c = \frac{R_{C_{60}} + R_p}{2\pi C_{ds}R_pR_{C_{60}}} \quad (2)$$

In the original experimental set-up,[1] the grid voltage was chosen (with no modulation applied) to maintain a tip apex-to-surface altitude of 14 Å and, with a 20-mV grid voltage, was reduced to 12 Å. This corresponds to a C_{60} tunnel junction resistance variation between 1 GΩ and 10 MΩ, respectively.[3] With a piezo electromechanical conversion factor of 1 Å per 10 mV, this yields the transconductance g = 0.5 μS and a static gain of 5. From **(1)**, the dynamic gain $G_d(\omega) = 2.45$ at low frequencies for a load resistance R_L much larger than 10 MΩ.

From (2), the cut-off frequency is controlled by the stray capacitance C_{ds} and by the load filtering capacitance C (FIG. 2a). C is somehow arbitrary and needs no further discussion here. Measurements and calculations of C_{ds} give a lower limit of 1 pF in a standard STM.[6,7] This gives f_c = 15 kHz from **(2)** with the experimental set-up seen in FIGURE 2a. This must be compared to the first resonance frequency of our piezo tube, which can be pushed towards the 100 kHz range,[7] and to the first soft vibration mode of C_{60} which is around 8 THz.[8] Consequently, the C_{60} amplifier bandwidth is limited by a large drain-source resistance and a large stray capacitance. Notice that in the standard CMOS technology, the limiting factor is the grid-source capacitance, which is negligible in our device due to the use of an electromechanical grid.

The energy required by our molecular triode to pass from its high to its low resistance states can be estimated from FIGURE 1 by calculating the work ΔW done by the tip apex to compress the C_{60} cage: $\Delta W = F\Delta z$, which gives 10^{-18} J since a Δz of 2 Å results in a 5-nN force applied to the cage. This places the C_{60} triode four orders of magnitude below the switching energy of the best semiconductor devices realized so far and equivalent in performance to an RSFQ superconducting device.[9] Nevertheless, this value is still three orders of manitude higher than kT. In2 at room temperature.[10] We note that R_p is also a dissipating element in the circuit. In the low-resistance state, the maximum drain-source current is 10 nA, which gives 2 nW of dissipation.

III. GAIN AND BANDWIDTH OPTIMIZATION

What limits the cut-off frequency and the gain of our first prototype is the very low electronic transparence of the molecule used. In this section, we demonstrate how this can be improved by a better optimization of the set point of the device. One guideline we have used is the resistance-force curve (FIG. 1). The original set point was chosen between 1 GΩ and 10 MΩ. The advantage of working in this resistance

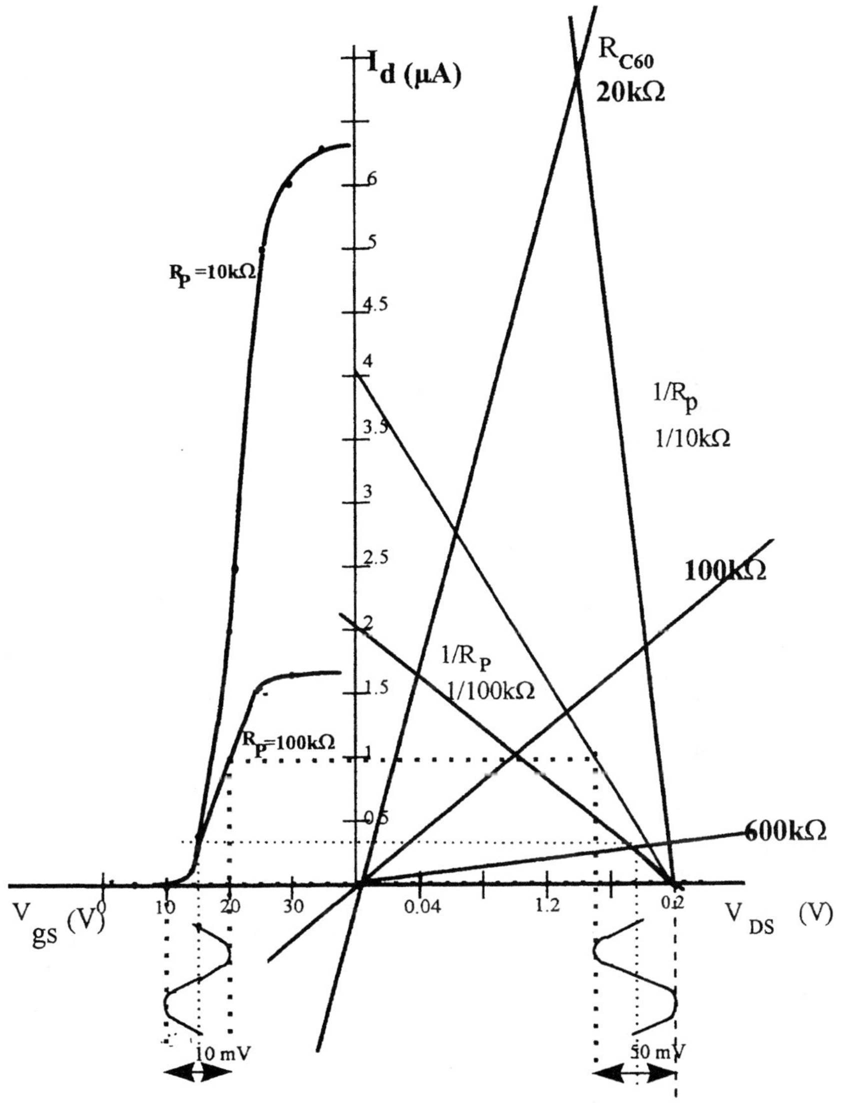

FIGURE 3. Load characteristics of a C_{60} triode for two differents values of R_p. Notice that the chosen set point of 11 Å is not optimum from the point of view of linearity of the $I_d = f(V_{gs})$ characteristics. As a consequence the output of the amplifier is distorted.[1] However, the C_{60} cage deformation must be limited to avoid the C_{60} escape from the tip apex. The electromechanical conversion factor chosen here is 1 Å per 10 mV.

range is the requirement of a moderate compression of the C_{60} cage. For example, a 10-MΩ resistance sets the tip-to-surface distance at 12 Å, which is 0.3 Å lower than the known tip-to-surface distance for an electrical contact on a C_{60} molecule.[3] As a consequence, the stability of a C_{60} under the tip apex surface junction is maintained and consequently, the amplification effect is reliable.[1]

Nevertheless, this set point does not completely exploit the intrinsic electronic properties of a C_{60} molecule. For example, when the tip apex-to-surface altitude is settled between 11.5 Å and 10.5 Å, R_p can be greatly reduced below 10 MΩ because of the higher electronic transparence of a C_{60} molecule that is achieved.[3] Notice also that a tip altitude of 10.5 Å on C_{60} is the limit to ensure no escape of the molecule from the tip apex. Starting with the equivalent circuit shown in FIGURE 2b, standard transistor-like characteristics can be plotted for the C_{60} triode (FIG. 3). Choosing R_p = 100 kΩ yields a new cut-off frequency of 3.4 MHz which is more than two orders of magnitude higher than achieved using our original set-up. There still remain three modes of resonance in the device: the soft vibration mode of C_{60}, around 8 GHz, the electronic cut-off frequency due to C_{ds}, now around 3 MHz, and the *z*-piezo grid resonance frequency around 100 kHz. Now the grid and not the molecule limits the working bandwidth of the triode. This can be optimized by reducing the dimensions of the grid, as described in the next section. Notice that the dynamic gain doubles, reaching 4.3 as a result of the change in the transconductance, *g*, obtained through the shifting of the set point of the device.

The energy required to pass from the high 6 MΩ to the low 100 kΩ C_{60} resistance regime remains with an associated switching energy of 10^{-18} J. Here our new set point requires a permanent voltage applied to the *z*-piezo to keep the tip apex-to-surface altitude around 11 Å. This opens interesting questions regarding the noise margin in such a device (the precision on the *z*-piezo) and also about the long-term stability of a molecule under prolonged current intensities of 0.28 μA when 56 nW are dissipated in the electrodes.

IV. PLANAR TECHNOLOGY

Fabrication on the surface of a semiconducting material was the key to go from the first point contact transistor to the most advanced integrated circuits.[11] Following this path, the planarization of the C_{60} amplifier must permit a separation between the grid and the drain and to reduce the grid size. Furthermore, C_{60} is not the only molecule presenting an electronic response to a mechanical deformation.[12] Therefore, a planar version of the C_{60} triode may also facilitate the use of a more sensitive molecule, not accessible to compression with an STM tip.

A planar version of the set-up shown in FIGURE 2a is relatively straightforward to design if the grid is kept electromechanical.[13] The "tip apex-C_{60}-surface" junction is planarized in a "metal-C_{60}-metal" nanojunction fabricated on top of an insulating material and a microcantilever grid is positioned on the nanojunction (FIG. 4) While this planar design is quite clear, the technical steps to fabricate it are nontrivial. They are discussed point by point in this section, indicating those issues that are currently the greater technological challanges.

The Coplanar Source-Drain Nanojunction

Many versions of a coplanar nanojunction approaching the design shown in FIGURE 4 have been proposed in recent years: unsupported platinium electrodes grown at the end of a supported 25-nm coplanar nanojunction,[14] break junctions made by

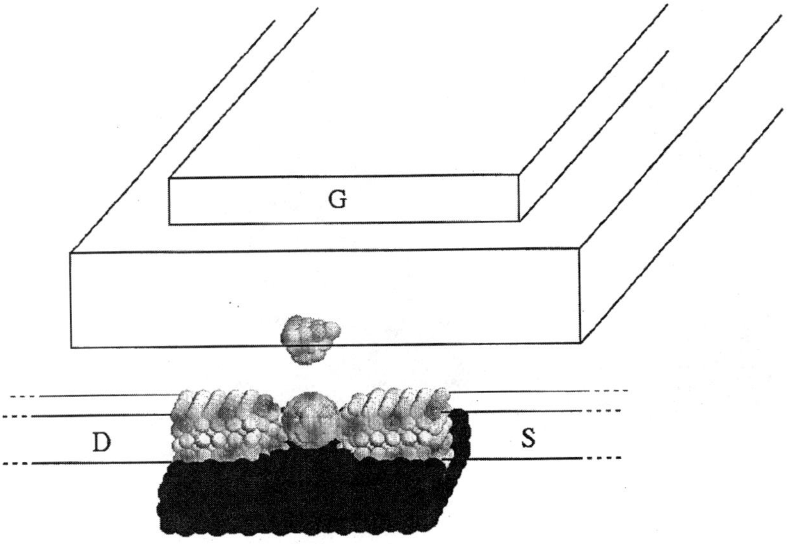

FIGURE 4. Model planar version of the C_{60} triode. The electromechanical grid is now separated from the drain. The grid is a cantilever equiped with an insulating tip apex and actuated by a piezo layer grown on top. Source and drain are at the ends of the electrodes separated by an insulating gap 1.5 nm in length. Notice the large dimensions of the cantilever compared to the "metal-C_{60}-metal" junction, indicating the need for a drastic size reduction of the cantilever.

controlling the break of a mesoscopic metal wire using piezoelectric drivers[15] as well as standard *e*-beam nanolithography coplanar metal-vacuum-metal nanojunctions[16] or embeded metal-insulator-metal nanojunctions.[17] Apart from the STM and the break junction, where the size of the vacuum gap is controlled by steps of a few picometers, the other techniques produce gap sizes in the 5-nm range. This is too large for a C_{60} molecule, which requires less than a 2-nm gap, but it is in principle small enough for conjugated oligomers to be employed.[18]

We have chosen embedded nanojunctions for two reasons. First, the metal electrodes are almost at the same altitude as the surface itself. This is very convenient for imaging the planar junction using atomic-force microscopy (AFM). Between the electrodes of the nanojunction, in the 5-nm gap, the insulating part of the junction is readily accessible to an AFM tip. Consequently, trouble-shooting or remodifications of the junction become possible when an active molecule is positioned in it. Secondly, thermal dilatations of the junction are reduced because the metal is embedded in a trench.

The 5-nm AuPd coplanar nanojunction presented in FIGURE 5 was obtained with a 10% throughput using *e*-beam nanolithography based on the pixel jump technique.[17] Although the resist insulation is stopped further away than the gap size because of proximity effects, the resist is insulated further than the *e*-beam spot size.

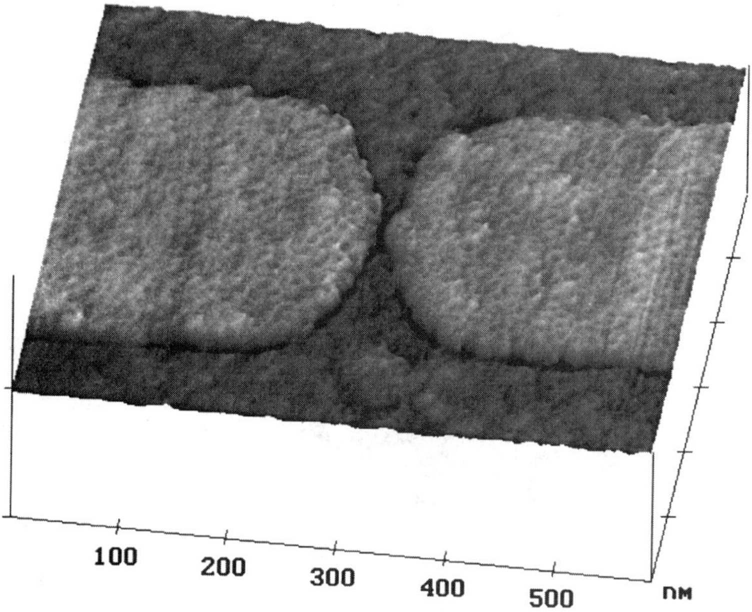

FIGURE 5. AFM constant force image of a 5-nm coplanar metal-SiO_2-metal nanojunction. The end of the two metal electrodes are imaged. The step height between the electrodes and the SiO_2 surface is only 3 nm and can be reduced practically to zero.[17]

By tuning these effects rather than attempting to suppress them, very small and reproducible gaps were fabricated (FIG. 5).

Positionning the Active Molecule in the Gap

Molecules can be manipulated one by one on a clean surface using STM.[19] For a surface containing nanojunctions prepared with a standard nanolithography process, STM manipulations are currently difficult to perform. They must be conducted using an AFM because of the insulating character of the substrate. This surface must also be cleaned at the atomic scale due to inherent contaminations of standard lithographic processes. Silanization is a good option for AFM manipulation on SiO_2. With such a surface preparation, we have demonstrated that nanofibers, 2 nm in diameter, can be manipulated one at a time, cut in pieces,[20] and brought on the nanojunction. The same manipulations can almost be realized with a carbon nanotube.[21] But this has not yet been extended to a single oligomer with a van der Waals diameter below 1 nm. This is because the shape of the tip apex at the end of a cantilever is very difficult to control in AFM.

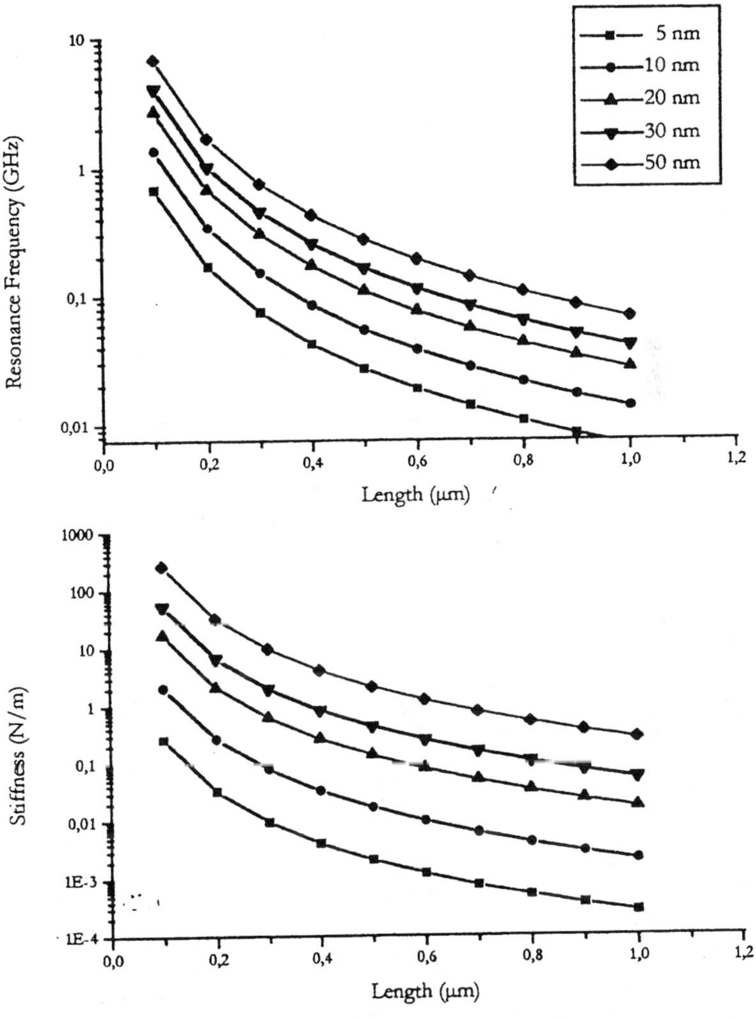

FIGURE 6. Variation of (*top*) the resonance frequency and (*bottom*) the stiffness of a cantilever as a function of its length for different thicknesses and a width $w = 50$ nm. The size reduction increases resonance frequency but also stiffness.

While those manipulation techniques are (or will be) very practical to fabricate exploratory devices, they are not suited for mass production. One approach recently proposed to solve this problem is nanoprinting, which had principally been explored for embossing.[22] Understanding the wetting of a surface at the nanometer scale remains an open subject which is crucial for nanoprinting of molecular structures.[23] The development of a nanoprinter may be one of the keys to mass production of devices at the nanoscale.

The Cantilever Grid

In the design shown in FIGURE 4, a tip cantilever is proposed as a grid. Therefore, the final surface extension of the planar amplifier will depend on how far the size reduction of a cantilever can be pushed. Interestingly, a size reduction is generally accompanied by an increase of the frequency of the first cantilever resonance mode. This provides good scaling characteristics for the device.

Starting with a free cantilever, the resonance frequency of its first resonance mode for a length L and a thickness b is given by[24]:

$$f_0 = (b/2\pi L^2)\,(Y/\rho)^{1/2} \tag{3}$$

with Y the Young modulus and ρ the density of the material used to fabricate the cantilever. For silicon, $Y = 1.69 \times 10^{11}$ N/m^2 [25] and $\rho = 2320$ kg/m^3. The variations of f_0 as a function of L and b for a $w = 50$ nm in width in a silicion cantilever are presented in FIGURE 6(*top*). The GHz regime is readily obtained for a cantilever with L approaching 0.2 µm and for a thickness in the range of 30 nm. The grid resonance frequency is now well above the 3 MHz electrical cut-off frequency of the amplifier discussed in the previous sections. f_0 must be also modified to take into account the mechanical charge due to the C_{60} molecule under the tip apex of the cantilever and the charge resulting from the piezoelectric layer which would be deposited on top of the cantilever to control its deflection.

The first mechanical charge is the C_{60} molecule. When physisorbed, its effective spring constant k_m is in the 10 N/m range when compressed by a few Å. A cantilever spring constant k_c of the same order of magnitude is required to ensure that the deflection induced by the piezo layer will not be damped by the molecule. Now, for free rectangular cantilever fixed at one end, k_c is given by[29]:

$$k_c = Yb^3w/4L^3 \tag{4}$$

FIGURE 6(*bottom*) shows k_c for different lengths and thicknesses of the cantilever. Since a small L provides a better k_c than a longer one, this determines the appropriate cantilever thickness. This follows the same trend as the dependence of f_0 upon L and b. Therefore, it is possible to mechanically adapt the cantilever to the mechanics of a single C_{60} and consequently, the final first resonance frequency f_{res} of the grid compressing C_{60} will be then close to f_0 because[27] $f_{res} = f_0\,(1 + k_m/k_c)^{1/2}$.

The second mechanical charge on the grid is the piezo top layer of the cantilever. We foresee that a layer of a thickness as low as 20 nm could be fabricated on a surface. This layer may also be organic. The first resonance frequency of a piezo-coated cantilever can be estimated considering this cantilever as a bimorph structure.[30] There is an upward shift of the overall resonance frequency which can be adjusted by restricting the extension of the piezo-coating along the cantilever or by choosing to increase the cantilever length (for example $L = 0.7$ µm, to maintain $f_{res} = 1$ GHz).

An important issue is the electromechanical conversion factor of the grid. The piezo-tube of the amplifier prototype had a conversion factor of 1 Å per 10 mV. Size reduction scales well[31] and the condition that the cantilever length does not become so short for the piezo deflection activity to fight against the stiffness is not a problem. Therefore, a "compromise" exists between the resonance frequency, stiffness,

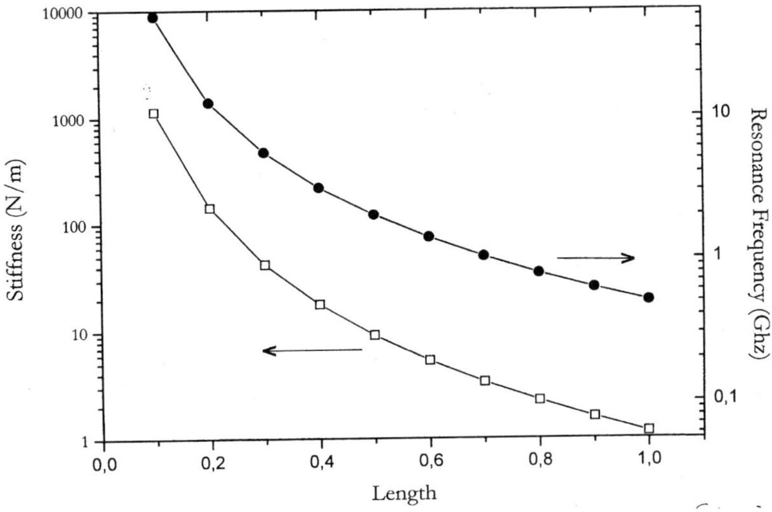

FIGURE 7a. Variations of the resonance frequency and stiffness of a bimorph cantilever fabricated from 20 nm of silicon and 20 nm of PZT for a width w = 50 nm. Compared to the variations seen in FIGURE 6, the resonance frequency is larger, but the stiffness is too.

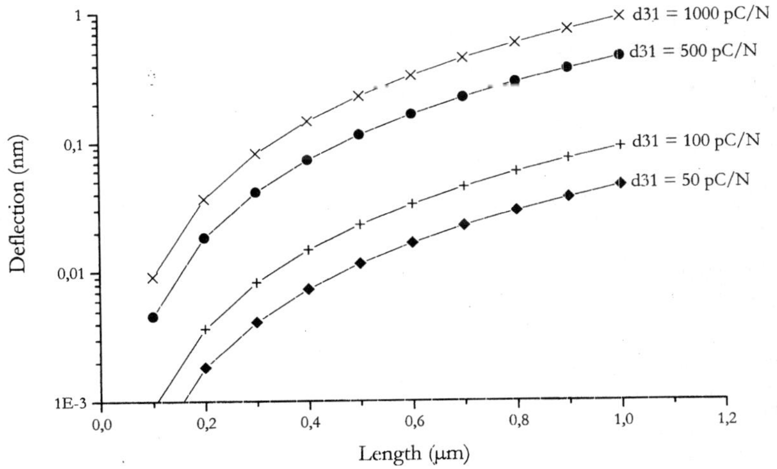

FIGURE 7b. Variations of the deflection amplitude of a bimorph cantilever as a function of length for a 1-mV bias voltage. Four values of the d_{31} piezo coefficient have been used, suggesting the difficulty in obtaining a deflection larger than 0.1 nm for a very size-reduced cantilever.

and the conversion factor of the piezo. For example, a 20-nm PZT layer on top of a standard silicon cantilever leads to a 1 Å per 1 mV conversion factor for $L = 0.7$ μm as shown in FIGURE 7b.[32] This increase of the conversion factor directly benefits the amplifier gain. In its optimized circuit version, we have now reached $G_d = 43$ for a 1 Å per 1 mV electromechanical conversion. In this condition, it would be possible to reduce the R_p resistance and consequently to increase the bandwith by the same factor.

The smallest cantilevers reported have a $L = 0.5$ µm and $w = 0.1$ µm for $b =$ 30 nm.[33] This is very close to the dimension proposed here for $f_{res} = 1$ GHz and for a good mechanical adaptation to C_{60}. What remains to be solved is how one equips a small cantilever with a very sharp insulating tip apex. Currently, AFM tips with an end radius of curvature as small as 2 nm have been fabricated,[34] which indicates that this issue is not an ultimate limitation to fabrication.

V. CONCLUSION

Assuming that the C_{60} tunnel junction resistance variation under compression is the same in a planar configuration as under an STM tip apex, we have shown that the characteristics of a planar C_{60} amplifier can be optimized to obtain useful device characteristics with some outstanding features:

(*1*) The surface occupation of the device is controlled by the ultimate dimensions of the top cantilever ($0.05 \times 0.7\ \mu m^2 = 0.035\ \mu m^2$). This cantilever must be aligned from the top on to C_{60} because of the cantilever support connected to the surface. It appears difficult to fabricate the nanojunction, the cantilever, the tip apex and to manipulate C_{60} in one stage;

(*2*) The dynamic gain of the amplifier can be higher than 40 using the proposed new set point for operation. This raises some questions on the noise and on the mechanical stability of the device since this high gain is obtained at the cost of a large C_{60} compression for a very low, 1-mV gate input.

(*3*) The bandwidth of the amplifier depends on the C_{60} electronic transparence and on the stray capacitance at the nanojunction. C_{ds} for a nanojunction of section 100 nm^2 is as low as 10^{-19} F. Using **(2)**, the intrinsic cut-off frequency of the triode will be in the THz range even for $R_p = 100$ kΩ. In this case, the cantilever will again limit the bandwidth in the 1 GHz range. The very low stray capacitance of the nanojunction will be very difficult to maintain in front of the external capacitance introduced by the connecting circuit. If this capacitance problem cannot be solved, the cut-off frequency will remain in the MHz range and new molecules with a higher electronic transparence need to be utilized.

Finally, the bandwidth (speed) of a molecular device is not so much limited by the intrinsic characteristics (electronics, mechanics) of the molecule used, but by the external driving circuit. Even if the metal-molecule-metal junction conductance can be as high as a quantum of conductance, the external wiring limits the device bandwidth to some hundred of MHz. Such embodiments have to consider the current and projected electrical characteristics of nanowiring. Our choice of a mechanical grid also limits the possiblility of size reduction. A mechanical cantilever can certainly be very much reduced below 50 nm in width and 0.2 µm in length. But its stiffness will increase so much that no piezo layer will be active enough to deflect it according to the molecular deformation required. This opens the search for a current induced or a field-effect gate at the molecular scale.

ACKNOWLEDGMENT

We wish to thank the MEL-ARI Nanowires Project for its support during this work.

NOTES AND REFERENCES

1. JOACHIM, C. & J.K. GIMZESWSKI. 1997. An electromechanical amplifier using a single molecule. Chem. Phys. Lett. **265:** 353–357.
2. JOACHIM, C. & J.K. GIMZESWSKI. 1995. Analysis of low-voltage J(V) characteristics of a single C_{60} molecule. Europhys. Lett. **30:** 409–414.
3. JOACHIM, C. & J.K. GIMZESWSKI.1995. Electronic transparence of a single C_{60} molecule. Phys. Rev. Lett. **74:** 2102–2105.
4. JOACHIM, C. & J. VINUESA. 1996. Length dependence of the electronic transparence (conductance) of a molecular wire. Europhys. Lett. **33:** 635–640.
5. LANDAUER, R. 1989. Can we switch by control of quantum mechanical transmission? Physics Today October: 119.
6. MICHEL, B., L. NOVTNY & U. DURIG, 1992. Low-temperature compatible I-V convertor. Ultramicroscopy **42–44:** 1647–1652.
7. MANALIS, S.R. *et al.* 1996. High-speed atomic force microscopy using an integrated actuator and optical lever detection. Rev. Sci. Instrum. **67:** 3294–3297.
8. CLOUGHERTY, D.P.& J.P. GORMAN. 1996. On the vibrational mode of C_{60}. Chem. Phys. Lett. **251:** 353–356.
9. BUNYK, P.I. *et al.* 1995. High-speed single-flux-quantum circuit using planarized niobium-trilayer Josephon junction technology. Appl. Phys. Lett. 66: 646–648.
10. LANDAUER R. 1988. Dissipation and noise immunity in computation and communication. Nature **335:** 779–784.
11. 1980. Special issue on Electronics. IEEE Spectrum. **17** (April).
12. DURIG, U. *et al.* 1992. Electronic and mechanical characterization of self-assembled alkanethiol monolayers by scanning tunneling microscopy combined with interaction-force-gradiant sensing. Phys. Rev. B **48:** 1711–1717.
13. We do not yet know how to make a field-effect grid on a single molecule even though proposals are numerous.
14. BEZRYADIN, A., C. DEKKER & G. SCHMID. 1997. Electrostatic trapping of single conducting nanoparticles between nanoelectrodes. Appl. Phys. Lett. **71:** 1273–1275.
15. ZHOU, C. *et al.* Microfabrication of a mechanically controllable break junction in silicon. Appl. Phys. Lett. **67:** 1160–1162.
16. DI FABRIZIO, E. *et al.* 1997. Fabrication of 5 nm resolution electrodes for molecular devices by means of electron beam lithography. Jpn. J. Appl. Phys. **36:** L70–L72.
17. ROUSSET, V. *et al.* 1995. Fabrication of buried co-planar metal-insulator-metal nanojunction with a gap lower than 10 nm. J. Phys. III (France) **5:** 1983–1989.
18. GOURDON, A. & H. TANG. 1998. Molecular landers. Ann. N.Y. Acad. Sci. This volume.
19. JUNG, T. *et al.* 1996. Controlled room-temperature positioning of individual molecules: molecular flexure and motion. Science. **271:** 181–184.
20. ONDARCUHU, T. & C. JOACHIM. 1998. Drawing a single nanofibre over hundreds of microns. Europhys. Lett. **42:** 215–218.

21. FALVO, M.R. *et al.* 1997. Bending and buckling of carbon nanotube under large strain. Nature **389:** 582–584.
22. CHOU, S.Y., et al. 1996. Imprint lithography with 25 nanometer resolution. Science **272:** 85–87.
23. ONDARCUHU, T. & C. JOACHIM. 1997. Unpublished material.
24. MEIROVITCH, L. 1967. Analytical Methods in Vibration. MacMillan Series in Applied Mechanics. F. Landis Ed. New York.
25. BRANTLEY, W.A. 1973. Calculated elastic constant for stress problems associated with semiconductor devices. J. Appl. Phys. **44:** 534–537.
26. TURNER, J.A. *et al.* 1997. High frequency response of atomic force microscope cantilevers. J. Appl. Phys. **82:** 966–972.
27. RABE, U. *et al.* 1996. Vibration of free and surface-coupled atomic force microscope cantilever: Theory and experiment. Rev. Sci. Instrum. **67:** 3281–3286.
28. ITOM, T. *et al.* 1996. Deflection, detection and feedback actuators using a soft excited piezo-electric Pb(Zr,Ti)O3 microcantilever for dynamic scanning force microscopy. Appl. Phys. Lett. **69:** 2036–2039.
29. SARID, D. 1991. Scanning force microscopy. Oxford University Press. Oxford.
30. YOUNG, W.C. 1990. Roark's Formulas for Stress and Strain. McGraw-Hill International Editions. New York.
31. DEVOE, D.L. & A.L. PISANO. 1997. Modeling and optimal design of piezoelectric cantilever microactuators. J. Micro. Elect. Mech. Sys. **6:** 266–275.
32. SMITHS, J.G. & W.S. CHOI. 1991. The constituent equations of piezoelectric heterogeneous bimorphs. IEEE Trans. Ultras., Ferro. Freq. Cont. **38:** 256–260.
33. BRUGGER, J. *et al.* 1996. Silicon micro/nanomechanical device Fabrication based on focused ion beam surface modification and KOH etching. International Conference on Micro and Nanoengineering. MNE 96. Glasgow, Scotland. September 1996. [Preprint].
34. MARCUS, R.B. et al. 1990. Formation of silicon tips with <1 nm radius. Appl. Phys. Lett. **56:** 236–239.

Atom Electronics: A Proposal of Atom/ Molecule Switching Devices

YASUO WADA[a]

Advanced Research Laboratory, Hitachi, Ltd., 2520 Akanuma, Hatoyama, Saitama 350-0395 Japan

ABSTRACT: This paper proposes very fast (more than Tera (10^{12}) Hz) and very small (less than a few nm) devices, atom relay transistors (ART), supra molecular atom relay transistors (SMART), and molecular single electron switching transistors (MOSES), which would supersede the MOSFET (metal-oxide-semiconductor field effect transistor) for future information processing. Their performances are evaluated on the basis of the characteristics necessary for integrated circuit devices and they are found to be the most promising candidates among the nanoscale devices including quantum devices and superconductor devices. Atom/molecule manipulation technology using the scanning tunneling microscope (STM) would be the key factor in achieving these devices. The basic technology developments toward the realization of ART, SMART, and MOSES are reported, including the beam-assisted scanning tunneling microscope (BASTM) for insulator observation, needle formation, and tip imaging (NFTI) for atom-scale evaluation of the STM tip apex, micromachine STM (μ-STM) for single molecule manipulation and the atom wire fabrication technology development on a silicon (100) hydrogen terminated surface.

INTRODUCTION

Advances in silicon ultra-large integrated circuit (ULSI) technology have been made possible by the downscaling of metal oxide semiconductor (MOS) transistors.[1] The trend curves of the MOS transistor downscaling and integration of large scale integrated circuits (LSIs) are shown in FIGURE 1. These curves indicate that device integration has increased by a factor of four every three years, while the minimum dimension has been reduced by a factor of 0.7 every three years. However, several physical and chemical limitations might prove to be obstacles against further microminiaturization beyond the 100 nm technology level.[2] Some of the major factors are:[3] (1) semiconductor limitation — the p-n junction depletion region width of about 20–30 nm cannot be reduced, because the built-in voltage cannot be scaled; (2) insulator limitation — the insulator tunneling phenomena (occuring below 4 nm for silicon dioxide) due to high electric field cannot be avoided; (3) metal limitation — the current density above 10^7 A/cm^2 cannot be sustained by conventional metal materials; (4) statistical error — a very small number of dopant atoms (about 10) would cause statistical fluctuation in a small doping area, such as a channel region, and lead to a fatal variation in device characteristics; and (5) switching energy — as the device dimensions become smaller, the switching energy is also reduced, and

[a]Address for telecommunication: Phone, 81/492-96-6111; fax, 81/492-96-6006; e-mail, wada@harl.hitachi.co.jp

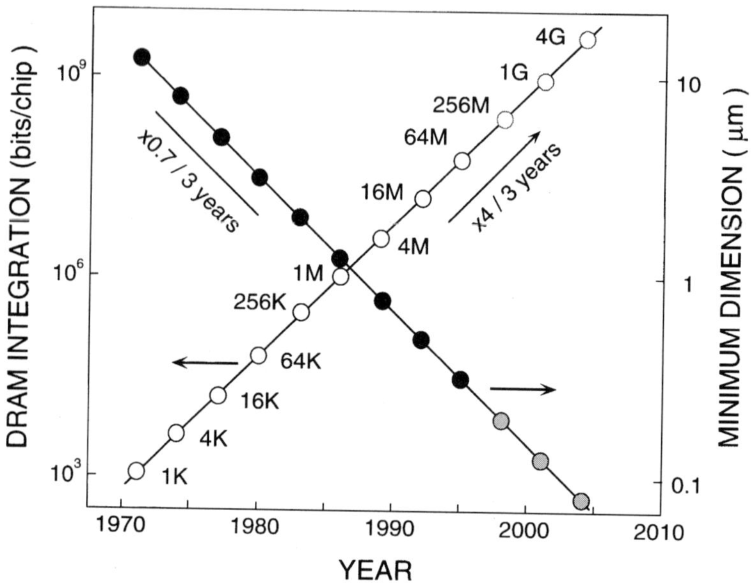

FIGURE 1. Trend curves of MOS transistor and DRAM LSI integration. Transistor size is shrunk 0.7 times every three years whereas integration quadruples every three years.

reaches the stage where quantum and thermal energy might cause errors in the device switching phenomena.

Innovative MOS transistor structures, it is claimed, make further downscaling possible,[4] though the improvement would be by a factor of two at most. The trend curves shown in FIGURE 1 also indicate that the minimum dimension of about 100 nm would probably be reached around the year of 2010, at the latest. Therefore, nanoscale devices, such as quantum effect devices,[5] have recently been attracting attention as candidates to replace silicon MOS transistors for future information processing integrated circuit devices. However, it is still uncertain whether these "quantum" devices will be able to replace MOS transistors, because the characteristics of these devices are not necessarily superior to those of MOS transistors, which are almost ideal as information processing devices, as will be shown later.

This paper proposes three atom size switching devices, the atom relay transistor (ART),[3] the supra molecular atom relay transistor (SMART)[6] and the Molecular single electron switching transistor (MOSES),[7] as candidate devices to supersede MOS transistors with dimensions well below a few nm, and an operation speed of more than 10^{12} (Tera:T) Hz. Operation principles, simulated device characteristics, and performance evaluation are shown, as well as the current technology development towards the realization of these devices.

2. CHARACTERISTICS NECESSARY FOR INFORMATION PROCESSING DEVICES

Generally, integrated circuit devices should fulfill the following requirements for high performance information processing[3]: (1) input and output (I/O) signal balance, the output of one device can directly drive the next device; (2) I/O signal isolation, input and output are isolated, in other words, at least three terminals are required in each device; (3) high switching speed, the switching speed of the device has to be much faster than that of the present devices; (4) dense integration capability, the separation between devices has to be very small, preferably the dimension should be the same as that of the minimum device size; and (5) fabricability, devices should be easy to fabricate, in terms of process technologies.

Other factors, such as operation margin, noise immunity, power dissipation, and reliability have to be taken into consideration in the integrated circuit design. Among these five factors, high switching speed has usually been the only criterion for evaluating newly proposed devices. However, it is essential to fulfill all five factors in a balanced manner. The reason why MOS transistors have been used in integrated circuits so far, and are expected to be used down to their limit of around 100 nm, is that they almost ideally fulfill the above five requirements. These requirements will be the basis for the development of any future information processing devices, even if computer architecture is changed. For example, massively parallel computers will operate faster, if the individual processor operates faster.

3. ATOM/MOLECULE SWITCHING DEVICES — PROSPECT FOR THE FUTURE

This section explains the operation principles of ART, SMART and MOSES and considers whether the proposed devices can fulfill the five requirements.

Atom Relay Transistor (ART)

Principle of Operation

The basic concept of ART[3] is schematically depicted in FIGURE 2, in which an individual atom is represented by a circle. In the figure, ART is depicted as consisting of an atom wire, a switching atom, a switching gate, and a reset gate. The operation principle is as follows. The atom wire is conductive (on) when the switching atom is set in the atom wire, and it is non-conductive (off) when the switching atom is removed from the atom wire.[3] The movement of the switching atom is made possible by the electric field supplied from the switching gate, and the first requirement of I/O balance for ART will be fulfilled. The on/off characteristics are simulated by using the tight binding method and the results are depicted in FIGURE 3, in which 250 atoms are arranged in a straight line with a separation of 0.2 nm. One electron is injected from the far left side of the atom wire with an electric field of 3×10^5 V/cm, and the electron density flowing through the atom wire is shown with a time interval of 6 fs. The results clearly indicate that ART exhibits on/off switching characteristics depending on the position of the switching atom, which is also ascertained by

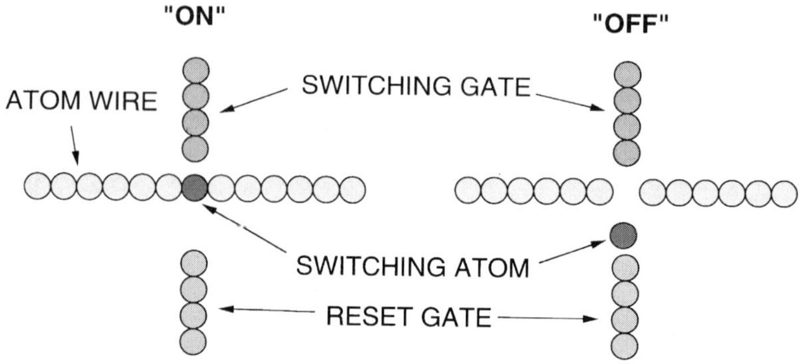

FIGURE 2. Basic configuration of atom relay transistor (ART),[3] consisting of atom wire, switching atom, switching gate and reset gate.

another simulation method.[8] The second requirement, I/O signal isolation, is also fulfilled, because ART is a three (four) terminal device.

Switching Speed

The switching speed of the switching atom from the "on" site to the "off" site, which corresponds to the switching frequency of ART, was simulated at around 30 THz by using the first principle method.[9] The transmission velocity of the electric field also allows the operation of ART at above 10 THz. Therefore the overall performance of ART should well exceed 10 THz. Thus the third condition, very high switching speed, has been fulfilled.

Memory Cell and Logic Circuits

The basic configuration of ART shown in FIGURE 2 suggests that ART can constitute a static memory cell by itself. In addition, ART offers dynamic operation us-

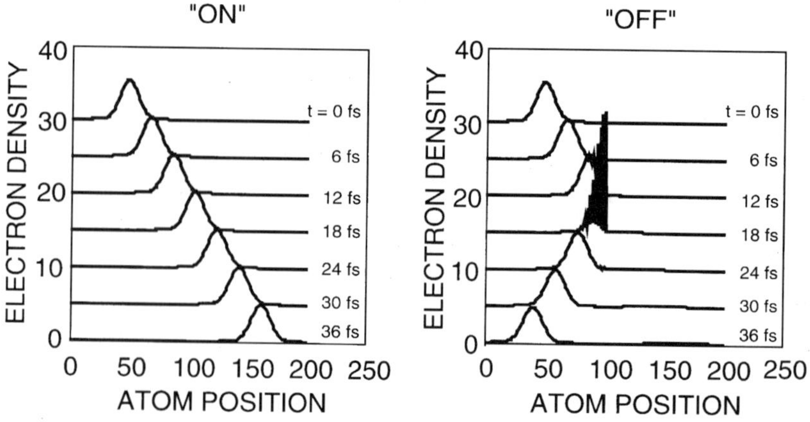

FIGURE 3. Simulated "on", "off" switching characteristics of ART[3] by tight binding method.

ing the self-relay configuration, in which the switching atom is removed from the atom wire by the electric field from the atom wire itself and electrons are trapped in the loop.[10] A dynamic memory cell uses only about 10^3 times of the atom area, and a 10^9 bit occupies only about 200 μm square of area. The trapped electron might disappear through leakage passes such as insulator leakage, tunneling through the gap or radiation of light. A lifetime evaluation of the trapped electron would also constitute a very interesting subject for scientific study. NAND and NOR gates were also designed,[3] and a von Neumann type computer can be constructed. The rule of thumb estimation of the area occupied by a 10^7 gate logic circuit and a 10^9 bit memory, which corresponds to almost the same integration as the state-of-the-art supercomputers, results in a 20 μm square area and 200 μm square area, respectively. Therefore the total area occupied by the ART supercomputer would be about 200 μm square, and even the longest signal delay on the chip would be well below the order of 10^{-12} s, which is short enough even for operation at more than 1 THz. Thus, the fourth condition, dense integration, would be fulfilled.

Fabrication Technologies of ART

Fabrication technologies of atom/molecule switching devices would be built using scanning probe technology, such as a scanning tunneling microscope (STM). Here, the current fabrication technology development is described toward the realization of atom electronics.

Beam Assisted STM (BASTM). In order to realize ART, it is necessary to manipulate individual atoms by, for example, STM, which can only observe and manipulate individual atoms on a conducting solid surface, because it requires a current flow between the tip and the sample. Atomic force microscope (AFM) allows observation of the insulator surface, but still not manipulation of individual atoms. Beam Assisted STM (BASTM) would be the most promising solution to this problem, because of its capability for observing insulators.[11] The operation principle is based on the activation of carriers in the insulator by an irradiating beam, such as electron, ion and photon beams. Thermally grown silicon dioxide layers have been successfully evaluated by using BASTM,[12] which confirms the possibility of manipulating atoms on insulators.

Needle Formation and Tip Imaging (NFTI). In order to manipulate individual atoms using STM, it is necessary to use tips with a very controlled apex. The needle formation and tip imaging (NFTI) method[13,14] delineates the three-dimensional tip apex image and *in situ* evaluation of the apex is made possible. The typical diameter of the nano needle measures about 1.5 nm, while the height can be controlled between few nm up to 20 nm, by selecting appropriate bias conditions and applied time. Using the nano needle as the probing tip, the STM tip apex structure was successfully delineated.[13] These results are the first direct experimental evidence of the theoretical and experimental hypothesis that an atomic protrusion on the tip apex is indispensable to obtain an atomic resolution during the STM observation of a surface.

Atom Wire Fabrication. The first step to demonstrate the operation of ART would be to measure the conductance of an atom wire. The atom wire can be formed on a silicon (100) substrate by the following method.[15,16] The silicon (100) surface was cleaned by vacuum annealing and exposed to an atomic hydrogen stream gen-

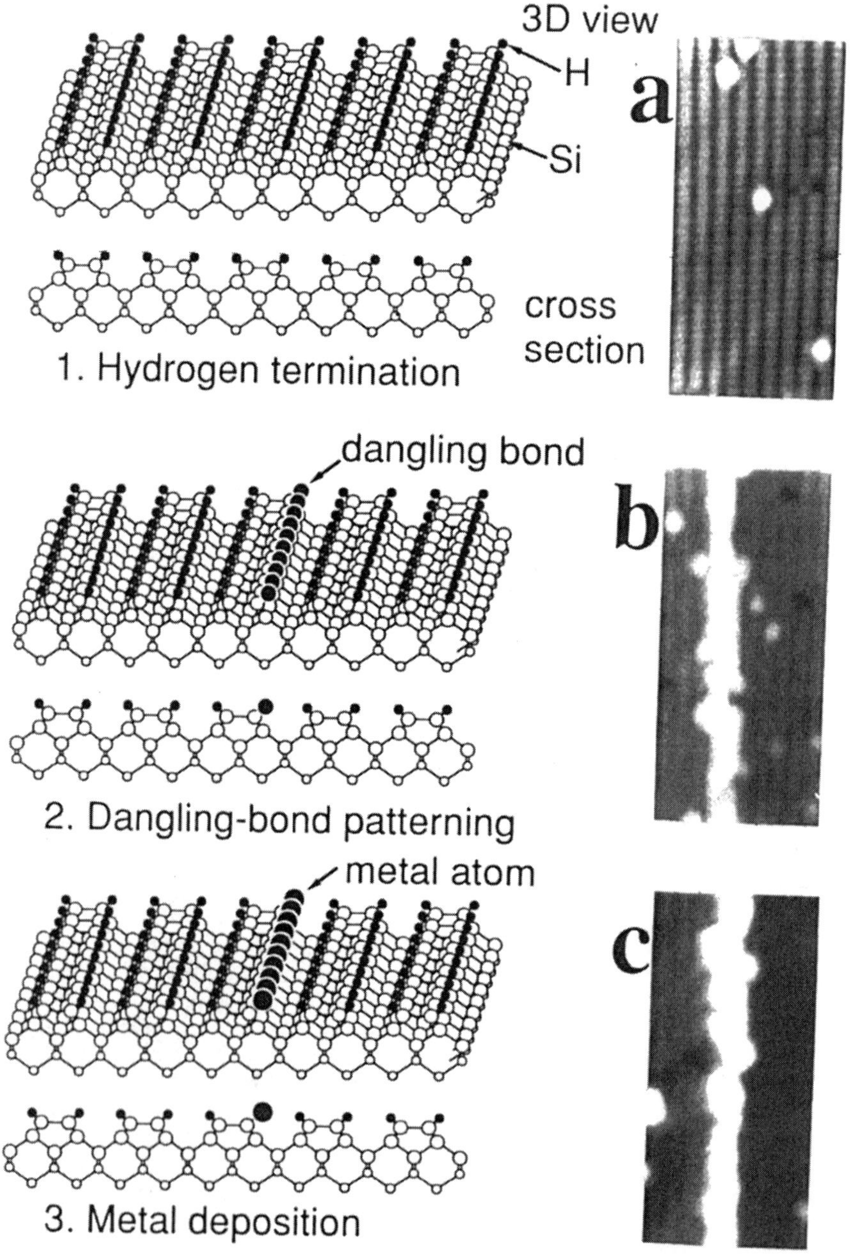

FIGURE 4. Atom wire fabrication procedures on hydrogen passivated Si (100) surface[16]: **(a)** hydrogen passivation of the Si (100) surface; **(b)** dangling bond formation by STM atom manipulation method; and **(c)** metal line formation on the dangling bond wire by a self-organization manner.

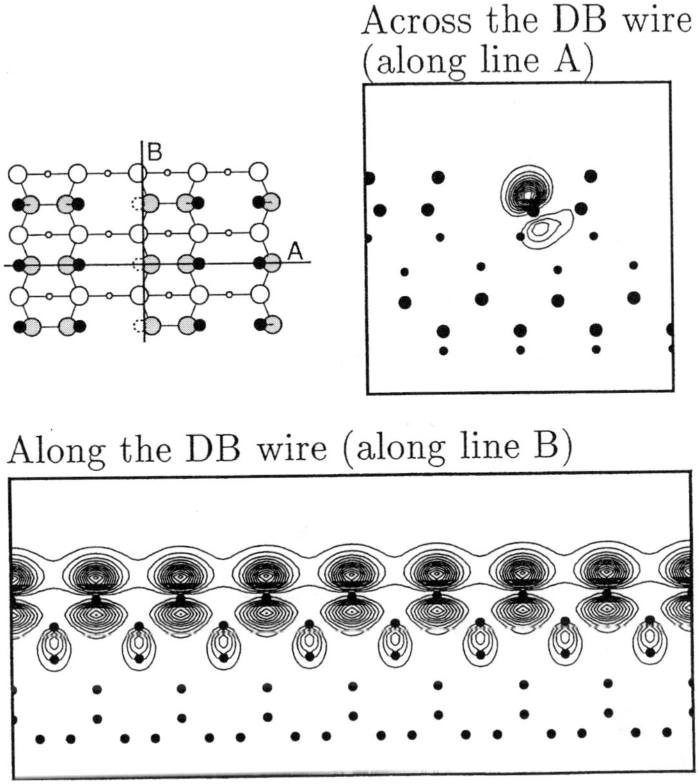

FIGURE 5. Electron density distribution of a dangling bond wire formed on a Si (100) -H surface simulated by the first principle method.[17]

erated by cracking hydrogen molecules using a hot (~2000 K) tungsten wire to form a hydrogen passivated surface as schematically shown in FIGURE 4(a). One hydrogen row was removed by an STM modification method to construct a dangling bond wire as shown in FIGURE 4(b). The dangling bond wire formed on the silicon (100) surface should be conductive, because of the high electron density predicted by the first principle simulation,[17] as shown in FIGURE 5. In the figure, the high electron concentration region along the dangling bond wire and the very sharp confinement of electron density across the wire are clearly depicted. Therefore, the wire can be conductive along the length direction, whereas it would be isolated from the adjacent wire perpendicular to the length direction. Scanning tunneling spectroscopy (STS) measurements have also ascertained the simulation results.[18]

By aligning metal atoms on the dangling bond structure and making the dangling bond wire into a metallic wire, the conducting wire would become more stable than the simple dangling bond wire. Therefore, the dangling bond wire shown in FIGURE 4(b) was filled with gallium metal atoms by evaporating an appropriate amount of gallium atoms, as schematically indicated in FIGURE 4(c). Because the activation energy for the diffusion of gallium atoms on the hydrogen passivated silicon surface is very small, in the order of 100 meV,[19] the gallium atoms stick to the dangling bonds selectively and spontaneously and form a gallium wire.

Connection to the Real World. The atom wire thus fabricated has to be connected to the real world in order to evaluate the electrical characteristics. One method is proposed[16] and the feasibility is demonstrated, in which the bonding pads were fabricated by through-the-mask evaporation method of metal layers together with the metal wiring. The final goal would be to connect the bonding pad, metal wiring, and the atom wire by thicker metal atom wire. Another method is to delineate pad and

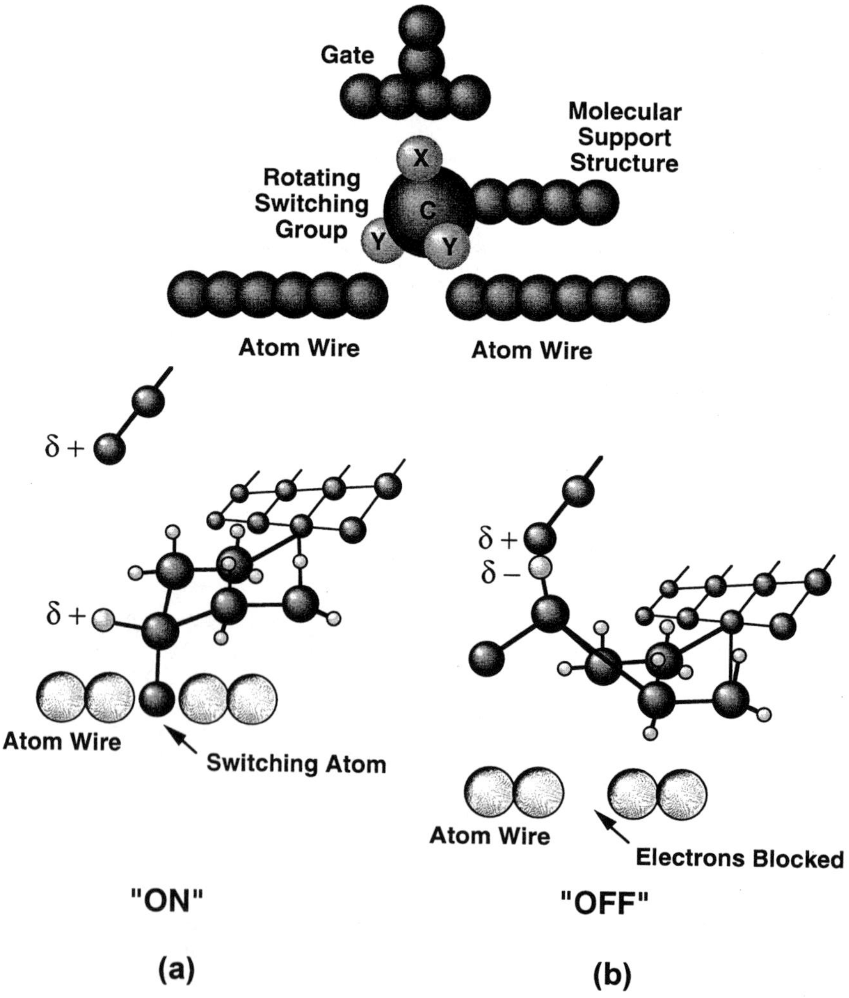

FIGURE 6. Schematic figure of molecular device version of ART, SMART, in which on/off characteristics are determined by the rotation of a molecule branch.[6]

wiring patterns by fine pattern lithography, such as e-beam lithography, and fabricate the atom wire between the metal wiring and connect them together. Thus the conductance of a gallium single atom wire formed on dangling bonds can be measurable if contact pads on both ends of the wire are probed by a conventional prober.[16]

Supra Molecular Atom Relay Transistor (SMART)

Recently, a molecular version of ART was proposed,[6] in which the switching operation is characterized by the molecular motion in the switching molecule, as schematically shown in FIGURE 6. The switching principle is that the switching molecule connects/disconnects the switching element in the molecular wires by rotating the small branch of the switching molecule. This structure would allow a switching speed in the THz range, deduced from the absorption spectra of the molecular rotation. In addition, this SMART configuration would offer vast kinds of molecules with various functions. Therefore, it should be possible to design and fabricate NAND molecules, NOR molecules and memory cell molecules, which would make von Neumann type very fast computers possible.

The connection between the molecules would be made possible by using the anchor atoms placed at the designated points on the surface, as will be explained in the later section. This idea is based on the experimental observation that molecules adsorb on the certain positions of the silicon surface, as shown in FIGURE 7,[20] where the corresponding adsorption structures are also schematically indicated. In addition, the combination of the atom wire concept described earlier as the ART configuration and the use of molecules as a switching device should provide another possible candidate for achieving the extremely high performance information processing. For example, the device region is fabricated by using the molecules, while the wiring portion is formed with the atom wires. The advantage of this concept would be in the "self-assembly" manner in formulating the circuit structures. In other words, if the wiring is fabricated so that the molecules attach spontaneously themselves to the designated positions of the wiring, the fabrication time would be shortened within a practical regime. This implies that by choosing the appropriate molecule and substrate combinations, SMART would be arranged to provide necessary functions.

Molecular Single Electron Switching Transistor (MOSES)

Molecular devices attracted much attention when they were proposed in 1980 by Carter.[21] However, the interest gradually faded away due to several causes such as the inaccessibility to individual molecules and the slow switching speed of molecules for an information processing device, because the switching is based on conformal changes. On the other hand, recent progress in single electron tunneling (SET) devices[22] have attracted attentions because of their potentially superior operation characteristics. The SET devices operate faster, at more than 1 THz, if the quantum dot size is reduced to 1 nm,[23] as will be discussed later. However, fabrication as well as precise size control, which is essential in fabricating integrated systems, should become almost impossible by the conventional physical fabrication technology in the nanometer range. Molecular systems should achieve almost per-

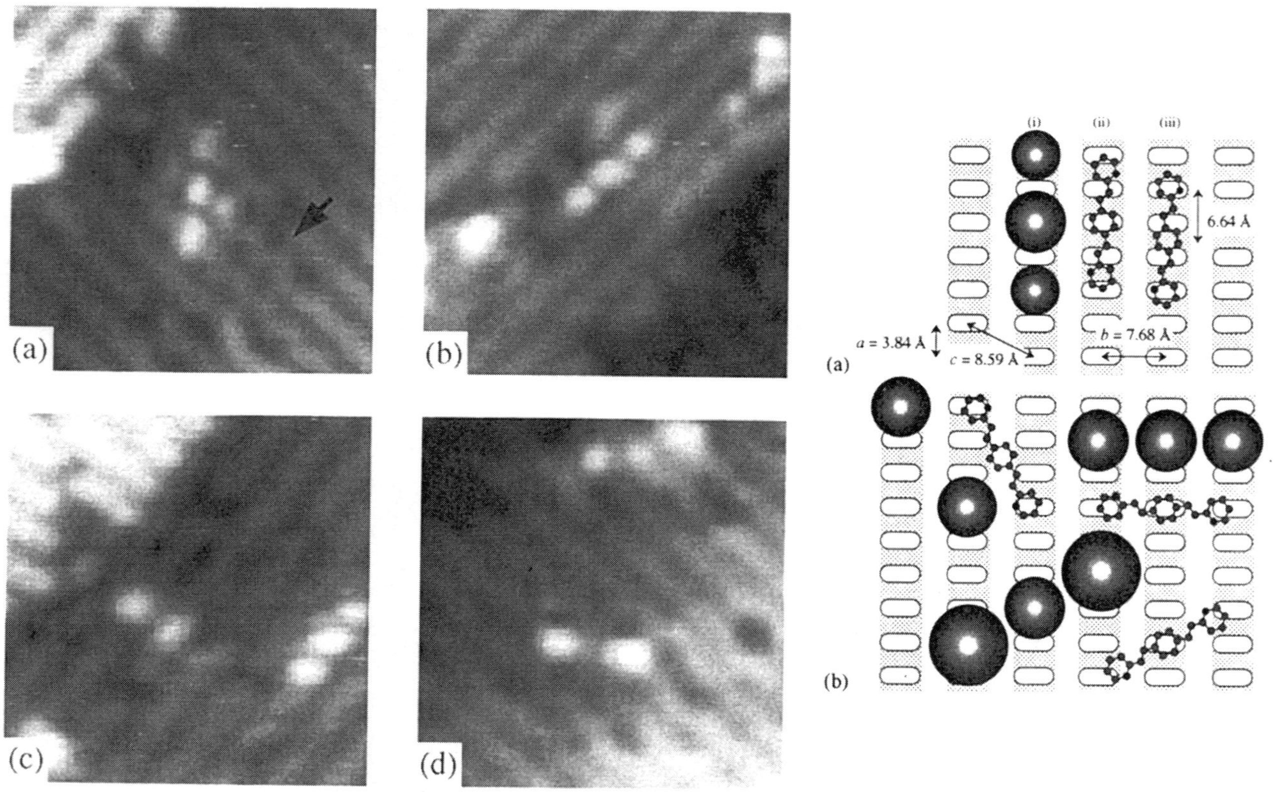

FIGURE 7. Example of selective molecular adsorption on silicon surface.[20]

fect dimensional control, because the dimension of a molecule is determined by the number of bonds within the molecule.

Principles of Operation

A molecular single electron switching transistor (MOSES) device consists of conducting polymer and insulating polymer, as schematically shown in FIGURE 8.[7] In the figure, polyacetylene and polyethylene molecules were chosen as typical conducting and insulating molecules, respectively. The insulator part acts as a tunnel barrier and the conducting polymer island as a quantum dot. The electron density of the polyacetylene-polyethylene-polyacetylene molecular structure was calculated by a molecular orbital (MO) method, and the results are also demonstrated in FIGURE 8. These results clearly indicate that the conduction electron density is negligible in the insulator (polyethylene) part, while it is high in the conductor (polyacetylene) part of the molecule. Recently, a relatively solid and straight chain molecule has been proposed as a candidate for the conducting polymer.[24] Therefore, these molecules might be suitable for realizing MOSES. In addition, three terminal MOSES devices should also be realizable by synthesizing appropriate molecular structures[25] with branches, dots, and gates.

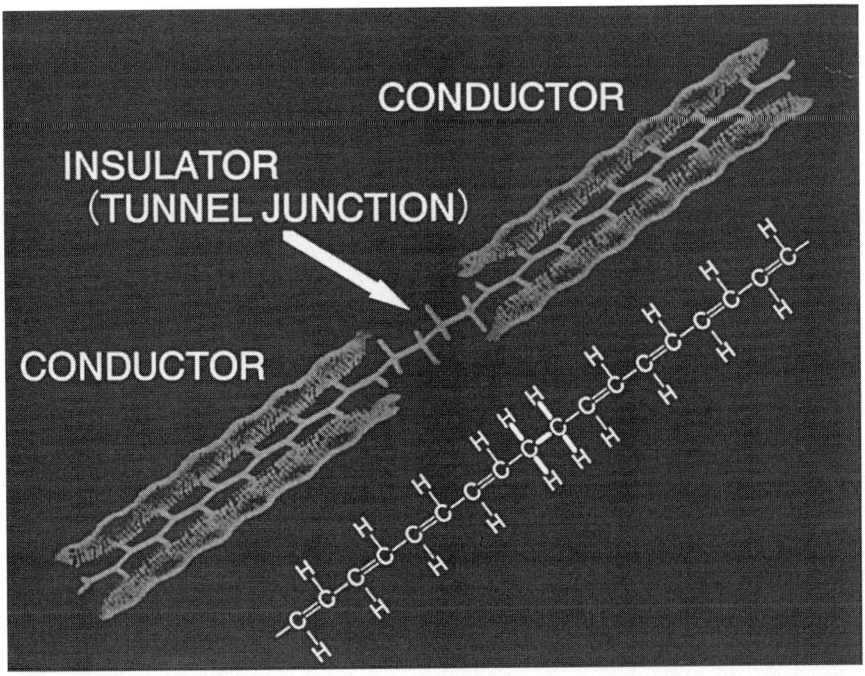

FIGURE 8. Schematic figure and simulated electron density of an example of MOSES.[7]

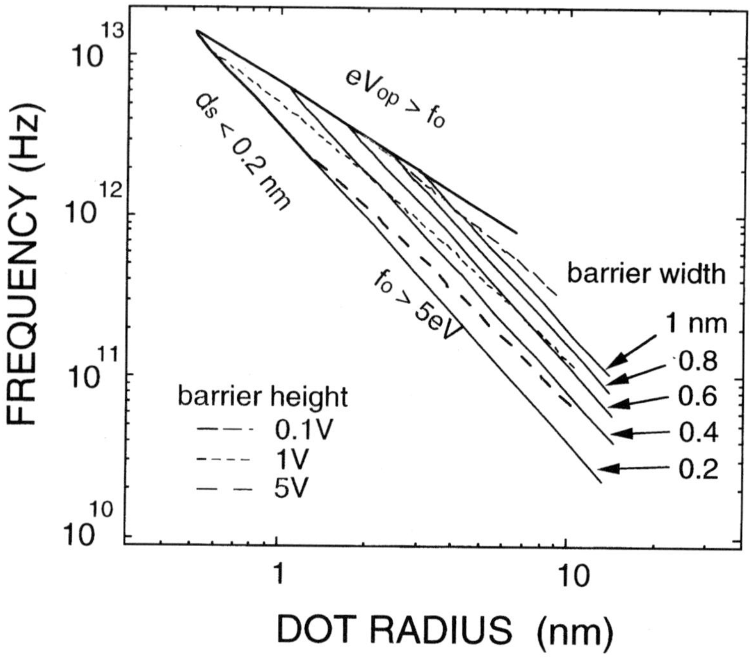

FIGURE 9. Simulated ultimate operation speed of SET devices.[23]

Operation Speed

The simulated ultimate operation speed of SET devices[23] indicates that the operation speed is in the order of 1 THz if the quantum dot size is less than 1 nm, and that the tunnel barrier width must also be very thin, less than 1 nm, as shown in FIGURE 9. For example, the size of a benzene ring, carbon double bond length, and carbon single bond length measure about 0.242 nm, 0.135 nm, and 0.153 nm, respectively. Other possible MOSES structure is based on the C60.[26] The diameter measures about 0.7 nm, and if C60 is used as the quantum dot, the operation speed should exceed several THz. In addition, C60 has many possible reaction sites, which offer lot of freedom in enabling the address of appropriate branches with appropriate characteristics. Therefore, gate capacitance can be made much larger than the tunnel barrier capacitance, which is essential to the high performance SET operation.[23] Thus, the size of the quantum dot in MOSES can be controlled by selecting the appropriate molecular design and molecules are likely to be the most appropriate material for realizing high performance SET devices.

Memory and Logic Circuits

Memory and logic circuits can be formed by using MOSES devices, in which a conventional SET device and circuit design[27] can be applied. Therefore, no new circuit design should be necessary. Appropriate molecular design should achieve the necessary circuit characteristics, including multiple output wiring.

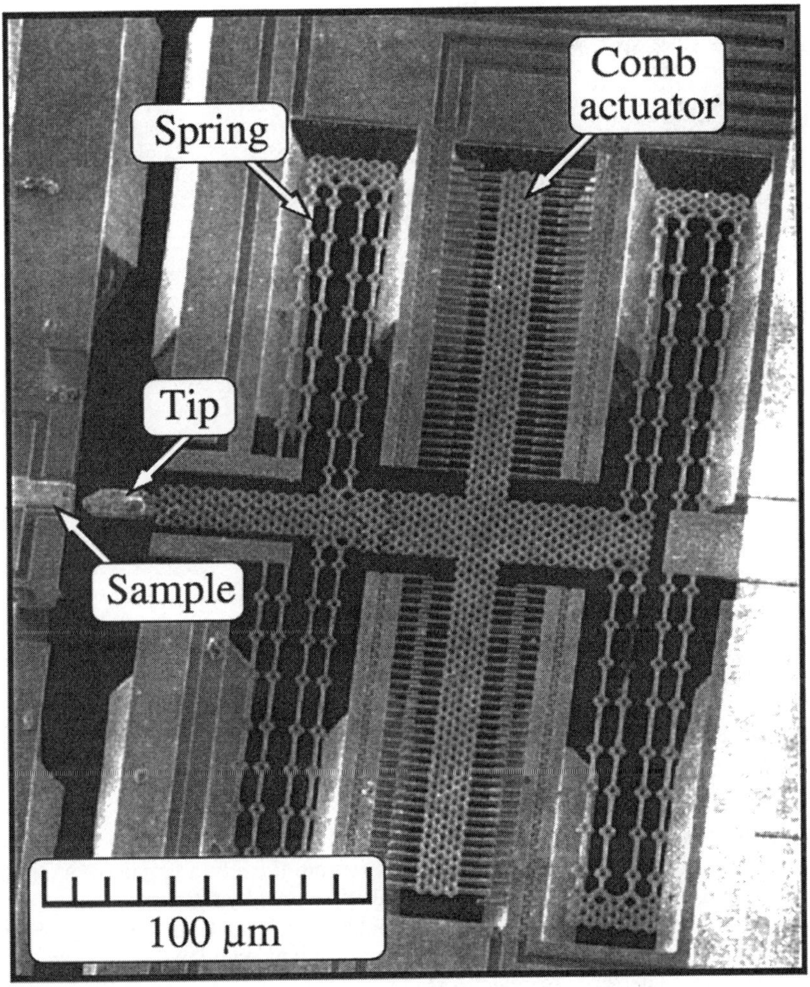

FIGURE 10. SEM micrograph of the fabricated micromachine STM (μ-STM).[30]

Measurement Technologies of MOSES Devices

The first major difficulty towards the realization of MOSES would be the access to individual molecules, as mentioned previously in this section. The possibility of measuring relatively large molecular systems, carbon nanotubes, are demonstrated using nano lithography[28] and STM.[29] These technologies might be applied to the access to single molecule, after intensive development. A novel micromachine scanning tunneling microscope (μ-STM) manipulation method was proposed, which is fabricated using a 0.4 μm ULSI fabrication technology.[30] A scanning electron microscope (SEM) micrograph of the fabricated μ-STM is shown in FIGURE 10. The

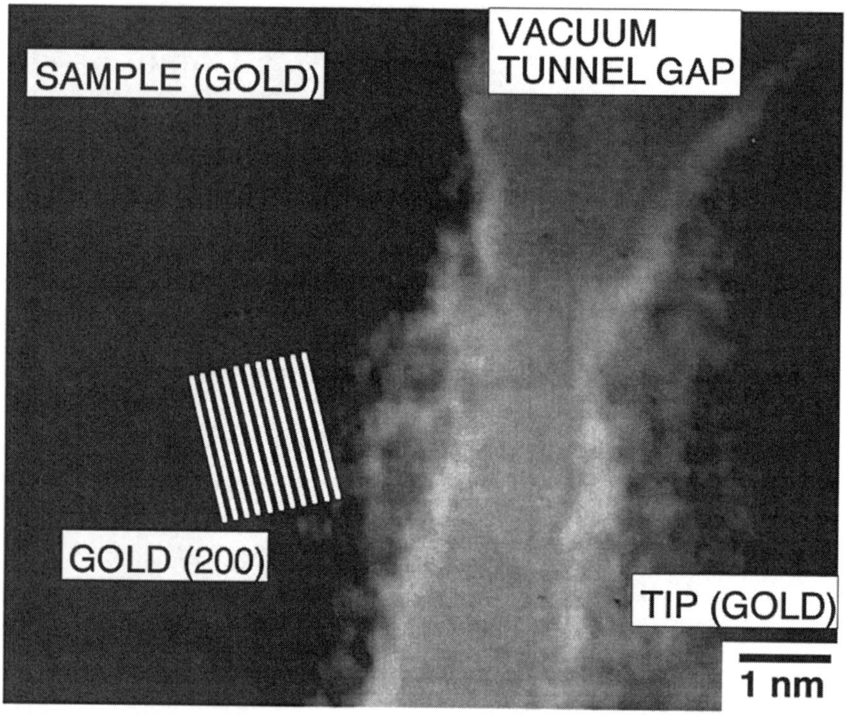

FIGURE 11. TEM micrograph of the vacuum tunneling gap between the tip and sample of the μ-STM.[30]

total μ-STM size measures about 200 μm square, but the size can be made smaller by using finer fabrication technologies. The vacuum tunneling gap between the μ-STM tip apex and the sample was successfully observed by a transmission electron microscope (TEM) (FIG. 11).[30] The two electrodes are made with gold, and thiol groups on both ends of the conducting molecule should easily make contacts to the gold electrodes. Therefore the electrical conduction characteristics of the molecule should be reliably measurable, and the high performance operation of MOSES should be demonstrable.

Recently, other possible technology to access to single molecule was demonstrated,[31] in which specially designed piezo-based STM was inserted in a TEM to also observe the vacuum tunneling gap. The single molecule manipulation might also be observable by polishing the present technology.

Another possible molecular addressing technology would be the silicon surface, in which the dangling bond connects the molecule quite firmly to the silicon.[20] If an appropriate atom, such as gold atom, is placed on the hydrogen passivated silicon surface, thiol molecule would adhere to the gold atom and form contacts. Other portion of the silicon surface than the gold atom would not react with the molecule, because of the very low surface energy of the hydrogen terminated region. Therefore, by removing hydrogen atoms from the designated position and placing appropriate

TABLE 1. Comparison of the performances of candidate information processing devices

	MOS	QUANTUM DEVICE	SUPER CONDUCTOR DEVICE	SET	MOLECULAR DEVICE	**ART**	**SMART**	**MOSES**
I/O BALANCE	○	△	○	○	△	○	○	○
I/O ISOLATION	○	○	△	○	△	○	○	○
SPEED	○	◎	◎	◎	△	◎	◎	◎
INTEG-RATION	○	○	○	○	◎	◎	◎	◎
FABRI-CABILITY	○	△	△	△	△	△	△	△

anchoring atoms there, it should be made possible to address molecules at an appropriate position, as schematically shown in FIGURE 12, in which gold atoms are used for the anchoring atoms of the MOSES molecule.

EVALUATION OF THE CANDIDATE DEVICES FOR THE FUTURE INFORMATION PROCESSING

The candidate switching devices for the next decade, including quantum devices, conventional single electron devices, conventional molecular devices, ART, SMART, and MOSES devices, were evaluated according to the five indispensable factors listed, and the results are summarized in TABLE 1. It is indicated that MOS transistors almost ideally fulfill all the requirements, and that ART, SMART, and MOSES are the most promising candidates to supersede them. The candidate nanoscale devices listed in TABLE 1, in addition to the historical information processing electron devices, were evaluated semi-quantitatively in terms of speed and integration capability, and the results are summarized in FIGURE 13. Human beings started electronic information processing using a mechanical relay. The very slow and unreliable operation of the mechanical relay (mechanical electronics) was soon replaced by a vacuum tube[32] (vacuum electronics). Then the invention of transistors[33] revolutionized the computer, and solid state electronics devices including integrated circuits[34] dominated electronic information processing. As indicated in FIGURE 13, the innovation, or paradigm shift (e.g., from mechanical electronics to vacuum electronics to solid state electronics), has been accomplished because the performances of the latter devices are superior to those of preceding ones by at least two orders of magnitude.

Now that solid state electronics devices are approaching their limits, candidate devices are expected to supersede them. Conventional molecular devices exhibit very high integration density, however, their operating speed is extremely slow and they cannot be the prime candidates for future devices. Quantum devices demonstrate better characteristics in terms of integration and switching speed, however, they do not exhibit orders of magnitude higher switching speed or integration than the present silicon MOS transistors. Single electron devices would have to be scaled down to less than 1 nm in order to obtain very high performance. Therefore ART,

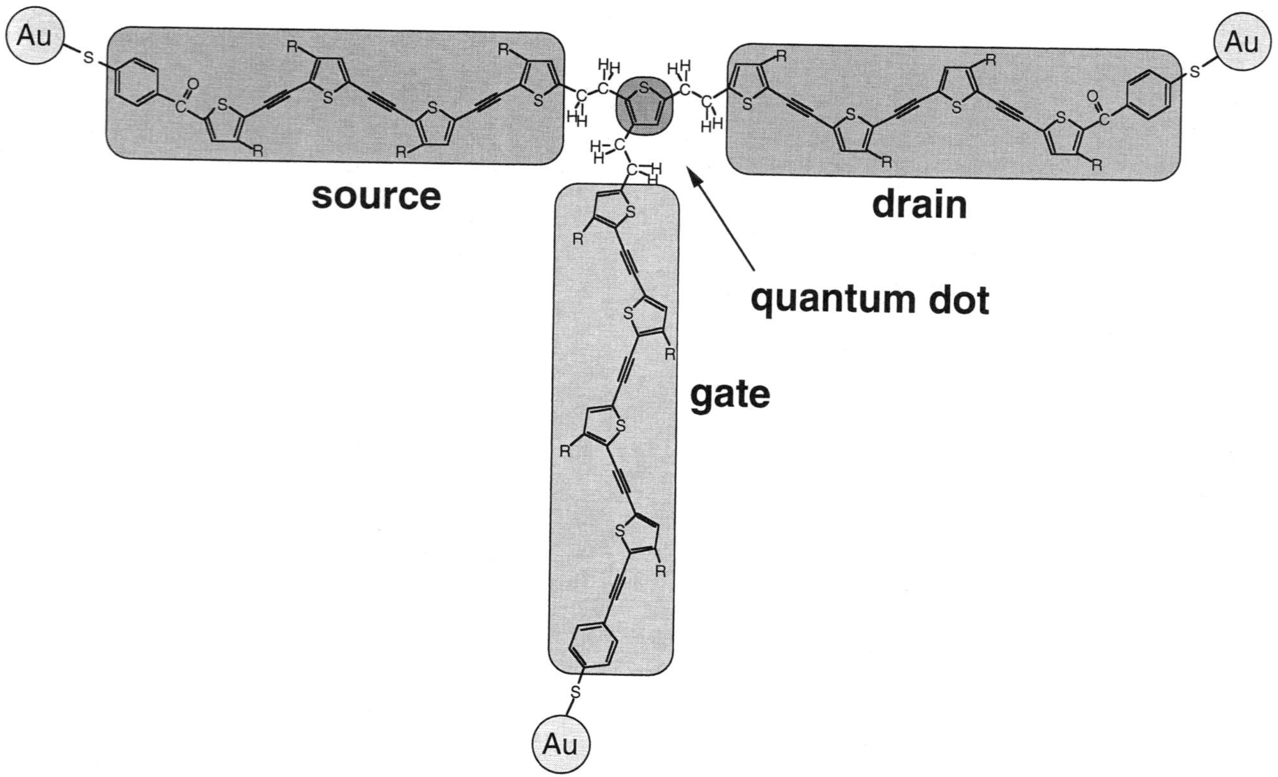

FIGURE 12. Schematic figure of positioning MOSES molecule on hydrogen passivated silicon surface with anchoring gold atoms.

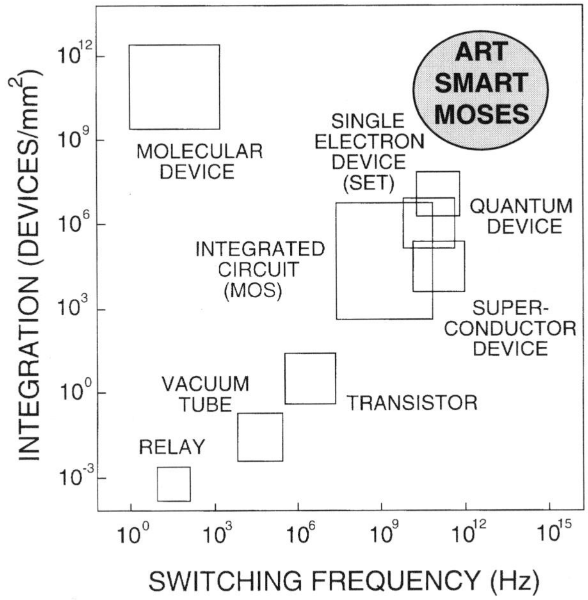

FIGURE 13. Comparison of the switching speed and integration density of historical and candidate information processing devices after the MOS transistor era.

SMART and MOSES are likely to be the most promising candidate devices for future electronic information processing integrated circuits, because of their potentially very fast operation speed and dense integration characteristics.

CONCLUSIONS

This paper proposed nanoscale switching devices, atom relay transistors (ART), supra molecular atom relay transistor (SMART) and molecular single electron switching transistors (MOSES), for the information processing in the next twenty years, with dimensions of only a few nm, and an operation speed of more than 10 THz. The switching characteristics of ART, SMART, and MOSES were demonstrated by simulation. Fundamental logic circuits such as NAND and NOR gates and memory circuits based on ART concept were proposed, which can integrate a supercomputer in an area 200 μm square, with 10^7 gates of logic circuit and 10^9 bits of memory, and operate at more than 1 THz. MOSES device structures were also demonstrated by simulation and very dense and fast operation characteristics were foreseen. New technologies indispensable for the fabrication of these nanoscale devices were demonstrated, including BASTM, NFTI, and μ-STM. Experimental results to fabricate atom wire structures were also demonstrated. Various candidate nanoscale devices were evaluated on the basis of the characteristics necessary for future integrated circuit devices, and ART, SMART, and MOSES devices were found to be the most promising candidates for future information processing integrated circuits.

ACKNOWLEDGMENTS

I would like to express my sincere gratitude to Dr. Shojiro Asai, Dr. Katsuki Miyauchi, and Dr. Michiharu Nakamura for their support and continuous encouragement throughout the work. I am also grateful to Dr. Akira Tonomura for valuable discussions. Dr. Tsuyoshi Uda, Dr. Satoshi Watanabe, and Dr. Takashi Goda are acknowledged for fruitful discussions on the structure and device performance simulation. I would also like to express my thanks to Dr. Tomihiro Hashizume, Dr. Hiroshi Kajiyama, Dr. Mark I. Lutwyche, Mr. Masayoshi Ishibashi, Ms. Nami Sugita, Mr. Seiji Heike, Dr. Ken Nakajima, and Mr. Taro Hitosugi for their contributions to the work. The ULSI Research Department and the Electron Device Department of the Central Research Laboratory are acknowledged for fabricating the μ-STM chips.

REFERENCES

1. DENNARD, R.H., F.H. GAENSSLEN, H.YU, V.L. RIDEOUT, E. BASSOUS & A.L. LEBLANC. 1974. Design of ion implanted MOSFETs with very small physical dimensions. IEEE J. Solid St. Circuits **SC-9:** 256–268.
2. SUGANO, T. 1990. VLSI Process Data Handbook. Science Forum Tokyo. 17–21; Keyes, R.W. 1993. The Future of the Transistor. Sci. Amer. **June:** 70–78.
3. WADA, Y., T. UDA, M.I. LUTWYCHE, S. KONDO & S. HEIKE. 1993. A proposal of nanoscale devices based on atom/molecule switching. J. Appl. Phys. **74:** 7321–7328.
4. FIEGNA, C., H. IWAI, T. WADA, M. SAITO, E. SANGIORGI & B. RICCO.1994. Scaling the MOS Transistor below 0.1 μm: Methodology, Device Structures, and Technology Requirements. IEEE Trans. Electron Devices **41:** 941–951.
5. AL'TSHULER, B.L., P.A. LEE & R.A. WEBB, Eds. 1991. Mesoscopic Phenomena in Solids. North-Holland. Amsterdam.
6. GOLDHARBER-GORDON, D., M.S. MONTEMERLO, J.C. LOVE, G.J. OPITECK & J.C. ELLENBOGEN. 1997. Overview of Nanoelectronic Devices. Proc. IEEE. **85:** 521–540.
7. WADA, Y. 1994. A proposal of molecular single electron devices. Trans. IEICE. **OME 93-54:** 31–36.
8. LANG, N.D. 1998. Conductance calculations for the atomic relay. Superlattices and Microstructures. **23:** 731–735.
9. VON ALLMEN, P. & K. HESS. 1995. Dynamics of an atomic switch computed by first-principles molecular dynamics. Phys. Rev. **B52:** 5243–5246.
10. WADA, Y. 1997. Atom electronics: A proposal of atom/molecule switching devices. Surf. Sci. **30:** 265–278.
11. HEIKE, S., Y. WADA, S. KONDO, M.I. LUTWYCHE, K. MURAYAMA & T. KURODA. 1994. Scanning tunneling microscope measurement of insulator surfaces. Appl. Phys. Lett. **64:** 1100–1102.
12. HEIKE, S., Y. WADA, S. KONDO, M.I. LUTWYCHE, K. MURAYAMA & T. KURODA. 1995. Evaluation of Thin silicon dioxide layers by beam assisted scanning tunneling microscope. Jpn. J. Appl. Phys. **34:** 1376–1380.
13. HEIKE, S., T. HASHIZUME & Y. WADA. 1995. *In situ* direct imaging of scanning tunneling microscope tip apex. Jpn. J. Appl. Phys. **34:** L1061–L1063.

14. HEIKE, S., T. HASHIZUME & Y. WADA. 1996. Nano-needle formation on silicon (111) surface using a scanning tunneling microscope Tip. J. Appl. Phys. **80:** 4182–4188.
15. LYDING, J.W., T.-C. SHEN, J.S. HUBACEK, J.R. TUCKER & G.C. ABELN. 1994. Nanoscale patterning and oxidation of H-passivated Si (100)–2×1 surfaces with an ultrahigh vacuum scanning tunneling microscope. Appl. Phys. Lett. **64:** 2010–2012.
16. HASHIZUME, T., S. HEIKE, M.I. LUTWYCHE, S. WATANABE & Y. WADA. 1996. Interaction of Ga adsorbates with dangling bonds on the hydrogen terminated Si(100) surface. Jpn. J. Appl. Phys., **35:** L1085–L1088.
17. WATANABE, S., Y. ONO, T. HASHIZUME, & Y. WADA. 1996. Theoretical study of electronic structures of atomic wires on an H-terminated Si (100) 2×1 surface. Phys. Rev. **B54:** 17308–17311.
18. HITASUGI, T., T. HASHIZUME, S. HEIKE, S. WATANABE, Y. WADA & K. KITAZAWA. 1997. Tunneling spectroscopy study of dangling bond wire on Si (100)-H passivated surface. Jpn. J. Appl. Phys. **36:** L361–L364.
19. WATANABE, S., M. ICHIMURA, T. ONOGI, Y. ONO, T. HASHIZUME & Y. WADA. 1997. Theoretical study of Ga adsorbates around dangling bond wires on an H-terminated si surface: Possibility of atomic-scale ferromagnets. Jpn. J. Appl. Phys. **36:** 929–932.
20. NAKAJIMA, K., T. HASHIZUME, S. HEIKE, S. WATANABE, M. MITSUYA, T. IKEHARA, Y. WADA & T. NISHI. 1996. Study of initial stage of molecular adsorption on Si (100) by scanning tunneling microscopy. Jpn. J. Appl. Phys. **35:** L1360–1363.
21. HONG, F.T., Ed. 1988. Molecular Electronics. Plenum Press. New York.
22. LIKHAREV, K.K. 1988. Correlated discrete transfer of single electrons in ultrasmall tunnel junctions. IBM J. Res. Dev. **32:** 144–158.
23. LUTWYCHE, M.I. & Y. WADA. 1994. Estimate of the ultimate performance of the single-electron transistor. J. Appl. Phys. **75:** 3654–3661.
24. TOUR, J.M. 1996. Conjugated macromolecules of precise length and constitution: organic synthesis for the construction of nanoarchitectures. Chem. Rev. **96:** 537–553.
25. ULMAN, A. 1991. An Introduction to Ultrathin Organic Films. Academic Press. San Diego, CA.
26. LI, J., T. YOSHIZAWA, M. IKUTA, K. KINBARA, K. NAKAHARA, T. HASAGAWA, K. KITAZAWA, M. HAYASHI, K. KINBARA, M. NOHARA & K. SAIGO. 1997. Novel copolyamides containing [60] fullerene in the main chain. Chemistry Lett.: 1037–1038.
27. YANO, K., T. ISHII, T. HASHIMOTO, T. KOBAYASHI, F. MURAI & K. SEKI. 1993. A room temperature single-electron memory device using fine-grain polycrystalline silicon. *In* Technical Digest. 541–544. IEEE International Electron Devices Meeting, IEEE. New York.
28. TANS, S.J., M.H. DEVORET, H. DAI, A. THESS, R. SMALLEY, L.J. GEERLIGS & C. DECKER. 1997. Individual single-wall carbon nanotubes as quantum wires. Nature. **386:** 474–477.
29. COLLINS, P.G., A. ZETTL, H. BANDO, A. THESS & R.E. SMALLEY. 1997. Nanotube nanodevice. Science **278:** 100–103.
30. LUTWYCHE, M.I. & Y. WADA. 1995. Observation of a vacuum tunneling gap in a transmission electron microscope using a micromechanical tunneling microscope. Appl. Phys. Lett. **66:** 2807–2810.
31. OHNISHI, H., Y. KONDO & K. TAKAYANAGI. 1998. Operation of scanning tunneling microscope in UHV transmission electron microscope. *In* Proceeding of International Centennial Symposium on Electron (ICSE). Cambridge. In press.

32. SPANGENBERG, K.R. 1948. Vacuum Tubes. McGraw-Hill, New York.
33. BRIDGERS, H.E., J.H. SCAFF & J.N. SHIVE, Eds. 1958. Transistor Technology. Nostrand. Princeton.
34. KILBY, J. 1976. Invention of the integrated circuit. IEEE Trans. Electron Devices. **ED-23:** 648–654.

Scanning Tunneling Microscopy (STM) on Physisorbed Chemical Groups of Individual Immobilized Molecules

YISHAY MANASSEN[a] AND DOV SHACHAL

Department of Chemical Physics, Weizmann Institute of Science Rehovot 76100, Israel

ABSTRACT: Most scanning tunneling microscopy (STM) studies of adsorbed molecules are performed either on molecules that are chemically bound to the surface or are part of a molecular monolayer. In the first case, the chemical reaction with the surface changes the surface local density of states. In the second one, the electrons tunnel through a physisorbed molecule. In the study described in this paper STM imaging is performed on isolated molecules immobilized on a clean semiconductor surface where atomic resolution is observed. This is an intermediate case.

The adsorption behavior of diene molecules on Si(111)7 × 7 was investigated. The molecules adsorb by reaction of the double bonds on the surface. In the reaction, the C=C bond is opened to create two neighboring Si—C bonds. Two such C=C bonds are used to immobilize the chain to the surface. The adsorption chemistry is affected not only by the nature of the Si—C bond but also by steric limitations.

Performing STM imaging on such systems ensures that the position of the immobilized chemical group (in this case an aliphatic chain) could be clearly identified and that it was possible to image parts of the molecule that are not chemically bound to the surface. STM images of 1,13-tetradecadiene revealed a white protrusion in the location of the aliphatic chain. The increase in transmission is explained by an additional channel for tunneling: through bond tunneling. Long-range electron transfer through molecules is a widely known chemical and biochemical process. This observation opens the possibility to investigate this fundamental phenomenon in large details. The potential applications for molecular electronics are discussed.

Since its invention, the capability of the STM (scanning tunneling microscope) to image individual molecules on the surface fascinated many chemists. Although significant progress was made in the direction of molecular imaging, still, the vast potential of the STM in solving questions of interest for the scientific community interested in chemistry remained largely unexploited. One problem is that despite recent impressive successes, STM images of adsorbed molecules did not achieve the resolution and the contrast of STM images on clean conducting surfaces (in particular semiconductors). It is widely accepted that although the STM contrast can be explained using molecular orbitals analysis, the exact mechanism in which the tunneling electrons are transferred through the molecules is not completely understood.

The first STM experiments on molecular imaging were on relatively small molecules which were chemisorbed on the surface. The contrast achieved in these systems was due to changing the surface LDOS (local density of states). This includes

[a]Present address: Department of Physics, Ben Gurion University of the Negev, P. 0. Box, 653, Beer Sheva, 84105, Israel.

studies of NH_3 adsorbed on Si(111)7 × 7,[1] O_2 on Si(111)7 × 7,[2,3] O_2 on GaAs[4] and C_2H_4 on Pt(111).[5] In these experiments it was normally difficult to see any molecular feature and only the affected LDOS is distinguishable. When larger molecules were adsorbed on the surface, the molecular features were, many times, distinguishable. In such systems, due to the strong chemical bond between the molecules and the surface, and the large size of the molecule, the LDOS includes a relatively large contribution from the molecular orbitals, and the STM image also reflects molecular features. Such studies were performed for example on adsorbed naphthalene,[6] azulene derivatives[7] copper phthalocyanine,[8] and benzene [coadsorbed with CO on Rh(111)].[9]

However, most STM molecular imaging studies were performed on two dimensional molecular monolayers (Delamarche and Michel,[10] Walba and colleagues,[11] and Patrick and colleagues[12] are only few examples). The hindered diffusion of each molecule as a result of the interaction with its neighbors in the monolayer causes immobilization of the molecule even if it is not chemically bonded to the surface. A recent elegant demonstration to the importance of this immobilization in observing good contrast was the study on adsorbed benzene (in a submonolayer coverage) near a step.[13] In this study, it was possible to observe good contrast of the adsorbed molecules only near a step, where the molecular diffusion is hindered by the step. As the distance from the step increased, the resolution of the molecular image deteriorated.

In the attempt to develop new STM chemical applications, there have been many studies on adsorbed monolayers that tried to demonstrate the capability of the STM to distinguish between different functional groups in physisorbed molecules. It was shown that the STM is quite effective in resolving alkyl chains on the surface and that the aromatic ring appears much brighter than the aliphatic chain.[14,15] In contrast STM studies on substituted alkanes on graphite indicated that the OH, COOH, and Cl functional groups can not be distinguished from the methylene group,[16–18] but the S atoms were easily resolved in STM images of thiols and disulfide compounds. Amide and anilide groups were also easily distinguishable. [18–20]

The question of the precise mechanism of the STM operation on physisorbed molecules is still not resolved. The molecules are normally imaged at bias voltages far from the HOMO (highest occupied molecular orbital) and the LUMO (lowest unoccupied molecular orbital) so tunneling through these orbitals (resonant tunneling) is excluded. One suggestion was that the LDOS is affected by the presence of the molecules, and that interaction of the HOMO and the LUMO with the surface is sufficient to perturb the surface LDOS.[14,15] This is supported by the agreement between the observed images and the HOMO and LUMO also in physisorbed molecules that form a monolayer such as alkyl cyanobiphenyls (on graphite).[15] Another suggestion was that the molecule changes the surface local work function and that a molecule with larger polarizability has a greater affect on the work function.[21] However neither of these mechanisms can explain the high transmission probability found for monolayers made of relatively long molecules or other systems where the tip sample separation is larger than 1 nm.[22,23] In order to explain these observations, two channels for tunneling were proposed, through space and through bond tunneling.[24] In the through space channel, the electrons tunnel from the tip to the sample without interaction with the molecule. In the through bond channel, the molecular orbital couples the tip and the surface. All orbitals whose tail reaches the Fermi level can contribute to this coupling (nonresonant tunneling). This is similar to the contrast mechanism observed in Xe atoms on Ni(110):[25] Although the Xe atomic level is far above the Fermi level, still the tail of this level at the Fermi level is enough to increase the tunneling probability and to create the STM contrast. This is connected to the more general problem of understanding long distance electron transfer. It is known for a long time[26–29] that such electron transfer processes must be mediated by the covalent bonds and the noncovalent interactions between the donor (of the electron) and the acceptor. Taking into account the role of these mediations gives a tunneling matrix element which is indeed exponential, but decays much slower with distance, enabling electron transfer over distances of several nanometers. Long-range electron transfer is a fundamental part of many basic biochemical reactions such as photosynthesis.[30] In a molecule, electron-electron interactions can have a significant effect on the chemistry of this molecule. Such interactions exist in biradicals like benzyne and ethlyne, or in molecules that have two nitrogen lone pairs such as pyrazine. The through bond coupling has an affect on the measurable chemical parameters of the molecule, such as the basicity, solvation energies, UV absorption spectra, etc.[31]

The intramolecular electron transfer was studied experimentally using several different techniques[32–34] (these references are only few examples). It was found that in many cases, where a complicated molecule serves as a bridge between the donor and the acceptor, the attenuation of the tunneling transmission probability is proportional to the number of bonds rather than the distance between the donor and the acceptor. One experiment[33] showed that the transmission probability decays one order of magnitude per two C—C bonds separating the donor and the acceptor. This is much slower than the decay of the transmission probability of vacuum tunneling as a function of distance. A very interesting characteristic of long-range electron transfer is its unique dependence on the thermodynamic drive of the process. When the transfer rate was measured as a function of the difference in electron affinity between the donor and the acceptor, it was found that while initially the transfer rate increased with an increase of the difference in electron affinity, increasing this difference even more (namely increasing the thermodynamic driving force), will *decrease* the transfer rate. This inversion is in agreement with theoretical predictions.[26] In addition, the transmission probability strongly depends on the orientation of the intervening bonds.[35]

When a molecule is strongly bound to the surface the STM can distinguish between molecular features. However, normally, such a well-defined and strong surface-molecule bond does not exist. This is why STM imaging on *physisorbed* molecules attracted such attention. As mentioned above, the key for a successful observation of physisorbed molecules is the immobilization of the molecule. All reported STM works proposed to perform such immobilization by imaging a molecular monolayer. However, first of all, a molecule in a monolayer, due to the strong interaction with neighboring molecules, is probably affected by this interaction, and there is always the question as to what extent the electrical properties of such a molecule, as reflected in the STM image or any other measurement are really the same as those of an isolated single physisorbed molecule. A second problem is that the degree of immobilization provided, even by a molecular monolayer, is not enough to bring the resolution of the STM image on such systems to the resolution routinely available on clean semiconductor surfaces. This is a major motivation behind the study which is described in this paper.

We immobilized a chemical group on a surface where atomic resolution is achievable. The STM measurement is made on an isolated molecule, which parts of it are not chemically bound to the surface. Thus the measurement is performed on a physisorbed functional group, which is isolated. The possibility to perform STM measurements on immobilized physisorbed chemical group on an atomic resolution surface, we believe, can lead to an improved resolution and reliability in STM molecular imaging. Questions of interest to a broad scientific community, such as electron transfer through molecules and molecular conductance, can now be investigated quantitatively with the STM. The possible molecular electronics applications will be mentioned at the end of this paper. Another motivation of this study was to show how the STM can be used to solve questions regarding the surface chemistry of more complicated molecules. In our work we have demonstrated how steric factors and noncovalent electrostatic interaction with the surface affects the chemical reaction. Such questions are quite fundamental in surface chemistry. In heterogeneous catalysis, for example, the main role of the catalyst is to provide steric matching between the surface sites and the reacting molecules.

EXPERIMENT

We decided to investigate the surface chemistry of bifunctional molecules, in which the two reacting groups are separated by an aliphatic chain. The studies were performed on a Si(111)7 × 7 surface. The reacting group was a double bond. Ethylene adsorbs non-dissociatively at room temperature.[36] The binding sites are the dangling bonds. The 7 × 7 unit cell is made of several layers. The dangling bonds in the external layers are the adatoms. Two types of adatoms exist: center and corner adatoms. FIGURE 1 shows a unit cell of the Si(111)7 × 7 with the corner adatoms as large filled dark circles and the center adatoms as empty circles. The atoms in the second layer, the rest atoms appear as small filled dark black circles. When ethylene reacts with the 7 × 7 surface, the C=C double bond is opened to create a C—C single bond and two C—Si bonds are formed, one with an adatom and a second one with a neighboring rest atom. In the empty state images,[36] the surface sites reacted with ethylene appear as vacancies at the position of the reacted adatom. The distances between the different adatoms in the 7 × 7 unit cell are shown in FIGURE 1. We have taken[37,38] four different dienes with an aliphatic chain with lengths that enable the reaction between two double bonds at the edge of the molecule and the different neighboring adatom–rest atom pairs. By changing the length of the chains in the diene we can change the fitting between the two reacting bonds and the binding sites on the surface. This changes both the kinetics of the reaction and the degree of strain that exists at the different Si—C bonds. One hope was that this difference in strain will be reflected in the contrast in the STM images, but so far we were not able to find such changes.

The experiment was performed on a custom-made UHV-STM. The course approach of this STM is an inchworm stepper motor capable of delivering a 10 mm motion at a resolution of 2 nm. A piezoelectric tube is glued on the inner part of the

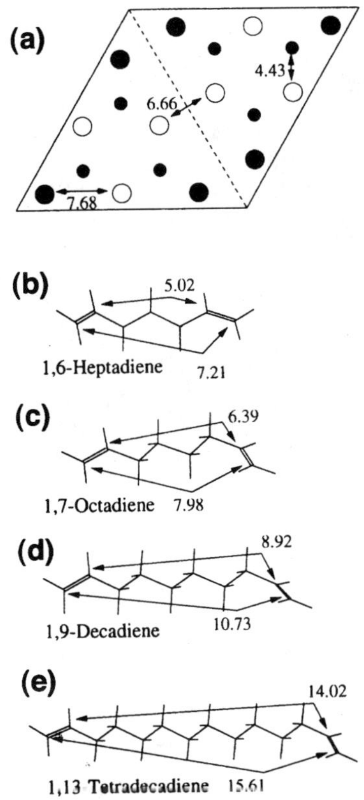

FIGURE 1. **(a)** The 7 × 7 unit cell with typical distances between nearest neighbors adatom pairs and between nearest neighbors adatom–restatom pairs. **(b)–(e)** The different diene molecules chemisorbed and imaged in this study, with the distances between the reacting groups.

inchworm. The vibrational isolation was based on a two-stage suspension spring with eddy current damping.[39] The system includes a custom-made electronics and a custom made STM-software. The experiment was performed under UHV conditions with a base pressure of 2×10^{-10} torr. The Si(111) wafer was outgassed and heated to remove the oxide layer and to induce the formation of the 7 × 7 reconstruction. After an hour, for cooling down, the molecules were dosed through a leak valve from a gas manifold. The images were taken at sample bias of +2V and tunneling current of 1 nA.

FIGURE 2 shows an STM image of a Si(111)7 × 7 exposed to 1,7-octadiene (molecule **c** in FIG. 1). In this molecule the distance between the double bonds is approximately the distance between neighboring adatoms. Indeed the molecules appear as dark pairs. Three types of pairs were identified in the image: a center-center pair, a center-corner pair, and a bridging pair (between two center adatoms in the two halves of the unit cell). The ratio between the number of center-center pairs and of center-corner pairs was 3:1[37] while the number of bridging pairs was much smaller. By taking a diene molecule shorter by one carbon atom (1,6-heptadiene **b** in FIG. 1) the ratio between the number of center-center and center-corner pairs is increased to 7:1. Also in this case, the number of bridging pairs was much smaller. Clearly, the

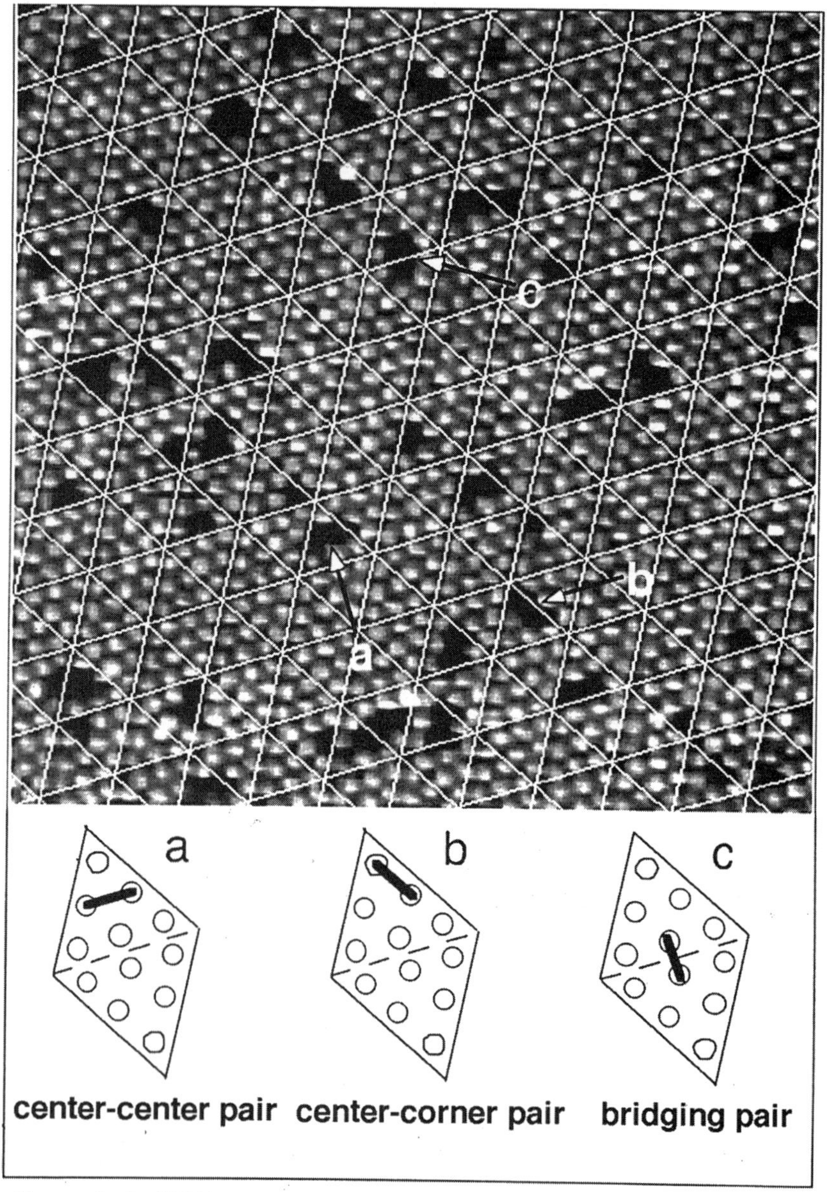

FIGURE 2. An STM image of Si(111)7 × 7 + 1,7-octadiene. The different possible pairs of reacted adatoms are marked and shown schematically below.

only factor that changes the chemistry of the molecule is the different steric matching due to a slightly different chain length.

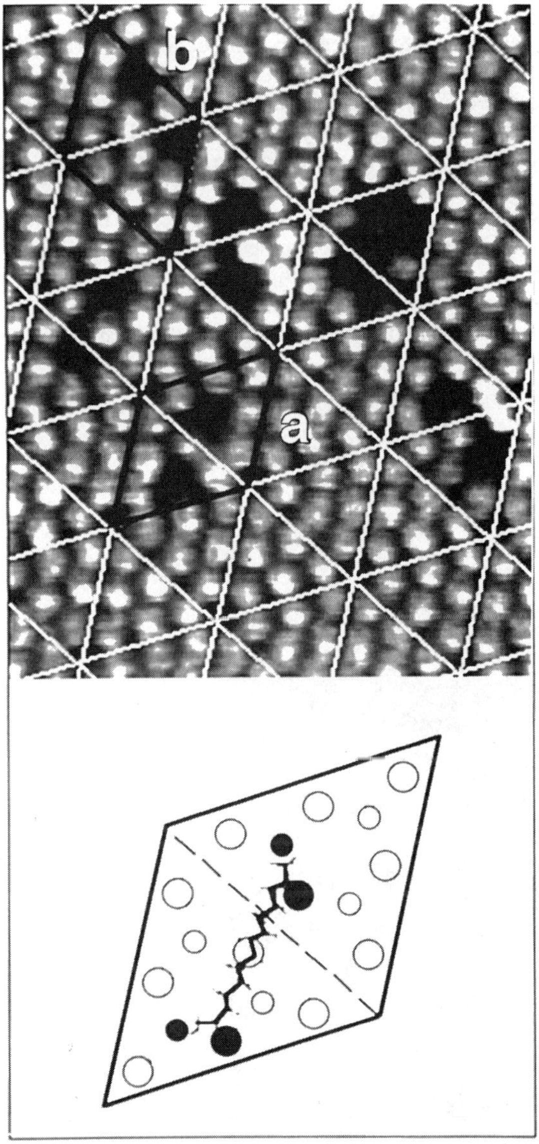

FIGURE 3. **(a)** An STM image of Si(111)7 × 7 + 1,13-tetradecadiene. **(b)** Schematic diagram representing a possible binding configuration of the molecule on the surface.

In the 1,7-octadiene and 1,6-heptadiene molecules, the chemical reaction is largely restricted to within one half of the unit cell (very few bridging pairs). Increasing the length of the aliphatic chain (1,9-decadiene) gives a much larger number of bridging pairs.[38] Different types of such pairs were identified.

A longer molecule, 1,13-tetradecadiene (molecule **e** in FIG. 1) was also adsorbed on a Si(111)7 × 7 surface.[37,40,41] Some of the images reveal pairs of reacted adatoms separated by unreacted ones (FIGS. 3 and 4), while in others a white protrusion appears between the two dark (reacted) adatoms (FIG. 5). The size of the long molecule causes a reaction with one end of the molecule situated in the faulted half, and the other end in the unfaulted half of the 7 × 7 unit cell.

Each half contains six adatoms, and the molecule can be bound in several ways. Looking at a number of such adsorbed molecules, it was possible to distinguish between two types of molecule-surface complexes: one (like molecules **a** and **b** in FIG. 4) where most of the aliphatic chain is positioned above the unfaulted half, and the second type where it is located above the faulted half (such as molecule **c** in FIG. 4).

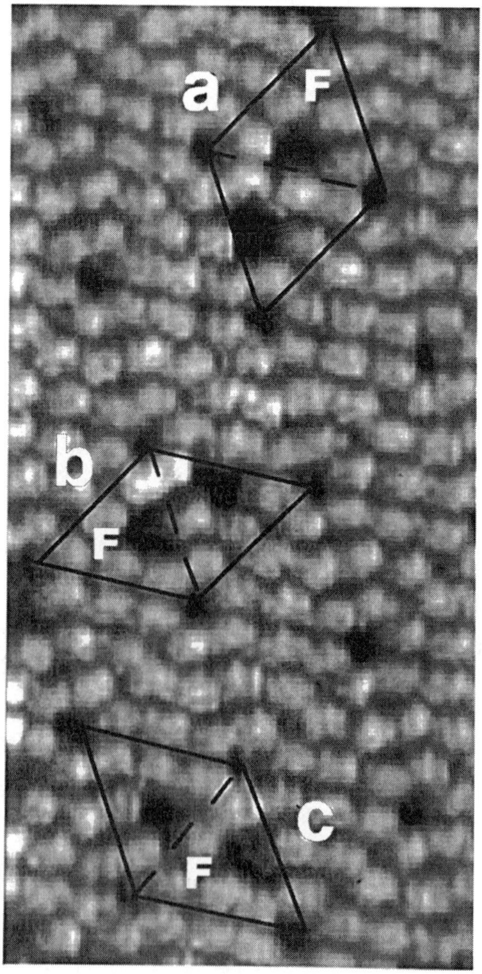

FIGURE 4. Three molecules of the 1,13-tetradecadiene on Si(1 11)7 × 7. The faulted and unfaulted halves of the unit cell are marked on the image.

Looking at the positions of the aliphatic chain in 172 adsorbed molecules, reveals a preference of 2:1 for the chain to reside over the unfaulted half. In both cases one Si—C—C—Si complex is formed in the faulted half of the unit cell, while the second Si—C—C—Si is formed in the unfaulted half. In addition, counting all the possible ways in which the molecule can be bound to the Si(111)7 × 7 unit cell gives equal number of configurations in which the aliphatic chain resides either above the faulted or the unfaulted half.

It was thus clear that there is a certain noncovalent interaction between the aliphatic chain and the 7 × 7 unit cell which is different in the two halves of the cell. We have made a calculation based on a molecular modeling program[42] to evaluate the energy of an aliphatic chain located 2.5 Å above either of the two halves of the 7 × 7 unit cell. The calculation was simplified by the possibility of representing each half by partial charges at the positions of the dangling bond as was shown by Brommer and colleagues.[43] We found that the difference in the effective electrostatic interaction between the chain and either of the two halves was 30 kcal/mol.

The electrostatic-noncovalent interaction with the surface leads to the preference of the molecule to adsorb on the Si(111)7 × 7 unit cell such that a larger fraction of the chain is physisorbed above the unfaulted half.[41]

The white protrusion that appears between the dark reacted adatoms (FIG. 5) in many of the images[40] can only be interpreted as an additional channel for tunneling, due to the presence of the aliphatic chain–through bond tunneling. The height of the protrusion is 0.4–0.7 Å. On Si(111)7 × 7 at bias voltage of 2 V and current of 1 nA, the transmission ($\exp[-2\kappa d]$) is 1.2×10^{-5},[44] which corresponds to a resistance of the order of $10^9\,\Omega$. The size of the protrusion measured in the experiment corresponds to an increase of the transmission of $1.4 \times 10^{-5} - 3.2 \times 10^{-5}$.[40] From the theoretical point of view, through bond tunneling is treated as a tunneling problem with a potential barrier with one or more potential wells. However this observation may also be linked to the question of the conductance of a single molecule. We recall that there have recently been several studies that investigated both theoretically and experimentally the question of the conductance of molecules.[45–48] In one of these works, the conductance of a single C_{60} molecule adsorbed onto a gold surface was measured. In this work the tip was accurately positioned such that an electrical contact was made. The measured resistance was 55 MΩ. Within certain limits of the tip sample bias (−0.2 to 0.2 V) the STM tip-C_{60}-molecule-surface tunnel junction obeys Ohm's law.[46] A theoretical work,[48] which calculated the conductance of a single alkane molecule connecting two gold electrodes, came out with a value of transmission of 10^{-12} for an alkane with 18 carbon atoms, and of 10^{-8} for an alkane of 10 carbon atoms. The abovementioned experimental observation[33] on through bond electron transfer verified that the transmission probability decreases one order of magnitude for each two C—C bonds. This is in agreement with the theoretical results. Although the calculation for the electron transmission through one methylene group was not performed, an extrapolation of the results of this calculation should give a transmission of the order of 10^{-3}. This, together with the measured resistance of the C_{60} molecule, leads to the conclusion that the largest factor in the resistance of the tip-chain-surface junction is actually at the tip-chain and the chain-surface interfaces, and not the resistance of the chain itself. By decreasing the tip molecule distance it should be possible to reach the limit where the current is determined only by the molecular conductance in a similar way as the conductance of the C_{60} molecules was measured by point contact.

As can be seen from the figures, not all the molecules give the white protrusion which is a result of through bond tunneling. (In FIGURE 5, molecule **d** does not have such a protrusion). The spatial distribution of the molecules with and without a protrusion is random (namely, tip switching effects are excluded). The appearance and disappearance of the protrusion is due to some local characteristics of the surface imaged. A possible explanation is that since the transmission probability in through bond tunneling is strongly dependent on the orientation of the mediating group and since the limited freedom of the chain creates small fluctuations in the orientation of the aliphatic chain relative to the tunneling electrons, this may lead to a random variation of the transmission probability and to the appearance and disappearance of the white protrusion in a random sequence.

It should be emphasized that the long-range decay of the transmission probability of through bond tunneling (as compared with vacuum tunneling) put a principle significant limitation on the resolution which can be achieved in STM molecular imaging when performed in the constant current mode. The slower the decay, the lesser will be the resolution. For this reason, better results might be observed many times when the STM is operated at the constant height mode when the contrast is based on the difference in conductance of the chemical groups (for example, an aromatic group compared with an alkyl group).

POSSIBLE APPLICATIONS FOR MOLECULAR ELECTRONICS

The capability to immobilize a certain well-defined group (unreacted) on the surface might open new possibilities for the field of molecular electronics.[49,50] Any real molecular device of the future will most probably be made from a "smart" molecule

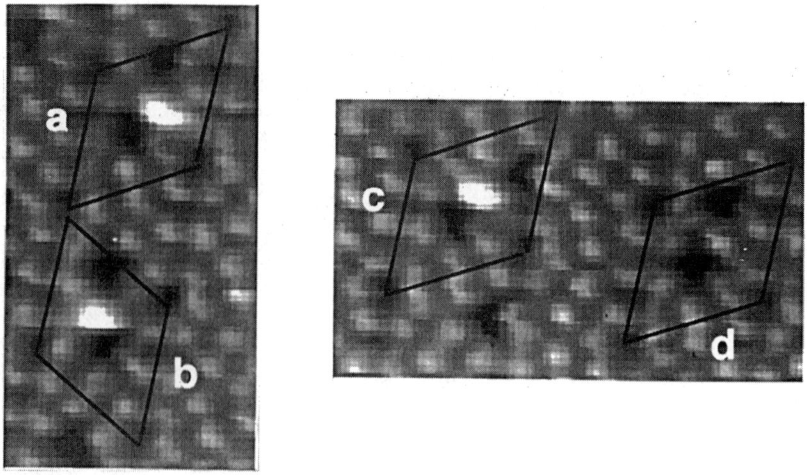

FIGURE 5. Si(111)7 × 7 + 1,13-tetradecadiene.

on a surface when the flow of information to and from the molecule will be controlled by an STM system. Molecules may serve as wires, switches, rectifiers, and memory elements for data storage. However, in order to create a functioning device, the different groups were brought together in a supermolecular assembly. This approach, however, has serious control problems. A possible different approach is the incorporation of the different functions on a single molecule. (The more functions, the better). These may include functions for surface binding, molecular transmission, rectification, etc. For this approach, the capability to immobilize different unreacted functional groups on an atomic resolution surface has several serious advantages. The assignment of the exact position of the chemical groups is straightforward. The tip could be used to study and exploit their electronic properties separately.

We foresee the future steps in this direction as follows: (1) It will be possible to verify whether the contrast of different chemical groups observed in STM molecular imaging of molecular monolayers is indeed correct also in the same chemical groups when they are physisorbed on the surface in isolated immobilized molecules. (2) We shall investigate incoherent electron transfer through different chemical groups and the dependence on different parameters such as the distance from or the size and the electronegativity of the electron acceptor which is immobilized on the chain. (3) We hope that it will be possible to measure the electrical resistance of different functional groups. This observation could be reliable, since the molecule is both physisorbed and immobilized on the surface. By arriving to point contact between the tip and the molecule, and between the molecule and the surface, the resistance can be measured and may be a sensitive finger print to the identity of the group. By immobilizing electron acceptors and molecular switches[51] on the chain it may be possible to investigate their electrical properties.

ACKNOWLEDGMENTS

This work was supported by the Minerva Foundation, Munich, Germany and the Basic Research Foundation administered by the Israeli Academy of Sciences and Humanities. One of us (Y.M.) is an incumbent of the Lilian and George Lyttle Career Development Chair.

REFERENCES

1. Avouris, P. & R. Wolkow. 1988. Phys. Rev. Lett. **60:** 1049.
2. Avouris, P., I.W. Lio & F. J. Bozso. 1991. J. Vac. Sci. Technol. B **9:** 424.
3. Pelz , J.P. & R.H. Koch. 1991. J. Vac Sci. Technol. B **9:** 774.
4. Stroscio, J.A., R.M. Feenstra & A.P. Fein. 1987. Phys. Rev. Lett. **58:** 1668.
5. Wand, T.A., T. Michely, R.J. Behm, C. Hemminger & G. Comsa. 1991. Appl. Phys. A **53:** 414.
6. Hallmark,V.M., S. Chiang, J.K. Brown & Ch. Wöll. 1991. Phys. Rev. Lett. **66:** 48.
7. Hallmark, V.M., S. Chiang, K.P. Meinhart, & K. Hafner. 1993. Phys. Rev. Lett. **70:** 3740.
8. Lippel, P.H., R.J. Wilson, M.D. Miller, Ch. Wöll & S. Chiang. 1989. Phys. Rev. Lett. **62:** 171.

9. OHTANI, H., R.J. WILSON, S. CHIANG & C.M. MATE. 1988. Phys. Rev. Lett. **60:** 2398.
10. DELAMARCHE, E. & B. MICHEL. 1996. Thin Solid Films **273:** 54.
11. WALBA, D.M., F. STEVEN, D. C. PARKS & M. D. WAND. 1995. Science **267:** 1144.
12. PATRICK, D.L., V.J. CEE & T.P. BEEBE, JR. 1994. Science **265:** 231.
13. STRANICK, S.J., M.N. KAMNA & P.S. WEISS. 1994. Science **266:** 99.
14. NEJOH, H. 1990. Appl. Phys. Lett. **57:** 2907.
15. SMITH, D.P.E., J.K.H. HORBER, G. BINNIG & H. NEJOH. 1990. Nature **344:** 41.
16. MCGONICAL, G.C., R.H. BERNHARD, Y.H. YEO & D.J.J. THOMSON. 1991. J. Vac. Sci. Technol. B **9:** 1107.
17. RABE , J.P. & S. BUCHHOLZ. 1991. Science **257:** 424.
18. VENKATARAMAN, B., G.W. FLYNN, J.L. WILBUR, J.P. FOLKERS & G.M. WHITESIDES. 1995. J. Phys. Chem. **99:** 8684.
19. RABE, J.P., S. BUCHHOLZ & L. ASKASKAYA. 1993. Synth. Met. **54:** 339.
20. TAKEUCHI, H., S. KAWAUCHI & A. AKAI. 1996. Jpn. J. Appl. Phys. **35:** 3754.
21. SPONG, J.K., H.A. MIZES, L.A. LACOMB, JR., M.M. DOVEK, J.E. FROMMER & J.S. FOSTER. 1989. Nature **338:** 137.
22. HORBER, J.K.H., C.A. LANG, T.W. HENSCH, W.M. HECKL & H. MOHWALD. 1988. Chem. Phys. Lett. **145:** 151.
23. MICHEL, B., G. TRAVAGLINI, H. ROHRER, C. JOACHIM & M. AMEREIN. 1989. Z Phys. B **76:** 99.
24. JOACHIM, CH. & P. SAUTET. 1990. *In* Scanning Tunneling Microscopy and Related Methods, R.J. Behm, N. Garcia & H. Rohrer, Eds. Kluwer. Dordrecht.
25. EIGLER, D.M., P.S. WEISS, E.E. SCHWEIZER & N.D. LANG. 1991. Phys. Rev. Lett. **66:** 1189.
26. MARCUS, R.A. 1956. J. Chem. Phys. **24:** 966.
27. HOPFIELD, J.J. 1974. Proc. Natl. Acad. Sci. USA **71:** 3640.
28. BERATAN, D.N., J.N. ONUCHIC & J.J. HOPFIELD. 1985. J. Chem. Phys. **83:** 5325.
29. BERATAN, D.N., J.N. ONUCHIC & J.J. HOPFIELD. 1987. J. Chem. Phys. **86:** 4488.
30. DIESENHOFFER, J., 0. EPP, K. MIKI, R. HUBER & H. MICHL. 1985. Nature **318:** 618.
31. HOFFMANN, R. 1973. Acc. Chem. Res. **4:** 1.
32. LELAND, B.A., A.D. JORAN, P.M. FELKER, J.J. HOPFIELD, A.H. ZEWAIL & P.B. DERVAN. 1985. J. Phys. Chem. **89:** 5571.
33. CLOSS, G.L., L.T. CALCATERRA, N.J. GREEN, R.W. PENFIELD & J.R. MILLER. 1986. J. Phys. Chem. **90:** 3673.
34. OHTA, K., G.L. CLOSS, K. MOROKUMA & N. J. GREEN. 1986. J. Am. Chem. Soc. **108:** 1319.
35. PAULSON, B.P., L.A. CURTISS, G. BAL, G.L. CLOSS & J.R. MILLER. 1996. J. Am. Chem. Soc. **118:** 376.
36. PIANCASTELLI, M.N., N. MOTTA, A. SGARLATA, A. BALZAROTTI & M.DE. CRESCENZI. 1993. Phys. Rev. B **48:** 17892.
37. SHACHAL, D., Y. MANASSEN & E. TER-OVANESYAN. 1997. Phys. Rev. B **55:** 9367.
38. SHACHAL, D. & Y. MANASSEN. 1997. J. Vac. Sci. Technol. B **15:** 1310.
39. PARK, S.-I. & C.F. QUATE. 1987. Rev. Sci. Instrum. **58:** 2004.
40. SHACHAL, D. & Y. MANASSEN. 1997. Chem. Phys. Lett. **271:** 107.
41. SHACHAL, D. & Y. MANASSEN. 1998. Appl. Phys. A. **66:** S1229.
42. [Insight 11, Biosym Technologies, San Diego]
43. BROMMER, K.D., M. GALVAN, A. DAL PINO, JR. & J.P. JOANNOPOULUS. 1994. Surf. Sci. **314:** 57.
44. CHEN, C.J. & R.J. HAMERS. 1991. J. Vac. Sci. TechnoL B **9:** 503.
45. BUNN, L.A., J.J. ARNOLD, M.T. CYGAN, T.D. DUNBAR, T.P. BURGIN, L. GONES, D.L. ALLARA & P.S. WEISS. 1995. Science **271:** 1705.

46. JOACHIM, C., J.K. GIMZEWSKI, R. SCHLITTER & C. CHAVY. 1995. Phys. Rev. Lett. **74:** 2102.
47. JOACHIM, C. & J.K. GIMZEWSKI. 1995. Europhys. Lett. **30:** 409.
48. JOACHIM, C. & J.F. VINUESA. 1996. Europhys. Lett. **33:** 635.
49. AVIRAM, A. & M.A. RATNER. 1974. Chem. Phys. Lett. **29:** 277.
50. AVIRAM, A. 1992. Int. J. Ouan. Chem. **42:** 1615.
51. FARAZDEL, A., M. DUPUIS, E. CLEMENTI & A. AVIRAM. 1990. J. Am. Chem. Soc. **112:** 4206.

Nanoscopic Molecular Memories

K. MATSUSHIGE,[a] H. YAMADA, H. TADA, T. HORIUCHI, AND X.Q. CHEN

Department of Electronic Science and Technology, Graduate School of Engineering, Kyoto University, Kyoto 606-8501, Japan

ABSTRACT: Scanning probe microscopy (SPM) was utilized to form novel nanometer-scale memory units in organic molecular layers. One method is based on the conductivity changes induced in evaporated *n*-paraffin films by imposing electric pulses. The scanning tunneling microscope (STM) image revealed clearly the formation of high-conductive area, the spot size of which is as small as 1 nm, suggesting the creation of the extremely high-density molecular memory over 10^{14} bit/cm^2. Moreover, the analysis with the aid of a computer simulation demonstrated that the extraction of H atoms for *n*-paraffin molecules by applying electric pulses results in the changes in the electric states as well as geometric conformation. Next, ferroelectric organic materials were utilized to construct another type of reversible molecular memory. Here, the nanometer-scale polarization switching in ferroelectric (VDF/TrFE) films was achieved with the SPM technique by applying a positive or a negative electric voltage to the film spin-coated on Pt substrate, and actually, several local domains with different polarities could be formed even within a small crystalite of about 200 nm in length. Finally, the possibility of creating the novel molecular memories of the next generation is discussed.

INTRODUCTION

The present industrialized world owes its significant development to the great advancement of Si technology or LSIs. Since the invention of a transitor, the electric computer has advanced its performance by reducing the sizes of the operaton unit, as shown schematically in FIGURE 1. For the past thirty years, the sizes have shrunk year by year (miniaturization) and so the integration has increased at the rate of about 4 times/3 years (Moore's law). However, the further integration based on the present miniaturization principle may no longer be possible because of several reasons including the limitations of fabrication techniques and quantum mechanics problems. Therefore, some revolutionary techniques should be developed in the near future to overcome this situation. Several new ideas have been proposed and some attempts have been made already.

The combination of molecular electronics and recent nanotechnology, called here *molecular nanoelectronics* (MNE), may be one of the possible ways of developing the electric devices of the next generation, as seen in FIGURE 1 and described below. In the 1980s, the Carter's proposal[1] on molecular electric devices (MED) inspired the possibility of building organic electric devices on a nanoscopic level and with novel functions, replacing conventional silicon-based inorganic devices. Since then,

[a]Address for correspondence: Professor Kazumi Matsushige, Director of Kyoto-University Venture Business Laboratory, Graduate School of Engineering, Kyoto University, Yoshida-honmachi, Sakyo-ku, Kyoto 606-8501, Japan. Phone, +81-75-753-5308; fax, +81-75-753-7570; e-mail, matusige@kuee.kyoto-u.ac.jp.

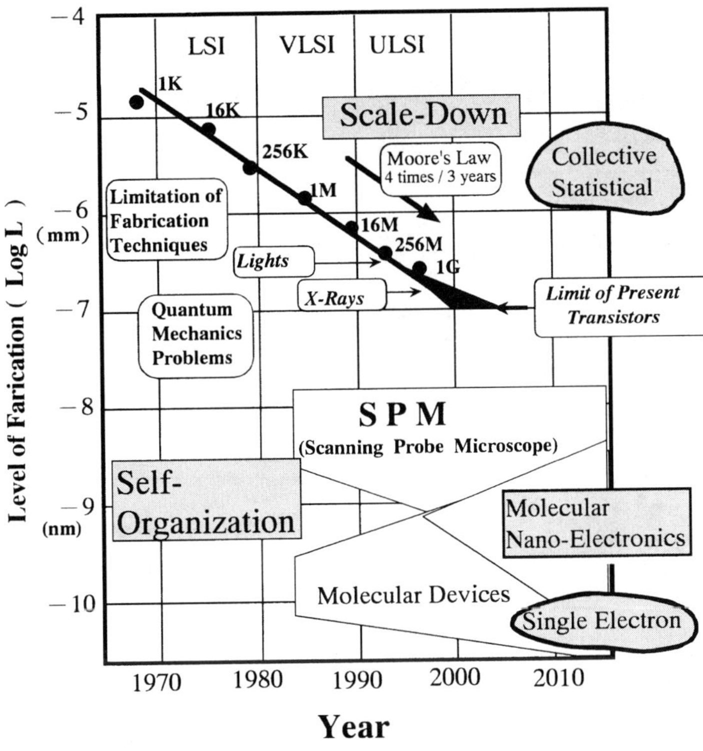

FIGURE 1. Advancement of LSI and the prospect of molecular nanoelectronics.

various ideas on molecular as well as electric circuit designs have been proposed and considered. However, the issues on how to produce such nanoscale devices and how to recognize the electric states on a single or a small group of molecules have remained as practical difficulties. Meantime, the STM developed by G. Binning and H. Rohrer[2] was revealed to have the capability of visualizing atoms and molecules. In addition, various nanoscopic tools such as the atomic-force microscope (AFM) were developed and it was proven that the SPM technique can be utilized as nanoscopic tools for individual manipulation of atoms and molecules.[3] The combination of these advancements in MED as well as nanotechnology propels the realization of molecular electric devices in nanoscale (MNE).

Thus, MNE has the strong possibility of inducing the paradigm shift from the present scale-down principle to a bottom-up one, including the drastic changes in the treatment of electrons involving the device operation process from collective static treatment to a single electron phenomenon. In this paper, as one of the simplest and the most promising cases in MNE devices, we demonstrate the capability of writing

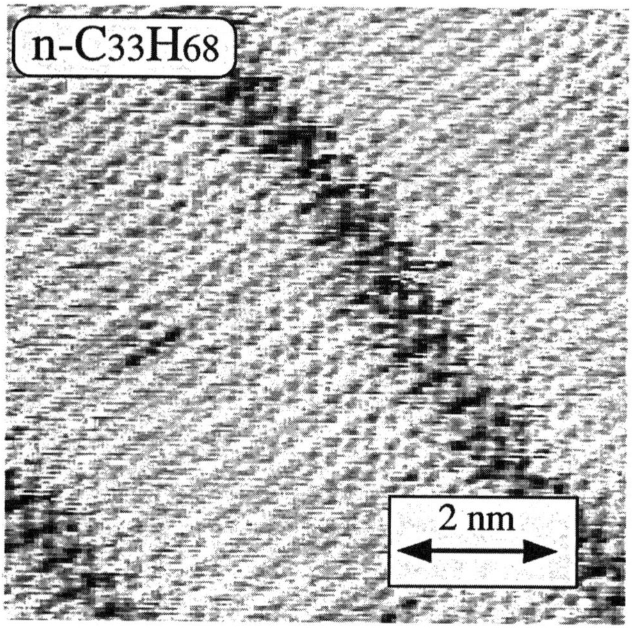

FIGURE 2. STM image of *n*-paraffin molecule (n-$C_{33}H_{68}$) on HOPG substrate.

information into molecular layers in nanoscopic scale and reading them by using the SPM technique. The further possibility of constructing such novel molecular memories is also discussed.

MOLECULAR MEMORY BASED ON INSULATOR-CONDUCTOR TRANSFORMATION

Reorganization of Molecules by Electric Pulses

As already reported, the STM can image organic molecules directly with an atomic resolution for some simple molecules such as *n*-alkanes[4] and liquid crystals.[5–7] FIGURE 2 shows the STM image of *n*-paraffin molecule, *n*-tritriacontane (n-$C_{33}H_{68}$), prepared by a vacuum evaporation method on the HOPG (highly oriented pyrolytic graphite) substrate, which had been cleaved and baked at 150° for 5 hr to get smooth and clean surfaces.[8] It is obvious that the molecules arrange quite regularly, probably in a mono layer, forming two-dimensional crystal different from the usual three-dimensional ones. Also, the molecular arrangement and crystal domains are revealed to be affected by the substrate crystal of HOPG, namely by an epitaxial effect.

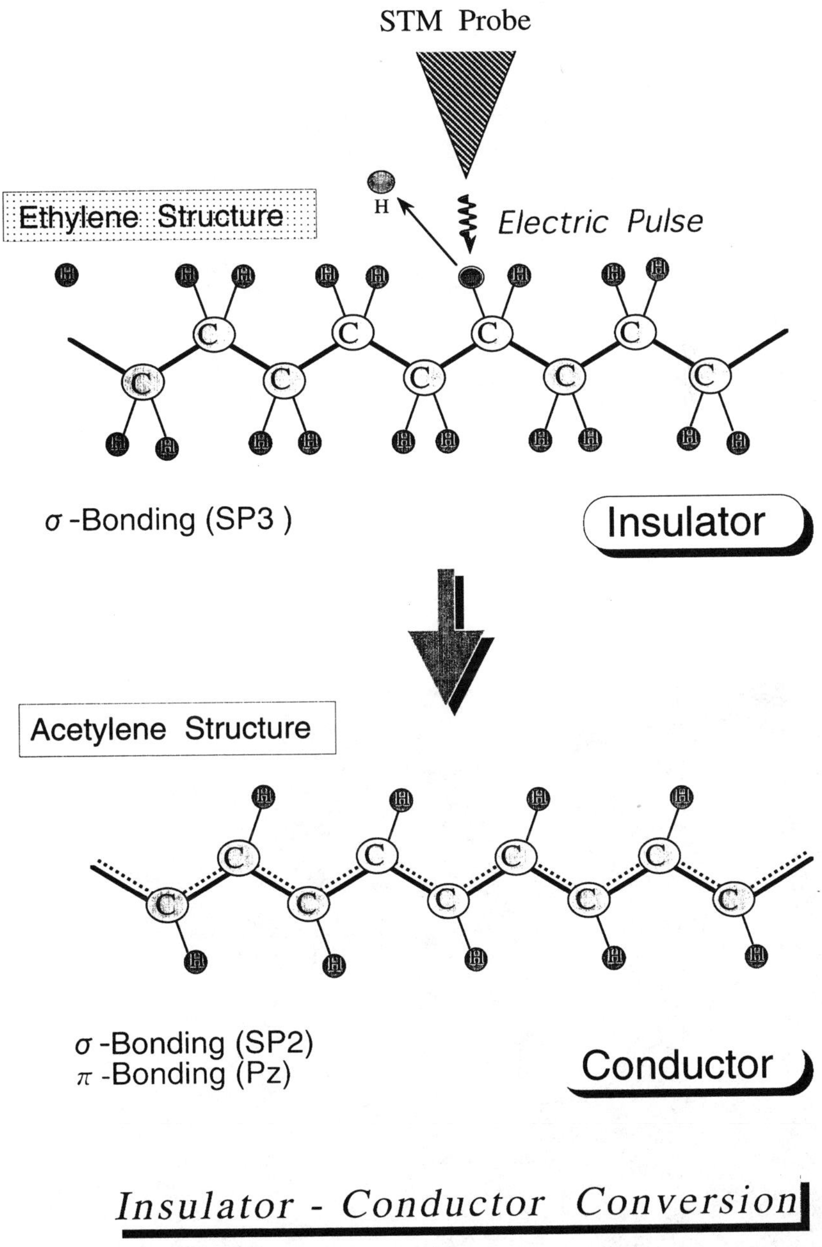

FIGURE 3. Schematic illustration for the removal of hydrogen atoms from ethylene molecules, inducing the insulator-to-conductor transformation, by the application of electric pulses.

The fact that the organic molecules can be imaged with the high-resolution of atomic level gives us expectation of the capability of modyfying the molecular structure using the STM tip. One of the most significant achievements by applying electric field will be the electric transformation from insulator to conductor by simple molecular reconstruction, as schematically shown in FIGURE 3. When the hydrogen atoms in *n*-paraffin with ethylene structure are removed by the action of electric pulses imposed on the STM tip, the structure changes to an acetylene structure. While the ethylene structure composed of σ-bonding (sp^3 orbital) with localized electrons is an insulator, the acetylene structure includes the π-bonding (p_z orbital) electrons with nonlocalized characteristics. Therefore, the removal of two hydrogen atoms from each ethylene unit results in the induction of the insulator-to-conductor transformation.[9]

The possibility of such molecular reconstruction was examined by applying electric pulses with various energies to evaporated *n*-paraffin films. The surface features of the samples that received such electric pulses were examined with STM to detect the induced changes. When the film was monolayer, any change was admitted for the pulses with relatively small amplitudes, while the molecular arrangement in the two-dimensional paraffin crystals was destroyed for the pulses with high amplitude. This is probably due to a situation in which the interaction between the molecules is too strong and larger than that between the molecules and the substrate graphite crystals.

On the other hand, the sample was much thicker, namely, double or multi-layer, and apparent changes were observed for the STM images, as shown in FIGURE 4. In

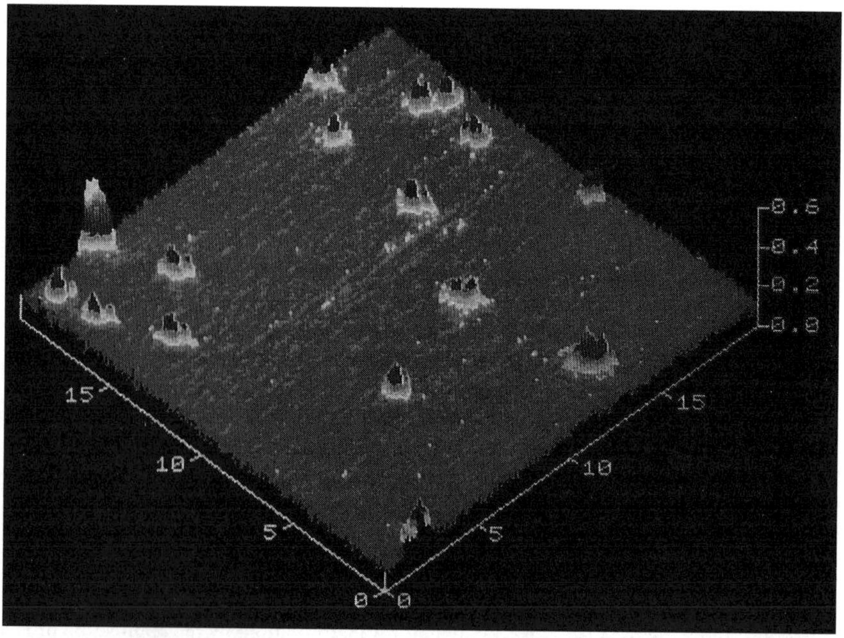

FIGURE 4. STM image of evaporated paraffin film observed after the application of electric pulses, showing the formation of highly conductive spots.

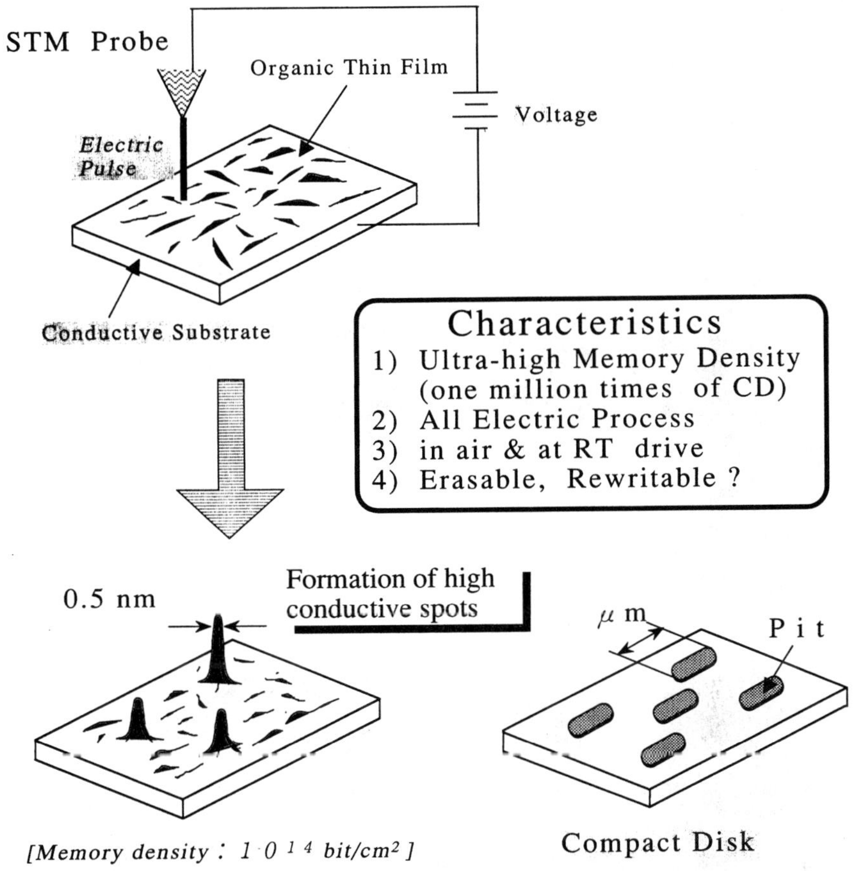

FIGURE 5. The formation of ultra-high density molecular memory with comparison to a compact disk.

this case, the electric pulse of 0.3 μm in width and 2.0 V was superimposed on the STM probe with the bias voltage of 0.5 V. In this image, the high-conductive area was restricted at the places where the pulses had been applied.

FIGURE 5 shows the schematic view of molecular memory thus produced, with the comparison to the present commercial CD (compact disk). Since the created memory units were extremely small, with the dimension of about one millionth in area, this molecular memory can be regarded as the one-million-times denser one. Moreover, this memory has several advantages such as all electric operation in writing and reading. Although there are various difficulties and the methods to be developed, the present primitive experiments promise us the nanoscopic and high-density molecular memories created by these principles demonstrated here.

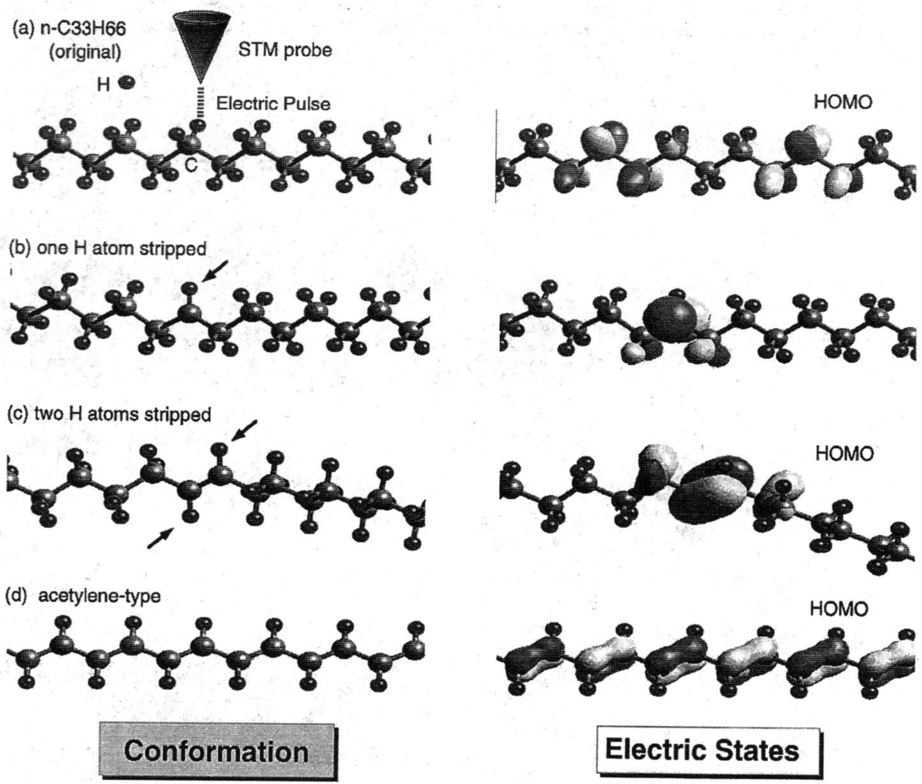

FIGURE 6. Geometrical conformation and electric states of original n-$C_{33}H_{68}$ molecule, the molecule of one H atom stripped, that of two H atoms stripped by application of electric pulses, and acetylene-type molecules reduced by a computer simulation.

Computer Calculation and Analysis

In order to foresee how the reconstruction of molecular structures discussed above induces the changes in electronic states as well as the conformational structures in *n*-paraffin molecules, a computer calculation and its analysis were conducted by using a workstation (Silicon Graphics, O2 R10000) with a commercially available program (Molecular Simulation, Cerius-2). FIGURE 6 shows the idea of the structural modification of *n*-paraffin with ethylene structure by electric pulses applied with an STM probe. The molecular structures were optimized by molecular mechanics calculation and semi-emperical molecular orbital calculation PM3/MOPAC. Amsterdam Density Functional (ADF) program based on density funtional theory was also used to evaluate the molecular orbitals for greater precision.

FIGURE 6(a) shows the molecular structure of original *n*-tritriacontane molecules. All carbon atoms connect covalently to the neighbors. The energy gap between highest occupied molecular orbital (HOMO) and lowest occupied MO (LUMO) was calculated to be 7.0 eV. FIGURE 6(b) shows the molecular structure optimized by PM3 calculation after being stripped of one hydrogen atom from the molecule. The molecule bends slightly at the carbon atom which lost the hydrogen. A new molecular orbital resulting from the lack of uniformity is generated between the HOMO and LUMO, as shown at the right side of FIGURE 6(b). Further stripping of the hydrogen atom from the neighboring carbon atom produces a π-bonding between two carbon atoms as shown in FIGURE 6(c). This new π-bonding produces HOMO and LUMO with a gap energy of 5.0 eV. Then, subsequent stripping of hydrogen atoms from each carbon atom produces a π-conjugated molecule, as shown in FIGURE 6(d). In this case, the HOMO-LUMO energy gap was 1.7 eV in the present calculation.

Molecular Quantum Well Structures

Furthermore, when the acetylene molecules are created regularly at the positions with a suitable separation by the method described above, we may be able to con-

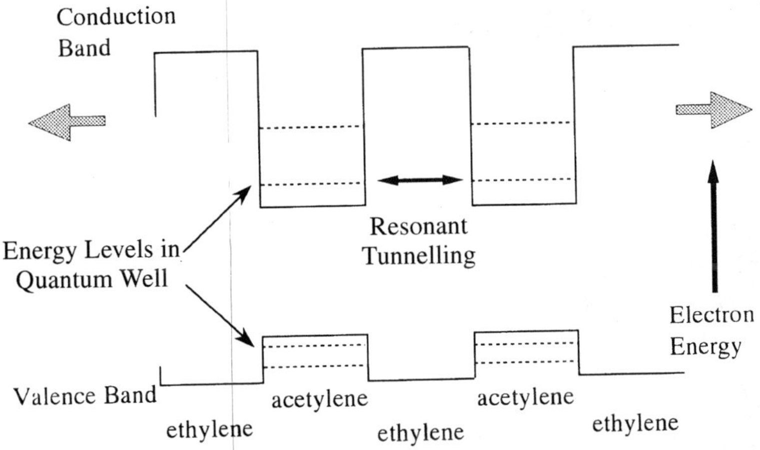

FIGURE 7. The energy band structure with quantum wells in the ethylene-acetylene super lattice.

struct ethylene-acetylene "super lattice" in the molecularly thin films.[9] Since the energy structures in these insulator and conductive molecules are quite different, the energy band structure with quantum well's may be formed, as shown in FIGURE 7. If the separation between the wells, that is, the width of the array of ethylene molecules, is narrow enough, the electrons in the wells can tunnel through the potential barrier upon the application of electric field and resonantly under some conditions. Moreover, the capacity of the conductive region may be quite small and so transportation of electrons is influenced greatly by the electronic state of the region, accompanying some single-electron tunneling (SET) phenomenon. Thus, the construction of the super lattice and the control of the lattice spacings provides characteristic and artificial energy band structures, by which notable electric functions such as molecular SET and various high quality devices can be created.

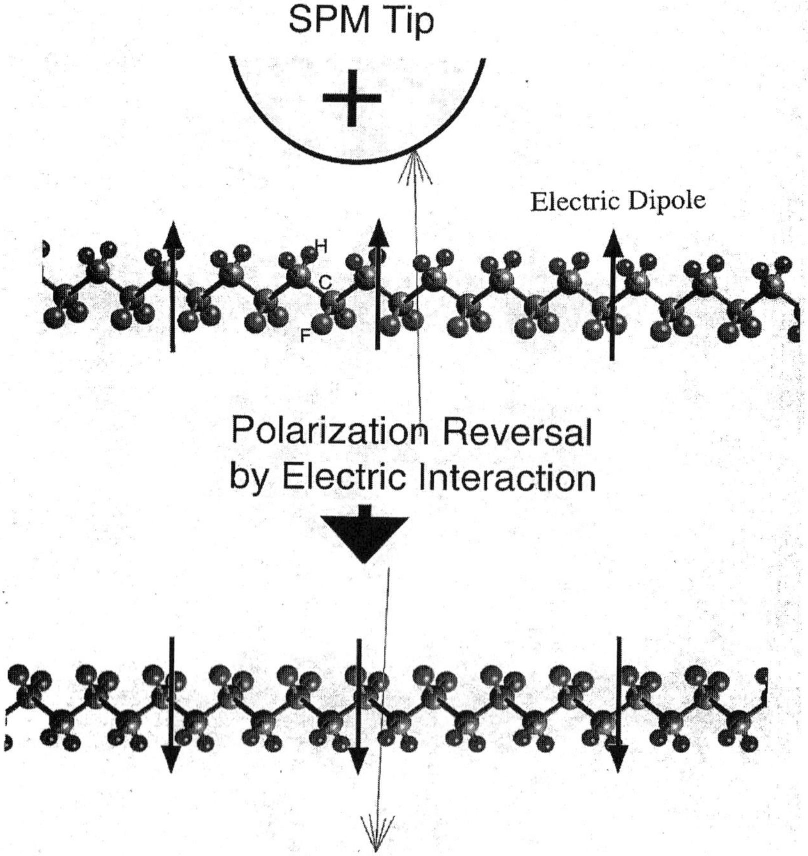

FIGURE 8. The structure of polyvinylidene fluoride and the electric dipoles. The lower molecule shows the rotated molecule upon the polarization reversal by electric interaction between the dipoles and the electric field from SPM tip.

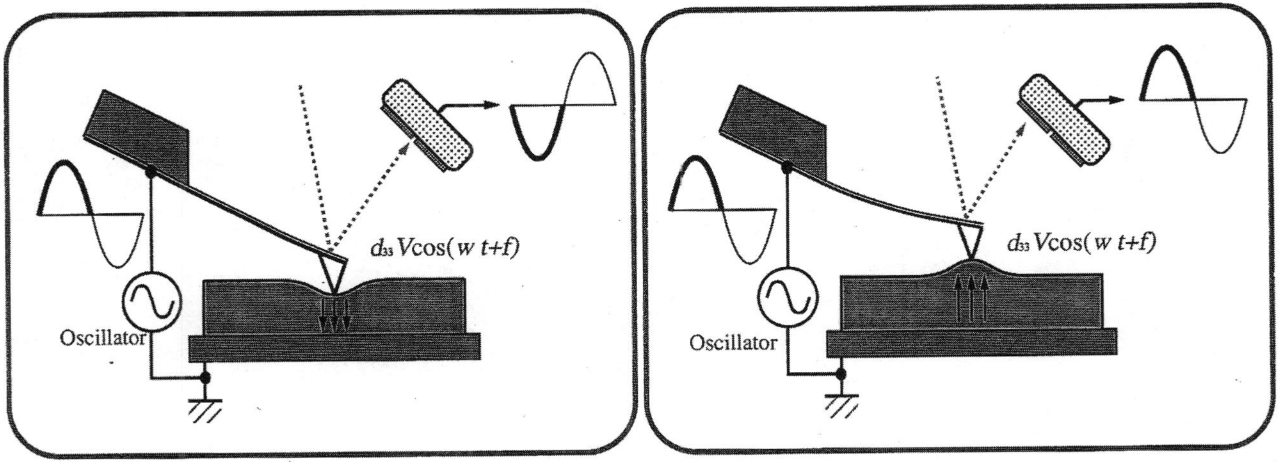

Positive Polarization *Negative Polarization*

FIGURE 9. The methods of detecting the locally formed domains with positive or negative polarization.

NANOSCALE CONTROL AND DETECTION OF ELECTRIC DIPOLES IN ORGANIC MOLECULES

Next, ferroelectric materials were utilized to construct another type of high-density molecular memory. As is well known, molecules contain lots of information in the geometrical conformation and electrical characteristics. Among them, electric dipoles may be one of the possible information units if the direction of the dipoles can be controlled in a nanometer scale. Here we utilized the ferroelectric nature to form nanometer scale memories by controlling the electric dipoles with SPM technique.

Experimental Procedure

The sample used in this experiment is copolymers of vinylidene fluoride and trifluoroethylene with the composition ratio of 73 to 27, which reveals most clearly the ferroelectric characteristics. As simply shown in FIGURE 8, polyvinylidene fluoride has electric dipoles perpendicular to the molecular axis originated from the large difference in the electron affinity of H and F atoms. When the SPM tip with opposite polarity is approached to the polar molecules, the electric force interaction between the tip and molecules will cause the polarization switching with molecular axis 180-degree rotation.

The examined sample was the copolymers deposited on Pt substrate by spin-coating method. The Pt substrate was sputtered on silicon dioxide. The sample was thermally annealed at 130°C for 1 hour after the deposition to increase its crystallinity.

The polarization alignment and switching were conducted by applying electric pulses to the film sample by using an Au-coated conductive AFM tip. The write and erase procedure was done by changing the polarity of the applied electric pulses. On the contrary, the ferroelectric domains were detected by utilizing piezo response, as schematically shown in FIGURE 9. When we locate the tip on a polarization domain by applying the oscillating weak electric field, piezoelectric vibration is caused and the tip in contact with the film is consequently vibrated, where the direction of the tip deflection depends on the nature of the polarization. The vibration is detected using the optical lever method and then the signal is demodulated by lock-in amplifier so that both in-phase and quadrature phase signals, that is, both amplitude and phase information can be obtained. In this set-up the oscillation frequency used must be located between the bandwidth of the feedback loop and the resonant frequency of the cantilever. When the tip is scanned, it follows the topography by feedback electronics but does not follow the oscillation. Then both topographic image and piezoelectric response can be simultaneously taken. From the detected phase information, we can determine the polarity directions in the domains. In this experiment, the gold-coated tips and cantilevers had a spring constant of 0.75 N/m and a resonant frequency of 88 kHz, and the modulation signal applied to the tip had a frequency of 70 kHz and a voltage of 0.5 Vrms.

Observation of the Polarized Domains

FIGURE 10 shows topographic and piezoelectric response images of the ferroelectric copolymer film deposited on Pt substrate, observed after application of the

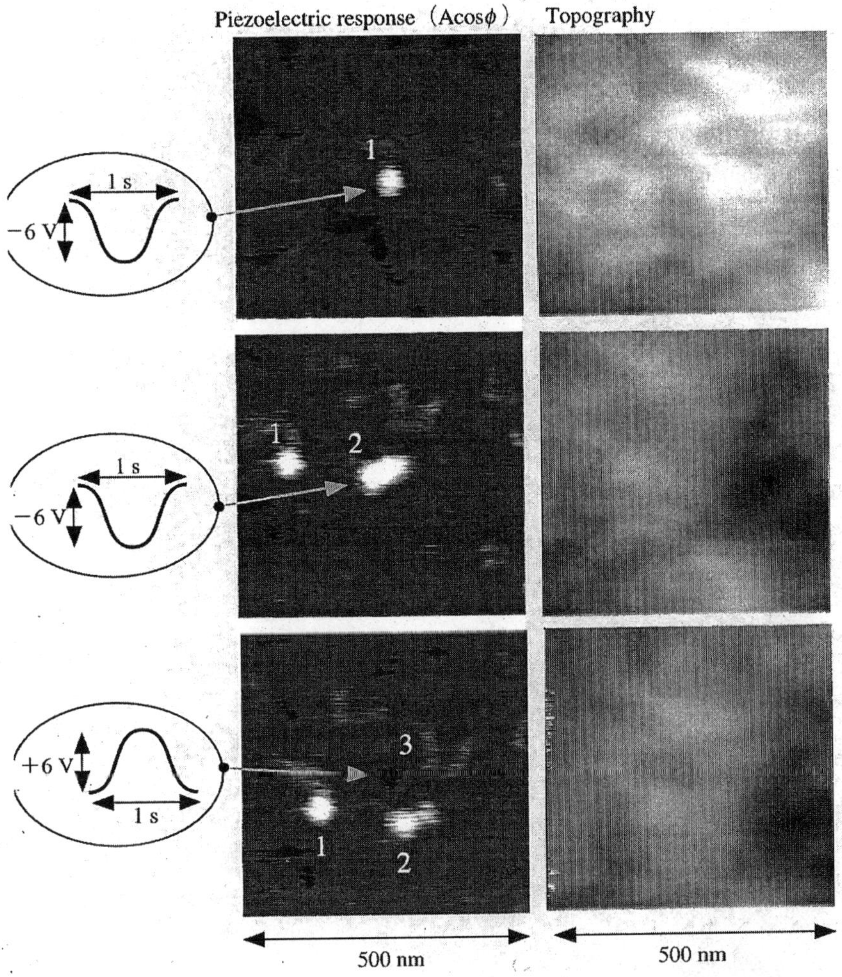

FIGURE 10. Topographic and piezoelectric response images of the samples on which electric pulses with opposite polarities were applied beforehand locally, showing the formation of nanoscopic polarized regions.

electric pulse of −6 V and +6 V with a duration of 1 second.[10] As shown clearly in these images, the bright or dark spots with high piezoelectric activities were created at the portion where the negative or positive electric pulse was imposed, while the corresponding topographic image taken simultaneously did not show any change in its feature. This fact thus suggests that the pulse application can actually form locally the ferroelectric domains as small as several tens nm in diameter. Here, it should be mentioned that the size of the domains is governed mainly by the degree of the concentration of electric field around the AFM tip, and therefore much smaller domains may be able to be formed if AFM tips with sharper points were employed. In a separate experiment, the temporal stability of the polarization domains formed by the application of electric pulses was examined, and it was proven that the ferroelectric local domains can exist keeping their piezoelectric activities for a suitably long period of time.[11] In addition, when the second electric pulse with the opposite polarity was applied at the once-formed domains, reverse piezoelectric responses were observed, thus suggesting the formation of the ferroelectric domains with an opposite polarization direction, namely, the reversibility. By utilizing these evidences and imposing electric pulses with positive and negative polarities during the scanning of

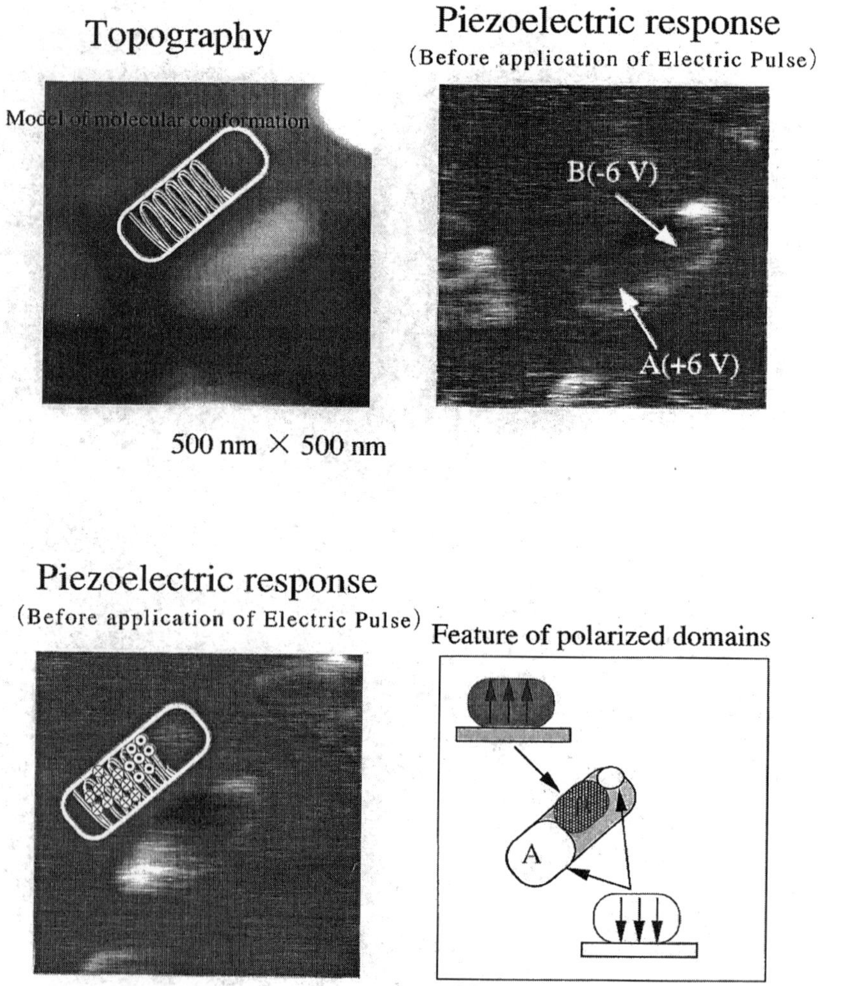

FIGURE 11. Topographic and piezoelectric response images for the crystallite samples on which different electric pulses were applied locally, showing the formation of several nanoscopic polarized regions.

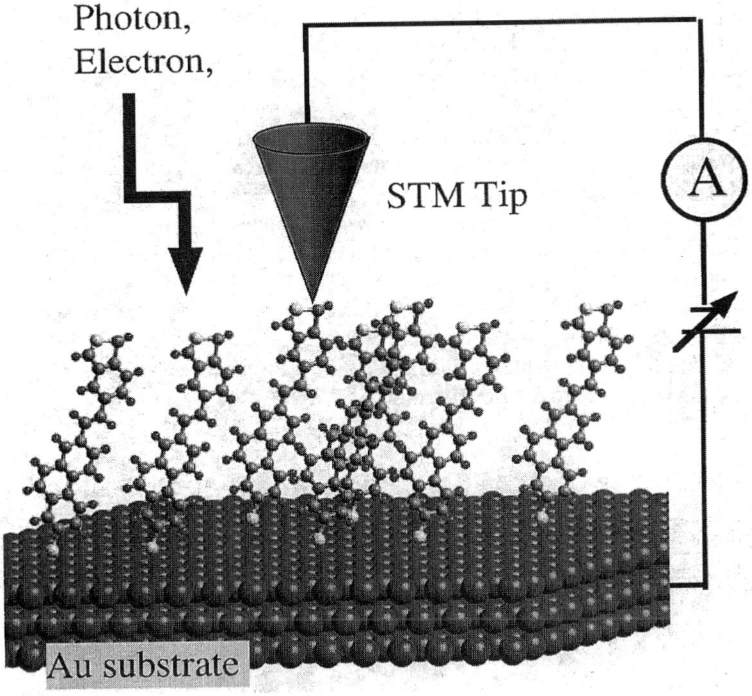

FIGURE 12. Schematic illustration of a novel molecular information system consisting of the combination of functional molecules and STM.

the AFM tip, it was revealed that one can write the binary information in series, suggesting that the novel molecular memory devices with nanometer-scale information units, and so extremely high storage density, can be created based on the write/read method for the molecules with ferroelectric nature.

FIGURE 11 demonstrated the formation of multi-domains in a crystallite. By annealing the copolymer sample, the crystallite with the dimension of about several hundred nanometers can be grown. Although the molecules are generally well organized in the crystallite, several domains can be formed within the crystallite by imposing electric pulses with opposite polarities. This suggests that the molecules within the crystallite have still rotational capability probably due to the existence of some defects or of folded parts in the lammelae crystals.

FUTURE MNE DEVICES

As is well known, molecules themselves contain much information in their chemical structures, geometric conformations, electric states, and so on. Recently, Andres *et al.*,[12] by conducting I/V measurement, observed at room temperature the Coulomb staircase in self-assembled molecular nanostructures composed of double-ended aryl dithiols and Au substrate. These findings stimulate one to create further novel molecular devices. If we can draw the information from each molecule itself, memory units of several bits can be stored in a single molecule. One of the possible ways to achieve this is schematically shown in FIGURE 12. Some functional molecules with a thiol unit at the chain end can arrange regularly and self-standingly on the Au (111) surface because of the chemical interaction between the Au atoms and thiol units, namely, self-arranged monolayer (SAM). Information, such as electric state and conformation, can be changed by inclusion of electrons and holes or photon energy and stored vertically in such a small area. By making spectroscopic I/V measurement for each molecule by using the STM tip, one can get various kinds of information stored within the molecules.

As discussed above, the combination of organic molecules and recently developed various SPM techniques will allow exploration of novel molecular memories and further contribute to the development of MNE of the next generation.

ACKNOWLEDGMENT

This work was partly supported by a grant-in-aid from the Ministry of Education, Science, Sports and Culture, Japan and by the Kyoto University-Venture Business Laboratory (KU-VBL) project. Also, we thank Dr. S. Taki of Kyushu University and Mr. H. Tanaka for their cooperation during this work.

REFERENCES

1. CARTER, F.L., ed. 1982. Molecular Electric Devices. Marcel Dekker, New York.
2. BINNIG, G. *et al.* 1982. Surface studies by scanning tunneling microscopy: Phys. Rev. Lett. **55:** 57–60.
3. EIGLER, D.M. *et al.* 1991. An atomic switch realized with the scanning tunneling microscope [letter]. Nature **352:** 600–603
4. MCGONIGAL, G.C. *et al.* 1991. 0bservation of highly ordered, twodimensional *n*-alkanes and *n*-alkanol structures on graphite: J. Vac. Sci. Technol. B **9:** 1107–1110.
5. FOSTER, J.S. & FROMMER. 1988. Imaging of liquid crystals using a tunneling microscope [letter]. Nature **333:** 542–545.
6. IWAKABE, Y. *et al.* 1991. Correlation between bulk orderings and anchoring structures of liquid crystals studied by scanning tunneling microscopy. Jpn. J. Appl. Phys. **30:** 2542–2546.
7. MATSUSHIGE, K. *et al.* 1993. Scanning tunneling microscope observation of a polar liquid crystal and its computer simulation: Jpn. J. Appl. Phys. **32:** 1716–1721.
8. TAKI, S. *et al.* 1993. Direct observation of the growth process of organic crystals by scanning tunneling microscopy. J. Crystal Growth **131:** 13–16.
9. MATSUSHIGE, K. & S. TAKI. 1994. The possibility of nanoscopic molecular archtecture with novel electric function: Mol. Cryst. Liq. Cryst. **247:** 203–213.
10. TANAKA, H. *et al.* 1997. Nanoscopic polarization domains in ferroelectric organic thin films: Appl. Phys. A. Submitted for publication.
11. MATSUSHIGE, K. *et al.* 1997. Nano-scale control and detection of electric dipoles in organic molecules: Proceedings of the Fifth Foresight Conference on Molecular Nanotechnology. In press.

12. ANDRES, R.P. et al.1996. Coulomb staircase at room temperature in a self-assembled molecular nanostructure. Science. **272:** 1323–1325.

Molecular Photodiodes with Langmuir-Blodgett Films Driven by Scanning Probe Microscopy

MASAMICHI FUJIHIRA[a]

Department of Biomolecular Engineering, Tokyo Institute of Technology, 4259 Nagatsuta, Mdori-ku, Yokohama 226-8501, Japan

ABSTRACT: Molecular photodiodes were fabricated with Langmuir-Blodgett (LB) films containing A-S-D triads for charge separation and H antenna molecules. Here, A, S, and D stand for electron acceptor, sensitizer, and electron donor moieties of the triad, and H for light harvesting moiety of the antenna, respectively. These functional amphiphiles mainly consisting of hydrocarbon (HC) can be phase-separated from fluorocarbon (FC) amphiphiles in mixed monolayers of HC and FC to form small domains. The phase-separated structure of the HC-FC mixed monolayers was studied with various scanning probe microscopes (SPM). SPM can be also used to detect the photo-induced charge separation in the molecular photodiodes.

INTRODUCTION

The highly oriented molecules in thin organic films such as Langmuir-Blodgett (LB) films and self-assembled monolayers (SAM)[1] are essential for some molecular functions. In the past 20 years, photo-electric conversion using such molecular assemblies and its application to molecular photodiodes[2–11] have been one of the main subjects of our laboratory. In the natural photosynthetic reaction center, solar energy is first captured by light harvesting antenna pigments, then excited energy harvested by the pigments is funneled to a special pair of the charge separation unit by energy migration and energy transfer, and finally an excited electron-hole pair in the special pair can be separated via electron and hole transfer by electron acceptors and donors located appropriately across the membrane.[12,13]

This molecular mechanism of photo-electric conversion has been simulated by using an artificial mixed monolayer consisting of light harvesting H and A-S-D triad amphiphilic molecules.[4–11] A, S, and D stand for electron acceptor, sensitizer, and electron donor moieties of the triad, respectively. The sensitizer moiety of the triad can absorb light energy by itself, but also can accept the excited energy of H molecules in the mixed monolayer. The A, S, and D moieties are linked covalenty in the triad and expected to be positioned in this order across the monolayer due to an amphiphilic property of the triad. Upon excitation of the S moiety followed by intramolecular electron and hole transfer to the A and D moieties, respectively, the charge can be separated unidirectionally across the monolayer in the same way as in the natural photosynthetic reaction center.[12]

[a]Address for telecommunication: Phone, 81/45-924-5784; fax, 81/45-924-5817; e-mail, mfujihir@bio.titech.ac.jp

The LB technique is suitable for making such highly oriented molecular assemblies. But, detailed structures of LB films had not been clarified until these films were studied recently by scanning probe microscopes (SPMs).[8,11,14–18] The SPMs, such as scanning tunneling microscope (STM), atomic force microscope (AFM), friction force microscope (FFM), scanning Maxwell-stress microscope (SMM), and scanning near-field optical microscope (SNOM), are useful to characterize the organic thin films on sub-micron to molecular scales. In order to realize true molecular devices, we have to reduce the size of our LB photodiodes not only in thickness (z-direction) but also in the x–y plane.[8,11] Micro processing by AFM[8,11] was applied previously to fabricate molecular devices in nanometer scales in lateral dimensions (x–y plane).

In the present paper, first we will describe how to fabricate artificial photosynthetic reaction center in nanometer scales by making use of phase separation in mixed monolayers of hydrocarbon (HC) and fluorocarbon (FC) amphiphiles.[17,18] The phase separated structures were studied by SPMs such as AFM, FFM,[16–18] SMM,[8,11] and scanning near-field optical/atomic force microscopy (SNOM-AFM)[8,14] as well as a conventional local surface analysis by SIMS. The model anionic and cationic HC amphiphilic compounds were used for the present study of phase separation for simplicity in place of the H light harvesting antenna and the A-S-D charge separation triad. The intense fluorescence due to cationic cyanine dyes with two long alkyl chains observed from HC islands by a conventional fluorescence microscopy and SNOM-AFM[14] indicates that the HC dyes preferentially dissolved in the HC islands in the phase-separated system. The success in fluorescence measurements of the nanostructure demonstrates also that SNOM-AFM can be used to access nano-devices from macro-scale external devices in optical modes.

We will also describe observation of charge separation in LB films under laser light illumination by SMM.[8] In the LB films, linear A-S-D triad amphiphilic molecules oriented unidirectionally in two ways with the D tail of the triad towards the air or towards the substrate plate. The A, S, and D moieties are expected to be located in this order or reversed order across each alternate monolayer in the LB films. By photo-excitation of the S moiety, the electron and hole can be separated unidirectionally, but with opposite directions in the LB films. The resulting photo-induced changes in dipole moments of the highly oriented triad molecules were detected as the surface potential changes of gold electrodes covered with the LB films by SMM. The results described here indicate that SMM can be used as one of electronic access tools to the nano-devices from the macro-scale external circuit. Finally, a dramatic increase in photo-electric conversion efficiency by using A-S-D-D′ will be presented.[19]

FABRICATION OF MOLECULAR PHOTODIODES

How to prepare an LB film is as follows. First, amphiphilic molecules are spread on an aqueous subphase, then compressed by a barrier while monitoring its surface-pressure with Willhelmy plate, and finally transferred onto a substrate as an LB film.

Before deposition of the LB film, a $\pi - A$ isotherm, namely a surface-pressure (π) versus area per molecule (A) curve is measured. During the compression, the

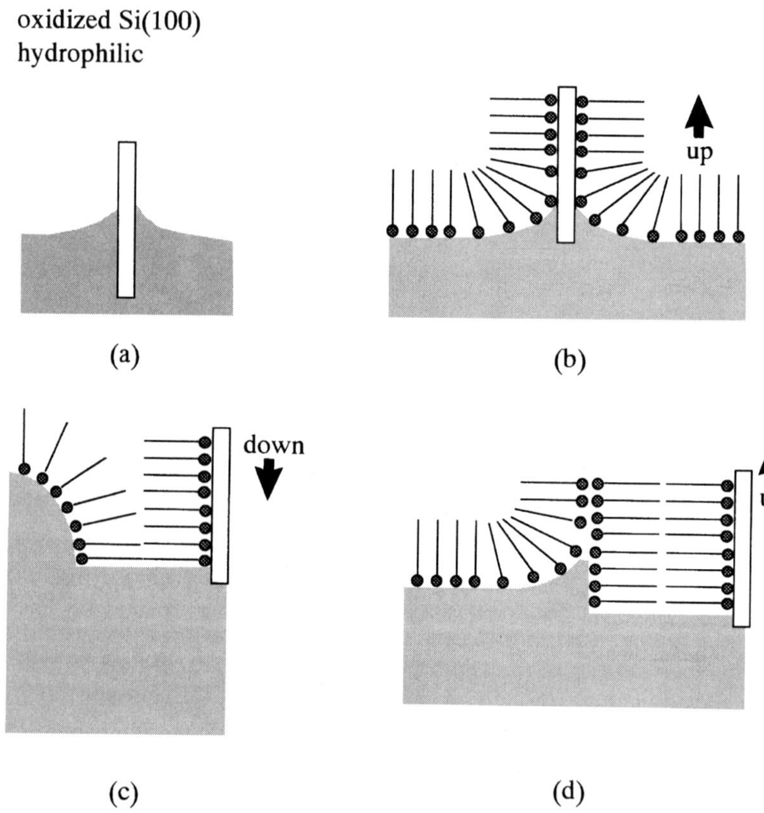

FIGURE 1. Preparation of a Langmuir-Blodgett multi-layer film on a hydrophilic substrate.

state of the monolayer is changing from gas phase, liquid phase, and finally to solid phase. When the surface pressure increases beyond a critical value, then the monolayer collapses into a multilayered film. Usually the monolayer in the solid or liquid phase is transferred onto the substrate.

FIGURE 1 illustrates preparation of a Langmuir-Blodgett multi-layer film on a hydrophilic substrate. In most cases, oxidized silicon(100) wafers were used as the flat hydrophilic substrate for SPM study of LB films. A surface of the silicon substrate can be cleaned and made hydrophilic by treatment in an ozone cleaner. The hydrophilic substrate is dipped in the subphase when the water surface is clean (FIG. 1(a)). After the spreading and the compression of the monolayer, the monolayer is trans-

TABLE 1. Invention of Friction Force Microscopes (FFM)

Mate, C.M., G.M. McClelland, R. Erlandsson & S. Chiang. 1987. Atomic-scale friction of a tungsten tip on a graphite surface. Phys. Rev. Lett. **59:** 1942–1945.
Meyer, G. & N.M. Amer. 1990. Simultaneous measurement of lateral and normal forces with an optical-beam-deflection atomic force microscope. Appl. Phys. Lett. **57:** 2089-2091.
Marti, 0., J. Colchero & J. Mlynek. 1990. Combined scanning force and friction. microscopy of mica. Nanotechnol. **1:** 141–144.

ferred onto the substrate during withdrawal (FIG. 1(b)). After the monolayer transfer, the surface of the substrate becomes hydrophobic, because long alkyl chains of the amphiphile orient toward the air. As shown in FIGURE 1(c and d), during the following dipping and withdrawal, two monolayers are transferred with orientations of the amphiphilic molecules shown in FIG. 1(d). For measurements of charge separation in LB films by SMM, semitransparent gold electrodes were used as the hydrophilic substrate.

For a hydrophobized substrate, such as a silicon wafer with a hydrogen terminated surface by HF treatment, the transfer process starts from this second step and even numbers of multi-layers are deposited.

In some cases for the SPM study of monolayers, chemisorbed monolayers on gold(111) or oxide surfaces are also used. A thiol compound with a long alkyl chain in solution is known to be adsorbed on a gold or silver surface and to form a closely packed stable monolayer. Surfaces of oxides are also known to be covered by a stable monolayer when the oxide plate is dipped in a solution containing a silanization reagent with a long alkyl chain. These types of monolayers are called "self-assembled monolayers" and abbreviated as SAMs.[1]

Since the invention of AFM by Binnig, Quate, and Gerber in 1986, a variety of force measurement modes of AFM cantilevers have been proposed. But in the present studies, the beam deflection method using a quadrant photo detector and a red laser diode has been used. As cantilevers with an AFM tip, Olympus micro fabricated Si_3N_4 triangular or rectangular cantilevers are used. The radius of the tip apex is found to be about 20 nm by SEM.

CHARACTERIZATION OF ORGANIC THIN FILMS USING SPMS

Friction Force Microscopy (FFM)

Friction force microscopes have been invented as early as 1987 and 1990 as listed in TABLE 1. In our case, we detected the lateral force with beam deflection using a quadrant photodetector according to the method described in 2 and 3.

As illustrated in FIGURE 2, by using the beam deflection method with the quadrant photodetector, we can measure the change in the lateral force under a constant normal force during scans parallel to the direction shown by the arrow. Here, it is important to note that what we keep constant as the normal force is a force due to the bending of the cantilever. If the adhesive force, like water capillary force, changes from one place to the other, the effective normal load can be changed during the scans. For quantitative analysis of the change in observed lateral force, recording of

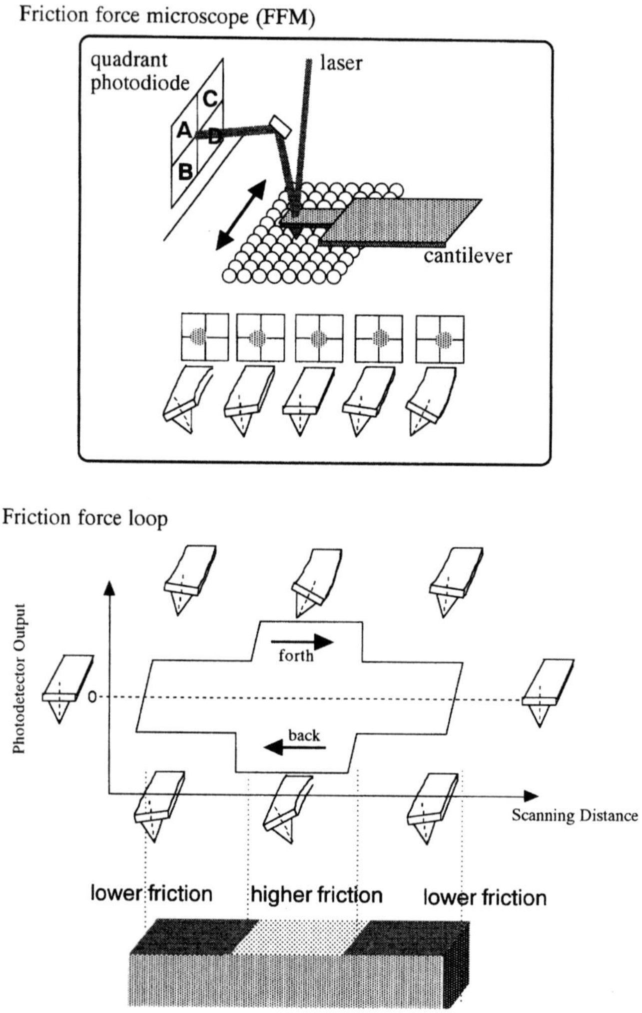

FIGURE 2. Beam deflection method with the quadrant photodetector.

friction loop is recommended in order to avoid the problem how to determine zero friction level in friction force coordinate.

The first application of FFM to the study of monolayers was demonstrated by Basel's group.[20] In their case, four layers of cadmium arachidate monolayers were deposited on HF-treated hydrophobic silicon wafer. In the AFM topographic image, they can clearly see three different levels corresponding to the substrate surface, the surface of the first bilayer, and the top surface of the two bilayers. In FFM image,

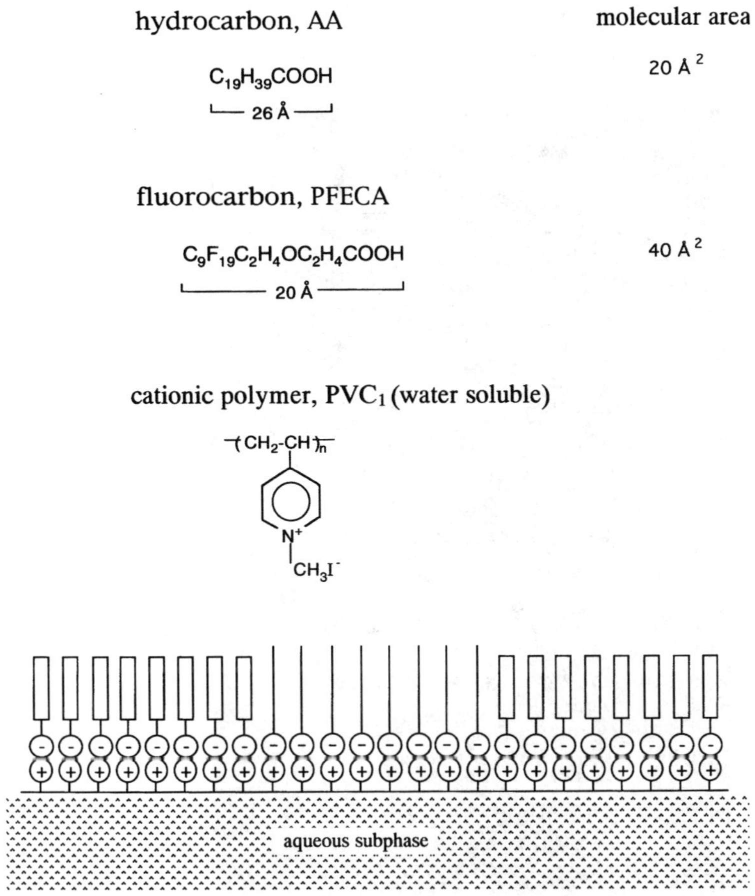

FIGURE 3. A mixture of two amphiphiles on an aqueous subphase dissolving a cationic polymer. At the bottom of the figure, the expected structure of the phase-separated monolayer on the aqueous subphase is illustrated.

the surface of the silicon substrate has higher friction than those of the LB films, but the two levels in the LB film show the same frictions.

When they were measuring the above friction, the author brought a sample to Basel to characterize its surface with FFM. Our sample was expected to have two phase separated domains, because our monolayer consisted of HC and FC amphiphiles. We spread a mixture of these two amphiphiles on an aqueous subphase dissolving a cationic polymer (FIG. 3). In the bottom, the expected structure of the phase-separated monolayer on the aqueous subphase is illustrated (a "side-by-side" structure). From the difference in molecular lengths, HC domains were expected to be higher than FC domains.

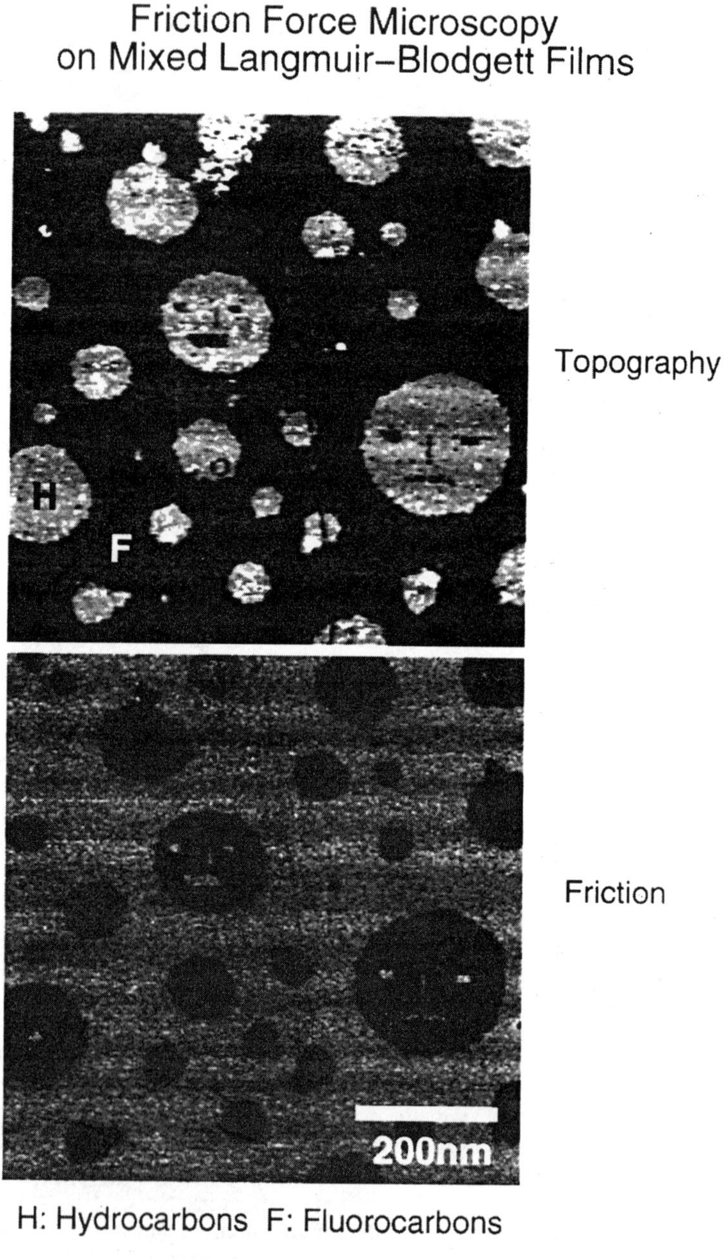

FIGURE 4. FFM of a monolayer film of a 1:1 mixture of PFECA and AA (2.4 × 2.4 μm^2). **(a, top)** Topography shows circular islands. "Smileys" were written with the atomic force microscope. The bottom level is at the same height as the sea of fluorocarbons, which shows that the hydrocarbons sit on the top of the fluorocarbons. **(b, bottom)** Friction force map.

As we would expect, the phase-separated structure of the mixed monolayer transferred on an oxidized silicon wafer is clearly seen in its AFM topography and FFM image.[21] Again it was confirmed that the different chemical species show different frictions and the higher friction was observed on the lower domains. But the difference in topographic height between the two domains in AFM is much higher than the difference in length between the two amphiphiles.[22]

After scratching islands part (FIG. 4), it was found that the depth of the scratched part was the same as the difference in height of two domains, and in addition the friction of the bottom of the scratched part was the same as that of the surrounding sea area. Taking account of the fact that single component HC monolayer was scratched,

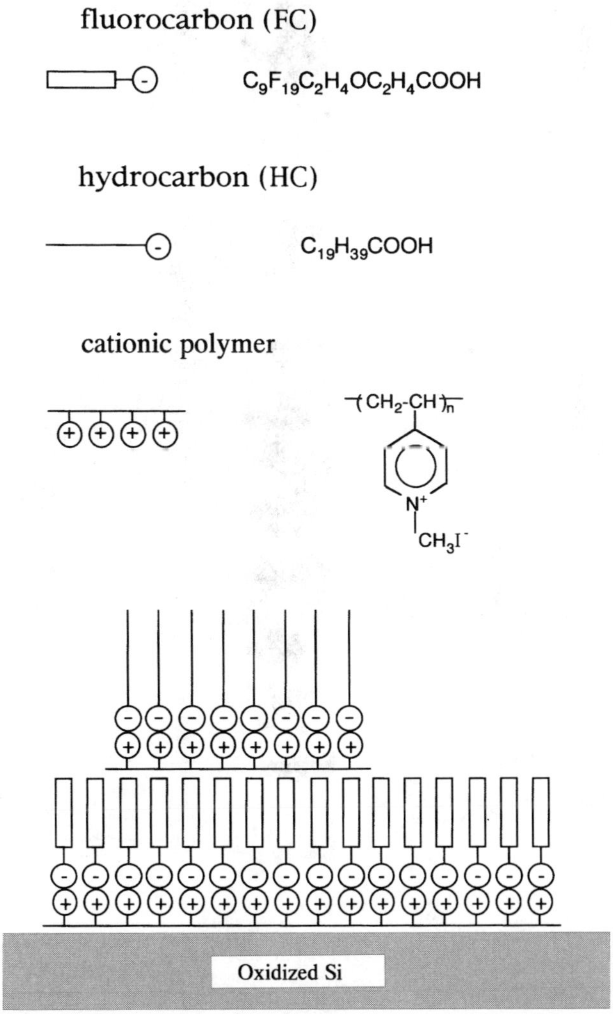

FIGURE 5. Proposed phase-separated monolayer with "on top" structure.

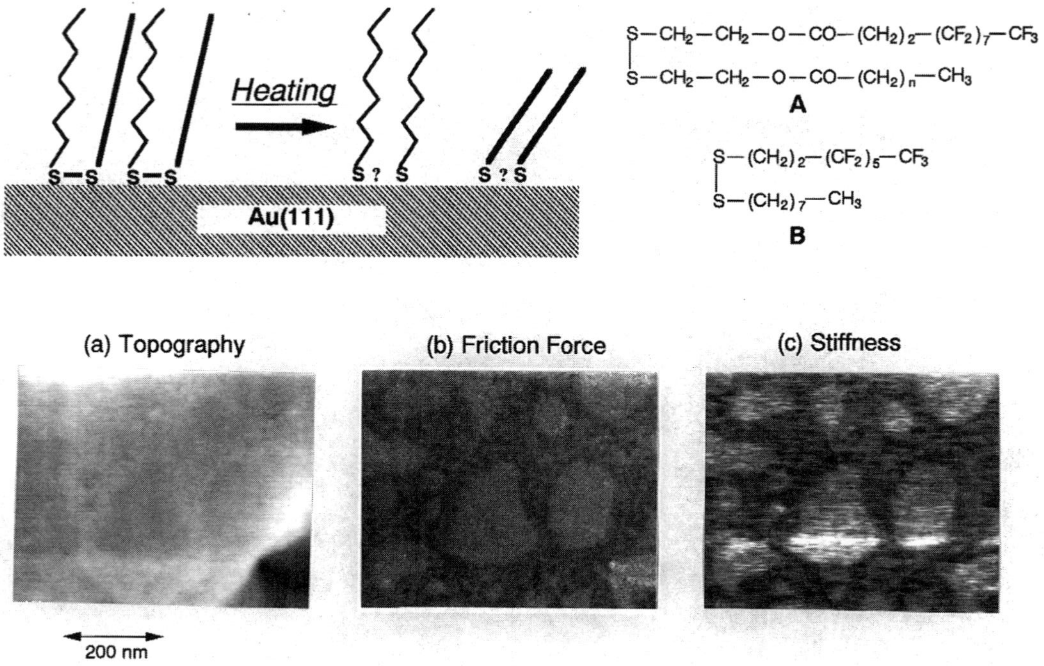

FIGURE 6. Phase-separated structure in **(a)** AFM, **(b)** FFM, and **(c)** stiffness images.

but the corresponding FC monolayer was not scratched, we proposed that the phase-separated monolayer has an "on-top" structure (FIG.5). Namely, the HC monolayer sits on the FC monolayer.

Recently FFM was applied to the study of cleavage of dissulfides in the SAM on gold(111).[23] A SAM of an asymmetric disulfide A was studied by Ringsdorf and his coworkers,[24–26] but we studied the phase separation of HC and FC of another SAM of a thermally more stable asymmetric disulfide B.[23] After heating of SAM B for 8 hr at 100°C, we observed the phase separated structure as shown in AFM, FFM, and stiffness images of FIGURE 6. This fact indicates that an S—S bond of B in its SAM was cleaved and the cleaved HC and FC adsorbates diffused by heating.

Force-Distance Curves in AFM and Pulsed-Force-Mode AFM (PFM-AFM)

In addition to FFM, a force-distance curve in AFM contains a lot of information of sample properties. In FIGURE 7, typical force-distance curves during expansion and retraction of *z*-piezo are shown in a *dashed* and a *solid line*, respectively. Topography and friction are measured under a low normal force. Under high normal force information of viscoelastic properties can be obtained, while from the jump-off force an adhesive force between the tip and sample surfaces is measured. If we can measure force-distance curves at high frequency, we can map the viscoelestic and adhesive properties in the *x*–*y* plane.

Pulsed-force-mode AFM (PFM-AFM) was proposed by Marti and his coworkers for this purpose.[27] In FIGURE 8, the cantilever is modulated in *z*-direction, but the sample can be modulated. By the use of sinusoidal modulation, much higher fre-

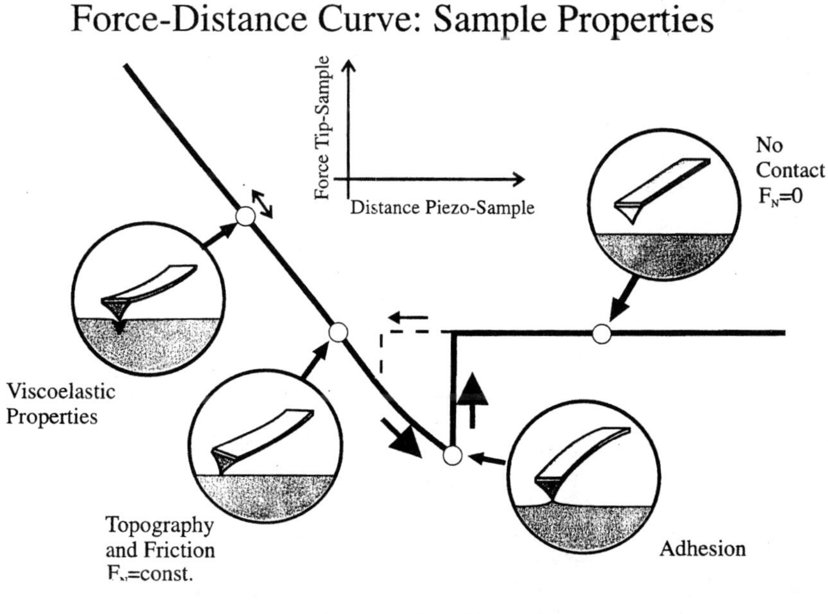

FIGURE 7. Sample properties of force-distance curve.

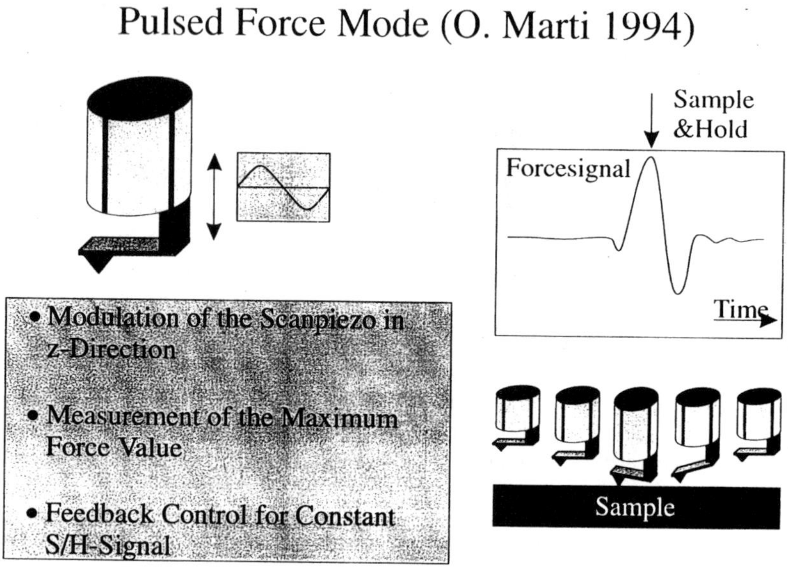

FIGURE 8. Pulse-force mode.

quencies can be used than the force-distance curve measurement with triangular modulation.[28] The maximum force value recorded by sample and hold circuit during one cycle was used as the feedback control signal for topography mapping. The viscoelastic properties and the adhesive force were obtained by recording the sample and hold force signals at different timings.

The curve in FIGURE 9 is a typical force signal-time curve measured in air. In our case, the sample is modulated in *z*-direction rather than the cantilever.[28] By the viscoelastic property of air, damping of oscillation of cantilever with a resonance frequency after jump-off is clearly observed. The damping in water was much faster than this and reached steady state within one cycle with the resonance frequency. From the force signals at S/H No. 2 and S/H No. 3 before jump (*inset*), one can measure weak interaction force between the tip and sample surfaces. Again the maximum force at S/H No. 1 is used for the feedback control.

A silica surface partially covered with an alumina thin film was prepared by the procedure as follows. (1) Water droplets containing SiO_2 beads were placed on a quartz substrate plate. The diameter of the beads was 3.5 μm. A concentration of the beads was ca. 2 wt.%. The substrate was dried in a desiccator for 2 days. (2) Aluminum was vacuum vapor-deposited on the bead covered quartz plate. The pressure and the deposition rate were 4.0×10^{-6} Torr and 3 nm s^{-1}, respectively. (3) The Al film deposited on the bead covered quartz plate was oxidized to Al_2O_3 by leaking oxygen into the evaporation system. Oxidation of Al produced a transparent film. (4) The silica beads together with the Al_2O_3 film on them were removed in water by ultrasonication. (5) Finally, in order to oxidize the Al completely and for the result-

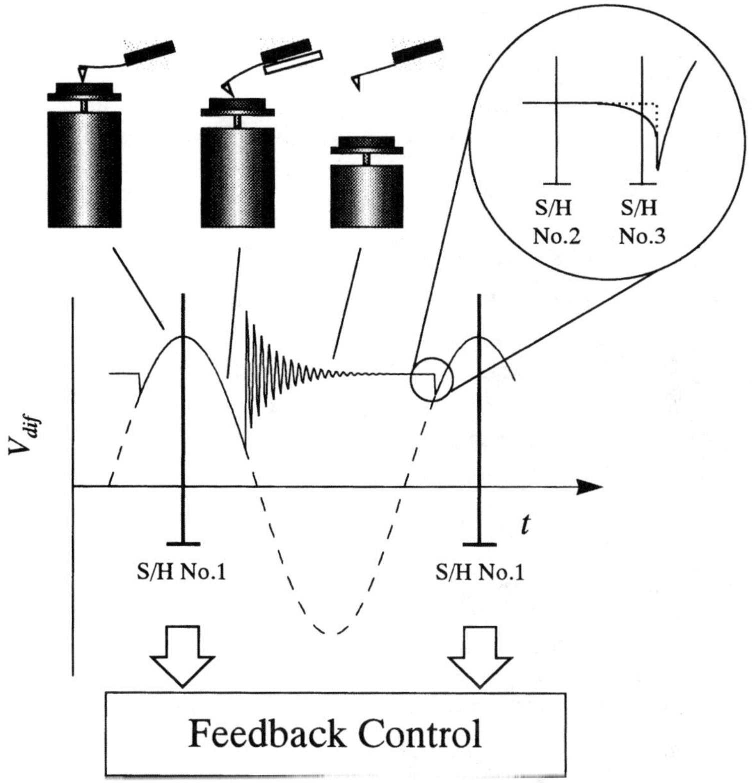

FIGURE 9. Typical force signal-time curve measured in air.

ing Al_2O_3 to adhere well to the SiO_2 substrate, the sample was further annealed at ca. 500°C in air overnight.

A topographic image recorded with the maximum force feedback in an aqueous solution of pH 8.6 is shown in FIGURE 10(a). The uncovered silica surface is seen as craters with a constant depth. The double-layer force between the silicon nitride tip and the sample surface was measured by the difference between the force signals at No. 2 and No. 3 in FIGURE 9 and is mapped in FIGURE 10(b). The isoclectric points of silicon nitride, silica, and alumina were measured previously by my group and are known to be about 6, 2, and 10, respectively. Therefore, the tip and the silica surface were charged negatively, while the alumina surface charged positively at this pH. The repulsive force observed on the craters and the attractive force on the alumina covered surrounding are consistent with the above expectation.

Scanning Maxwell-stress Microscopy (SMM)

As described above (FIG. 5(a)), the structure of the monolayer was the "on top" structure in the polyion complexed mixed monolayer of HC and FC. In order to con-

TABLE 2. Studies of high-resolution potentiometry Electrostatic Force Microscopy (Kelvin probe microscope)

Y. Martin, D.W. Abraham & H.K. Wickramasinghe. 1988. Appl. Phys. Lett. **52:** 1103.
J.M.R. Weaver & D.W. Abraham. 1991. J. Vac. Sci. Technol. **B9:** 1559.
M. Nonnenmacher, M. O'Boyle & H.K. Wickramasinghe. 1991. Appl. Phys. Lett. **58:** 2921.
M. Nonnenmacher, M. O'Boyle & H. K. Wickramasinghe. 1992. Ultramicroscopy **42-44:** 268.
Scanning Maxwell Stress Microscope
H. Yokoyama, K. Saito & T. Inoue. 1992. Mol. Electronics. Bioelectronics **3:** 79.
Scanning surface potential microscope
M. Fujihira, H. Kawate & M. Yasutake. 1992. Chem. Lett. 2223.
M. Fujihira & H. Kawate. 1994. Thin Solid Films **242:** 163.
M. Fujihira & H. Kawate. 1994. J. Vac. Sci. Tech. **B12:** 1604
M. Yasutake, D. Aoki & M. Fujihira. 1996. Thin Solid Films **273:** 279.

firm the "on top" structure, we also studied this LB film by electrostatic force microscopy. FC monolayer is known to have large surface potential because the dipole moment of terminal CF_3 group is much larger than that of CH_3 group due to the high electronegativity of the fluorine atom.

In TABLE 2, the studies of high resolution potentiometry by electrostatic force microscopy are listed. Initial electrostatic force microscopes included the Kelvin-probe force microscope were developed by the group of Wickramasinghe in IBM

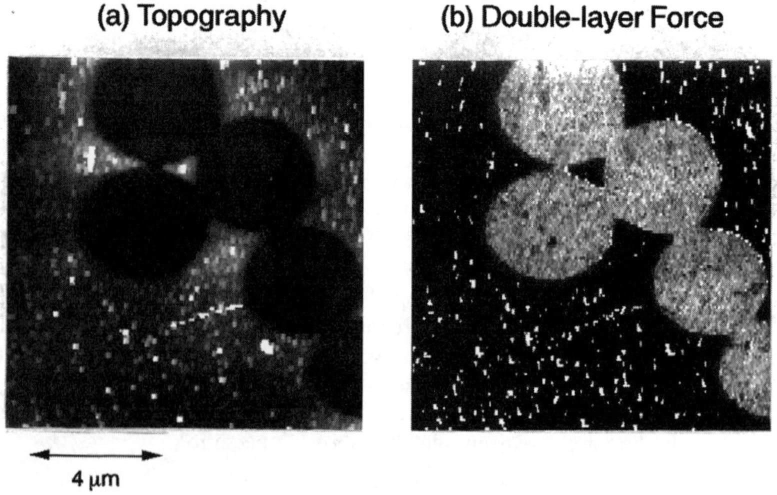

FIGURE 10. **(a)** Topographic image recorded with the maximum force feedback in an aqueous solution. **(b)** Mapping of the double-layer force between the silicon nitride tip and the sample surface.

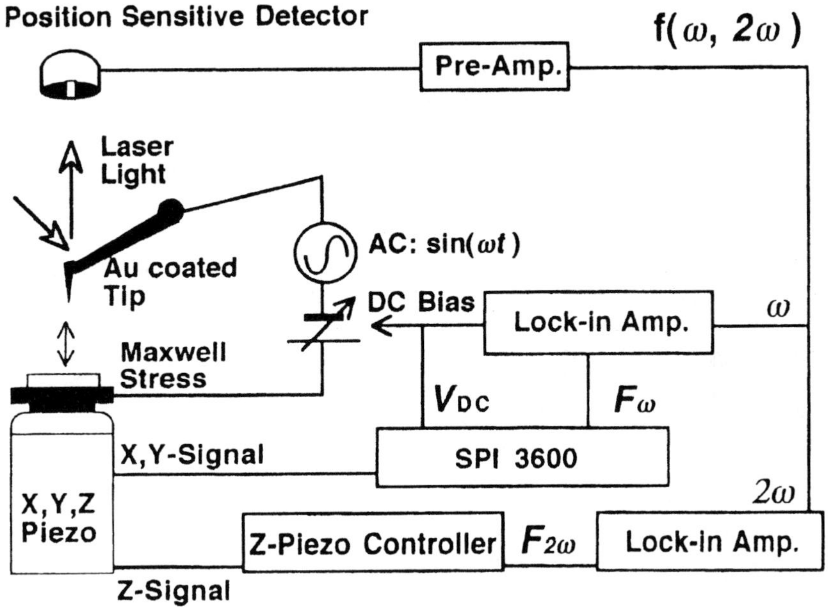

FIGURE 11. Operational diagram of SMM.

Yorktown Heights. Later, Yokoyama and his coworkers at Electrotechnical Laboratory, Tsukuba modified the method and named the method "Scanning Maxwell-stress Microscopy (SMM)." We used these methods for the study of organic thin films as listed at the bottom of TABLE 2.

In FIGURE 11, an operational diagram of SMM is shown. The electrostatic force between the tip and sample under application of AC voltage and DC offset is described[29] by

$$F = -(1/2)(dC/dz)[V_{DC}^2 + 2V_{DC}V_{AC}\sin(\omega t) + (V_{AC}^2/2)(1 - \cos(2\omega t))] \quad (1)$$

From the result of the HC-FC phase-separated mixed monolayer, the potential detection limit by the present SMM is determined to be below 1 mV.[30]

In FIGURE 12(a), a topographic image observed by the SMM is shown. In FIGURE 12(b), a potential image of the phase-separated monolayer is shown. The potential difference between the islands and the sea part is found to be about 50–60 mV.[30] The potential difference is consistent with the "on top" structure.

Scanning Near-Field Optical/Atomic Force Microscopy (SNOM-AFM)

FIGURE 13 shows how the super-resolution of optical microscopy can be obtained. If the size of the aperture and the distance between the aperture and the sample surface is much smaller than one wavelength, interaction of the confined photon in the near-field region with the surface of the sample gives the super-resolution. The

TABLE 3. Recent progress in scanning near-field optics (SNOM)

D.W. Pohl, W. Denk & M. Lanz. 1984. Appl. Phys. Lett. **44:** 651.
A. Lewis, M. Isaacson, A. Harootunian & A. Murray. 1984. Ultramicroscopy **13:** 227.
U.Ch. Fischer. 1985. J. Vac. Sci. Technol. **B3:** 386.
D. Courjon, K. Sarayeddine & M. Spajer. 1989. Opt. Commun. **71:** 23.
R.C. Reddik, R.J. Warmack & T.L. Ferrell. 1989. Phys. Rev. **B39:** 767.
K. Lieberman, S. Harush, A. Lewis & R. Kopelman. 1990. Science **247:** 59.
E. Betzig and J.K. Trautman. 1992. Science **257:** 189.
N.F. van Hulst, M.H.P. Moers, O.F.J. Noordman, T. Faulkner, F.B. Segerink, K.O. van der Werf, B.G. de Grooth & B. Bölger. 1992. Proc. SHE **1639:** 36.
M. Fujihira, H. Monobe, H. Muramatsu, & T. Ataka. 1994. Chem. Lett. 657.
H. Muramatsu, T. Ataka, H. Monobe, & M. Fujihira. 1995. Ultrarnicroscopy **57:** 141.

idea was proposed by Synge as early as 1928,[31] but we had to wait for the realization of such experimental conditions until the invention of STM using a piezo-scanner under computer control.

The recent progress in SNOM since the invention of STM is listed in TABLE 3. For the distance control, STM current, evanescent field itself, or shear force has been used. But in our apparatus, the distance control was achieved by the non-contact mode AFM or the contact mode AFM.[32,33]

In FIGURE 14, operation of our SNOM with AFM control[32,33] is illustrated. The bent optical fiber and a sharpened tip end are used as a cantilever for AFM and a probe for SNOM, respectively. In this case, the bimorph is used for the separation control for the non-contact mode AFM.

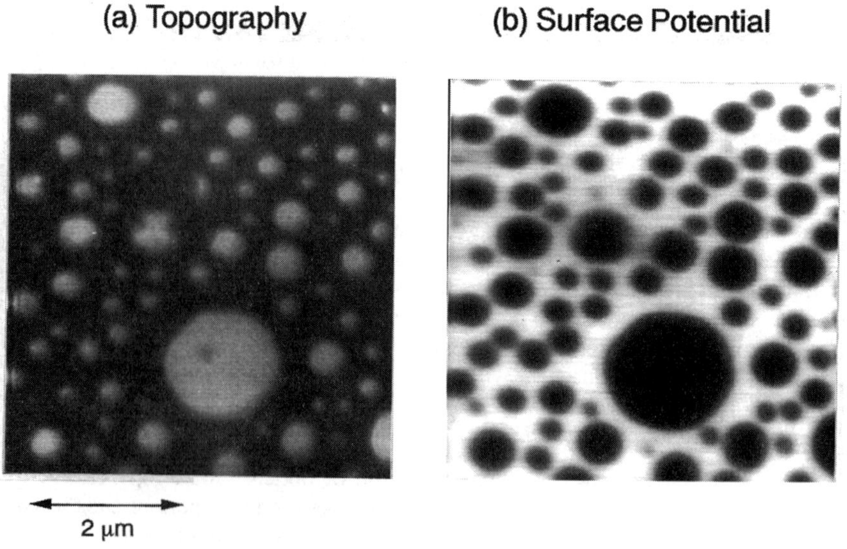

FIGURE 12. (**a**) Typical topographic and (**b**) surface potential image observed by SMM.

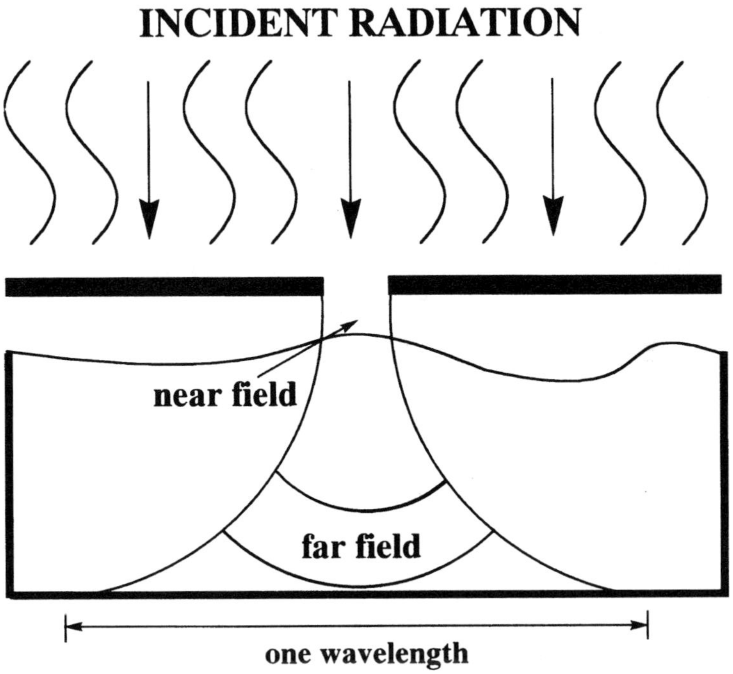

FIGURE 13. A schematic diagram of scanned aperture microscopy showing the potential of achieving high resolution by restriction of the sample to the "near field."

When we use the bent optical fiber with a diameter of 125 μm as a cantilever for AFM, the spring constant is calculated to be about 1000 N/m. Therefore such bent optical fiber cannot be used for the contact mode AFM. We reduced the spring constant down to about 10 N/m by decreasing the diameter by chemical etching.[34] In this way, we can show experimental observation of the liquid-solid phase separation in a monolayer of a fluorescent lipid[35] by SNOM with the contact mode AFM. The structural formula of the lipid and the $\pi - A$ isotherm of the pure monolayer of the lipid are shown in FIGURE 15. When the surface pressure was kept at near 5 mN/m, the solid and the liquid phase coexist. Therefore the monolayer was transferred at this surface pressure.

In FIGURE 16, the contact mode AFM topographic and the FFM image observed simultaneously are shown in **(a)** and **(b)**, respectively.[36] Due to the lack of capacity for recording three images simultaneously in our apparatus, the SNOM fluorescent intensity image was recorded later **(c)**. It was found from comparison of these images that topographically higher solid phase shows the high friction and is nonfluorescent, while the liquid phase shows the lower friction and is fluorescent. This is the first observation of these three images with high lateral resolutions.

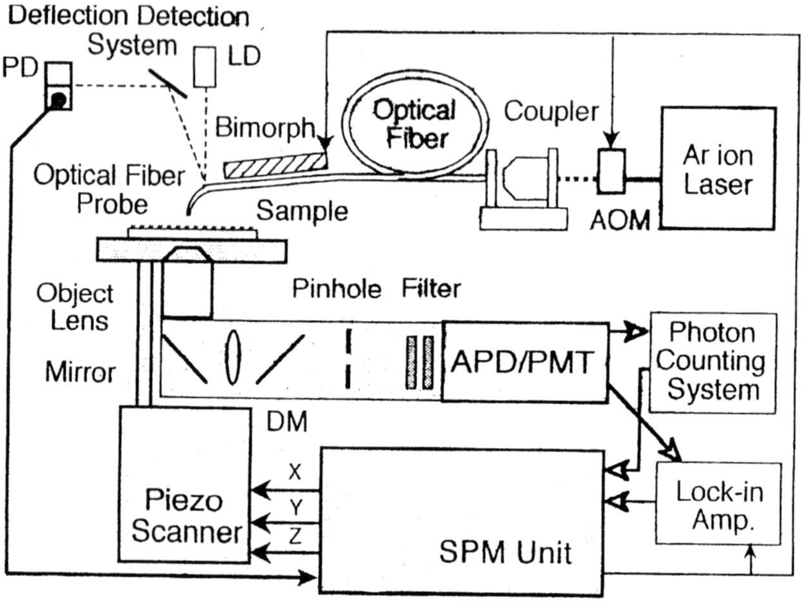

FIGURE 14. Scanning near-field optical atomic force microscope (SNOAM or SNOM-AFM).

SXX′M

Here, the conventional local analysis methods and scanning probe microscopies and spectroscopies will be compared. In the conventional local analysis, probing particles impinge on the sample surface and monitoring particles are collected in the far-field mode. The lateral resolution of these methods is limited by diffraction of the corresponding waves.[11] But,the SPM methods are all in the near-field mode and therefore the lateral resolution of the SPM methods is not limited by the diffraction. However, the concept of the probing and monitoring particles in the conventional far-field methods is also useful for the classification of the SPM methods.[11]

This is SXX′M. Here, X represents a probing particle from the tip or a probing interaction by the tip, while X′ represents a monitoring particle from the sample surface or a monitoring physical quantity as a result of the tip-sample interaction. In comparison with SPM or SXM, SXX′M calls one's attention more to *cause* and *effect*. For example, SNOM including Raman, IR, and fluorescence spectroscopies can be classified by a relationship between probing-monitoring photons. According to this idea, we are currently studying emitted photons and particles under frictional interaction between the tip and sample surfaces during FFM. The result will give us an insight into the atomic or molecular mechanism of triboemission and tribochemistry.

MOLECULAR PHOTODIODES DRIVEN BY SMM

Amphiphilic A-S-D triad molecules can be oriented unidirectionally in monolayers at the air/water interface. The resulting oriented monolayer assemblies are fabricated into thin films on electrode surfaces by LB technique. The A-S-D triads for charge separation together with light harvesting antenna molecules can be assembled into sub-µm island domains by making use of phase separation of mixed monolayers of HC and FC amphiphiles.[8] The structure and properties of the domains of HC-FC mixed monolayers were studied by SPMs as described above. Some of SPMs can be used not only to characterize the LB films, but also to drive the molecular devices. The photo-induced charge separation in unidirectionally oriented triad molecules embedded in alternate LB films was observed by SMM.[8]

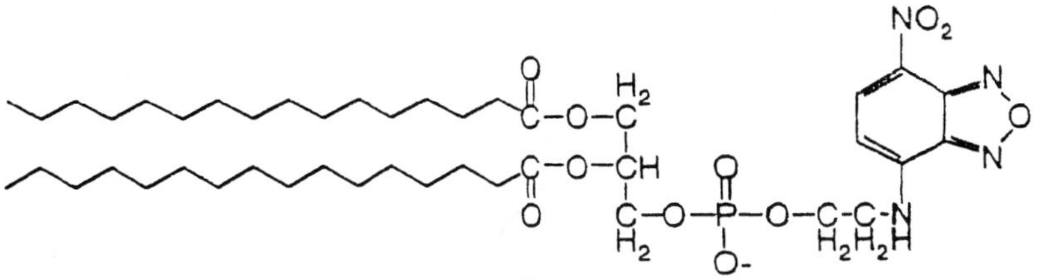

N-(7-nitrobenz-2-oxa-1,3,-diazol-4-yl)-1,2-dihexadecanoyl-*sn*-glycero-3-phosphoethanolamine (NBD-PE)

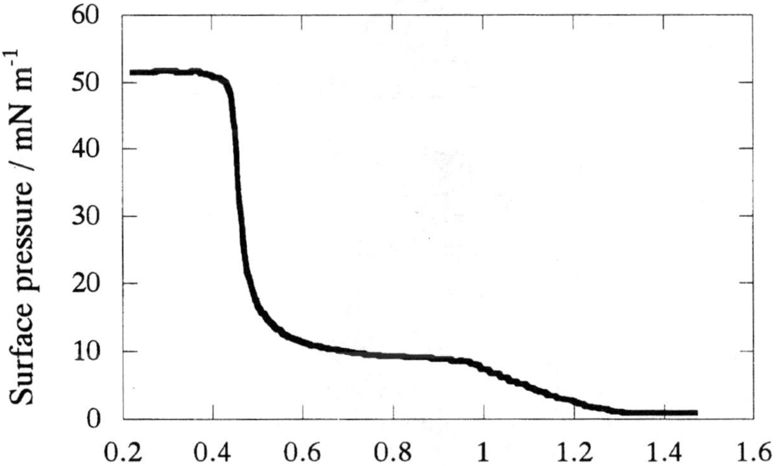

FIGURE 15. Structural formula of the lipid and the $\pi - A$ isotherm of the pure monolayer of the lipid.

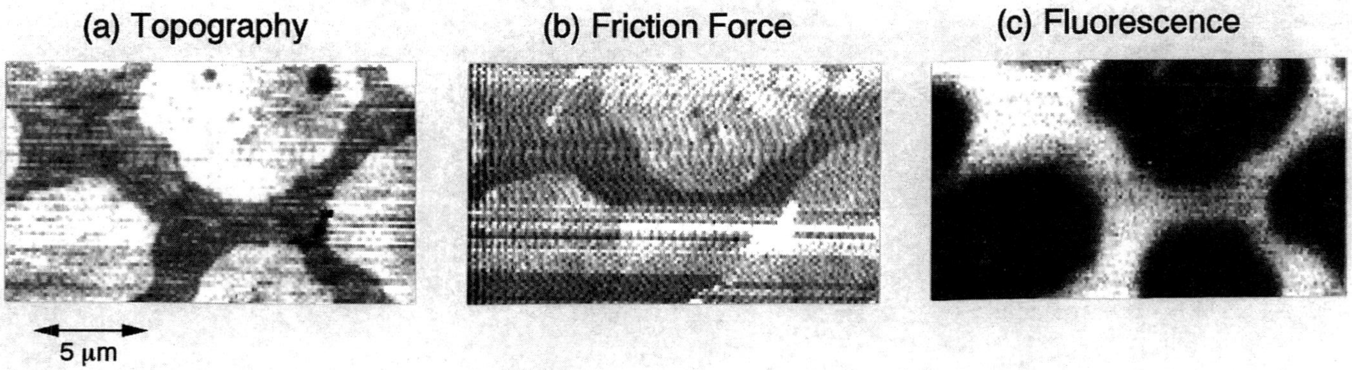

FIGURE 16. (**a**) Contact mode AFM topographic image, (**b**) FFM image, and (**c**) SNOM fluorescent intensity image.

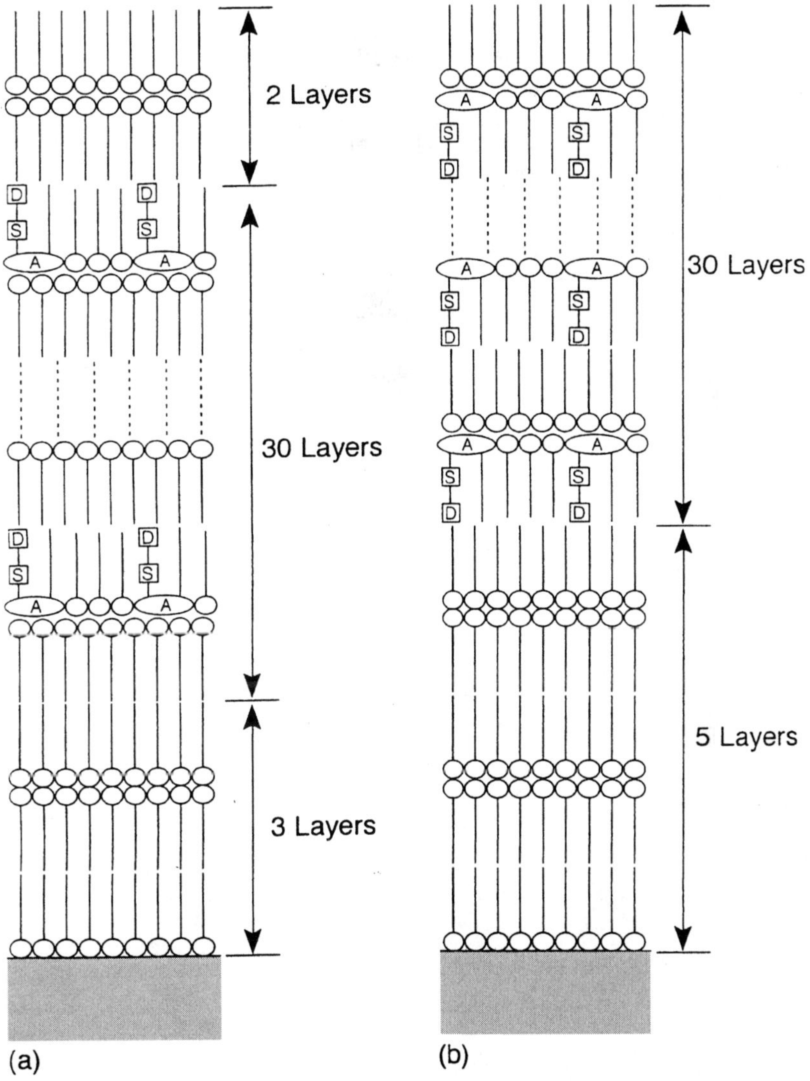

FIGURE 17. Unidirectionally oriented A-S-D triads in LB films. (**a**) D tails orient toward the air, while in (**b**) the D tails orient toward the substrate plate.

In FIGURE 17 are shown unidirectionally oriented A-S-D triads in LB films containing 15 alternate bilayers. In type a (**a**), the D tails of A-S-D triads orient towards the air, while in type b (**b**) the tails orient towards the substrate plate. The charge separation in these two types of LB films was observed with SMM by irradiating a He-Cd laser of 441.6 nm laser light step functionally in a total reflection mode.[8] The change in the surface potential by step illumination for ca. 2–3 s on (FIG. 18(a)) a type a alternate and (FIG. 18(b)) a type b alternate LB film shown in FIGURE 17 is shown in FIGURE 18 as well as on (FIG. 18(c)) a blank LB film. The direction of the change in surface potential shown in FIGURE 18 agrees with orientation of A-S-D triads in these two types of alternate LB films. In this way, SMM can be used to detect the charge separation in molecular photodiode.

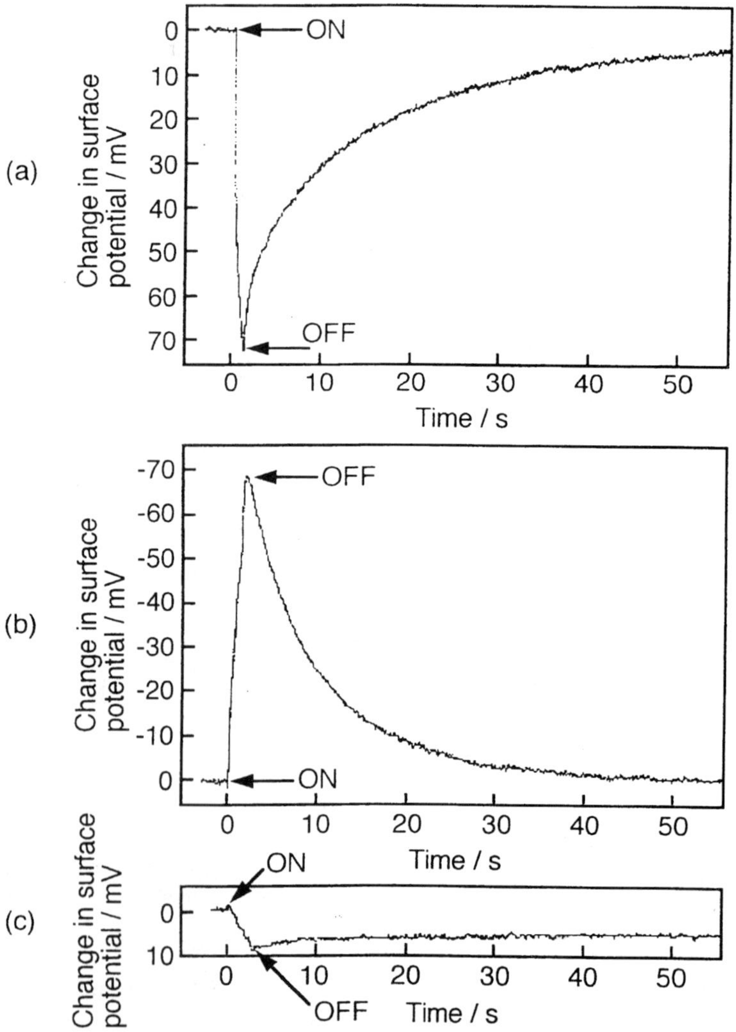

FIGURE 18. Change surface potential by step illumination for type a **(a)**, type **(b)**, and blank **(c)** LB films.

Recently, we succeeded in obtaining the increase in efficiency of photo-electric conversion by A-S-D molecular photodiodes by adding the second donor layer, D′. The ionization potential of D′ was smaller than that of the donor moiety of the A-S-D triad. The D′monolayer was deposited on top of the donor of the A-S-D monolayer. The photo-induced charge separation in the monolayer assemblies was observed by SMM. The addition of the second donor layer promoted the photo-induced charge separation due to the longer charge separation distance by the presence of the D′ monolayer and lateral diffusion of charges through the D′ monolayer.[19]

REFERENCES

1. ULMAN, A. 1991. Ultrathin Organic Films. Academic Press. San Diego, CA.
2. OSA, T & M. FUJIHIRA. 1976. Photocell using covalently-bound dyes on semiconductor surfaces. Nature **264:** 349–350. FUJIHIRA, M., N. OHISHI & T. OSA. 1977. Photocell using covalently-bound dyes on semiconductor surfaces. Nature **268:** 226–228.
3. FUJIHIRA, M. 1986. Modified electrodes. *In* Topics in Organic Electrochemistry. A.J. Fry & W. E. Britton, Eds.: 255–298. Plenum. New York, NY
4. FUJIHIRA, M., K. NISHIYAMA & H. YAMADA. 1985. Photoelectrochemical responses of optically transparent electrodes modified with Langmuir-Blodgett films consisting of surfactant derivatives of electron donor, acceptor and sensitizer molecules. Thin Solid Films **132:** 77–82.
5. FUJIHIRA, M. & H. YAMADA. 1988. Molecular photodiodes consisting of unidirectionally oriented amphipathic acceptor-sensitizer-donor triads. Thin Solid Films **160:** 125–132.
6. FUJIHIRA, M., M. SAKOMURA & T KAMEI. 1989. Simulation of the primary process in photosynthetic reaction centre by monolayer with charge separation and antenna molecules. Thin Solid Films **180:** 43–50.
7. FUJIHIRA, M. 1990. Photoelectric conversion by monolayer assemblies. Mol. Cryst. Liq. Cryst. **183:** 59–69.
8. FUJIHIRA, M., M. SAKOMURA, D. AOKI & A. KOIKE. 1996. Scanning probe microscopies for molecular photodiodes. Thin Solid Films **273:** 168–176.
9. FUJIHIRA, M. 1995. Photoinduced electron transfer in monolayer assemblies and its application to artificial photosynthesis and molecular devices. *In* Thin Films, A. Ulman, Ed. **20:** 239–277. Academic Press. Boston, MA.
10. FUJIHIRA, M. 1994. The photoinduced electron transfer and energy transfer in LB films. *In* Molecular and Biomolecular Electronics. R.R. Birge, Ed. **240:** 373–394. Adv. Chem. Series. American Chemical Society, Washington, DC.
11. FUJIHIRA, M. 1995. Study of thin organic films by various scanning force microscopes. *In* Forces in Scanning Probe Methods. H.-J. Güntherodt, D. Anselmetti & E. Meyer, Eds. **286:** 567–591. NATO ASI Series E: Appl. Sci. Kluwer Academic Publishers. Dordrecht, The Netherlands.
12. DEISENHOFER, J. & H. MICHEL. 1989. The photosynthetic reaction centre from the purple bacterium Rhodopseudomonas viridis. The EMBO Journal **8:** 2149–2170.
13. VAN GRONDELLE, R., J.P. DEKKER, T. GILLBRO & V. SUNDSTROM. 1994. Energy transfer and trapping in photosynthesis. Biochim. Biophys. Acta **1187:** 1–65.
14. FUJIHIRA, M. 1996. Fluorescence microscopy and spectroscopy by scanning near-field optical /atomic froce microscope (SNOM-AFM). In Optics at the Nanometer Scale: Imaging and Storing with Photonic Near Fields. NATO ASI Series, Series E: Applied Sciences. **319**. M. Nieto-Vesperinas and N. Garcia, Eds.: 205–221. Kluwer Academic Publishers. Dordrecht, the Netherlands.

15. FUJIHIRA, M. 1997. Friction force microscopy of organic thin films and crystals. In Nanotribology. NATO ASI Series, Series E: Applied Sciences. **330.** B. Bhushan, Ed.: 239–260. Kluwer Academic Publishers. Dordrecht, the Netherlands.
16. FROMMER, J. 1992. Scanning tunneling microscopy and atomic force microscopy in organic chemistry. Angew. Chem., Int. Ed. **31:** 1298–1328.
17. OVERNEY, R. M., E. MEYER, J. FROMMER, D. BRODBECK, R. LÜTHI, L. HOWALD, H.-J. GÜNTHERODT, M. FUJIHIRA, H. TAKANO & Y GOTOH. 1992. Friction measurements on phaseseparated thin films with a modified atomic force microscope. Nature **359:** 133–135.
18. MEYER, E., R. OVERNEY, R. LÜTHI, D. BRODBECK, L. HOWALD, J. FROMMER, H.-J. GÜNTHERODT, O. WOLTER, M. FUJIHIRA, H. TAKANO & Y GOTOH. 1992. Friction force microscopy of mixed Langmuir-Blodgett films. Thin Solid Films **220:** 132–137.
19. SAKOMURA, M., T.OONO & M. FUJIHIRA. The effect of second donor-layer on charge separation in LB films. Thin Solid Films. In press.
20. MEYER, E., R. OVERNEY, D. BRODBECK, L. HOWALD, R. LÜTHI, J. FROMMER & H.-J. GÜNTHERODT. 1992. Friction and wear of Langmuir-Blodgett films observed by friction force microscopy. Phys. Rev. Lett. **69:** 1777–1780.
21. OVERNEY, R. M., E. MEYER, J. FROMMER, D. BRODBECK, R. LÜTHI, L. HOWALD, H.-J. GÜNTHERODT M. FUJIHIRA, H. TAKANO & Y. GOTOH. 1992. Friction measurements on phase-separated thin films with a modified atomic force microscope. Nature **359:** 133–135.
22. MEYER, E., R. OVERNEY, R. LÜTHI, D. BRODBECK, L. HOWALD, J. FROMMER, H.-J. GÜNTHERODT, O. WOLTER, M. FUJIHIRA, H. TAKANO & Y. GOTOH. 1992. Friction force microscopy of mixed Langmuir-Blodgett films. Thin Solid Films **220:** 132–137.
23. ISHIDA, T., S. YAMAMOTO, W MIZUTANI, M. MOTOMATSU, H. TOKUMOTO, H. HOKARI, H. AZEHARA & M. FUJIHIRA. 1997. Evidence for cleavage of disulfides in the self-assembled monolayer on Au(111). Langmuir **13:** 3261–3265.
24. SCHÖNHERR, H. & H. RINGSDORF. 1996. Self-assembled monolayers of symmetrical and mixed alkyl fluoroalkyl disulfides on gold. 1. Synthesis of disulfides and investigation of monolayer properties. Langmuir **12:** 3891–3897.
25. SCHÖNHERR, H., H. RINGSDORF, M. JASCHKE, H.-J. BUTT, E. BAMBERG, H. ALLINSON & S. D. EVANS. 1996. Self-assembled monolayers of symmetrical and mixed alkyl fluoroalkyl disulfides on gold. 2. Investigation of thermal stability and phase separation. Langmuir **12:** 3898–3904.
26. SCHÖNHERR, H. & G. J. VANCSO. 1997. Lattice imaging of self-assembled monolayers of partially fluorinated disulfides and thiols on sputtered gold by atomic force microscopy. Langmuir **13:** 3769–3774.
27. ROSA, A., E. WEILANDT, S. HILD & O. MARTI. 1997. The simultaneous measurement of elastic, electrostatic and adhesive properties by scanning force microscopy pulsed-force mode operation. Meas. Sci. Technol. **8:** 1–6.
28. MIYATANI, T., M. HORII, A. ROSA, M. FUJIHIRA & O. MARTI. 1997. Mapping of electrical double-layer force between tip and sample surfaces in water with pulse-force-mode atomic force microscopy. Appl. Phys. Lett. **71:** 2632–2634.
29. FUJIHIRA, M. & H. KAWATE. 1994. Scanning surface potential microscope for characterization of Langmuir-Blodgett films. Thin Solid Films **242:** 163–169.
30. YASUTAKE, M., D. AOKI & M. FUJIHIRA. 1996. Surface potential measurements using the Kelvin probe force microscope. Thin Solid Films **273:** 279–283.
31. SYNGE, E.H. 1928. Suggested method for extending microscopic resolution into the ultra-microscopic region. Phil. Mag. **6:** 356–362.

32. MURAMATSU, H., N. CHIBA, T. ATAKA, H. MONOBE & M. FUJIHIRA. 1995. Scanning nearfield optic/atomic-force microscopy. Ultramicroscopy **57:** 141–146.

33. MURAMATSU, H., N. CHIBA, K. HOMMA, K. NAKAJIMA, T. ATAKA, S. OHTA, A. KUSUMI & M. FUJIHIRA. 1995. Near-field optical microscopy in liquids. Appl. Phys. Lett. **66:** 3245–3247.

34. MURAMATSU, H., N. CHIBA & M. FUJIHIRA. 1997. Frictional imaging in a scanning nearfield optical/atomic-force microscope by a thin step etched optical fiber probe. Appl. Phys. Lett. **71:** 2061–2063.

35. IVANOV, G. R., J.I. PETKOVA, Y OKABE, D. AOKI, H. TAKANO, H. KAWATE & M. FUJIHIRA. 1997. Scanning probe microscopy studies of aggregation in Langmuir-Blodgett films. Supramolec. Sci. **4:** 549–557.

36. FUJIHIRA, M., H. MONOBE, A. KOIKE, G. R. IVANOV, H. MURAMATSU, N. CHIBA, N. YAMAMOTO & T. ATAKA. 1998. Application of scanning near-field optical microscopy to thin organic film devices. Ultramicroscopy. **71:** 269–274.

Electronic Interaction between Submonolayer Molecular Films and Substrates of Different Nature

H.C. WOLF,[a] U. GÓMEZ, AND H. PORT

3 Physikalisches Institut, Universität Stuttgart, D–70550 Stuttgart, Germany

ABSTRACT: The electronic interaction between UHV-deposited organic molecules (MePTCDI) and ultraclean insulating, semiconducting, and metal substrates (quartz, Ag, Si) is investigated. By means of submonolayer coverage (0.01 monolayer) intermolecular interaction is suppressed and quenching of molecular excitation is studied directly by analyzing the fluorescence lifetime as a function of the distance between molecule and substrate. The quenching on Ag(111) can quantitatively be described by the classical image dipole theory (CPS), implying interaction with the charge carriers in the metal. Quenching on the semiconductor Si(111) is much less effective and is explained by creation of electron/hole pairs in the substrate.

INTRODUCTION

For realizing molecular electronic devices, the interface between the molecular counterpart and the world of conventional electronics is a crucial problem. In order to improve the understanding and possibly control the mutual interaction, we investigated the radiationless quenching of electronic excitation of molecules by insulators, semiconductors, and metals as substrates. The question addressed in our work is: How do different substrates influence the spectral behavior and lifetime of molecular fluorescence?

Since the first experiments[1] there have been a series of theoretical studies (reviews in Refs. 2 and 3) as well as experimental investigations, especially for mono-/multilayer molecular films in the excited triplet state and metallic substrates.[3] In the case of excited singlet state quantitative analysis of fluorescence lifetimes suffers from nonexponential fluorescence decays caused by competing intermolecular interactions within the molecular films.[4,5] The present work is concerned with the quenching of the excited singlet state of molecules which are adsorbed on surfaces in such a low concentration that intermolecular interaction is negligible.

We have chosen MePTCDI (*N,N′*-dimethylperylene-3,4:9,10-bis(dicarboximide)) as an example.[6,7] This molecule is large enough that it can be immobilized on the substrate at low temperatures,[7] and thus does not aggregate after evaporation like anthracene, which we have used in earlier experiments.[5,8] Moreover, MePTCDI has a higher fluorescence quantum yield (≈ 93%) than anthracene and is better suited for the present experiments, where quenching ratios as high as 1 : 1000 occur.

[a]Address for telecommunication: Phone, (++49) 711 685-5278; fax, (++49) 711 685-5281.

In this contribution it will be shown that the results for metal substrates are quantitatively understood within the framework of the classical CPS theory, whereas for semiconducting substrates a different quenching mechanism comes into play.

EXPERIMENTAL METHOD

All measurements presented in this paper were performed *in situ* under ultrahigh vacuum conditions ($< 10^{-9}$ mbar). Before loading the commercially available MePTCDI source material[9] into the UHV system, it was purified twice by gradient sublimation. The purified material is loaded immediately into the effusion cells, where it is further purified by heating to 300°C for 72 hr while keeping under UHV conditions. The organic films were grown by molecular beam deposition at a deposition rate of 1–2 Å/min in the case of mono-/multilayer films and at 0.1–0.2 Å/min in the case of submonolayer films. The film thickness was monitored by a quartz crystal oscillator calibrated by optical reflection spectroscopy. During film deposition the substrate temperature was kept at $T_S = 10$ K using a helium flow cryostate (Cryovac).

As metal substrates we used polished Ag(111) single crystals, cleaned in an UHV preparation chamber by standard UHV techniques (sputtering with Ar^+ ions at 500 eV and consecutive heating to 900 K). As semiconducting substrates we used polished Si(111) wafers, cleaned *ex situ* by applying a modified RCA procedure,[10] hydrogen terminated by wet etching in aequous solution of NH_4F,[11] and transferred into the UHV chamber under inert gas atmosphere immediately after etching. As insulating substrates we used polished quartz slides (Suprasil W), cleaned *ex situ* with acetone and methanol prior to heating at 500°C in the UHV chamber for several hours.

The optical experiments were carried out *in situ* at $T = 10$ K immediately after film deposition without intermediate warming-up to room temperature. For time-resolved fluorescence measurements we used a frequency-doubled actively mode-locked Nd:YAG laser system (532 nm) for photoexcitation and a narrow-band detection system with a subtractive double monochromator and time-correlated single-photon counting.

SUBMONOLAYER COVERAGE: INTERMOLECULAR ENERGY TRANSPORT EXCLUDED

By a systematic reduction of the coverage into the submonolayer regime the (statistically) averaged distance between the individual MePTCDI molecules is gradually reduced.

As a result energy transport within the organic film is suppressed successively, emission of the excimeric trapping centers 'E' is diminished at the expense of a higher energy monomeric emission 'M' (FIGURE 1). In the limit of ultrathin films (< 0.05 ML) the excimeric emission is totally absent and a pure monomeric fluorescence spectrum remains.

The corresponding monomeric fluorescence transients reflect this situation of isolated molecule behavior. In the limit of ultrathin films the transients are essential-

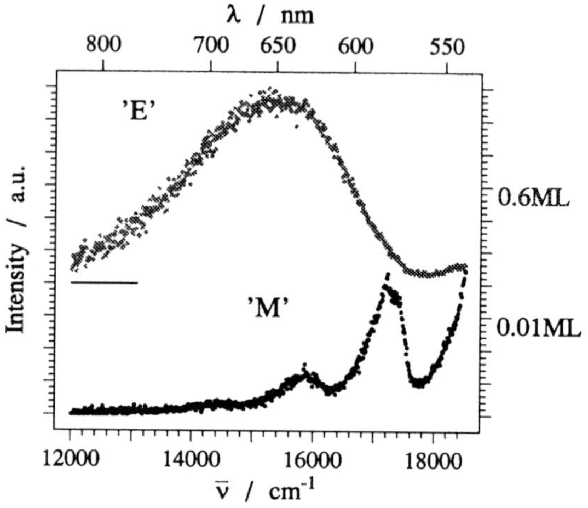

FIGURE 1. *In situ* quasi-cw emission spectra of submonolayer MePTCDI films on quartz at 10 K: excimeric 'E' for 0.6 ML, monomeric 'M' for 0.01 ML.

ly monoexponential and the decay time on noninteracting substrate (quartz) $\tau \approx$ 3.9 ns is the same as for the molecule in deoxygenated solution[7] (FIGURE 2b).

MOLECULE–SUBSTRATE INTERACTION

Substrate Variation

The interaction between the electronically excited MePTCDI molecules and substrates of different nature was analyzed by investigating the (monomer-type) emission of ultrathin films with submonolayer coverage deposited directly on metal, semiconducting, and insulating substrates. The identical shape and energetic position of the fluorescence spectra (FIGURE 2a) indicate that there is no specific effect of binding between the MePTCDI molecules and the various substrates.

In contrast to the spectral shape, the fluorescence decay is strongly influenced by the substrate (FIGURE 2b) because of nonradiative energy transfer to the substrate. On semiconducting substrates (Si(111):H) the molecular lifetime (τ_{Si} = 110 ps) is already reduced by more than one order of magnitude. On metal substrates (Ag(111)) the reduction of the lifetime is further enhanced, and from deconvolution of the decay a molecular lifetime $\tau_{Ag} \approx$ 2 ps is derived.

For the case of Si(111):H, we investigated the dependence of the molecular lifetime on the concentration of the free charge carriers in the semiconductor. To this end both dopant concentration and temperature were varied.

FIGURE 3 shows the fluorescence decay of ultrathin MePTCDI films on (n-type) Si(111):H wafers with dopant concentrations $N_D = 10^{13} \div 10^{16}$ cm^{-3}. No difference in the transients is observed upon doping. Moreover, additional experiments on p-type Si(111) proved, that also the type of doping does not influence the molecule–substrate interaction.

Upon varying the temperature in the range 10 ÷ 200 K, the charge carrier concentration of the Si(111):H substrates increases by more than eight orders of magnitude, but no effect on the fluorescence decays is observed.

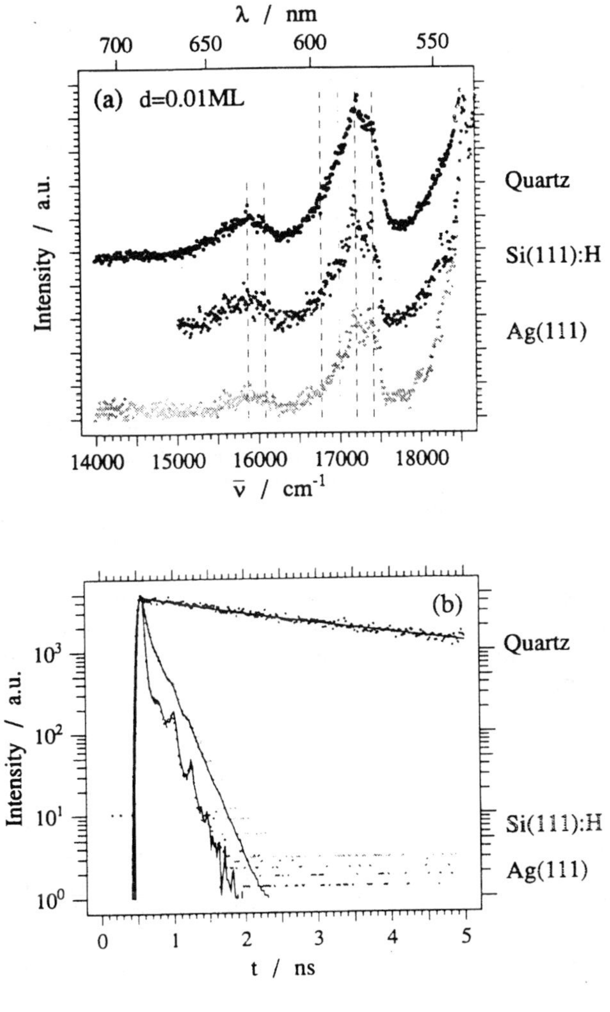

FIGURE 2. Ultrathin (d = 0.01 ML) MePTCDI films on substrates of different nature: insulator (quartz), semiconductor (Si(111):H) and metal (Ag(111)). **(a)** *In situ* quasi-cw fluorescence spectra. **(b)** Fluorescence transients; *solid lines*: mono-/biexponential fitting curves.

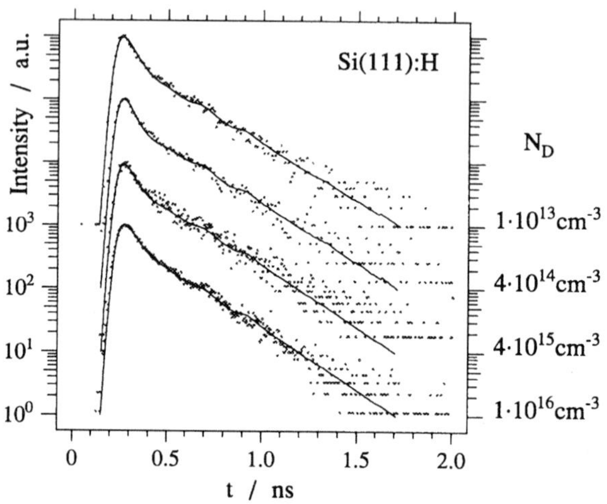

FIGURE 3. Fluorescence transients of ultrathin MePTCDI films on n-type Si(111):H; variation of the dopant concentration $N_D = 10^{13} \div 10^{16}$ cm^{-3}. *Solid lines*: Mono-/biexponential fitting curves.

These experiments show that the free-charge carriers (electrons as well as holes) in the Si(111) substrates are not responsible for the molecule–substrate interaction in the considered doping range $N_D \leq 10^{16}$ cm^{-3}.

It should be emphasized, that this type of experiment requires well-defined surface preparation, since surface contamination heavily influences the effective molecule-substrate interaction. In the example of FIGURE 4 it is shown that fluorescence quenching of the MePTCDI molecules is significantly reduced (τ = 273 ps) on Si(111) substrates without hydrogen termination. It is obvious, that the natural oxide layer (typ. ≥ 5 Å) on the Si(111) surface[12] — acting as an inert spacer layer — is responsible for the reduced energy transfer.

Distance Dependence

The drastic lifetime shortening effect described above (FIGURE 2b) has been analyzed as a function of the distance between the MePTCDI molecules and metal substrates (Ag(111)) in the limit of isolated molecule behavior, where intermolecular interaction can be neglected. The experiments were performed by using inert rigid spacer (Ar) layers of variable thickness (0–300 Å) to separate the MePTCDI submonolayers (0.01 ML) from the substrate.

The Ar layers were prepared by vacuum evaporation onto the substrate at T_S = 20 K to ensure smooth growth of spacer films. For quantitative analysis a well-defined thickness of the spacer films is essential. Subsequently the temperature was decreased to T_S = 10 K and the ultrathin MePTCDI film was deposited. Detailed temperature-dependent measurements proved that at low substrate temperatures between 10–30 K the film structure remained stable for several hours and the MePTC-

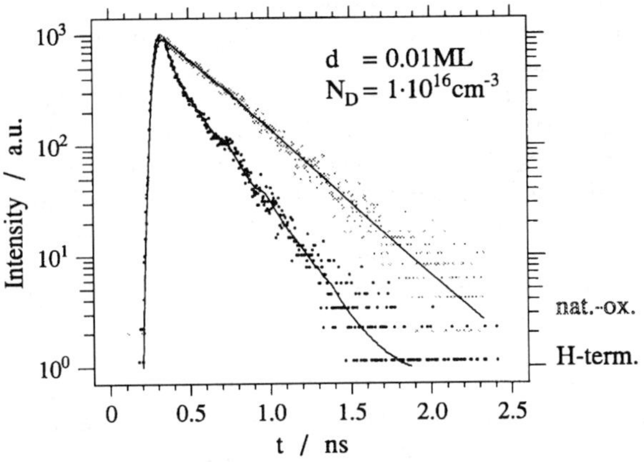

FIGURE 4. Fluorescence transients of ultrathin MePTCDI films on Si(111) substrates: H-terminated versus naturally oxidized Si(111) surface.

DI molecules did not penetrate into the spacer layers, whereas at higher temperatures penetration could be observed.

In FIGURE 5a the fluorescence decays of the MePTCDI molecules at various spacer thicknesses (0–300 Å) are shown. The individual transients deviate only little from an exponential decay and are modeled by mono-/biexponential decay functions. Approximating the decay by an average decay time $\bar{\tau}$ (FIGURE 5b), a monotonic fluorescence lifetime shortening of more than three orders of magnitude is observed when reducing the spacer thickness between 300 Å and 0 Å. The nearly exponential shape of the decay curves indicates a well-defined distance between the individual MePTCDI molecules and the Ag(111) substrate and reflects the homogeneity of the Ar spacer films.

DISCUSSION

The experimental approach of using ultrathin MePTCDI films with submonolayer coverage is appropriate for the direct comparison of the observed distance-dependent fluorescence quenching by dielectric substrates with theoretical predictions. Due to the suppression of competing intermolecular interaction processes the theoretical assumptions of isolated molecular dipoles is perfectly fulfilled.

Application of the classical CPS-theory[2] requires the knowledge of various parameters: (i) the dielectric constants of the substrates and spacer-layers at the appropriate dipole frequency (well known from the literature) and (ii) the intrinsic molecular fluorescence lifetime $\tau_0 = 3.95$ ns and fluorescence quantum yield $\Phi_F = 0.93$. Both quantities were measured in our group for a dilute deoxygenated solution of MePTCDI molecules in $CHCl_3$, and Φ_F was determined relative to anthracene as a fluorescence standard.[13] For the calculation of the distance-dependent molecular lifetimes within the CPS-model there are no further assumptions required.

For metallic substrates (Ag(111)) we found an excellent agreement between the measured molecular lifetimes and the CPS calculations in the full distance range of FIGURE 5b. Consequently, the interaction mechanism implied by CPS, which can be understood as a dipolar coupling between the excited molecules and the charge carriers in the metal substrate (in our case coupling to surface plasmons [nonresonant] and lossy surface waves[3]), provides a correct description of the coupling, even for small molecule-substrate distances (down to 20 Å). There is no experimental evidence requiring consideration of either higher multipolar interactions[14] or different elementary excitations in the substrate.[15]

Furthermore, from the excellent agreement with the parallel dipole case (FIGURE 5b), the orientation of the MePTCDI molecules can be deduced: as the S_0S_1 transition dipole moment of the MePTCDI molecules is directed along the long molecular axis, that very axis is oriented parallel to the substrate surface.

In contrast to Ag(111), the measured molecular lifetimes for MePTCDI on Si(111):H substrates are in agreement with the CPS predictions only for molecule-substrate distances >25 Å. For smaller distances the fluorescence quenching is clearly much less than predicted; in the limit $d \rightarrow 0$ when the molecules are deposited directly on the substrate, the deviation is more than one order of magnitude.

It is concluded that for semiconducting substrates the CPS description implying coupling of the molecular dipole near field components directly to free charge carriers in the substrate is not valid.

On the other hand, a different behavior is expected since the charge carriers in the semiconductor should not couple to the molecular dipole in the visible spectral range because of the plasma frequency in the FIR.

The latter aspect was confirmed in the experiments, demonstrating that a large variation of the charge-carrier density by changing dopant concentration and/or temperature does not influence the molecule–substrate interaction (FIGURE 3).

Therefore another mechanism must be responsible for the nonradiative fluorescence quenching on semiconducting substrates. Stavola, Dexter, and Knox (SDK)[16] discussed the possibility of dipole-induced generation of electron-hole pairs in the semiconductor and developed a semiclassical model for the distance-dependent quenching rate. For large molecule–substrate distances ($\gtrsim$25 Å) the SDK model reproduces the CPS results, whereas in the very near field ($\lesssim$25 Å) the fluorescence quenching predicted by SDK is much less than that given by CPS.

Our experiments of ultrathin MePTCDI films on Si(111):H substrates (as well as additional ones on multilayer films[17]) are in agreement with the SDK model. Therefore, in this case the generation of electron-hole pairs by a dipole-type interaction seems to be the dominant interaction mechanism between the electronically excited MePTCDI molecules and the semiconductor.

This paper concentrates on experiments in the limit of ultrathin MePTCDI films with submonolayer coverage in order to study the molecule–substrate interaction directly. A more detailed analysis considering competing intermolecular interaction in films of multilayer coverage will be published separately.[17]

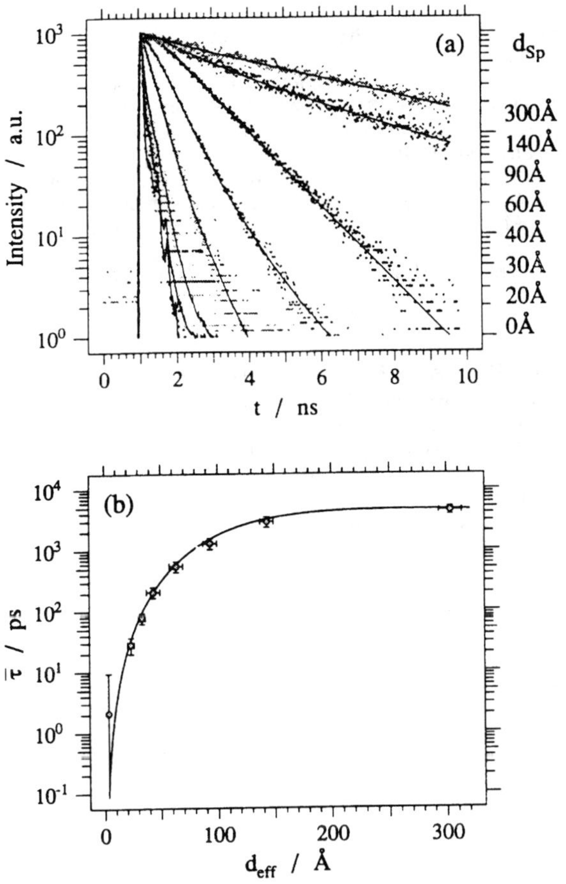

FIGURE 5. Distance-dependence of the direct molecule–substrate interaction: spacer separated MePTCDI molecules on Ag(111) substrates. **(a)** Fluorescence transients for different spacer thicknesses d_{Sp}. *Solid lines*: Mono-/biexponential fitting curves. **(b)** Averaged monomer lifetimes $\bar{\tau}$ deduced from **(a)** plotted as a function of the effective distance between the molecules and the substrate: $d_{eff} = d_{Sp} + 3.5$ Å. *Full line*: theoretical prediction (CPS) for parallel dipole orientation with respect to the surface.

ACKNOWLEDGMENT

Financial support by Deutsche Forschungsgemeinschaft and by Fonds der Chemischen Industrie is gratefully acknowledged.

REFERENCES

1. DREXHAGE, K.H. 1974. *In* Progress in Optics. E. Wolf, Ed. Vol. **XII:** 163–232. North-Holland. Amsterdam.
2. CHANCE, R.R., A. PROCK & R. SILBEY. 1978. Advan. Chem. Phys. **37:** 1–64.
3. WALDECK, D.H., A.P. ALIVISATOS & C.B. HARRIS. 1985. Surf. Sci. **158:** 103–125.
4. ALIVISATOS, A.P., M.F. ARNDT, S. EFRIMA, D.H. WALDECK & C.B. HARRIS. 1987. J. Chem. Phys. **86:** 6540–6549.
5. DAFFERTSHOFER, M., H. PORT & H.C. WOLF. 1995. J. Lumin. **66–67:** 208–212.
6. GRASER, F. & E. HÄDICKE. 1984. Liebigs Ann. Chem. **3:** 483–494.
7. GÓMEZ, U., M. LEONHARDT, H. PORT & H.C. WOLF. 1997. Chem. Phys. Lett. **268:** 1–6.
8. DAFFERTSHOFER, M., H. PORT & H.C. WOLF. 1995. Chem. Phys. **200:** 225–233.
9. HOECHST AG. Frankfurt/Main, Germany.
10. KERN, W. 1993. Handbook of Semiconductor Wafer Cleaning Technology. Noyes Publ. Park Ridge, NJ.
11. PIETSCH, G.J. 1995. Appl. Phys. A **60:** 347–363.
12. HAHN, P.O. & M. HENZLER. 1981. J. Appl. Phys. **52:** 4122–4127.
13. PARKER, C.A. & W.T. REES. 1960. Analyst **85:** 587–600.
14. CHANCE, R.R., A. PROCK & R. SILBEY. 1976. J. Chem. Phys. **65:** 2527–2531.
15. PERSSON, B.N.J. & N.D. LANG. 1982. Phys. Rev. B **26:** 5409–5415.
16. STAVOLA, M., D.L. DEXTER & R.S. KNOX. 1985. Phys. Rev. B **31:** 2277–2289.
17. GÓMEZ, U., H. PORT & H.C. WOLF. To be published.

The Effect of Electric Fields on Double-Well-Potential Molecules

ARI AVIRAM AND PAUL ROLAND

IBM Research Division, Thomas J. Watson, Research Center, P.O. Box 218, Yorktown Heights, New York 10598, USA

INTRODUCTION

The tunneling phenomenon in double-well-potential molecules emerges as the dominant feature in the design of molecular-electronic switches,[1] and plays a pivotal role in *molecular electronics* (ME). Of particular interest is the study of double-well-potential molecules in electric fields.[1–4] It was found that the *inverted effect* plays an important role in electric-field switching, since it is expected that there would be an applied electric field (F_{crit}), at which the electron transfer rate k, is at a maximum. This is apparent from the Marcus/Hush expression for electron-transfer rates,[5] for the condition where the exponent is equal to zero.

$$k = A\exp\left[\frac{-(\Delta G + \lambda)^2}{4\lambda k_B T}\right] \qquad A = \frac{2\pi}{\hbar}|T_{ab}|^2\left(\frac{1}{\sqrt{4\pi\lambda k_B T}}\right)$$

Here λ is the reorganization energy, T_{ab} the tunneling matrix element, and ΔG the standard free energy of the reaction. The free energy is directly related to the exothermicity of the electron transfer (the energy difference between the two states of the flexomer), which in turn is dependent on the applied electric field. The exponent in the Marcus/Hush expression can be written as $\lambda_{tot} - 2\mu_z F_z$, where μ_z is the dipole moment component parallel to the field (which includes field-induced contributions) and $2\mu_z F_z$ is the field-induced exothermicity.

When the exponent is set to zero, one can obtain the relationship between F_{crit} and the dipole moment[2,3]

$$F_{crit} = \lambda/2\mu$$

From a practical point of view the value of F_{crit} should be as small as possible. This can be obtained when molecules with large dipole moments are considered. Fields as low as 10^5 V/cm could conceivably switch certain double-well molecules. This is a remarkable conclusion since normally molecules are hardly affected by such small electric fields. Compared to this effect, the *Stark effect* is a minor perturbation. It seemed important to study this effect experimentally and this article presents our initial experimental results.

The bistable character of hemiquinones and their possible use as switches in molecular electronics has been described in several publications,[6] where the possibility of electric field switching of these mixed-valence molecules has also been proposed. A representative example of a hemiquinone is model compound **1** which is shown in FIGURE 1. The double-well character of the potential surface of these flexomers is

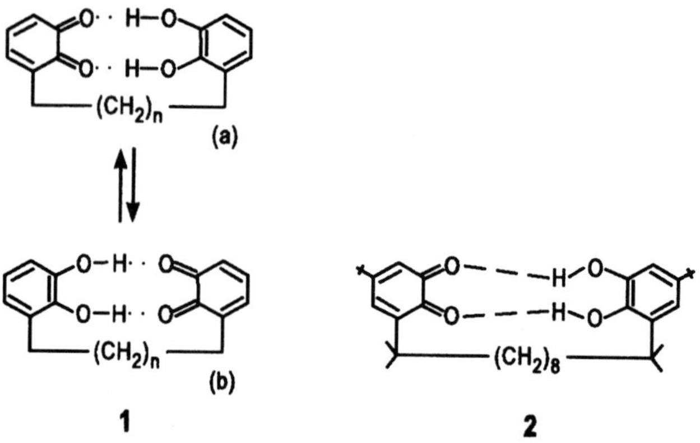

FIGURE 1. The model compound "hemiquinone" and the one reported in Ref. 1.

understood on the basis of theoretical studies, and was confirmed experimentally by NMR spectroscopy, following the synthesis of the model compound **2**.[6] However, the model compound **2**, which was synthesized to confirm the double-well character, wasn't suitable for the electric-field switching experiments since it was not designed for proper orientation in electric fields. For such an application the molecules have to be placed perpendicularly between metallic plates. Additionally, since the molecular properties are of interest, the preferred embodiment is a single molecular layer deposited perpendicularly on a metal surface. We designed a synthetic method aimed at producing monomolecular films of hemiquinones covalently bound to conductive surfaces. Our method was based in part on the surface silylation method developed by Murray,[7] by Haller,[8] and by Sagiv,[9] and on a coupling reaction developed in our laboratory. This method was published previously[10] and is repeated below with added details. Experiments of electric-field switching have been performed on hemiquionone monolayers and the results of these experiments are also presented below.

SYNTHESIS AND CHARACTERIZATION OF COMPOUNDS

The addition reaction of thiols to quinones is well known and has been reviewed.[11] However, the case of the *t*-butyl substituted *o*-quinones such as the quinone ring in compound **2**, has not been studied previously. We used the model compound 3,5-di-*tert*-butyl-*o*-quinone (DTBQ) and reacted it (i) with 3-mercaptopropionic acid to give **3**, and (ii) with ethanethiol to give **4**, by the reaction sequence shown in FIGURE 2. Both reactions produced quantitative yields at room temperature within minutes. Both reactions were carried out by dissolving the thiol in hexane and adding it to a solution of the quinone in hexane. The product precipitated within minutes and was collected and recrystallized from methylcyclohexane. The products were analyzed by elemental analysis, NMR, IR, and MS. This established that the site "4" on the

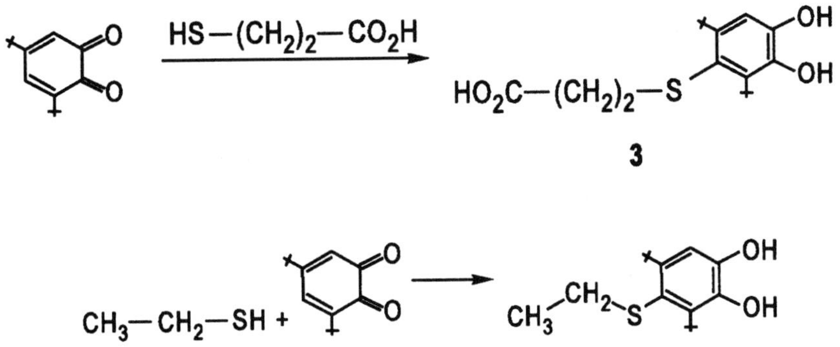

3

4

FIGURE 2. Reaction of DTBQ with mercaptans.

rings is accessible and very reactive with thiols. It was also determined that the thioether coupling product is the catechol **3** or **4** (depending on the thiol reactant) and that it is not oxidized by the reactant *o*-quinone. This was also verified by NMR spectroscopy: The room temperature NMR spectrum of a 1:1 molar mixture of DTBQ and the thioether catechol **4** is identical with the superposition of the two spectra of the individual components, suggesting that the proton exchange reaction is very slow at room temperature, and that the equilibrium is in favor of the unexchanged *o*-quinone/thiocathechol pair. NMR temperature-dependent line broadening spectroscopy was used to study quantitatively the kinetics of proton exchange and the thermodynamic equilibrium of the *o*-quinone-thiocathechol mixture.[6] These studies revealed that signals are broadened above room temperature and that exchange half life at 125°C is 60 msec.

SURFACE ANCHORING OF A FREE THIOL BRIDGE

Our scheme calls for binding a bridging molecule with a "free" thiol group to a metal surface. The purpose is to use this "free" thiol to subsequently anchor the hemiquinones by the thiol/quinone reaction described above. 3-mercaptopropyltrimethoxysilane was chosen for this purpose. It has a silane group that should bind to the metal and a thiol group that can be used to bind the quinone. Unexpectedly, this compound is fairly unreactive with silicon and aluminum surfaces, but does react with noble metals. However, this type of binding is in a way that is opposite to our need. The reaction of 3-mercaptopropyltrimethoxysilane with aluminum surfaces was studied extensively and conditions were established for proper silylation.

The following is the procedure that we used: The substrate was coated with an aluminum film of 1200 Å. The substrate was placed in a container that could be hermetically sealed which also contained two ports, both equipped with valves. These ports were designed to permit the input and output of gases. The chamber was placed in an oven at 75°C. The chamber was continuously purged with nitrogen through the

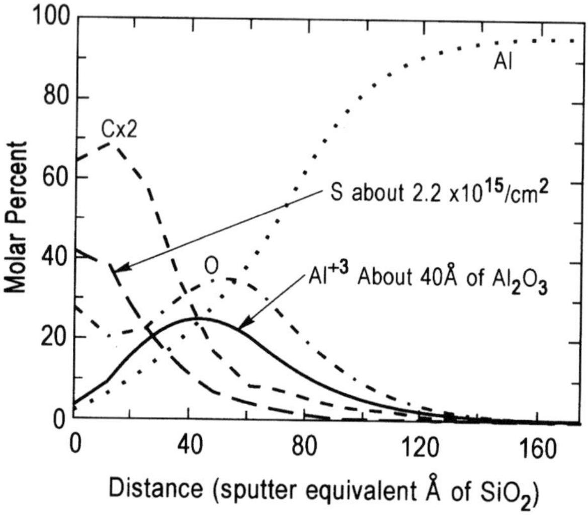

FIGURE 3. Auger spectra at different sputter depth for S, C, O, and Si. The concentration of S on the surface indicates a full monolayer coverage.

input port, while the effluent gas from the exit port was directed to an efficient fume hood. Simultaneously the nitrogen gas was passed through an additional chamber that contained 3-mercaptopropyltrimethoxysilane, thus collecting vapors from this chamber and transporting them into the surface reaction chamber. The liquid reagent cell was also placed in the oven. The reaction was continued for 18 hours and then the flow of reagent was discontinued, although the flow of pure nitrogen was maintained. After cooling, the substrates were taken out for further reactions or analysis.

Analysis of the silylated surfaces was carried out by Auger, XPS, and grazing angle IR spectroscopies, and the Auger spectrum of the silylated aluminum surface is shown in FIGURE 3. The figure clearly shows that sulfur was incorporated in the surface. All analytical data were consistent with monolayer incorporation.

PREPARATION OF HEMIQUINONE MONOLAYERS

The surface silylation provided a thiol group dangling from the surface and free to react with quinones in general. The surface reaction was first carried out with DTBQ to produce the corresponding 4-thiocatechol on the aluminum surface. The successful surface reaction was verified by comparing the XPS spectra of the above-mentioned surface (FIG. 4a) with the XPS spectrum of an aluminum surface coated with a thick layer of 4-ethylthio-3,5-di-*tert*-butylcatechol 4 (FIG. 4b), which was prepared by conventional chemical means. Comparison between the two spectra reveals striking similarities that suggest that the two surfaces are almost identical. Analogously, for preparation of hemiquinone monolayers on aluminum surfaces, the aluminum surface was first reacted with 3-mercaptopropyltrimethoxysilane, followed by reaction of the surface with the bisquinone, as shown schematically in FIG-

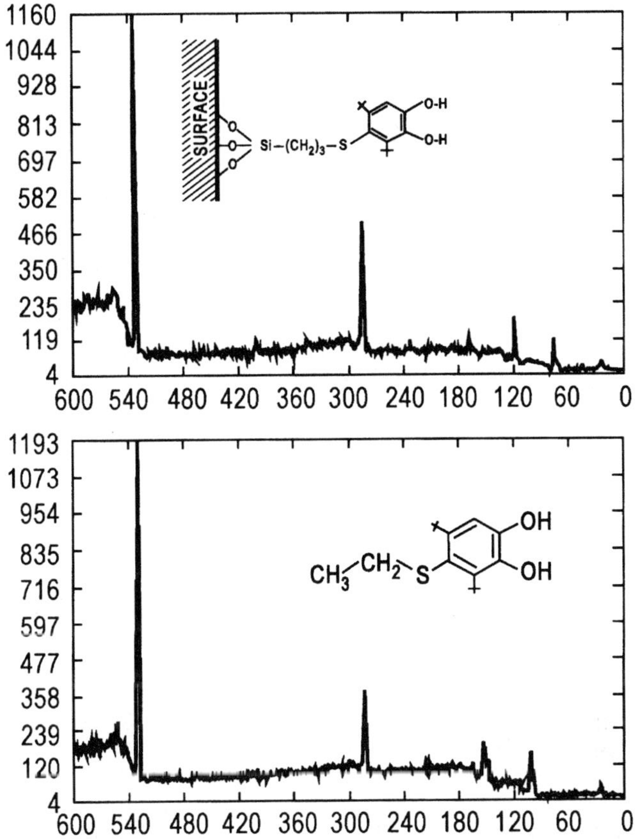

FIGURE 4. XPS spectra of **(top)** a surface containing a chemically bound catechol and **(bottom)** of a surface coated with 4-alkylthio-3,5-di-*tert*-butylcatechol,4. *Abscissa*: electron binding energy (eV), relative –; *ordinate*: photoelectron intensity (arbitrary units).

URE 5. This reaction produces a hemiquinone in one step since the ring that reacts with the mercapto group is transformed into a catechol. The reaction was also corroborated by Auger, XPS, and IR spectroscopies.

THE BICHROISM OF SURFACE HEMIQUINONES

The surface-anchored hemiquinone **5** is not symmetric and consequently it is bichroic. (The term bichroism is used here to emphasize that each flexomer should have a different spectrum.) The experimental bichroism in the visible region is shown in FIGURE 6. These spectra were taken on the two quinones that comprise halves of the two flexomers, one being the quinone derived by oxidation of com-

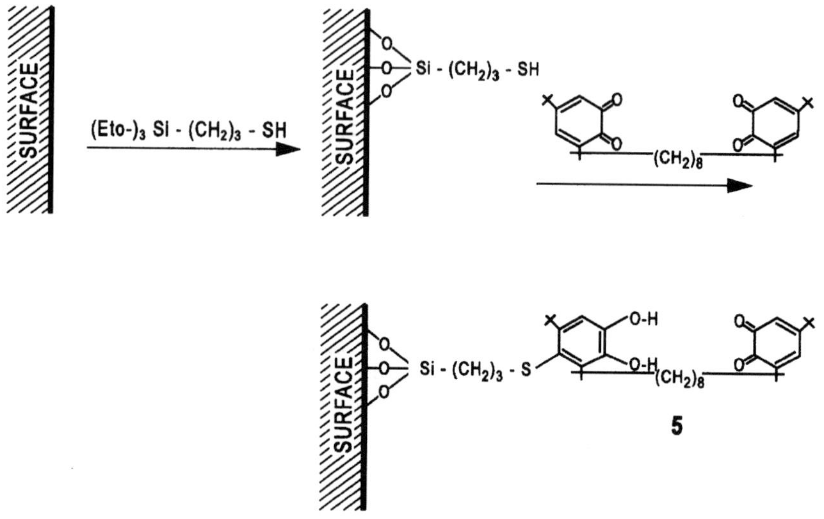

FIGURE 5. The reaction sequence that produces surface-anchored hemiquinones.

pound **4**, and the other DTBQ. Since the corresponding catechols are both white, they do not absorb in the visible, and consequently the absorption spectra of the two quinones is representative of the spectra of the two hemiquinones.

EXPERIMENTAL SETUP FOR E-FIELD SWITCHING

Two 10 cm × 10 cm mirrors were placed facing each other with a gap of 0.2 mm that was obtained with two spacers. These spacers had mirror-finish sides that were metallized with aluminum and acted as reflectors of light traveling inside the cavity. The optical path is shown schematically in FIGURE 7. A laser beam that enters the cavity on top is repeatedly reflected inside the cavity by the large mirrors and by the spacer mirrors. Then it emerges at the other end as shown. This cell was designed to increase the probability of optical detection.

One of the mirrors was covered with a hemiquinone monolayer as specified above and the big mirrors were fitted with electrical leads. The whole cavity was placed in a high vacuum chamber (10^{-7} Torr) to permit the application of intense electric fields.

The light beam used was a green He/Ne laser beam operating at 543 nm, very close to the absorption peak of the higher energy flexomer. The detection scheme consisted of two photodiodes that measured the intensity of the beam that passes through the cavity and a reference beam that did not. Both beams passed through an optical chopper. The outputs of the detectors were compared with the help of a lock-in amplifier.

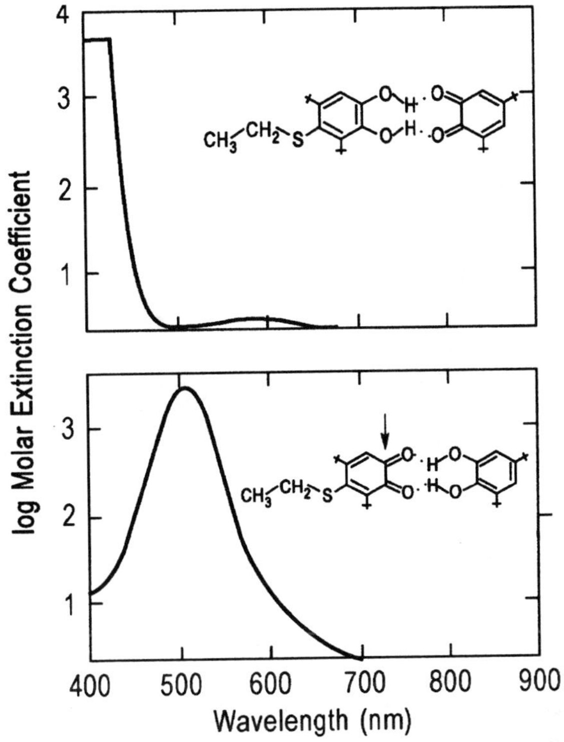

FIGURE 6. The bichroism of the two flexomers that represents the expected electrochromism of the hemiquinone monolayer.

RESULTS

When the cell and electro-optical instruments were in place, it was possible to perform the electric field switching experiments. In FIGURE 8 we present one of our results with an applied voltage of +/– 700 volts across the cavity that contained the monolayer. Under similar experimental conditions, cavities that didn't contain molecules produced no output signal. The figure shows two graphs plotted versus time. The bottom graph shows the polarity across the cell. The cell operated at three applied voltages: +700 V, 0 V, and –700 V. The top graph gives the output values of the optical detector in arbitrary units. There is a time correspondence between the two plots. Thus when the voltage is +700 V the detector signal is at a maximum. This is the *on state*. The *on state* persists at this level until the voltage is turned off. At this point the signal drops but doesn't return to 0. This is due to the fact that the cell is also a capacitor and while the applied voltage drops, the field still persists.

The optical signal returns to 0 only when the polarity of the applied voltage is reversed. Following the reversal of polarity, the applied voltage is reduced to 0. This

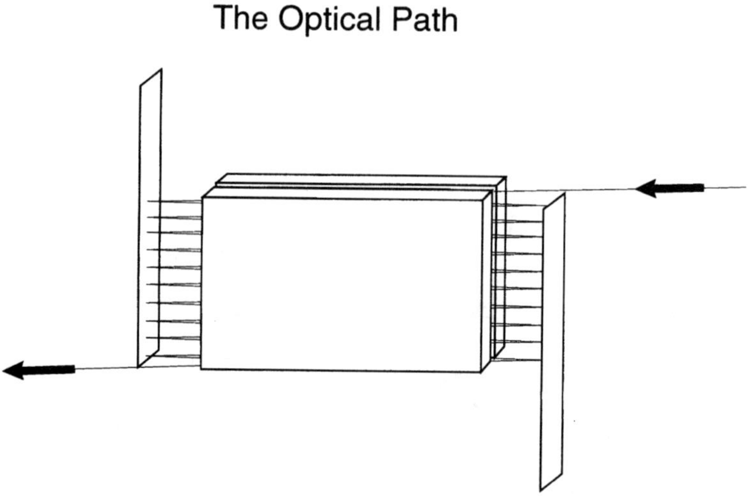

FIGURE 7. Schematic drawing of the experimental apparatus used for inducing electric-field switching of hemiquinones, simultaneous with grazing angle spectroscopy.

time no change in the optical detector signal is observed. The indication is that at the negative polarity the molecules are in their *off state* and 0 voltage doesn't alter this state. The process can be repeated over and over.

DISCUSSION

The Marcus/Hush equation provides the theoretical basis upon which the electric field switching of double-well-potential molecules stands. It predicts switching with fields that are in practical range. (For molecules with large dipole moments, switching field can be below 10^5 V/cm.) Thus it should be possible to observe such an effect.

The preparation of the monomolecular layer of hemiquinones on metallic surfaces opens the door to a variety of possible experiments aimed at verification of the electric-field switching theory of these molecular systems. It is worthwhile to point out that the surface synthesis produces a hemiquinone monolayer with all the dipoles aligned in parallel. The alignment is an integral part of the surface synthesis that produces hemiquinones with the thioether-bound catechol next to the surface. This provides a uniform starting point. The other feature of the surface-bound hemiquinone is that it is not symmetric. This property makes optical detection possible.

From the NMR data in Ref. 6 it is clear that temperature plays an important role in the proton exchange. The NMR data show that below −60°C the symmetric hemiquinone is completely stationary,[6] while the NMR results of thiol-substituted hemiquinones that resemble the surface-anchored hemiquinone **5** show that the proton shuttle begins at +60°C. The higher activation energy of the nonsymmetric flex-

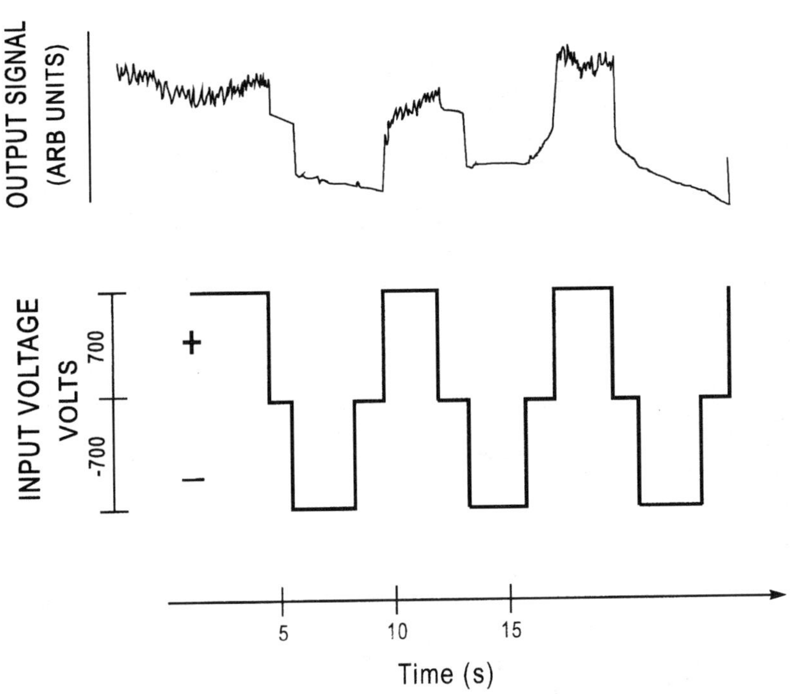

FIGURE 8. The experimental results of electric-field switching. The **bottom** graph indicates the polarity across the cell, which is +700 V, 0 V, and –700 V. The **top** graph shows the output signal of the optical detector starting with the *on state* and back to 0 *off state* when the voltage polarity graph is –700 V.

omer can be regarded as a difficulty, yet it provides some convenience. The difficulty arises from the fact that the field has to overcome a larger energy barrier than it would have to in the case of a symmetric double-well-potential molecule. On the other hand it makes it possible to do the switching experiment at room temperature (up to +60°C) because in this range there is no competition with the exchange that originates from thermal effects.

In examining the switching results in FIGURE 8 one ponders whether these results are due to signals obtained from the molecules or due to some artifact of the experimental setup. For instance it could have been argued that any geometrical changes of the mirrors could lead to a change in the intensity of the transmitted beam. Such a change could have come about from the forces of attraction imposed by the field. This particular concern can be put to rest by observing that the output signal intensity is asymmetric with respect to the electric-field polarity. Therefore we conclude that the optical signal output indicates that electric-field switching was obtained.

CONCLUSIONS

The results of this work show that it is possible to switch double-well-potential molecules by electric fields. To demonstrate this effect it was necessary to prepare surfaces containing hemiquinone molecules as monolayers aligned perpendicularly to the field and oriented with parallel dipoles. It was also necessary to develop a new surface-anchoring reaction that enabled the preparation of a monolayer of hemiquinone molecules on an aluminum surface. The results of this work are important because they confirm theories of electric-field switching of molecules and promise the possible technological application of this effect to molecular electronics.

ACKNOWLEDGMENTS

I wish to express my thanks to Dr. Robert Johnson for help with the NMR spectroscopy, to Dr. J. Clabes for the XPS spectroscopy, and H. Wildman for the Auger spectroscopy.

REFERENCES

1. AVIRAM, A. 1988. J. Amer. Chem. Soc. **110:** 5687.
2. FARAZDEL, A., M. DUPUIS, E. CLEMENTI & A. AVIRAM. 1990. J. Amer. Chem. Soc. **112:** 4206.
3. HUSH, N.S., A.T. WONG, G.B. BACSKAY & J.R. RIMERS. 1990. J. Amer. Chem. Soc. **112:** 4192.
4. TOUR, J.M., R. WU & J.S. SCHUM. 1991. J. Amer. Chem. Soc. **113:** 7604.
5. HUSH, N.S. 1961. Trans. Faraday Soc. **57:** 557; MARCUS, R.A. & N. SUTIN. 1985. Biochem. Biophys. Acta **811:** 265.
6. AVIRAM, A., P.E. SEIDEN & M.A. RATNER. 1983. *In* Molecular Electronics. F.L. Carter, Ed.: **5**. Marcel Dekker. New York.
7. CHIDSEY, C.E.D. & R.W. MURRAY. 1978. Science **231:** 25.
8. HALLER, I. 1978. J. Amer. Chem. Soc. **100: 8050.**
9. MAOZ, R. & J. SAGIV. 1984. J. Coll. Interf. Sci. **10:** 465.
10. AVIRAM, A. 1990. J. Molec. Electron. **4:** 247.
11. FINLEY, K.T. 1974. *In* the chemistry of Quinoid Compounds. S. Patai, Ed.: 877. John Wiley & Sons. New York.
12. This work was done in cooperation with R. Johnson of IBM Almaden.
13. This work was carried out by J. Clabes and F. Cardone in out laboratory.

Evolution of Strategies for Self-Assembly and Hookup of Molecule-Based Devices

D.L. ALLARA,[a] T.D. DUNBAR,[a] P. S. WEISS,[a] L.A. BUMM,[a] M.T. CYGAN,[a] J.M.TOUR,[b] W.A. REINERTH,[b] Y. YAO,[b] M. KOZAKI,[b] AND L. JONES, II[b]

[a]*Department of Chemistry, Pennsylvania State University, University Park, Pennsylvania16802, USA*
[b]*Department of Chemistry and Biochemistry, University of South Carolina, Columbia, South Carolina 29208, USA*

ABSTRACT: Strategies for self-assembling molecule-based devices are considered in terms of current chemical issues whose resolution appears critical to efficient connection and addressing of electronically active molecules between electrodes. We discuss issues related to the type and shape of the molecules, chemical bonding at junctions, molecular lengths and electrode gap matching, molecular alignment at electrodes, chemistry of deposited metal contacts, and the doping of molecular conductors. Examples of each of these aspects is given using fully conjugated molecules, constituted from rigid rod phenylene-ethynylene units, self-assembled onto metal and semiconductor surfaces.

INTRODUCTION

There has been significant interest for a number of years in utilizing the electronic properties of organic molecules for electronic and optoelectronic devices.[1–13] One of the most challenging aspects in this area is the use of single molecules as components of electronic circuits[1,2,6] to be applied to such purposes as computation. When an electrical potential is applied across two ends of a fully π-conjugated molecule, one can imagine the molecule acting as a wire with the electron flow carried by the π electrons. For an electronically asymmetric molecule, one can imagine diode-like behavior.[1,2,8] For asymmetric molecules with three or more distinct "arms," transistor-like behavior can be envisioned which would lead to amplification of input signals. Intensive efforts are now in place to synthesize appropriate molecules,[14,15] and recent experiments are beginning to show that single molecules can indeed act as "wires." Reed and co-workers have used a break junction configuration to measure the conductance of a single molecule chemisorbed between gold electrodes.[16] Bumm *et al.* have used scanning tunneling microscopy measurements to show that individual π-conjugated molecules, chemisorbed at gold electrodes via sulfur atoms, have enhanced conductivities compared to adjacent chemisorbed alkanethiolate chains.[17] Recently, Reed and coworkers also have demonstrated that properly structured molecules can act as resonant tunneling diodes.[18] Other experiments have shown that through-molecule conduction[19–21] and rectification[22,23] can occur with microensembles of molecules at electrodes, and theoretical calculations have been able to explain an interesting temperature-dependence of ensemble conduction in terms of molecular structure dynamics.[24] Related experiments of interest include the demonstration of the conducting properties of individual nanometer-scale objects such as carbon tubes,[25–27] conducting polymer filaments,[28] and semiconductor clusters.[29,30]

Why are molecules of interest as electronic components? The primary motivation stems from the general perception that the intrinsic properties of semiconductor materials may not be capable of forming the basis for the new generations of electronic devices envisioned in terms of feature sizes, operation speeds and architectures. In principle, molecules could meet some of these demanding requirements.

First, molecules are nanometer-scale objects which have extremely precise, well-defined structures. Huge numbers of these objects can be made at one time at the laboratory bench scale and all the objects will have indistinguishable chemical structures and properties.[15] Contrast this with typical electronic materials such as semiconductors. Such materials are composed of a semi-infinite periodic lattice of atoms. These structures give rise to remarkable electronic properties which are utilized as the basis for engineering complex devices, e.g., silicon for computer processors. However, when these seemingly perfect materials are confined as objects at the nanometer scale, the lattices are no longer infinite and the properties become controlled to an increasing extent by the surfaces and edges as the feature size decreases. Atoms at surfaces and edges are of higher energies, and consequently less stable, chemically and physically, than atoms in the bulk lattice. This often means that the properties of extremely small structures vary erratically as a result of the quirks in the ways in which the structures were formed and the environments in which they are placed. Most importantly, the electronic properties become difficult to control in the same precise way as is possible for the semi-infinite lattices. For example, if one were to use lithographic processing to make a large, periodic array of nanometer-scale quantum dots on a semiconductor wafer surface, there would be a distribution of responses over the array since the edges and surfaces of each object could never be made identical. In fact, it would be typical that some of the objects would not even function because of variability in the fabrication process.

Second, organic molecules can be chemically changed by well-known synthetic methods to produce different molecules with slightly altered chemical and physical properties. This allows an entirely new dimension in engineering flexibility that does not exist with the typical kinds of inorganic electronic materials. Such subtleties in molecular variations form the basis of the amazing structure/function relationships that propel the biochemistry of living organisms. For electronics applications, the multitude of electronic energy levels in molecules can be fine-tuned by simple variations in molecular structure, e.g., by changing substituents on aromatic rings in conjugated compounds. Moreover, derivatization of a molecule can lead to changes in the solubility without significant perturbation of the electronic structure, an advantage for improving the handling of compounds without changing the eventual device properties.

The case for developing molecule-based electronics is clear. But what are the challenges? One major problem lies in the fact that much of the highly sophisticated knowledge learned in semiconductor device processing does not transfer directly to assembling molecular components. Semiconductor devices are essentially made by lithographically "carving" out a large number of features in a monolithic chunk of semiconductor material. In contrast, molecules are made by synthetic chemistry methods. This means that ways need to be developed to incorporate and interconnect them into some pre-existing template structure, typically formulated at a low level of sophistication for a semiconductor-based circuit with prepared contacts for the molecules. While other architectures can be imagined, such as three-dimensional ones with novel network properties, the only practical architectural target at present is a planar array of prefabricated contacts on a chip. In fact, it has been the difficulty of meeting the elementary challenge of "hooking up" electronically active molecules between nanometer-scale contacts that has confounded the field of molecule-based electronics since its early days, more than two decades ago.[1–3]

Imagine that two metal contacts have been fabricated on a planar surface, e.g., of insulating SiO_2, with a spacing exactly the size of a desired molecule. How would one get the molecule positioned exactly between the gaps and then how would the "ends" be "attached?" There are two fundamental limiting approaches, single molecule manipulation and statistical assembly. The first is a brute force approach in which a molecular resolution tool such as the tip of a scanning tunneling microscope (STM) or an atomic force microscope (AFM) transfers each molecule to its desired location, one at a time. This approach, unfortunately, is impractical for engineering applications since it would take literally years to build a chip of the complexity of current PC processors. The second approach, which borrows from the way nature assembles living systems, utilizes a pre-designed system of contacts, molecules, and containing environment, e.g., a solution, in such a way that the desired structure self-assembles into place spontaneously, driven by the global minimization of free energy. A cartoon illustration of this process is given in FIGURE 1A. The figure also suggests how other nanometer-scale components, such as semiconductor clusters, can be incorporated as parts of the functional structures when the system is chemically primed in an appropriate way.[29,30] FIGURE 1B shows an example of hybrid processing in which a pre-self-assembled array of molecules in a microscopic pore is capped

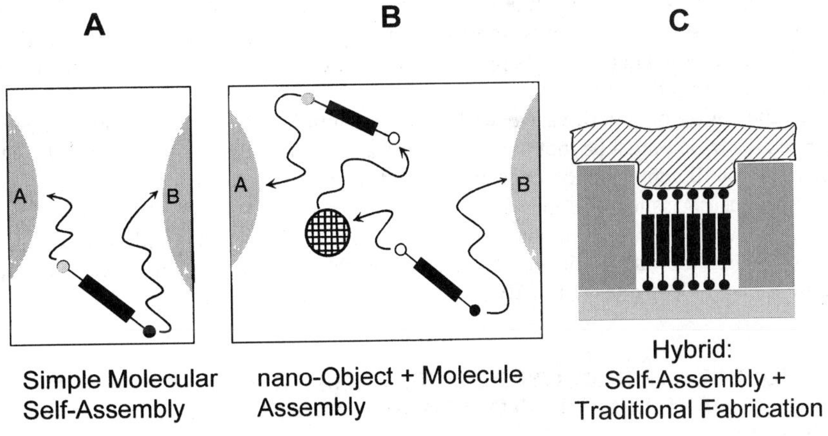

FIGURE 1. Schematic of two terminal molecule-based devices: **(A)** self-assembly of a single molecule; **(B)** self-assembly of a molecule/nano-object combination and **(C)** a hybrid device with a self-assembled ensemble of molecules contacted by a deposited top metal electrode.

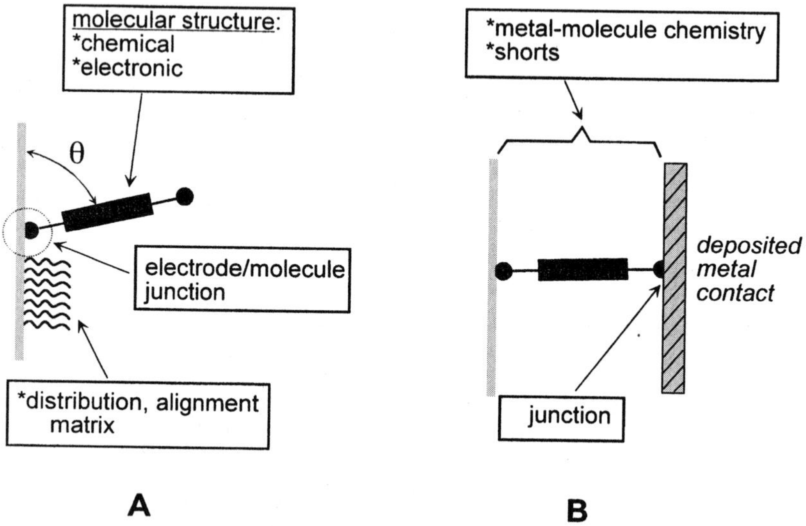

FIGURE 2. Schematic of molecules connected to electrode surfaces showing critical chemical-based aspects of their structure: **(A)** a chemisorbed molecule at an electrode and **(B)** a molecule chemisorbed at an electrode and covered by a deposited metal counterelectrode.

with a vapor-deposited metal electrode. Such "nano-pore" molecule-based devices have recently been reported.[19] The spontaneous character of the self-assembly process, in principle, allows high throughput fabrication of highly complex devices. Although problems can be expected, such as the appearance of inherent variations in the structure, the self-assembly approach has captured the interest of many workers in the device field because of its promise as a platform for building a new generation of molecule-based devices.

In the following sections we will discuss a number of the fundamental design challenges which must be met in order to self-assemble molecule-based devices. The discussion will center on the chemical aspects of forming functional structures. The specific design issues of concern are indicated in FIGURE 2. The functional structures include those prepared both by pure self-assembly (FIG. 2A) and by hybrid processes, which include traditional fabrication, such as vapor-phase deposition of top-contact metal electrodes (FIG. 2B).

A MOLECULAR STRUCTURE ISSUE: WHAT KINDS OF MOLECULES ARE USEFUL AND HOW BIG SHOULD THEY BE?

Types of Molecules: Electronic Activity and Shape

In general, to be capable of performing some useful electronic activity, molecules need a set of overlapping electronic states which connects two or more distant func-

TABLE 1. Summary of examples of fully conjugated, linear types of molecules useful as model systems in studies of molecule-based devices

Molecule	Notation	~Length, Å (for R=H)	Ref.
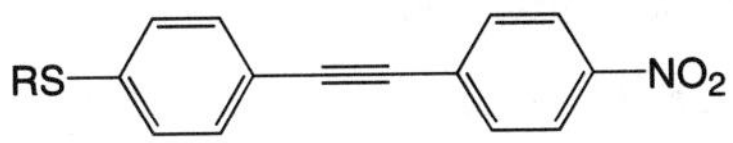	PE2(NO_2)	15.4	39
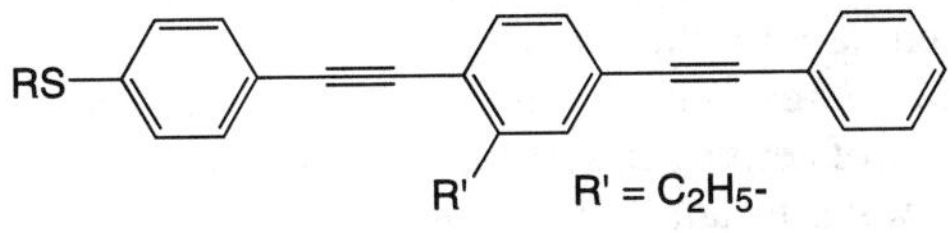	PE3	21.3	17
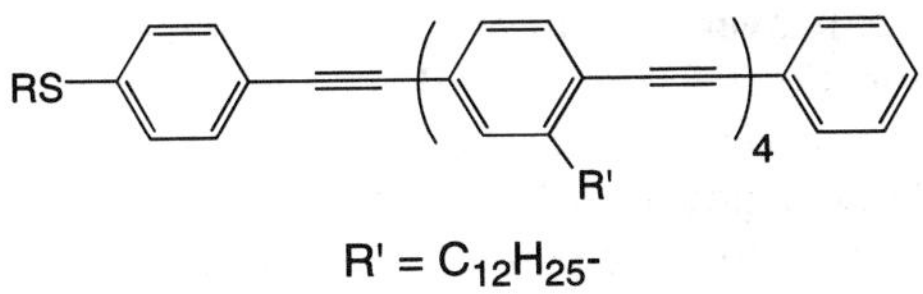	PE6	42.0	39
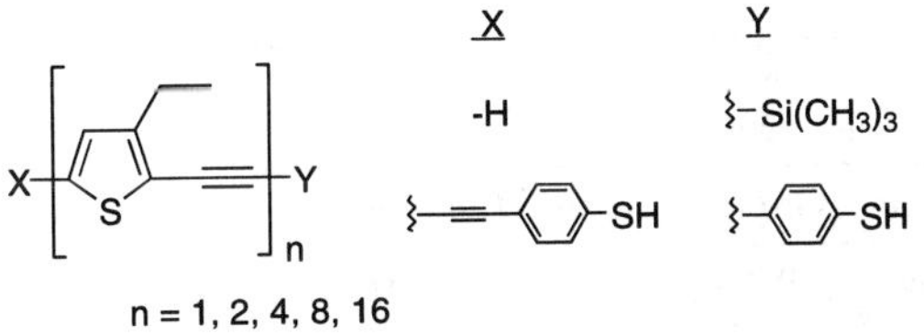	TEn	33.3 (for X = H-, Y = -C_6H_4SH, n = 4)	32

tional points in the molecule. A typical candidate would have a conjugated π orbital system on an extended σ-framework with terminal functional groups. Molecules of this type generally will be rigid with 1-, 2-, or 3-dimensional limiting shapes as depicted below. The typical example is a long, linear (~1-D) molecule with a fully conjugated π-electron system. Examples of recent interest in our own work are shown in TABLE 1. For reference, the approximate lengths are also given. These molecules all contain an acetylenic linkage. The triple bond imparts a local uniaxial symmetry to the π-electron clouds which reduces geometrical bottlenecks in the electron trans-

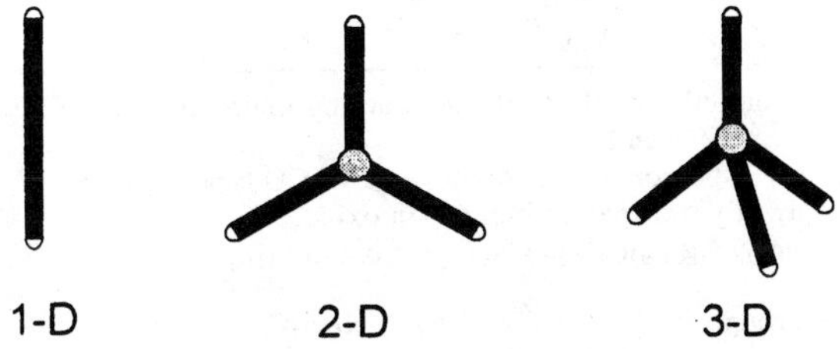

mission and forces the molecules into a useful linear shape. The syntheses of these molecules have been reported elsewhere.[31,32] Similar structural rigidity could be obtained using ladder oligomers in which multiple linked backbones span the molecule.

Terminal Functionality: "Alligator Clips"

Molecules which are to be connected between two electrodes in order to function as a "wire" or other electronic component require terminal functional groups that can form direct chemical attachments to the electrode surfaces. In this way, the appropriate electronic states of the system are all interconnected, viz., the electrode surface bands and those molecular states with similar energies and matching symmetries. In recent years the flurry of activity in self-assembly of monolayer films has produced a limited variety of ways to attach functional groups chemically to conductive surfaces. Selected examples are summarized in TABLE 2.

For gold electrodes, highly preferable over other metals because of the high conductivity and general inertness, formation of Au—S bonds can establish a stable junction. It has been estimated from experiments that the Au—SR bond strength is ~28 kcal/mol for R = alkyl groups. The compounds in TABLE 1 all show the presence of a terminal divalent sulfur derivative for forming bonds to Au surfaces. However,

TABLE 2. Selected summary of functional groups which form stable, direct bonds to bare[a] electrode surfaces. The entries are marked as yes = observation, no = not observed and not expected, and ? = not observed but a reasonable possibility

	Functional Group[d] Bonding (Literature reference)						
Electrode	RS-	RSe-	RTe-	RNC-[e]	RNH_2-	RCNS-[f]	RCH_2-
Metals							
Au	Yes[1,2,3,4]	Yes[5,6,7]	Yes[7]	Yes[8,9,10]	No	Yes[11]	No
Pt	Yes[12]	?	?	Yes[13]	?	?	No
Cu	Yes[14]	?	?	?	?	?	No
Ag	Yes[14]	?	?	?	?	?	No
Hg	Yes[15]	?	?	?	?	?	No
Oxides							
HTC[b]	Yes[16]	?	?	?	Yes[17]	?	No
ITO[c]	Yes[18]	?	?	?	?	?	No
Semiconductors							
Si	No	No	No	No	No	NO	Yes[19]
GaAs	Yes[20]	Yes[21]	Yes[21]	No	No	?	No
InP	Yes[22]	Yes[21]	Yes[21]	No	No	?	No

[a]Chemical bond formed directly to the conducting material with no intervention of a non-conducting layer, e.g., SiO_2 on Si.
[b]High-temperature superconducting oxide, e.g, YBCO type materials.
[c]Conducting, optically transparent indium tin oxide.
[d]R = organic moiety, e.g., an alkyl chain or aromatic ring.
[e]Isonitrile group.
[f]Isothiocyanate group.

—*continued*—

Table 2 — continued

1. R.G. Nuzzo and D.L. Allara. J. Am. Chem. Soc. 105, 4481 (1983).
2. R.G. Nuzzo, F. Fusco and D.L. Allara. J. Am. Chem. Soc. 109, 2358 (1987).
3. M.D. Porter, T.B. Bright, D.L. Allara and C.E.D. Chidsey. J. Am. Chem. Soc. 109, 3359 (1987).
4. Bain, C.D.; Troughton, E.B.; Tao, Y.T.; Evall, J.; Whitesides, G.M.; Nuzzo, R.G. J. Am. Chem. Soc. 1989, 111, 321–325.
5. Samant, M.G.; Brown, C.A.; Gordon, J.G. Langmuir 1992, 8, 1615–1618.
6. Dishner, M.H.; Hemminger, J.C.; Feher, F.J. Langmuir 1997, 13, 4788–4790.
7. Dunbar, T.; Reinerth, W.; Tour, J.; Allara, D., manuscript in preparation
8. Bain, C.D.; Evall, J.; Whitesides, G.M. J.Am.Chem.Soc. 1989, 111, 7155–7165.
9. Robertson, M.J.; Angelici, R.J. Langmuir 1994, 10, 1488–1492.
10. Henderson, J.I.; Feng, S.; Ferrence, G.M.; Bein, T.; Kubiak, C.P. Inorg.Chim.Acta 1996, 242, 115–124.
11. Han, W.H.; Li, S.M.; Lindsay, S.M.; Gust, D.; Moore, T.A.; Moore, A.L. Langmuir. 1996, 12, 5742–5744.
12. Shimazu, K.; Sato, Y.; Yagi, I.; Uosaki, K. Bull.Chem.Soc.Jpn. 1994, 67, 863–865
13. Hickman, J.J.; Zou, C.; Offer, D.; Harvey, P.D.; Wrighton, M.S.; Laibinis, P.E.; Bain, C.D.; Whitesides, G.M. J.Am.Chem.Soc. 1989, 111, 7271–7272.
14. Laibinis, P.E.; Whitesides, G.M.; Allara, D.L.; Tao, Y.T.; Parikh, A.N.; Nuzzo, R.G. J.Am.Chem.Soc. 1991, 113, 7152–7167.
15. Demoz, A.; Harrison, D.J. Langmuir 1993, 9, 1046–1050.
16. Chen, K.M.; Xu, F.; Mirkin, C.A.; Lo, R.K.; Nanjundaswamy, K.S.; Zhou, J.P.; Mcdevitt, J.T. Langmuir 1996, 12, 2622–2624
17. Chen, K.M.; Mirkin, C.A.; Lo, R.K.; Zhao, J.N.; Mcdevitt, J.T. J.Am.Chem.Soc. 1995, 117, 6374–6375.
18. Sato, Y.; Uosaki, K. Denki.Kagaku. 1994, 62, 1269–1275.
19. Linford, M.R.; Fenter, P.; Eisenberger, P.M.; Chidsey, C.E.D. J.Am.Chem.Soc. 1995, 117, 3145–3155.
20. Sheen, C.W.; Shi, J.X.; Mårtensson, J.; Parikh, A.N.; Allara, D.L. J.Am.Chem.Soc. 1992, 114, 1514–1515.
21. Mars, C.; Allara, D., unpublished result
22. Gu, Y.; Lin, Z.; Butera, R.A.; Smentkowski, V.S.; Waldeck, D.H. Langmuir 1995, 11, 1849–1851.

since the electronic states of the functional group-electrode interface may well influence the overall electrical response of the molecule-electrode system, it is desirable to have the option of varying the type of bonding in any given system in order to "tune" the overall performance. For example, optical spectroscopy data indicate that the Au-SR interface (R = alkyl) has localized electronic properties involving some degree of charge transfer.[34] Recent work has shown that stable Au-Se interfaces are possible with reports of the formation of organized monolayers during self-assembly of aryl[35] and alkylselenium[36] compounds. An example of an ordered alkaneselenolate SAM on Au is given by the STM image in FIGURE 3A for the case of a SAM formed by chemisorption of $(C_{12}H_{25}S)_2$.[37] FIGURE 3B shows an image of a co-assembled monolayer of S- and Se-terminated chains from a solution with the latter as the minority component. The image suggests that the two types of chains coexist with identical structural order on the overall SAM lattice. Further, since the two types of chains are equal in length, the difference in gray-scale contrast (due to sample-tip height differences at constant current) of the lattice points in the image implies that the electronic conductivity of RS- and RSe- are different. This difference would presumably be due to intrinsic differences in the junction states. Given the behavior of the Se-Au system, it follows that Te-Au chemistry should be quite suitable

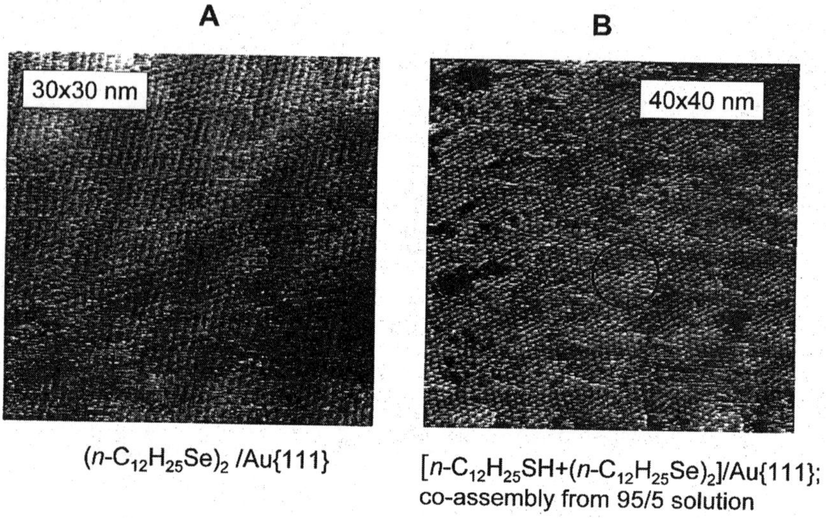

FIGURE 3. Constant current STM images of alkanethiolates and alkaneselenoates self-assembled on Au{111} surfaces: **(A)** monolayer prepared from dodecanediselenide and **(B)** mixed monolayer formed from a tetrahydrofuran (THF) solution of dodecanethiol and dodecanediselenide. In **B** the darker spots in the lattice, one shown within the circled area, are tentatively identified as the images of individual alkylselenoate molecules. Since the length of the S- and Se-based molecules are virtually identical, the darker intensity of the Se-molecule image indicates that the tip-substrate distance is lower and thus the Se-molecules exhibit lower tunneling currents.

for forming junctions also. Examples of possible chemical junctions for metals other than gold also are given in TABLE 2.

Electrodes other than metals are of interest, in particular, conducting oxides and common semiconductor device materials such as Si and III-V heterocompounds. Recent developments of self-assembly chemistry on several of these materials has provided some possibilities for useful molecule-electrode junctions and these are listed in TABLE 2. FIGURE 4A shows the IRS spectrum of a n-$C_{18}H_{37}S$- SAM on GaAs(100) formed by assembly from a thiol solution at room temperature.[38] The frequency positions of the symmetric and antisymmetric C-H stretching modes of the CH_2 units show that the chains exhibit a high degree of conformational ordering, indicative of a dense, highly organized monolayer. The observation of a 43° hexadecane contact angle is consistent with this structure. SAMs of conjugated molecules also can be formed. FIGURE 4B shows the IRS spectrum of a SAM on GaAs(100) formed from nitrobenzenethiol.[38] The presence of the characteristic -NO_2 stretching modes confirms the attachment of the aromatic ring to the GaAs surface.

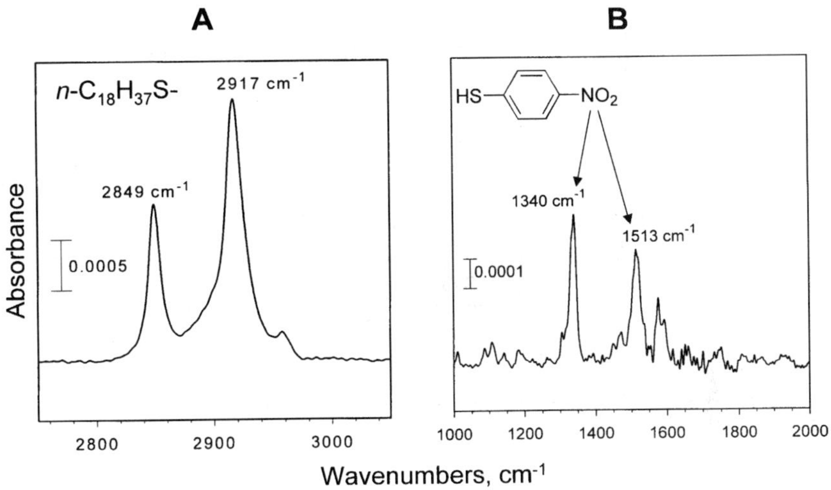

FIGURE 4. High-sensitivity transmission spectra of monolayers of thiol compounds self-assembled onto GaAs(100) from THF solution: **(A)** spectrum of an octadecanethiolate monolayer, which shows sharp peaks at 2849 and 2917 cm^{-1} characteristic of a high degree of conformational ordering and **(B)** spectrum of a nitrobenzenethiolate monolayer showing the presence of the symmetric and antisymmetric NO_2 group stretching modes. The formation of the monolayer in B demonstrates that it will be possible to self-assemble much longer conjugated molecules onto GaAs semiconductor surfaces.

Size Scales: Molecular Lengths vs. Electrode Gaps

Consider self-assembly of the simplest two-terminal device, as shown in FIGURE 5A. While it seems straightforward to come up with combinations of appropriate molecules, attachment chemistry and fabricated electrodes, significant practical issues arise in choosing the specific size of the device. The problem lies in the required match of the molecular length and the electrode gap. Starting from the 1–2-nanometer scale, the ease of synthesizing discrete molecules decreases sharply with molecular length while the ease of fabricating gaps increases rapidly with increasing spacing. On the basis of current state of the art in these two tasks, one can estimate that the optimum match exists currently at ~10–20 nm, as summarized graphically in the hypothetical plot in FIGURE 5B.

The Problem of Molecular Alignment at Electrodes During Self-Assembly

The design of a self-assembly process to connect a bifunctional molecule between two electrodes requires careful consideration of molecular orientation effects. In particular, one needs to build into the self-assembling "template" the driving force for the molecules to orient outward at the initially contacted electrode surface such that a high probability will exist of making a connection to the second. This is illustrated in FIGURE 6 for the case of rod-like molecules. The central difficulty is that the highly polarizable π-electron clouds of typical highly conjugated, planar mole-

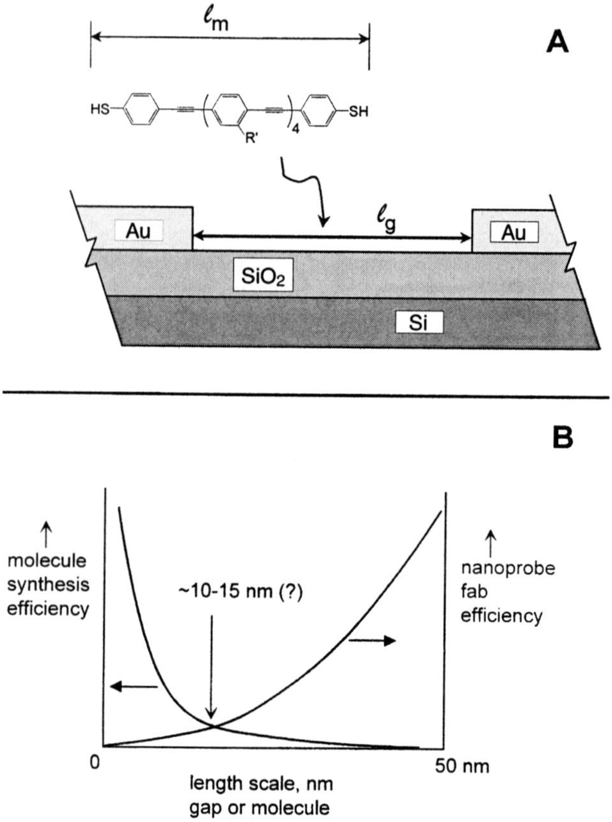

FIGURE 5. **(A)** Schematic showing the matching of the electrode gap distance with the molecular length in the assembling of a two-terminal device. **(B)** A qualitative plot illustrating the opposite trends in difficulty in synthesizing molecules and electrode gaps with controlled lengths.

cules induce substantial attractive dispersion forces towards polarizable metal surfaces. These interactions will force the molecules to lie flat, as shown in FIGURE 6A, unless other factors intervene to cause an outward orientation, as shown in FIGURE 6B.[39] Further, this problem is magnified as the size (length) of the molecule increases since the surface interaction energy per molecule increases in the flat configuration but remains constant in the end-on configuration. One can easily visualize approaching a limit of a highly extended graphite-like structure in which it would be virtually impossible to arrange the large individual units edge-on to a surface by simple physical adsorption processes. We find this trend experimentally. For example, for short, fully conjugated rod-like molecules with end-to-end length of $\ell_m \lesssim 2$ nm [e.g., PE2(NO_2) in TABLE 1] we find from IRS analyses that adsorption on Au{111} surfaces gives tilt angles of θ ~20 to 30° with respect to the surface normal, similar to θ for alkanethiolate chains.[40] In contrast, for long fully conjugated molecules such

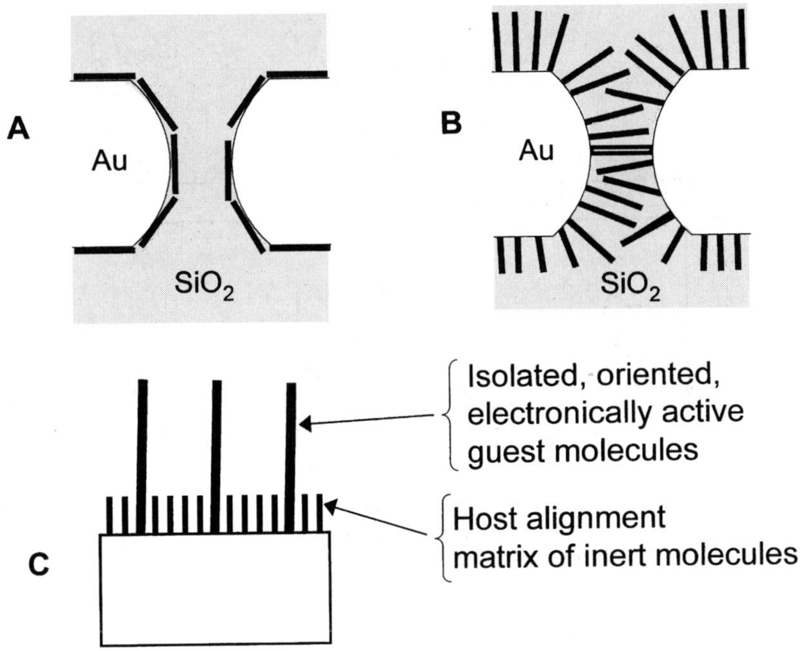

FIGURE 6. Schematic illustrating the issue of how molecular alignment at electrode surfaces affects self-assembly of contacts: **(A)** molecules lying flat at electrodes cannot form contacts; **(B)** molecules vertically oriented are ideal; and **(C)** alignment forced by insertion of guest molecules into a host matrix.

as PE6 and TE6, the molecules appear to adsorb with a significant fraction of the molecules lying close to flat to the surface. Once θ is known for a given molecule, then the ideal electrode gap distance would be $\ell_E = \ell_m \cos\theta$.

One recently developed strategy for overcoming this alignment problem involves self-assembling mixed composition monolayers in which guest electronically active molecules are dispersed, individually and/or in small bundles, in a host SAM of oriented, low dielectric, short alkanethiolate chains on a gold electrode surface.[17,39] This is illustrated in FIGURE 6C for a hypothetical case of completely vertical orientations on the electrode surface. In actual fact, as will be discussed later, the *average* tilt angles are always 30° or more in these systems. The basic idea is that the presence of the alkanethiolate chains prevents a "face to face" contact of the aromatic rings against the electrode surface. Further, when the alkanethiolate chains are arranged in an oriented lattice, one might expect the structure to induce a similar orientation in the guest molecules, analogous to a pencil placed in a stiff brush. We have examined two ways to make mixed composition SAMs, co-assembly and post-assembly insertion, as depicted in FIGURE 7. We consistently observe[41] that the insertion method gives superior results, both with respect to inducing outward alignment of the guest molecule and to maintaining order in the host alkanethiolate SAM

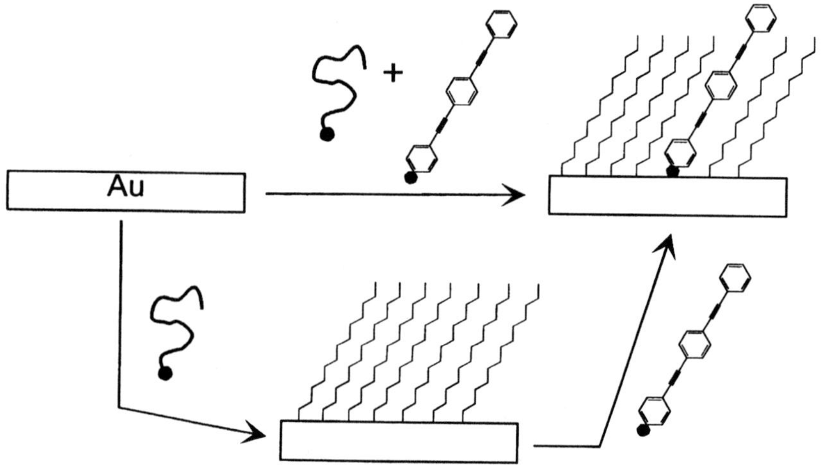

FIGURE 7. Schematic showing the self-assembly of a mixed monolayer of alkanethiolates and fully conjugated oligomers at an electrode via co-assembly and via sequential alkanethiolate host assembly followed by oligomer insertion.

matrix. Typical host SAMs are made by self-assembly from octanethiol, decanethiol, dodecanethiol, and tetradecanethiol solutions. For convenience, the respective SAMs are referred to as A8, A10, A12, and A14, where A represents alkanethiolate.

An example of the host-ordering effect is given by the simple case of the relatively short molecule PE2 (see TABLE 1) in a host A12 alkanethiolate SAM for which co-assembly degrades the alkanethiolate lattice whereas insertion maintains ordering.[39] FIGURE 8 shows an STM image of the ordering for a monolayer modified by insertion. Notice the well-resolved molecular lattice of the A12 chains. In contrast, images of corresponding samples made via co-deposition (data not shown) reveal no traces of molecularly resolvable lattices. A summary of IRS evidence for these effects is given in the plot in FIGURE 9, which shows the C-H stretching peak frequencies vs. the fraction of guest PE3 for the two deposition methods. The well-established correlation of increasing frequency of both the symmetric and antisymmetric mode peaks with increasing chain conformation disorder is used as a monitor for the host SAM ordering. The observed frequency increases with increasing guest content for the co-deposition method establishes the disordering effect on the host lattice. In contrast, the frequencies of the two peaks remain relatively unaffected by the presence of guest for the insertion method.

The effects of insertion in a host alkanethiolate SAM can be readily seen for insertion of the long PE6 molecule (~40 Å) into a shorter A8 SAM. From analysis of IRS data, the molecules in the pure PE6 SAM appear nearly randomly oriented while in the inserted state the average tilt is ~20–30°, as summarized in cartoon form, with the molecular lengths made approximately to scale, in FIGURE 10. It is interesting to see the strong orientation effect exerted by a relatively thin host SAM layer.

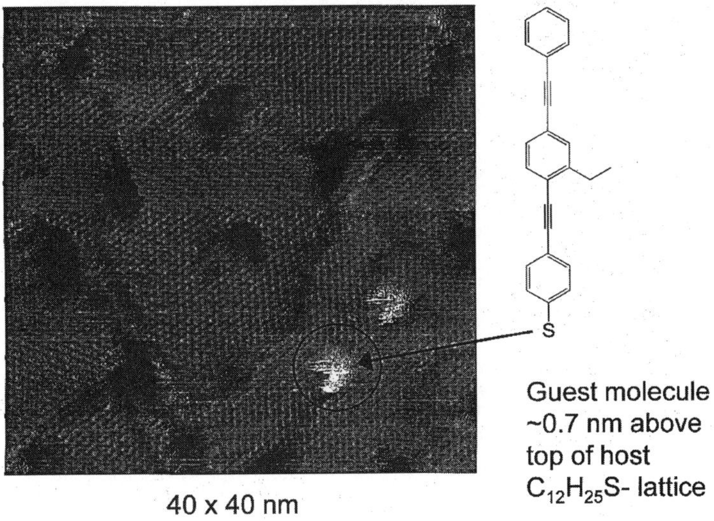

FIGURE 8. Constant-current STM image of guest PE2 molecules inserted into a host dodecanethiolate SAM on Au. The white spot in the circled region is identified as the location of a guest molecule for which the STM tip height is ~0.7 nm higher than the top of the alkanethiolate lattice. The higher distance at constant current indicates that the guest has a significantly enhanced electronic conduction compared to the host.

The course of the insertion process can be controlled accurately by regulating the solution concentration of the guest molecules and the immersion time of the host SAM. For example, we observe that for fixed concentrations of dilute solutions (~millimolar and lower in tetrahydrofuran) of $PE(NO_2)$, the relative initial insertion rates into A8, 10, 12, and 14 alkanethiolate host SAMs on Au are 3.4, 1.6, 1.2, 1.0, respectively, with an absolute rate of ~21% insertion per hour for the A8 SAM.

Other strategies are possible to force molecular alignment. One involves using 3-D-shaped molecules in which the intrinsic design forces a unique surface orientation under self-assembly conditions. One potential candidate is a "3-legged" tetrahedral-shaped molecule in which three equal-length linear legs are terminated by S- groups for attachment to a Au surface. In the case of an ideal three-point attachment, the remaining leg will be forced to align perpendicular to the substrate surface plane. An example based on a PE-based oligomeric structure with a central Si atom is shown in FIGURE 11. When a gold surface was exposed to solutions of two types of these molecule, converted to the thiol forms *in situ*, surface films self-assembled as shown by the infrared spectra shown in the figure. However, in this first trial, the films were slightly thicker than a theoretical monolayer. This indicated that other modes of chemisorption in addition to the perfect three-point attachment mode also occurred. The approach is promising and further work is in progress.

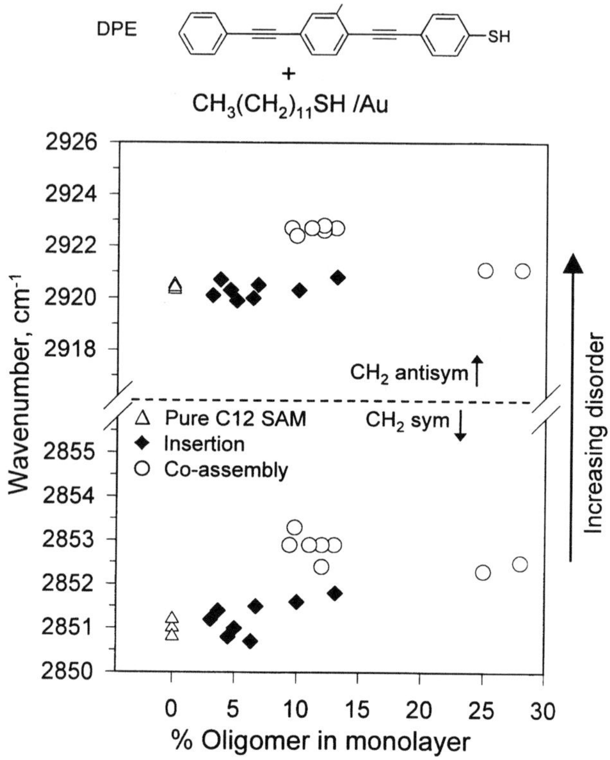

FIGURE 9. Infrared frequency positions of the symmetric and antisymmetric C-H stretching modes of the alkanethiolate molecules in a host matrix on gold into which PE3 molecules have been either inserted or co-adsorbed to different extents. The constant frequency trends for the insertion case show that the host matrix is undisturbed, whereas the changes observed in the co-adsorbed case indicate disruptive perturbations in the host matrix.

Selective Attachment at Designated Electrodes

Perhaps one of the most challenging design issues faced in utilizing self-assembly for making molecule-based devices with sophisticated functions is the question of how to create an assembly template which will direct the attachment of designated site on a multifunctional molecule to a selected target electrode in an electrode array. One of the pieces of this puzzle may lie in our observation that steric effects can control the insertion rates of guest molecules into host SAMs at the electrodes. In the process of learning how to solubilize large molecules by adding constituent alkyl side chains, we discovered that the location and length of the side chain could regulate the insertion into host SAMs. This effect suggests a method for blocking of a pre-selected part of a molecule as illustrated in FIGURE 12. The basic idea is to locate a long side arm at a distance away from the electrode attachment group which is

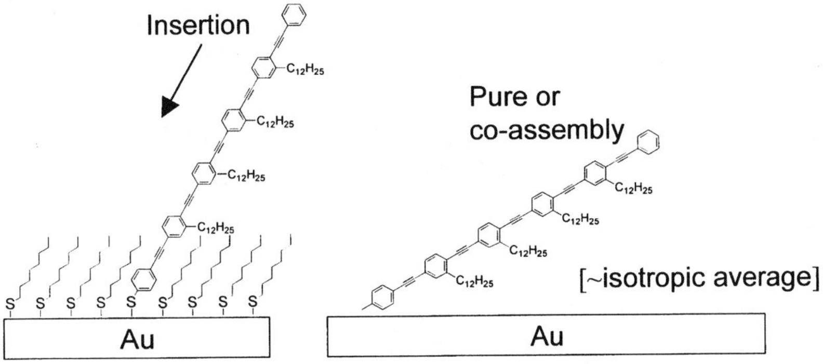

FIGURE 10. Schematic of the tilt angles adopted by the PE6 molecules in short alkanethiolate host SAMs when the guests are inserted and co-adsorbed. The tilt angles shown are representative of the actual average values deduced from infrared spectra. The large average tilt in the co-adsorption case likely indicates that the molecules are distributed with a wide variety of orientations approaching an isotropic distribution.

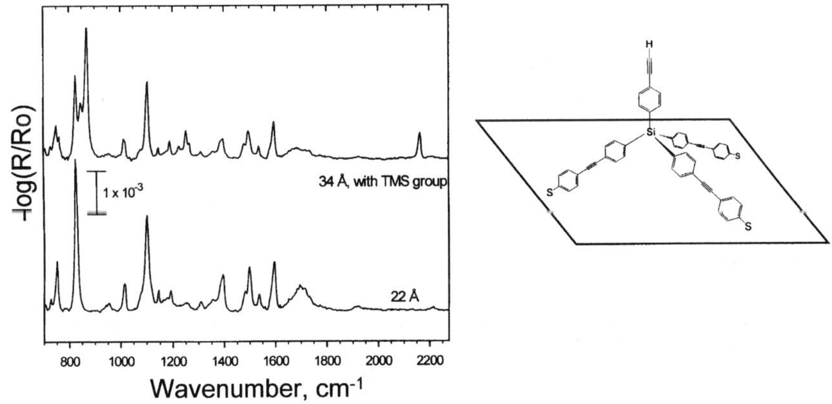

FIGURE 11. Illustration of the chemisorption of a "caltrops" tri-attached molecule on a surface. The infrared spectra are for the molecule shown and its analogue terminated by a trimethylsilyl group. The spectra show that the molecules are attached to the gold surface with a general vertical orientation according to the picture.

shorter than the height of a host SAM matrix on a pre-selected electrode. In principle, this situation could prevent insertion of a side arm if the host matrix were unyielding to the incoming arm. Data to support this simple steric effect are given in TABLE 3, where the relative insertion rates of several molecules with varying structures are shown. The data indicate that the presence of a long $C_{12}H_{25}$- side chain located near the end of the molecule considerably slows insertion into a host SAM of thickness greater than the arm-terminal group distance. One possibility for implementing this strategy to a three-terminal device would be to assemble SAMs on all

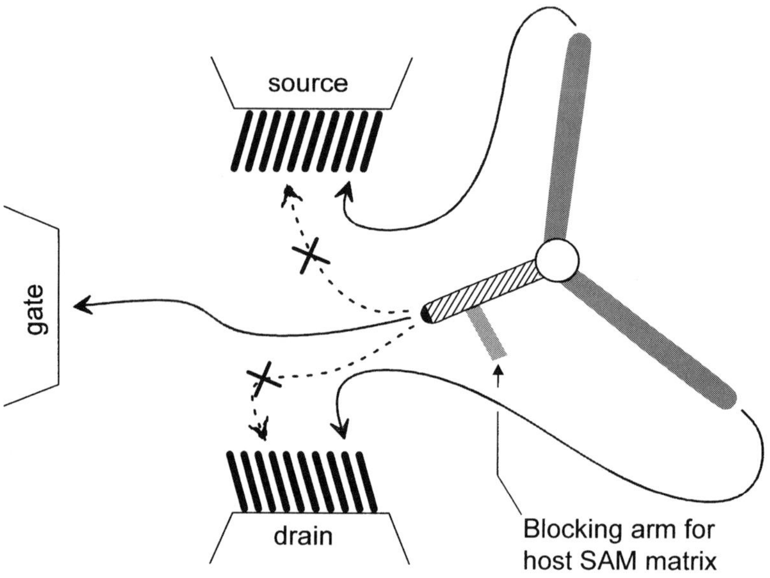

FIGURE 12. Cartoon illustrating the basis of a strategy to control the specific configuration of a tri-functional molecule chemisorbed in a three-electrode structure. The figure illustrates how the presence of a long side group on one of the arms of the molecule can help to direct attachment of that arm to the bare electrode.

three electrodes and then to desorb the SAM from a selected electrode to give the structure in FIGURE 12. Porter and coworkers[42] have shown that under solution conditions application of a negative potential to a gold electrode can cause desorption of an alkanethiolate SAM. After selective desorption, exposure to a solution of the appropriate trifunctional molecule for short times would lead to three-point attachment only when the blocked arm is connected to the bare electrode.

Deposited Metal Contact Chemistry

There is considerable interest in molecule-based devices in which a contact is made by the traditional methods of metal deposition using either physical vapor or chemical vapor depositions. For example, in a recent report the electrical response of small molecules was probed using a micropore type of device, represented by the structure in FIGURE 1C, where the top contact was made via a metal vapor deposition.[19] In cases of reactive metal atoms, for example, Ti, Al, or Cr, the character of the metal/organic junction will depend critically upon the detailed chemical reactions which occur during deposition and the diffusive transport of the metal through the molecular layer. In one limit, the metal atom reactivity could be so great as to decompose the molecules, for example, forming metal carbides and oxides. However, in the other limit, where the reactivity is negligible, one may face a new problem in which poor molecular packing could allow metal atom diffusion through the layer to form an electrical short. It is thus clear that significant care must be taken in the

TABLE 3. Rates of insertion of selected linear oligomeric molecules into an A12 alkanethiolate SAM host matrix on gold

Molecule	% inserted in selected time	
	1 hr	10 hr
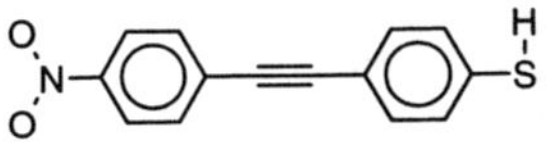	4.1	8.8
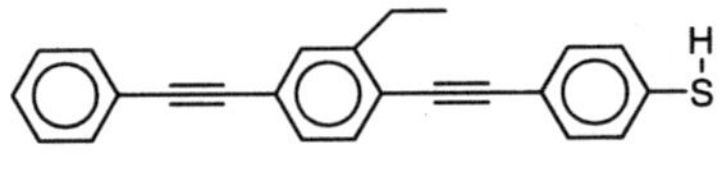	5.1	—
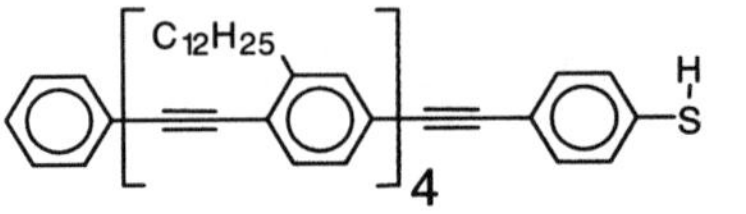	0.0	0.0[a]
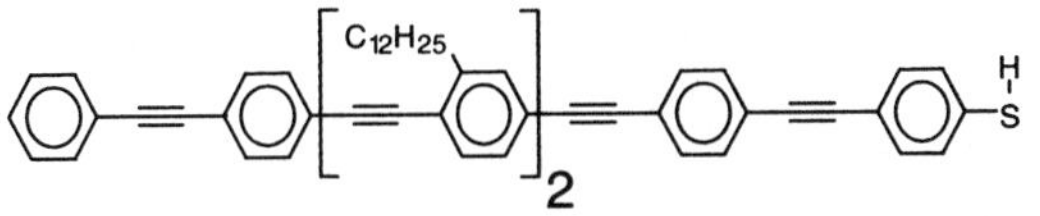	4.5	—

[a]No insertion until 50 hr.

design of the metal/organic contact. An example of the type of chemistry that could occur is given in FIGURE 13, which summarizes the XPS results from vapor deposition of Al metal onto a pure SAM of the nitro-terminated compound PE2(NO_2) on Au. The data show that for <1 Å of Al deposited, an AlN_XO_Y interlayer forms, while for greater coverages, the metal begins to deposit as metallic clusters which coalesce in the final stages to form a uniform metallic film. Most importantly, analysis of the C1s core level data shows no evidence for formation of metal carbides, thus implying that after the degradation of the NO_2 group to form an interlayer, the remainder of the molecule stays chemically intact.

Doping of "Molecular Wires"

It is typical for electroactive polymers that high electrical conductivity is not realized unless strong electron donors or acceptors, e.g., K metal or AsF_5, respectively, are incorporated into the polymer matrix to provide new electronic states for conduction along the polymer backbones.[43] For example, the doping of polyphenylene by either AsF_5 or K metal can increase its conductivity by well over 10 orders of magnitude.[43] Similarly, such doping may be useful for controlling the conductivity of discrete-sized, fully conjugated oligomeric molecules. As a prelude to the fabrication of doped molecule-based devices, we have been exploring the reactivity of SAMs of fully conjugated molecules with alkali metals, which are strong electron donors. An example of a readily reduced adsorbate molecule is given by PE2(NO_2). Exposure of both the pure SAM and mixed SAMs formed via insertion of the oligomer into host alkanethiolate SAMs to K vapor under UHV conditions results in for-

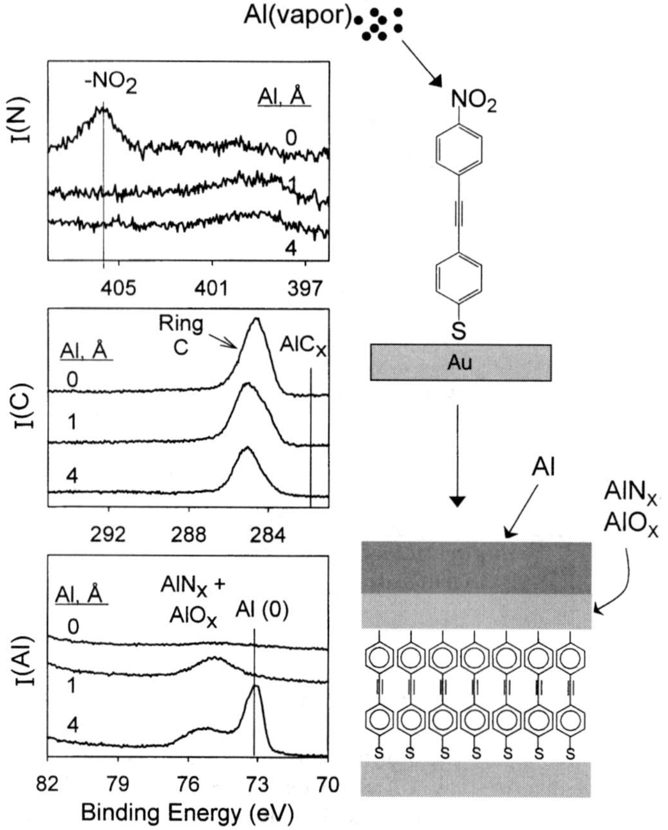

FIGURE 13. Illustration of the chemical effects of vapor-deposited Al atoms on the terminal groups of a SAM of PE2(NO_2) molecules on Au. The XPS N1s, C1s, and Al2p core level data on the left show that the nitro groups react with the Al to form and AlN_XO_Y interlayer.

mation of 1-electron reduced states at low K coverages with formation of 2-electron reduction following at higher coverages. The IRS spectra in FIGURE 14 reveal the dramatic changes in the molecular vibrations after K vapor exposure. These changes signal the appearance of the reduced states. The electronegativity of the NO_2 groups obviously helps to drive the reductions. However, initial results indicate that other fully conjugated molecules, such as the PEn type in TABLE 1, also appear susceptible to alkali metal doping.

CONCLUSIONS AND PROSPECTS

The recent demonstrations of direct electronic measurements of conjugated molecules and nanometer-scale inorganic objects (e.g., clusters and nanotubes) offer ex-

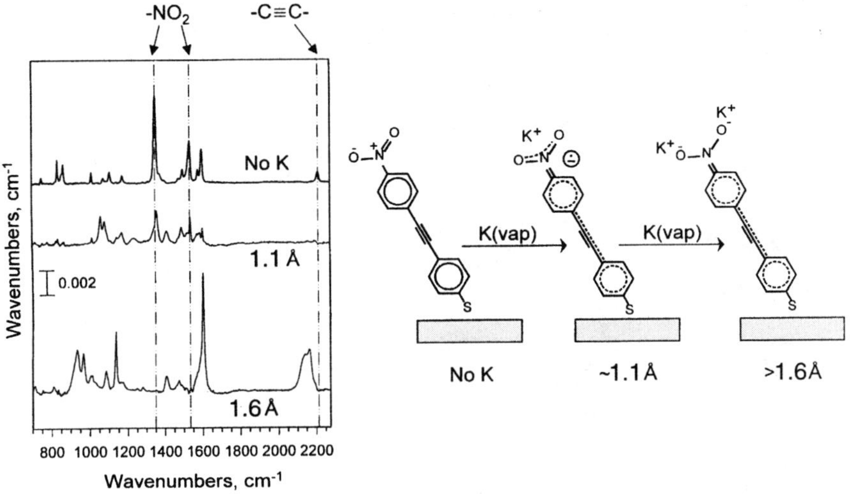

FIGURE 14. A summary of the results of stepwise vapor deposition of K metal onto a PE2(NO_2) SAM on Au. The infrared spectral changes in the NO_2 symmetric and antisymmetric modes, as well as the aromatic ring and acetylene group modes, give evidence for the sequential formation of the single and double electron injection states.

citing prospects for new generations of electronic devices. Practical fabrication of these devices will require the primary use of self-assembly methods in which the driving force for the desired device configuration will be built into the chemical characteristics of the molecules and the self-assembling process. In order to design these processes efficiently, a number of critical chemical issues will need to be addressed, including optimal electrode/chemical attachments; control of molecular orientation at electrodes; chemical bonding mechanisms at electrodes; and methods for selective chemical attachments in asymmetric, multi-terminal devices. Further, manipulation of conductivity via chemical doping should be explored as well as the chemical and electrical interface effects of deposited metal electrodes in hybrid devices using ensembles of molecules. All these aspects are new to the traditional-device community, but a sustained effort to explore these, and related, chemical issues will be a key step in the development of molecule-based electronics.

ACKNOWLEDGMENTS

We acknowledge support of this work by the Defense Advanced Research Projects Agency, the National Science Foundation, and the Office of Naval Research.

REFERENCES AND NOTES

1. AVIRAM, A. 1974. Molecular Rectifiers. Chem. Phys Lett. **29:** 277–283.

2. CARTER, F.L., Ed. 1982. Molecular Electronic Devices. Marcel Dekker. New York; CARTER, F.L., Ed. 1987. Molecular Electronic Devices II. Marcel Dekker. New York.
3. AVIRAM, A. 1988. Molecules for memory, logic, and amplification. J. Am. Chem. Soc. **110:** 5687–5692.
4. HONG, F.T. 1989. Molecular Electronics: Biosensors and Biocomputers. Plenum Press. New York.
5. MIRKIN, C.A. & M.A. RATNER. 1992. Molecular electronics. Annu. Rev. Phys. Chem. **43:** 719–754.
6. AVIRAM, A. 1992. A strategic plan for molecular electronics. Int. J. Quant. Chem. **42:** 1615–1624.
7. KRIEGER, YU.G. 1993. Molecular electronics: Current state and future trends. J. Struct. Chem. **34:** 896–904.
8. WALDECK, D.H. & D.N. BERATAN. 1993. molecular electronics: Observation of molecular rectification. Science **261:** 576–577.
9. BIRGE, R.R. 1994. Molecular and Biomolecular Electronics. American Chemical Society. Washington, D.C.
10. HONG, F.T. 1994. Molecular electronics: Science and technology for the future. IEEE Engineering in Medicine and Biology **13:** 25–32.
11. GUST, D. T.A. MOORE & A.L. MOORE. 1994. Photosynthesis mimics as molecular electronic devices. IEEE Engineering in Medicine and Biology **13:** 58–66.
12. MUJICA, V., M. KEMP, A. ROITBERG & M. RATNER. 1996. Current-voltage characteristics of molecular wires: Eigenvalue staircase, Coulomb blockade, and rectification. J. Chem Phys. **104:** 7296–7304.
13. JOACHIM, C. 1991. The conductance of a single molecule? New J. Chem. **15:** 223–229.
14. MULLER, C.J., B.J. VLEEMING, M.A. REED, J.J. LAMBA, R. HARA, L. JONES & J.M. TOUR. 1996. Atomic probes: A search for conduction through a single molecule. Nanotechnology **7:** 409–411.
15. TOUR, J.M. 1996. Conjugated macromolecules of precise length and constitution. Organic synthesis for the construction of nanoarchitectures. Chem. Rev. **96:** 537–553.
16. REED, M.A., C. ZHOU, C.J. MULLER, T.P. BURGIN & J.M. TOUR. 1997. Conductance of a molecular junction. Science 278: 252–254.
17. BUMM, L.A., J.J. ARNOLD, M.T. CYGAN, T.D. DUNBAR, T.P. BURGIN, L. JONES, D.L. ALLARA, J.M. TOURS & P.S. WEISS. 1996. Are single molecular wires conducting? Science **271:** 1705–1707.
18. REED, M.A. Personal communication.
19. ZHOU, C., M.R. DESHPANDE, M.A. REED, L. JONES & J.M. TOUR. 1997. Nanoscale metal self-assembled monolayer metal heterostructures. Appl. Phys. Lett. **71:** 611–613.
20. SLOWINSKI, K., R.V. CHAMBERLAIN, C.J. MILLER & M. MAJDA. 1997. Through-bond and chain-to-chain coupling. Two pathways in electron tunneling through liquid alkanethiol monolayers on mercury electrodes. J. Am. Chem. Soc. **119:** 11910–11919.
21. ANDRES, R.P., T. BEIN, M. DOROGI, S. FENG, J.I. HENDERSON, C.P. KUBIAK, W. MAHONEY, R.G. OSIFCHIN & R. REIFENBERGER. 1996. "Coulomb staircase" at room temperature in a self-assembled molecular nanostructure. Science **272:** 1323–1325.
22. MARTIN, A.S., J.R. SAMBLES & G.J. ASHWELL. 1993. Molecular rectifier. Phys. Rev. Lett. **70:** 218–221.
23. METZGER, R.M., B. CHEN, U HOPFNER, M.V. LAKSHMIKANTHAM, D. VUILLAUME, T. KAWAI, X.L. WU, H. TACHIBANA, T.V. HUGHES, H. SAKURAI, J.W. BALDWIN, C. HOSCH, M.P. CAVA, L. BREHMER & G.J. ASHWELL. 1997. Unimolecular electri-

cal rectification in hexadecylquinolinium tricyanoquinodimethanide. J. Am. Chem. Soc. **119:** 10455–10466.
24. SEMINARIO, J.M., A.G. ZACARIAS, J.M. TOUR. J. Am. Chem. Soc. In press.
25. ODOM, T.W., J.L. HUANG, P. KIM, & C.M. LIEBER. 1998. Atomic structure and electronic properties of single-walled carbon nanotubes. Nature **391:** 62–64.
26. WILDOER, J.W., L.C. VENEMA, A.G. RINZLER, R.E. SMALLEY & C. DEKKER. 1998. Electronic structure of atomically resolved carbon nanotube. Nature **391:** 59–62.
27. TANS, S.J., M.H. DEVORET, H.J. DAI, A. THESS, R.E. SMALLEY, L.J. Geerligs & C. Dekker. 1997. Individual single-wall carbon nanotubes as quantum wires. Nature **386:** 474–477.
28. WU, C.G. & T. BEIN. 1994. Conduction of polyaniline filaments in a mesoporous channel host. Science **264:** 1757–1759.
29. KLEIN, D.L., R. ROTH, A.K. LIM, A.P. ALIVISATOS & P.L. MCEUEN. 1997. A single-electron transistor made from a cadmium selenide nanocrystal. Nature **389:** 699–701.
30. PERSSON, S.H.M., L. OLOFSSON, L. HEDBERG, D. SUTHERLAND & E. OLSSON. Personal communication.
31. TOUR, J.M., L. JONES, D.L. PEARSON, J.J.S. LAMBA, T.P. BURGIN, G.M. WHITESIDES, D.L. ALLARA, A.N. PARIKH & S.V. ATRE. 1995. Self-assembled monolayers and multilayers of conjugated thiols, alpha,omega-dithiols, and thioacetyl-containing adsorbates. Understanding attachments between potential molecular wires and gold surfaces. J. Am. Chem Soc. **117:** 9529–9534.
32. PEARSON, D.L. & J.M. TOUR 1997. Rapid synthesis of oligo(2,5,-thiophene ethynylene)s with thioester termini: Potential molecular scale wires with alligator clips. J. Org. chem. **62:** 1376–1387; JONES, L., J.S. SCHUMM & J.M. TOUR. 1997. Rapid solution and solid phase syntheses of oligo(2,5,-thiophene ethynylene)s with thioester termini: Molecular scale wires with alligator clips. Derivation of iterative efficiences on a polymer support. J. Org. Chem. **62:** 1388–1410.
33. DUBOIS, L.H. & R.G. NUZZO. 1992. Synthesis, structure, and properties of model organic surfaces. Annu. Rev. Phys. Chem. **43:** 437–463.
34. SHI, J. B. HONG, A.N. PARIKH, R.W. COLLINS & D.L. ALLARA. 1995. Optical characterization of electronic transitions arising from the Au/S interface of self-assembled *n*-alkanethiolate monolayers. Chem. Phys. Lett. **246:** 90–94.
35. DISHNER, M.H., J.C. HEMMINGER & F.J. FEHER. 1997. Scanning tunneling microscopy characterization of organoselenium monolayers on Au(III). Langmuir **13:** 4788–4790.
36. SAMANT, M.G., C.A. BROWN & J.G. GORDON. 1992. Formation of an ordered self-assembled monolayer of docosaneselenol on gold(111). Structure by surface x-ray diffraction. Langmuir **8:** 1615–1618.
37. CYGAN, M.T., T.D. DUNBAR, J.J. ARNOLD, L.A. BUMM, L. JONES, II, J.M. TOUR, P.S. WEISS & D.L. ALLARA. Manuscript in preparation.
38. MARS, C.A. T.D. DUNBAR & D.L. ALLARA. Manuscript in preparation.
39. CYGAN, M.T., T.D. DUNBAR, J.J. ARNOLD, L.A. BUMM, N.F. SHEDLOCK, T.P. BURGIN, L. JONES, II, D.L. ALLARA, J.M. TOUR & P.S. WEISS. 1988. Insertion, conductivity and structures of conjugated organic oligomers in self-assembled alkanethiol monolayers on Au{111}. J. Am. Chem. Soc. **120:** 2721–2732.
40. DUNBAR, T.D., M.T. CYGAN, J.M. TOUR, P.S. WEISS & D.L. ALLARA. Manuscript in preparation.
41. WEISS, P.S. L.A. BUMM, T.D. DUNBAR, T.P. BURGIN, J.M. TOUR & D.L. ALLARA. 1998. Probing electronic properties of conjugated and saturated molecules in self-assembled monolayers. This volume.

42. WIDRIG, C.A., C. CHUNG & M.D. PORTER. 1991. The electrochemical desorption of *n*-alkanethiol monolayers from polycrystalline Au and Ag electrodes. J. Electroanal. Chem. 310: 335–539.
43. SKOTHEIM, T.A., Ed. 1986. Handbook of Conducting Polymers, Vols. 1&2. Marcel Dekker. New York.

Index of Contributors